D1811004

METHODS IN MOLECULAR BIOLOGY

For further volumes:
http://www.springer.com/series/7651

Cytoskeleton

Methods and Protocols

Third Edition

Edited by

Ray H. Gavin

Brooklyn College, Brooklyn, NY, USA

Editor
Ray H. Gavin
Brooklyn College
Brooklyn, NY, USA

ISSN 1064-3745 ISSN 1940-6029 (electronic)
Methods in Molecular Biology
ISBN 978-1-4939-3123-1 ISBN 978-1-4939-3124-8 (eBook)
DOI 10.1007/978-1-4939-3124-8

Library of Congress Control Number: 2015947776

Springer New York Heidelberg Dordrecht London

Printed on acid-free paper

Humana Press is a brand of Springer
Springer Science+Business Media LLC New York is part of Springer Science+Business Media (www.springer.com)

Preface

It has been 15 years since the publication of the first edition of *Cytoskeleton Methods and Protocols*, and in reflection, I am mindful of the many advances in our understanding of the cytoskeleton over the last four decades. The earliest images of the cytoskeleton revealed a complex network of *seemingly* interconnected arrays of various fibers, some of which appeared filamentous, and others appeared tubular. Synergy of cytoskeleton fibers was not widely accepted, and the notion was slow to gain acceptance. However, convincing evidence for synergistic action among cytoskeleton fibers as well as more detailed knowledge about their structure and function continually emerges. The dramatic interplay of cytoskeleton components is aptly illustrated in the image of a HeLa cell (Fig. 1). Continuing advances in cell imaging reflect refinements in various forms of microscopy and have significantly expanded our understanding of cytoskeleton structure and function. Technological advances have led to development of magnetic tweezers, in vitro reconstructions, and improved proteomics, all of which have become prominent experimental tools in cytoskeleton research. Many of these new tools are the focus of the third edition of *Cytoskeleton Methods and Protocols.*

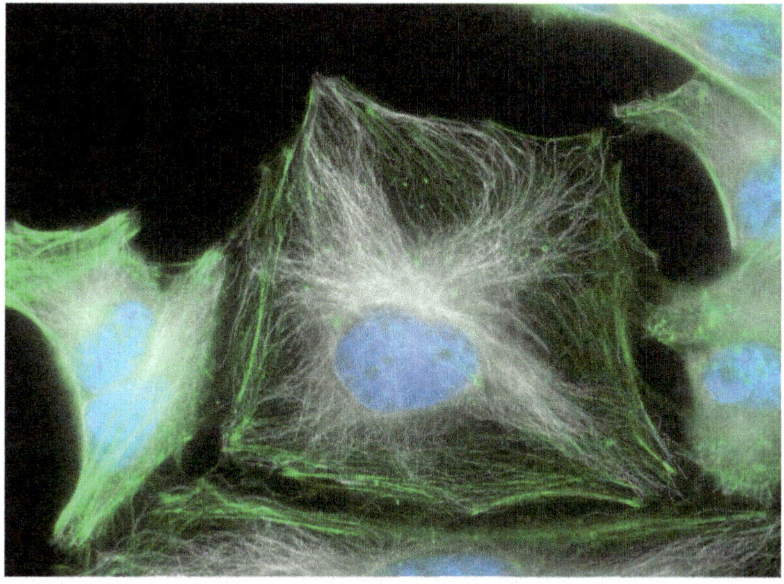

Fig. 1 The cover illustration shows a HeLa cell with microtubules (*white*), actin filaments (*green*), and DNA (*blue*). Photo courtesy Omar Quintero

The first two editions of this book aimed to provide the investigator with protocols based on diverse experimental models. The third edition continues the focus on experimental models that are useful for investigating various aspects of cytoskeleton structure and function. Animal, plant, protist, and fungal models highlight 24 chapters that provide detailed protocols for live and fixed-cell imaging, dynamics of cytoskeleton components, cell and organelle motility, and genetics and proteomics. Protocols in each chapter are up-to-date menus organized in a useful step-by-step format appropriate for novice and established investigators. Each chapter is equipped with a highly valuable notes section that provides a troubleshooting guide and helpful, and often unpublished, technical information aimed at ensuring success with implementation of the protocols.

I thank John Walker for help with the review of manuscripts. The third edition of *Cytoskeleton Methods and Protocols* belongs to more than 60 internationally renowned experts in their fields who contributed chapters, and I thank them for submitting manuscripts in a timely manner and for their great professionalism.

Brooklyn, NY, USA ***Ray H. Gavin***

Contents

Contributors

CHRIS AMBROSE • *Department of Biology, University of Saskatchewan, Saskatoon, SK, Canada*
DEREK A. APPLEWHITE • *Department of Biology, Reed College, Portland, OR, USA*
ISABEL VAN AUDENHOVE • *Nanobody Laboratory, Department of Biochemistry, Faculty of Medicine and Health Sciences, Ghent University, Ghent, Belgium*
NICOLAS BIAIS • *Department of Biology, Brooklyn College of the City University of New York, Brooklyn, NY, USA; The Graduate Center of the City University of New York, New York, NY, USA*
KATHRIN BLAAS • *Functional Plant Biology, Institute of Botany, University of Innsbruck, Innsbruck, Austria*
ELISON B. BLANCAFLOR • *Plant Biology Division, The Samuel Roberts Noble Foundation, Inc., Ardmore, OK, USA*
ISTVAN R. BOLDOGH • *Department of Pathology and Cell Biology, College of Physicians and Surgeons, Columbia University, New York, NY, USA*
DERRICK BRAZILL • *Department of Biological Sciences, Hunter College, New York, NY, USA; The Graduate Center, City University of New York, New York, NY, USA*
KATHERINE CELLER • *Department of Botany, The University of British Columbia, Vancouver, BC, Canada*
CHRISTINE A. DAVIS • *Department of Biology, University of Richmond, Richmond, VA, USA*
JULIA DYACHOK • *McDermott Center for Human Growth & Development, University of Texas Southwestern Medical Center, Dallas, TX, USA*
MANSOUR EL-MATBOULI • *Department for Farm Animals and Veterinary Public Health, University of Veterinary Medicine, Vienna, Austria*
MIKI FUJITA • *Department of Botany, The University of British Columbia, Vancouver, BC, Canada*
JAN GETTEMANS • *Nanobody Laboratory, Department of Biochemistry, Faculty of Medicine and Health Sciences, Ghent University, Ghent, Belgium*
MICHAEL GOTESMAN • *Department of Biology, Technion – Israel Institute of Technology, Technion, Haifa, Israel*
ERIC R. GRIFFIS • *Wellcome Trust Centre for Gene Regulation and Expression, College of Life Sciences, University of Dundee, Dundee, UK*
QI HE • *Department of Biology, Brooklyn College, City University of New York, Brooklyn, NY, USA*
MARY ELLEN HEAVNER • *Biochemistry Program, The Graduate Center, City University of New York, New York, NY, USA*
KLAUS HERBURGER • *Functional Plant Biology, Institute of Botany, University of Innsbruck, Innsbruck, Austria*
RYO HIGUCHI-SANABRIA • *Department of Pathology and Cell Biology, College of Physicians and Surgeons, Columbia University, New York, NY, USA*
ANDREAS HOLZINGER • *Functional Plant Biology, Institute of Botany, University of Innsbruck, Innsbruck, Austria*

LINDA A. HUFNAGEL • *Department of Cell and Molecular Biology, University of Rhode Island, Kingston, RI, USA*
EIKO KAWAMURA • *Western College of Veterinary Medicine, University of Saskatchewan, Saskatoon, SK, Canada*
ROBERTA S. KING • *Department of Cell and Molecular Biology, University of Rhode Island, Kingston, RI, USA*
CHRISTINA KING-SMITH • *Department of Biology, Saint Joseph's University, Philadelphia, PA, USA*
SPENCER KUHL • *W.M. Keck Dynamic Image Analysis Facility, Department of Biological Sciences, University of Iowa, Iowa City, IA, USA*
GEORGE M. LANGFORD • *Syracuse University, Life Sciences Complex, Syracuse, NY, USA*
BIN LI • *Biomaterials and Cell Mechanics Laboratory, Orthopedic Institute, Soochow University, Suzhou, Jiangsu, China*
DANIEL F. LUSCHE • *W.M. Keck Dynamic Image Analysis Facility, Department of Biological Sciences, University of Iowa, Iowa City, IA, USA*
SIMON MENANTEAU-LEDOUBLE • *Department for Farm Animals and Veterinary Public Health, University of Veterinary Medicine, Vienna, Austria*
EMILIA LAURA MUNTEANU • *Department of Biology, Brooklyn College of the City University of New York, Brooklyn, NY, USA; Herbert Irving Comprehensive Cancer Center, Columbia University, New York, NY, USA*
SHOICHIRO ONO • *Department of Pathology, Emory University, Atlanta, GA, USA*
ANA PAEZ-GARCIA • *Plant Biology Division, The Samuel Roberts Noble Foundation, Inc., Ardmore, OK, USA*
GIOVANNI J. PAGANO • *Department of Cell and Molecular Biology, University of Rhode Island, Kingston, RI, USA*
KARUPPAIAH PALANICHELVAM • *Department of Biotechnology, Kalasalingam University, Krishnankoil, Tamil Nadu, India*
LIZA A. PON • *Department of Pathology and Cell Biology, College of Physicians and Surgeons, Columbia University, New York, NY, USA; Confocal and Specialized Microscopy Shared Resource, Herbert Irving Comprehensive Cancer Center, Columbia University, New York, NY, USA*
OMAR A. QUINTERO • *Department of Biology, University of Richmond, Richmond, VA, USA*
CHRISTOPHER ROBLODOWSKI • *Department of Biological Sciences and Geology, Queensborough Community College, City University of New York, Bayside, NY, USA*
LABIB ROUHANA • *Department of Biological Sciences, Wright State University, Dayton, OH, USA*
LUÍS CARLOS SANTOS • *Department of Biology, Brooklyn College of the City University of New York, Brooklyn, NY, USA; Icahn School of Medicine at Mount Sinai, New York, NY, USA*
MICHAEL SCARPATI • *Biology Program, The Graduate Center, City University of New York, New York, NY, USA*
AMANDA SCHERER • *W.M. Keck Dynamic Image Analysis Facility, Department of Biological Sciences, University of Iowa, Iowa City, IA, USA*
SHANEEN SINGH • *Biochemistry Program, The Graduate Center, City University of New York, New York, NY, USA; Department of Biology, Brooklyn College, City University of New York, Brooklyn, NY, USA; Biology Program, The Graduate Center, City University of New York, New York, NY, USA*

Roger D. Sloboda • *Department of Biological Sciences, Dartmouth College, Hanover, NH, USA; The Marine Biological Laboratory, Woods Hole, MA, USA*
David R. Soll • *W.M. Keck Dynamic Image Analysis Facility, Department of Biological Sciences, University of Iowa, Iowa City, IA, USA*
Tatyana Svitkina • *Department of Biology, University of Pennsylvania, Philadelphia, PA, USA*
Theresa C. Swayne • *Confocal and Specialized Microscopy Shared Resource, Herbert Irving Comprehensive Cancer Center, Columbia University, New York, NY, USA*
Junichi Tasaki • *Department of Biological Sciences, Wright State University, Dayton, OH, USA*
Chihiro Uchiyama-Tasaki • *Department of Biological Sciences, Wright State University, Dayton, OH, USA*
Sven K. Vogel • *Department of Cellular and Molecular Biophysics, Max Planck Institute of Biochemistry, Martinsried, Germany*
Edward Voss • *W.M. Keck Dynamic Image Analysis Facility, Department of Biological Sciences, University of Iowa, Iowa City, IA, USA*
James H.-C. Wang • *MechanoBiology Laboratory, Departments of Orthopaedic Surgery and Bioengineering, University of Pittsburgh School of Medicine, Pittsburgh, PA, USA*
Geoffrey O. Wasteneys • *Department of Botany, The University of British Columbia, Vancouver, BC, Canada*
Deborah J. Wessels • *W.M. Keck Dynamic Image Analysis Facility, Department of Biological Sciences, University of Iowa, Iowa City, IA, USA*
Eliza Wiech • *Biology Program, The Graduate Center, City University of New York, New York, NY, USA*
Selwyn A. Williams • *Department of Biology, Miami Dade College, Miami, FL, USA*
Torsten Wöllert • *Syracuse University, Life Sciences Complex, Syracuse, NY, USA*
Cheol-Min Yoo • *Plant Biology Division, The Samuel Roberts Noble Foundation, Inc., Ardmore, OK, USA*
Guangyi Zhao • *MechanoBiology Laboratory, Departments of Orthopaedic Surgery and Bioengineering, University of Pittsburgh School of Medicine, Pittsburgh, PA, USA*

Part I

Imaging the Cytoskeleton in Live and Fixed Cells

Chapter 1

Long-Term Live Cell Imaging of Cell Migration: Effects of Pathogenic Fungi on Human Epithelial Cell Migration

Torsten Wöllert and George M. Langford

Abstract

Long-term live cell imaging was used in this study to determine the responses of human epithelial cells to pathogenic biofilms formed by *Candida albicans*. Epithelial cells of the skin represent the front line of defense against invasive pathogens such as *C. albicans* but under certain circumstances, especially when the host's immune system is compromised, the skin barrier is breached. The mechanisms by which the fungal pathogen penetrates the skin and invade the deeper layers are not fully understood. In this study we used keratinocytes grown in culture as an in vitro model system to determine changes in host cell migration and the actin cytoskeleton in response to virulence factors produced by biofilms of pathogenic *C. albicans*. It is clear that changes in epithelial cell migration are part of the response to virulence factors secreted by biofilms of *C. albicans* and the actin cytoskeleton is the downstream effector that mediates cell migration. Our goal is to understand the mechanism by which virulence factors hijack the signaling pathways of the actin cytoskeleton to alter cell migration and thereby invade host tissues. To understand the dynamic changes of the actin cytoskeleton during infection, we used **long-term live cell imaging** to obtain spatial and temporal information of actin filament dynamics and to identify signal transduction pathways that regulate the actin cytoskeleton and its associated proteins.

Long-term live cell imaging was achieved using a high resolution, multi-mode epifluorescence microscope equipped with specialized light sources, high-speed cameras with high sensitivity detectors, and specific biocompatible fluorescent markers. In addition to the multi-mode epifluorescence microscope, a spinning disk confocal long-term live cell imaging system (Olympus CV1000) equipped with a stage incubator to create a stable in vitro environment for long-term real-time and time-lapse microscopy was used. Detailed descriptions of these two **long-term live cell imaging** systems are provided.

Key words Keratinocytes, *Candida albicans*, Cell migration, Actin cytoskeleton, Live cell imaging, Environmental control chambers, Biofilms, Real-time and time-lapse observations

1 Introduction

1.1 Cell Migration and Biofilm Formation

Cell migration is a fundamental process by which cells move in a directed fashion from one location to another most often in response to an extracellular chemical signal [1]. Migration requires proper coordination and interaction between the cytoskeleton and cell surface adhesion plaques that serve as attachment sites to the

Ray H. Gavin (ed.), *Cytoskeleton: Methods and Protocols*, Methods in Molecular Biology, vol. 1365,
DOI 10.1007/978-1-4939-3124-8_1, © Springer Science+Business Media New York 2016

extracellular matrix or surrounding cells. Regulation of cell migration is a complex process which involves cell signaling molecules and their receptors, signaling pathways that facilitate cell polarization and protrusion, and substrate adhesion and release resulting in the coordinated movement of single cells or cell sheets [2]. Local activation of different Ras GTPases and in particular the Rho subfamily GTPases, Rac, Cdc42 and Rho, play essential roles in cell migration and are necessary for the dynamic reorganization of actin cytoskeleton at the leading edge [3, 4]. Cell migration is essential for cell differentiation, embryonic development and organogenesis, homeostatic processes such as immune responses to infections, as well as wound and tissue repair. Defects in the regulation of cell migration contribute to pathological conditions including tumor angiogenesis and metastasis, chronic inflammation such as atherosclerosis, asthma and arthritis, and brain abnormalities [5, 6].

Defects in cell migration have been shown to be caused in some cases by pathogens that secrete virulence factors that function to hijack the actin cytoskeleton, thereby altering the cell migration pattern and leading to serious infections and morbidity. One such pathogen is *Candida albicans*, a fungal pathogen that forms biofilms and is responsible for infections in infants and adults. *C. albicans* is commonly present in the environment and interacts with epithelial cells in the human host as a normal commensal but can be triggered to become an invasive pathogen. Fungal infections start with adhesion of *C. albicans* yeast cells to epithelial cells followed by penetration into individual cells, and/or degradation of interepithelial cell–cell adhesions and passing between epithelial cells [7, 8]. The pathogenicity of *C. albicans* is heightened when biofilms are formed. The molecular mechanism of biofilm formation differs among different classes of microorganisms but the stages are similar. The formation of biofilms occurs in response to environmental and physiological factors such as cell–cell recognition, cellular responses to various stresses, nutritional indicators, and antibiotic resistance [6, 9]. Many human pathogens cause biofilm-mediated diseases such as dental plaque, gingivitis, urinary tract infections and infections of catheters and implanted medical devices.

C. albicans forms biofilms through a complex process that involves four stages: (a) attachment and colonization of yeast cells to the substrate, (b) initiation of biofilm development by cell–cell adhesion and proliferation of yeast cells, (c) maturation by growth of pseudo-hyphae and hyphae and secretion of extracellular matrix and (d) dispersion of individual cells from the biofilm to colonize new locations. Hyphal formation and adhesion to surfaces are pivotal for biofilm formation, which depend on the transcription regulators Rob1, Brg1, Ndt80, Efg1, Tec1, and Bcr1 [10, 11].

Biofilm development in *C. albicans* and responses of epithelial cells, such as changes in cell migration and the actin cytoskeleton, can be monitored and characterized in vitro by long-term live cell imaging techniques including phase-contrast, differential interference contrast (DIC), spinning disk confocal microscopy, and high magnification imaging such as total internal reflection fluorescence microscopy (TIRF) and super-resolution microscopy. Phase-contrast or DIC microscopy is commonly used in combination with fluorescence microscopy in the presence of vital dyes or fluorescent fusion proteins to visualize living cells and cellular components, and therefore dynamic processes can be observed and analyzed quantitatively as they occur. Furthermore, fluorescence microscopy can be used to image fixed cells and tissues stained with specific antibodies to reveal internal structures and organization of cells [12]. These advanced fluorescence microscopic techniques and super-resolution imaging methods have become powerful tools to investigate the mechanism and dynamics of many of the proteins involved in cell migration.

Long-term live cell imaging studies in vitro require that the living cells be maintained in a natural state [13]. Cells exhibit normal migration behavior when they are cultured under the specific conditions that maintain their biological environment. In mammals, the most important parameters are a temperature of 37 °C, a pH of 7.4 that is controlled by a bicarbonate buffer at a 5 % CO_2 concentration, and physiological concentrations of salts and nutrients. In order to achieve biologically relevant results, it is crucial to maintain these conditions on the microscope stage during long-term live cell imaging experiments [14].

In this chapter we describe two multimode imaging systems, a wide-field epifluorescence and a spinning disk confocal microscope in combination with a cooled CMOS (complementary metal-oxide semiconductor) or CCD (charge-coupled device) camera, for long-term live cell imaging of human oral keratinocytes (OKF6/TERT-2) [15] and green monkey kidney cells (COS-7) in the presence of fungal biofilms. These imaging systems were controlled by commercially available software that also contains programs for analyzing real-time and time-lapse sequences of migrating epithelial cells.

1.2 Long-Term Live Cell Imaging Microscopy System Components

Long-term live cell imaging systems for cell migration studies require high-resolution optics and cameras capable of real-time image acquisition. Imaging systems that use phase contrast optics provide basic information, however, enhanced structural details and molecular interactions can be revealed by DIC and/or fluorescence microscopy, where reporter proteins like GFP and its variants are used to tag specific proteins. For fluorescence imaging, several options are available including standard epifluorescence optics or more sophisticated designs such as the laser scanning confocal, spinning disk confocal, TIRF, and multi-photon microscopy. A technique such as TIRF provides super-resolution through its ability to image the dynamic behavior of single molecules, such as cytoskeletal F-actin during cell migration [14].

To provide optimum conditions for long-term live cell imaging, it is important to use a microscope that is designed for **focus or focal plane stability** during the entire period of observation, which may last for 24–48 h. Temperature fluctuations are the primary causes of instability and must be controlled to maintain the **focus stability** for long-term live cell imaging systems. Therefore, it is important to prevent temperature fluctuations, and thereby prevent focal drift. It helps to have an optical focus system such as Perfect Focus (Nikon), Ultimate Focus (DeltaVision), Definite Focus (Zeiss), or Zero Thermal Drift Compensation (Olympus), which corrects any change in position of the coverslip relative to the objective using lasers or light emitting diodes (LED) systems.

For optimal optical quality, it is essential to use glass coverslips or glass bottom dishes for high magnification and high-resolution real-time and time-lapse microscopy. It is also important to prevent **condensation** of water on the optical components. Humidity in ambient air can lead to condensation on surfaces that are placed in the optical path and thereby cause light scattering, which will diminish the image quality. Environmental chambers that include independently controlled, heated glass lids and active humidity control will prevent the formation of condensation on lids of cell culture vessels.

Another important consideration for monitoring cell migration efficiently is to choose the right **magnification** based on the in vitro assay and the amount of structural detail that is needed for analysis of cell migration. Low magnification objectives (4× or 10×) can be sometimes more beneficial because of less focus drift and a higher number of cells that can be monitored during cell migration or wound healing experiments resulting in a larger data set for analysis. Furthermore, proper **illumination** that minimizes **photobleaching** and maintains high image quality for later image analysis is essential. Excitation light that leads to photobleaching can damage living cells and compromise the results of long-term live cell imaging experiments when using fluorescence-based imaging techniques. To avoid cell stress and phototoxicity, it is important to reduce the intensity and duration of the excitation laser or metal halide lamp during long-term live cell imaging. Image acquisition settings of the camera, such as binning or gain can be adjusted and special low-light cameras such as electron multiplying charge-couple device (EM-CCD; *see* below) can be used to improve the signal-to-noise ratio of low intensity fluorescence signals and cell viability. When selecting **time intervals** for image acquisition of time series during long-term live cell imaging, the following three parameters should be considered: the ability (1) to track single cell movements, (2) to identify cell type and morphological changes during cell migration, and (3) to visualize the dynamic changes of the actin cytoskeleton in the migrating cell.

1.2.1 Long-Term Live Cell Imaging with Multi-mode Epifluorescence Microscopy

The configuration of the epifluorescence light microscope used for long-term live cell imaging of migrating epithelial cells is shown in Fig. 1a. The inverted research microscope, e.g., Zeiss Axio Observer Z1, is equipped with a motorized Z-focus, motorized reflector and objective turrets, and a motorized condenser to ensure precise and fast switching between different optical components. The instrument is placed on a vibration isolation table for maximum stability. The microscope base has a high wattage metal halide light source (e.g., PhotoFluor LM75 or X-Cite 120) for

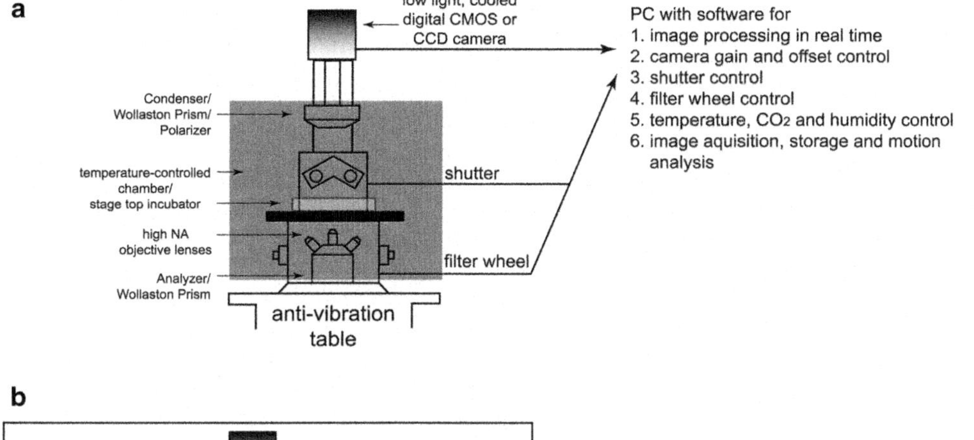

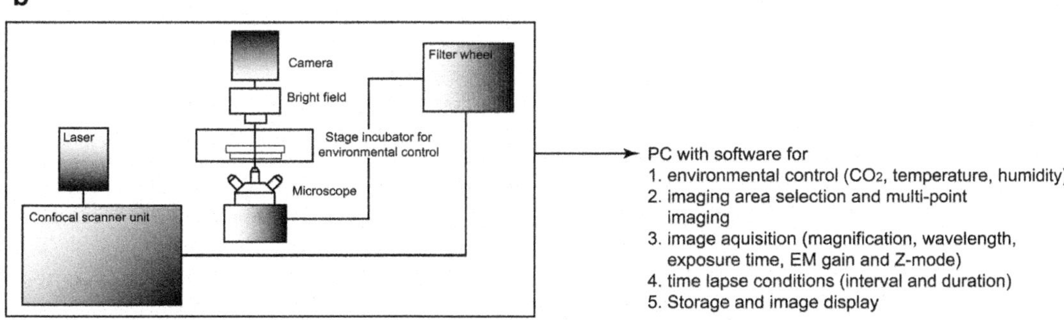

Fig. 1 Schematic diagrams of two types of imaging systems for long-term live cell observation. (**a**) High resolution light microscope system for fluorescence intensified and differential interference contrast (DIC) microscopy. An inverted light microscope placed on a vibration isolation table for stabilization and equipped with a digital high-resolution complementary metal-oxide semiconductor (CMOS) camera and an environmental specimen control chamber, which is required for long-term live cell imaging. (**b**) Diagram of a spinning disk confocal imaging system designed by Olympus for long-term live cell imaging. The CV1000 consists of up to three excitation laser wavelengths (405, 488, 561 nm), a dual Nipkow spinning disk and a stage incubator with temperature and humidity control. The microscope unit contains a red LED illumination light source for bright-field imaging, a back-illuminated electron-multiplying charge-couple device (EM-CCD) camera, up to six objective lenses, a auto *X*–*Y* stage and a motorized *Z*-axis control

epifluorescence microscopy. Microscope objectives were selected based on the appropriate numerical aperture, working distance and magnification required for imaging fine details of the specimen. Oil immersion objectives (40× or 63×) with high numerical aperture (1.3–1.4 NA) and a short working distance for high lateral and axial resolution are used routinely for capturing the actin cytoskeleton in migrating epithelial cells. High numerical aperture objectives are optimal because they collect more light resulting in better resolution and brighter images in fluorescence microscopy. Most importantly for long-term live cell imaging, the microscope stage is equipped with a live cell imaging chamber/incubator or a controlled environment microscope incubator to maintain the desired temperature and humidity (Fig. 1a).

Single images or time series of the actin cytoskeleton in migrating fluorescently labeled epithelial cells are captured with a CCD (e.g., Hamamatsu C4742-98-24NR) or CMOS camera such as the Hamamatsu Orca-Flash 4.0 LT (Fig. 1a and *see* below). These cooled, low-light level detection cameras provide high resolution (2048×2048 pixels) images and the speed required to capture images in real time (framing rate of 30 frames/s) at full resolution. These cameras are suitable for a wide range of image applications such as brightfield and epifluorescence, ratiometric Ca^{2+} imaging, and live cell and 3D imaging. Additional features of the CMOS camera include USB 3.0, high quantum efficiency and a wide dynamic range. PC-based imaging software, e.g., MetaMorph (Molecular Devices) (IPLab Imaging from BD Biosciences can also be used) is used to control image acquisition and processing (e.g., exposure time, background correction, saturation, 2D or 3D convolution) and camera settings (e.g., gain and offset levels, image size and position). The captured images or time-lapse sequences are then stored on external hard-disk drives (HDD) or online backup devices such as a server-based flash storage. The imaging software contains motion analysis tools to quantify migration of epithelial cells using object tracking and kymograph function for distance, time and velocity measurements.

1.2.2 Long-Term Live Cell Imaging with Spinning-Disk Confocal Microscopy (Olympus CV1000)

Spinning disk confocal microscopy allows acquisition of images at real time or even faster with a minimum of specimen illumination. These specific features make this fluorescent technique suitable for high-speed 3D imaging of living specimen and dynamic events such as cell migration. The principle is based on a Nipkow spinning style disk (pinhole disk) that contains thousands of pinholes and a second rotating disk (lens disk) that contains the same number of focusing microlenses. Each illuminated pinhole is focused on the specimen plane by the objective and emitted fluorescence is collected and visualized after passing back through the pinhole. The spinning disk confocal microscope illuminates multiple points in the specimen simultaneously compared to a single point at a time with laser scanning confocal microscopy.

We used the Olympus CV1000, a spinning disk confocal microscope (Fig. 1b), which combines an environmental control and imaging system for long-term time lapse imaging of multiple cell locations (multi-point imaging). The CV1000 contains a Yokogawa spinning disk scanning unit and up to three excitation laser lines (405, 488 and 561 nm) that can be integrated for either confocal or epifluorescence. LED light is used for brightfield illumination and image acquisition is performed with a high-sensitivity back-illuminated EM-CCD camera (512×512 pixels). The stage incubator holds multiple inserts for fixed or live samples (e.g., single or multiple 35 mm dishes, multi-well plates or glass slides) while environmental conditions such as temperature, humidity and CO_2 gas concentration are controlled by software. Finally, the Olympus CV1000 provides an autofocus system that uses laser light to avoid focus drift by comparing images taken in reference to the glass surface of the culture vessel.

1.3 Cameras for Long-Term Live Cell Imaging

Fluorescence microscopy remains one of the best methods to visualize and track dynamic molecular components in living specimen. However, to capture images of fluorescent molecules in cells without photobleaching and photodamage, require high-sensitivity cameras that are designed for long-term time-lapse imaging. Determining the most suitable camera for a specific application is an important exercise because no single imaging sensor possesses the full range of specifications to meet all requirements for recording fluorescence in living cells.

Cooled charge-coupled device (CCD) cameras are most widely used for live cell imaging and different applications of fluorescence microscopy such as TIRF and real-time confocal microscopy. In CCD image detectors photodiodes collect photons that are converted into voltage levels, which are then amplified and translated into intensity values. Cooled CCD cameras provide a low-light level detection with a high signal-to-noise ratio and high resolution, a wide dynamic range, and high quantum efficiency and spatial resolution. Even though these cameras are most often used for FRET, ratio and color time-lapse imaging, they are also suitable for high-resolution applications such as TIRF and real-time confocal microscopy. CCD cameras can be obtained from Hamamatsu (Orca-3 and Orca-R2), Photometrics (CoolSnap) or Princeton Instruments (PIXIS), Andor (Clara), and Qimaging (QIClick). The disadvantages of CCD's are that they are expensive, need a higher data rate and consume more power than CMOS sensors (*see* below). Dual CCD cameras have been designed for simultaneous acquisition of dual wavelengths with different intensities, which makes them unique for FRET and ratio imaging techniques such as Ca^{2+} and membrane potential measurements. Several manufacturers, e.g., Andor Technology, Princeton Instruments, Stanford Computer Optics and Hamamatsu also provide intensified CCD (ICCD's) cameras that are equipped with an image intensifier mounted in front of the CCD sensor for very low light-level detection down to a single photon at high framing rates. These digital cameras offer high signal-to-noise ratios below that of slow-scan CCD cameras, which are designed specifically for low-light applications. However, ICCD cameras show limitations in spatial resolution and have a reduced dynamic range and quantum efficiency compared to EM-CCD cameras (*see* below) [16].

Electron multiplying charge-couple device (EM-CCD) cameras were first introduced by Andor Technologies in early 2000. An EM-CCD camera does not contain an image intensifier as ICCD's but uses an electron-multiplying CCD sensor that multiplies the signal electrons prior to the addition of read-out noise. EM-CCD cameras combine high quantum efficiency and resolution, and a rapid framing rate to capture images at very low light levels resulting in shorter exposure times. EM-CCD's offer the best signal-to-noise ratio of any camera technology, which will require lower fluorophore concentrations and power levels. Applications include imaging of protein-protein interactions, real-time spinning disk confocal and single molecule TIRF microscopy. Scientific EM-CCD cameras can be obtained from Hamamatsu (ImageEM), Andor Technology (iXon), Photometrics (Evolve) or Raptor Photonics (Falcon and Kite) [17, 18].

Complementary metal oxide semiconductor (CMOS) cameras have replaced many CCD cameras since these detectors provide a faster framing rate, have low power consumption and are more cost-effective than CCD's. CMOS imaging sensors have pixels composed of a photodiode and an amplifier that converts light energy into voltage. However, these sensors do not perform very well at low-light applications. However, second generation CMOS cameras that are built with improved sCMOS sensors, offer a higher signal-to-noise ratio, rapid framing rates (30–100 frames/s), a wide dynamic range, high quantum efficiency and resolution, and a large field of view (e.g., 2048 × 2048 pixel for the Hamamatsu Orca-Flash 4.0 LT). These features allow image acquisition of rapid structural changes in cells and resolution of fine structural detail such as the actin cytoskeleton. Further applications for sCMOS imaging include single molecule detection, TIRF and super-resolution microscopy, and light sheet microscopy. sCMOS cameras are available from Andor Technology (Neo and Zyla), Hamamatsu Corporation (Hamamatsu Orca-Flash 4.0 LT), Photonics USA (xSCell), ThorLabs (DCC), and Qimaging (Rolera Bolt) [18].

Finally, modern video cameras remain appropriate for imaging cells in brightfield but are not recommended for live cell fluorescence microscopy. They can be used for conventional immunofluorescence imaging and FISH however. Most of them have a fixed video rate of 30 frames per second and are useful in low to moderate light applications. Both color and monochrome video cameras have comparable signal-to-noise ratios and dynamic ranges.

1.4 Environmental Control for Long-Term Live Cell Imaging

For long-term live cell imaging of mammalian cells, the observation chamber becomes a critical part of the microscope apparatus and must function to maintain conditions as close to the natural biological state as possible. Therefore, it is important to control the temperature, pH and concentrations of CO_2, and nutrients while cells are being observed on the microscope stage to obtain biologically relevant results. Modern long-term live cell imaging systems allow one to control all of these critical parameters and create a long-term stable environment.

The simplest systems are **microscope stage warmers for microscope slides and petri dishes/6-well plates**. They provide localized temperature control during short-term live cell imaging and these systems work well if humidity and CO_2 control are not required (Fig. 2a). Microscope stage warmers are available from Omano (SW99), AmScope (TCS-100), Lec Instruments (Lec 916) and Okolab. These temperature control systems are designed to fit most brands of microscopes and vary in size and shape. Okolab for instance offers thermal glass or metal plates for inverted and upright microscopes to maintain the temperature of the living specimen. To reduce temperature loss during the imaging process, many stage warmers are equipped with insulating pads to isolate the heated platform from the microscope stage.

Better options are **live-cell perfusion imaging chamber** systems that can be obtained in either an open or closed chamber configuration (Fig. 2b). Open chamber systems are similar to petri dishes and provide easy access to the cells whereas closed or sealed chambers protect the cells from any environmental disturbances. The three primary methods to perfuse these chambers are: gravity flow, manual injection using a syringe or mechanical pumps (motorized syringe or peristaltic). Gravity flow is most difficult to control as the formation of air bubbles may block the laminar flow of the cell culture medium and deprive cells of fresh medium. For better control of perfusion of cell culture chambers, a syringe or peristaltic pump is used to generate a constant flow rate [19]. Open dish systems are offered from Bioptech (Delta T) or Bioscience Tools (Flow-Petri) and more advanced closed perfusion systems are available for upright (FSC3 from Bioptech) and inverted microscopes (FSC2 from Bioptech, TC-MIS from Bioscience Tools or CV-30 from Warner Instruments).

The best option is a **stage top incubator** (e.g., ibidi, Tokai Hit, Bioscience Tools, Warner Instruments or Okolab) that provides precise temperature, humidity and CO_2 control, an ideal way to regulate the incubation and observation environments (Fig. 2c).

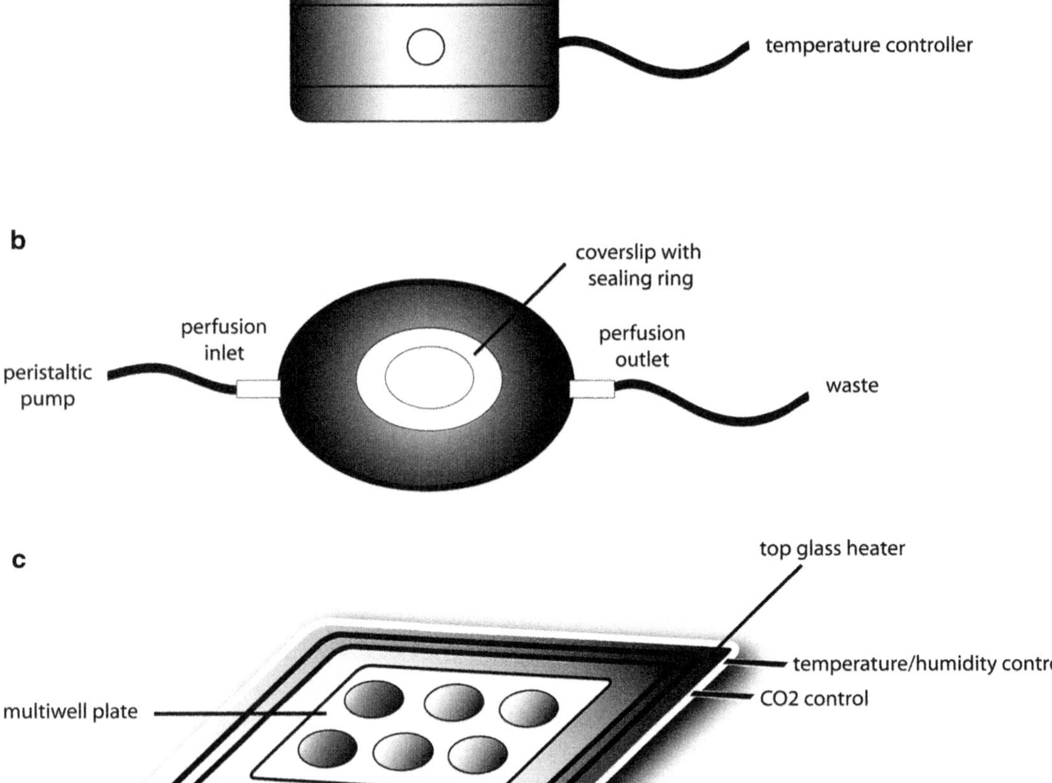

Fig. 2 A schematic diagram of a commercially available microscope temperature control stage slide warmer (**a**), live cell perfusion imaging chamber (**b**), and microscope stage top incubator (**c**) for maintaining the biological environment required for the observation of living cells during long-term live cell imaging. The stage slide warmer (**a**) consists of a metal top plate that is suitable for standard dimension microscope slides. Closed circular perfusion chambers (**b**) are sealed to protect the specimen and to maintain the environmental settings. Peristaltic pumps or motorized syringes are required to deliver medium or perfusate to the perfusion chamber. Stage top incubators (**c**) represent advanced live cell imaging chambers that maintain proper humidity levels and are equipped with carbon dioxide and temperature injectors

Most stage top incubators provide temperature sensors in the heated glass lid and in the chamber body, perfusion holes for inlet and outlet tubes and fit most microscope platforms and stages. In addition, temperature and gas management software modules are often available to control a digital gas mixer and thereby regulate gas concentrations (CO_2 and O_2) during long-term live cell imaging. Complex microfluidic systems for medium exchange or drug administration are also available from suppliers such as EMD Millipore (CellAsic), Bioscience Tools (TC-MIW) or Tokai Hit. These systems combine a syringe pump, a perfusion chamber and a CO_2 incubator into a single unit. Gas and temperature controls are connected directly to the cell culture chamber and can be mounted on any inverted microscope stage [20].

"Cage incubators or microscope enclosures" (e.g., Okolab, Zeiss and Nikon) are the most complex and sophisticated units to maintain all required environmental conditions around the microscope stage for long-term live cell imaging over a period of several days. Alternatively, a cell incubator with a built-in imaging system such as the incubator fluorescence microscope (VivaView FL) from Olympus or the BioStation IM/CT from Nikon Instruments are available to perform long-term time-lapse imaging of epithelial cells [21].

2 Materials

2.1 Cell Culture

1. Human oral keratinocytes (OKF6/TERT-2) (Harvard Skin Disease Research Center & BWH/Partners) ([15]; *see* **Note 1**) and African green monkey kidney cells (COS7) (ATCC CRL-1651, Rockville, MD).
2. Keratinocyte serum free medium (K-SFM) (Invitrogen, Carlsbad, CA) supplemented with bovine pituitary extract (BPE) and 0.2 ng/ml epidermal growth factor (EGF). BPE and EGF are supplied with K-SFM.
3. Dulbecco's modified eagle medium (DMEM) (Invitrogen) supplemented with 10 % fetal bovine serum (FBS) (Atlanta Biologicals, Flowery Branch, GA).
4. Penicillin/streptomycin (100×) from cellgro.
5. Calcium chloride ($CaCl_2$) is dissolved at 1 M in molecular grade water (cellgro, Mediatech, Inc., Herndon, VA) and filter sterilized before storage at 4 °C.
6. Sterile phosphate buffered saline (PBS) (1×) from cellgro.
7. Solutions of 0.05 and 0.25 % trypsin and ethylenediamine tetra-acetic acid (EDTA) from Gibco (Invitrogen, Carlsbad, CA).
8. Freezing media: 20 % DMSO (Sigma-Aldrich, St Louis, MO) in 1 mL K-SFM for OKF6/TERT-2 cells, and 5 % DMSO in 1 mL DMEM for COS-7 cells.
9. 10 cm plastic cell culture dishes (Falcon, Beckton Dickinson Labware, Franklin Lakes, NJ).
10. T-75 flasks (75 cm^2, Corning Incorp., Corning, NY).
11. 15 mL centrifuge tubes (Fisher Scientific, Pittsburgh, PA).
12. 45 mm bottle top filter for rapid sterile filtration (EMD Millipore, Billerica, MA).

2.2 C. albicans Culture

1. *Candida albicans* wild-type strain SC5314 [22].
2. Yeast nitrogen base (YNB) medium: 1.7 g YNB from Difco (Voigt Global Distribution Inc., Lawrence, KS), 5 g ammonium

sulfate (Sigma-Aldrich), 9 g glucose (Sigma-Aldrich), 10 mL of 0.02 % biotin (Sigma-Aldrich) are dissolved in 1 L of distilled water and filter sterilized.

3. YNB plates: All supplements for YNB medium (*see* above) except for biotin and 15 g agar are dissolved in 1 L of distilled water and autoclaved. After the medium is cooled down until hand-hot 10 mL of 0.02 % biotin is added and plates are poured.

2.3 Transient Cell Transfection

1. Optifect transfection reagent (Invitrogen) (*see* **Note 2**).
2. Opti-MEM reduced serum medium supplemented with HEPES (4-(2-hydroxyethyl)-1-piperazineethanesulfonic acid), 2.4 g/L sodium bicarbonate, and L-glutamine (Gibco).
3. pAcGFP1-actin vector (Clontech Laboratories Inc., Mountain View, CA) or LifeAct-TagRFP (ibidi, Planegg/Martensried, Germany).
4. 35 mm glass bottom dishes (MatTek, Ashland, MA).

2.4 Long-Term Real-Time and Time-Lapse Live Cell Imaging

1. 35 mm glass bottom dishes (MatTek, Ashland, MA).
2. Immersion oil for microscopy, Immersol 518N (Carl Zeiss Inc., Thornwood, NY)
3. Zeiss Axio Observer Z1 motorized microscope equipped with differential interference optics (DIC), a 40×/1.3 NA Plan Apo oil objective lens and filter sets for FITC, TRITC, and DAPI fluorescence (Carl Zeiss Inc.).
4. PhotoFluor-LM 75 light source for epifluorescence microscopy (Chroma Technology Corp., Bellows Falls, VT).
5. Orca-Flash 4.0 LT C11440-42U digital CMOS camera (Hamamatsu Photonics, Bridgewater, NJ).
6. Apple Mac Pro 3.2GHz Quad-Core Intel Xeon with 6 GB RAM (random access memory) and 1 TB hard disk drive (HDD) (Apple Inc., Cupertino, CA).
7. Stage top incubator with dish attachment for 35 mm dishes (UNIV-D35), temperature controller and built-in gas mixer (Tokai Hit CO., Japan).
8. MetaMorph image acquisition and analysis software (Molecular Devices, Downingtown, PA).

2.5 Long-Term Real-Time and Time-Lapse Live Cell Imaging Using Incubator Spinning Disk Confocal Microscope (Olympus CV1000)

1. 35 mm glass bottom dishes (MatTek).
2. Olympus CV1000 Incubator spinning disk confocal microscope (Olympus America Inc., Center Valley, PA) equipped with a motorized stage incubator for one 35 mm glass bottom dish, an back-illuminated EM-CCD camera (512 × 512 pixels), a dry 10×/0.3 NA (Standard), and a 60×/1.25 NA and 100×/1.3 NA oil objective lenses (*see* **Note 3**).

3. 1 W LED (625 nm) for transmitted light.
4. Late model PC computer with fast processor and most random access memory (RAM).
5. Image acquisition and environmental control software (Olympus).

2.6 Data Analysis

1. MetaMorph image acquisition and analysis software (Molecular Devices).

3 Methods

3.1 Cell Culture

3.1.1 Culture of OKF6/TERT-2 Cells

1. Combine the supplements (1.089 mL BPE and 0.2 ng/mL EGF) and 0.3 mM $CaCl_2$ with the keratinocyte serum free medium (K-FSM), filter sterilized with a 45 mm bottle top filter and store at 4 °C.
2. Maintain the oral mucosal cell line, OKF6/TERT-2 keratinocytes in 10 mL K-FSM at 37 °C in a 5 % CO_2 in air atmosphere. Grow the cells in 10 cm plastic dishes and change the medium every other day before microscopic observation.
3. Passage OKF6/TERT-2 keratinocytes when they reached ~30 % confluence (~$7–9 \times 10^5$ cells) with 0.5 % trypsin/EDTA. Remove the K-SFM from the cell culture dish and wash cells with 10 mL PBS. Add 2 mL trypsin/EDTA solution to the cells and incubate them for 2–10 min at 37 °C until the cells are rounded up and detached spontaneously from the cell culture dish. Collect cells with 8 mL K-SFM to inactivate the trypsin and transfer them to a 15 mL centrifugation tube. Centrifuge the cells at 75 g for 5 min in a clinical centrifuge. Aspirate supernatant to remove trypsin before adding fresh medium and seeding in culture dishes. Split the OKF6/TERT-2 cells at a 1:3 ratio (~1×10^5 cells), which provides experimental cell cultures that will reach ~30 % confluence after 2 days. Cryopreserve cells in freezing medium containing 20 % DMSO (v/v) followed by long-term storage in liquid nitrogen.

3.1.2 Culture of COS-7 Cells

1. Remove 50 mL DMEM from 500 mL bottle, then add 50 mL FBS (10 %) and filter sterilized with a 45 mm bottle top filter, store at 4 °C.
2. Culture COS-7 cells green monkey kidney cells in T-75 flasks containing 12 mL complete Dulbecco's modified eagle medium (DMEM) at 37 °C in a 5 % CO_2 in air atmosphere.
3. Passage cells when they reached ~90 % confluency. Remove the medium from the flask and trypsinize cells for 5–15 min with 2–3 mL 0.25 % Trypsin/EDTA at 37 °C. Add 7–8 mL of complete growth medium and centrifuge the cells at for 3 min.

Aspirate the supernatant and resuspend cell pellet in 12 mL of fresh DMEM. Split cells at between 1:4 and 1:8. Replace the culture medium 2–3 times per week before processing cells for live cell imaging. The cells are cryogenically frozen in growth medium supplemented with 5 % (v/v) DMSO.

3.2 Transfection of OKF6/TERT-2 and COS-7 Cells with Actin Expression Vectors

1. For transient transfections using either the pAcGFP1-actin vector or the LifeAct-TagRFP vector, plate cells in 35 mm glass bottom dishes 1 day before transfection. Importantly, grow the cells in culture medium without antibiotics 1 day before transfection to prevent cell death.
2. The following protocol used to transfect epithelial cells was optimized in our laboratory.
3. To prepare the DNA-Optifect complexes, add 4 μL (~4 μg) of pAcGFP1-actin DNA or LifeAct-TagRFP DNA to 246 μL of Opti-MEM reduced serum medium and gently mix the solution. Dilute 17 μL of Optifect transfection reagent with 233 μL Opti-MEM, mix gently and incubate for 5 min at room temperature. Combine the diluted DNA with the diluted Optifect solution and incubate the mixture for 20 min at room temperature.
4. Add the DNA-Optifect complexes (500 μL total volume) to 2 mL of culture medium containing the OKF6/TERT-2 cells while rocking the 35 mm glass bottom dish. Replace the cell culture medium after 4 h and incubate the cells for a period of 24 h at 37 °C prior to live cell imaging.

3.3 Coculture of *C. albicans* with Epithelial Cells

1. Inoculate *C. albicans* cultures in 10 mL of YNB medium in an Erlenmeyer flask.
2. Grow cultures of *C. albicans* overnight in a rotary shaker at 250 rpm at 30 °C (*see* **Note 4**).
3. Pellet the overnight *C. albicans* cultures in a tabletop centrifuge and resuspend the pellet in 10 mL K-SFM (OKF6/TERT-2) or 10 mL DMEM (COS-7).
4. Replace the cell culture medium of the 35 mm glass bottom dish with 2 mL of either K-SFM or DMEM medium containing ~8.5×10^6 *C. albicans* cells and incubate the dish for 5 min at 37 °C in a 5 % CO_2 in air atmosphere.
5. Wash cell cultures twice with 2 mL PBS to remove any unbound *C. albicans* and add 2 mL medium (K-SFM or DMEM) to the 35 mm glass bottom dish.
6. Incubate 35 mm glass bottom dishes for defined time intervals (e.g., 0, 1, 2, and 3 h) at 37 °C in a 5 % CO_2 in air atmosphere followed by live cell imaging (*see* **Note 5**).

3.4 Long-Term Live Cell Imaging

3.4.1 Multi-mode Epifluorescence Microscopy

1. Grow and transfect epithelial cells in 35 mm glass bottom dishes followed by coculture with *C. albicans* as described above (*see* Subheadings 3.2 and 3.3).
2. Turn on the temperature controller of the stage top incubator and lens heater at least 30 min before starting live cell imaging to ensure focus stability.
3. Install the lens heater to the Plan-Apochromat 40× objective and place a drop of immersion oil on top of the objective lens. Insert the 35 mm glass bottom dish into the dish attachment of the stage top incubator on the microscope stage. Place the sterilized sensor lid, which contains the sterilized feedback sensor for the temperature control unit, on the 35 mm glass bottom dish. Ensure that the sensor tip is inserted into the cell culture medium. Attach the top heater and the top heater protector to the stage top incubator.
4. Image cells during coculture with *C. albicans* over periods of 1 h at 37 °C. Actin filament dynamics in migrating cells is determined by epifluorescence microscopy. Microscope control, camera settings and image acquisition are done with MetaMorph imaging software.
5. Prior to epifluorescent illumination, locate the focal plane with DIC microscopy to focus the specimen and to prevent photobleaching of the fluorescent specimen. When the cells are in focus, perform Köhler illumination (the field diaphragm is focused by adjusting the height of the condenser) to provide even illumination over the specimen field.
6. Use epifluorescence microscopy to detect positively transfected cells and to monitor actin filament dynamics in migrating cells during coculture with wild-type *C. albicans* (Fig. 3). Switch the light from the ocular to the camera with the beam splitter and acquire either single fluorescence images or time-lapse sequences with the MetaMorph imaging software and store them directly on the computers HDD or an external HDD for later analysis (Fig. 3).

3.4.2 Spinning Disk Confocal Long-Term Live Cell Imaging

1. Turn on the temperature control at least 3 h or preferably overnight before the beginning of imaging. This will allow all components of the microscope and the stage incubator to reach 37 °C to prevent focus drift. Set the desired CO_2 concentration (usually 5 %) using the Environment dialog box.
2. Apply a drop of immersion oil to the 100× objective lens and place the 35 mm glass bottom dish in the suitable attachment of the stage incubator. Insert the attachment into the spinning disk confocal microscope.

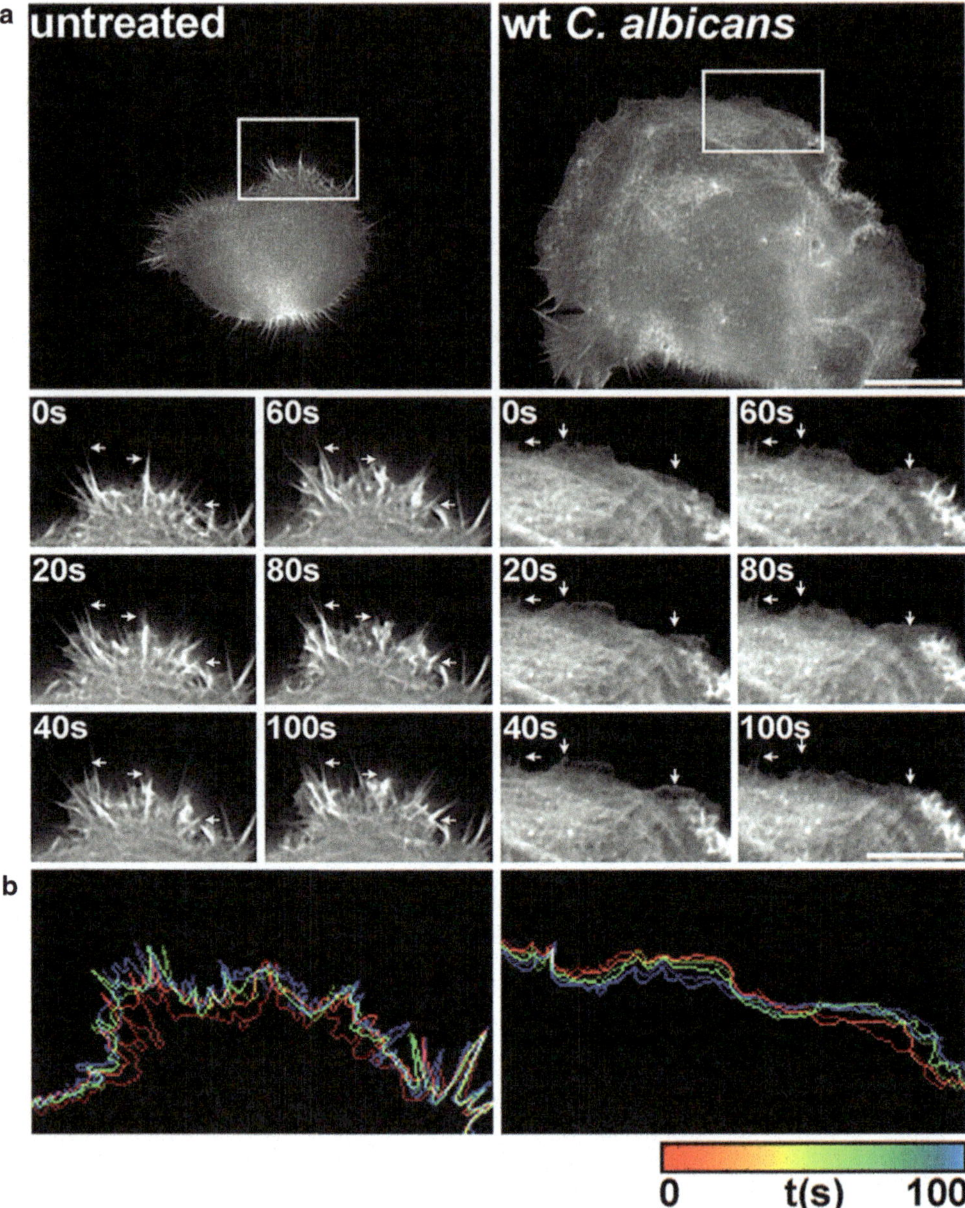

Fig. 3 Fluorescence micrographs of the actin filament dynamics in migrating human oral keratinocytes (OKF6/TERT-2) in the presence of *C. albicans* biofilms observed with fluorescence-intensified microscopy. (**a**, *top row*) Single frames from fluorescence time-lapse movies of OKF6/TERT-2 cells that were transiently transfected with mCherry-LifeAct and grown for 24 h. Untreated OKF6/TERT-2 cells (*left panel*) and OKF6/TERT-2 cells cocultured in the presence of biofilms formed by wild-type *C. albicans* (*right panel*) were observed with a Zeiss Axio Observer Z1 microscope using a 40× oil-immersion objective as described in Subheading 1.2.1. (**a**, *bottom panel*) Micrographs (0–100 s) of the actin containing structures and actin filament dynamics at the cell periphery (*see boxed regions* of the fluorescent micrographs in the *top row*). *Arrows* indicate regions of filopodial (untreated, *left panel*) or lamellipodial (treated with wild-type *C. albicans*, *right panel*) protrusion and retraction cycles. (**b**) Activity maps of the leading edge of migrating OKF6/TERT-2 cells shown in (**a**). *Color* encodes time points of leading edge position. *Bars* represent 20 μm

3. Determine the cell distribution and identify transfected cells within the cell population in the 35 mm dish using the Map view dialog box.
4. For imaging of the actin cytoskeleton within the region of interest (ROI) choose the Magnification dialog box to select a suitable objective lens that offers the best resolution.
5. Set the wavelength, excitation power, exposure time, electron-multiplying (EM) gain of the camera and Z-mode using the Measurement dialog box from the control panel (*see* **Note 6**).
6. Select the size of the recording area by choosing single or multiple fields within the ROI. Multiple imaging areas that differ in size can be recorded successively within the selected ROI.
7. Set the time-lapse conditions such as interval and duration for the live cell imaging using the Time Lapse Condition dialog box and start recording.

3.4.3 Motion Analysis

1. We have successfully used the MetaMorph imaging and analysis software for detailed studies of cell migration in response to pathogenic *C. albicans* in our laboratory (*see* **Note 7**).
2. Adjust brightness and contrast of the time-lapse image series using the Adjust Digital Contrast dialog box from the Display menu or use the Auto Enhance function to increase the appearance of fine details of the actin cytoskeleton by changing the grayscale or color level.
3. Calibrate the pixel-to-distance conversion factor to determine the protrusion velocities using a reference image (e.g., calibrated stage micrometer). Use the Calibrate Distance dialog box from the Measure menu to change the calibration settings.
4. Analyze the protrusion velocity of the leading edge in migrating cells from fluorescence time-lapse image series by generating kymographs (Fig. 4a, b). Choose the Kymograph dialog box from the Stack menu to create pixel intensity values along a user-defined line region for each image of a time-lapse series (Fig. 4a) (*see* **Note 8**).
5. Place a single 1 pixel width line using the point-to-point line drawing function at the margins of individual cells along the direction of movement on the first image of the time-lapse series (Fig. 4a) [23]. Note that the line width can be changed by increasing the number of pixels to capture structures of interest that move or oscillate laterally off the single 1 pixel line.
6. For data interpretation of the actin filament dynamics the activity maps and kymographs of the leading edges in migrating cells in untreated cultures or cells cocultured with wild-type *C. albicans* are compared (Fig. 5).

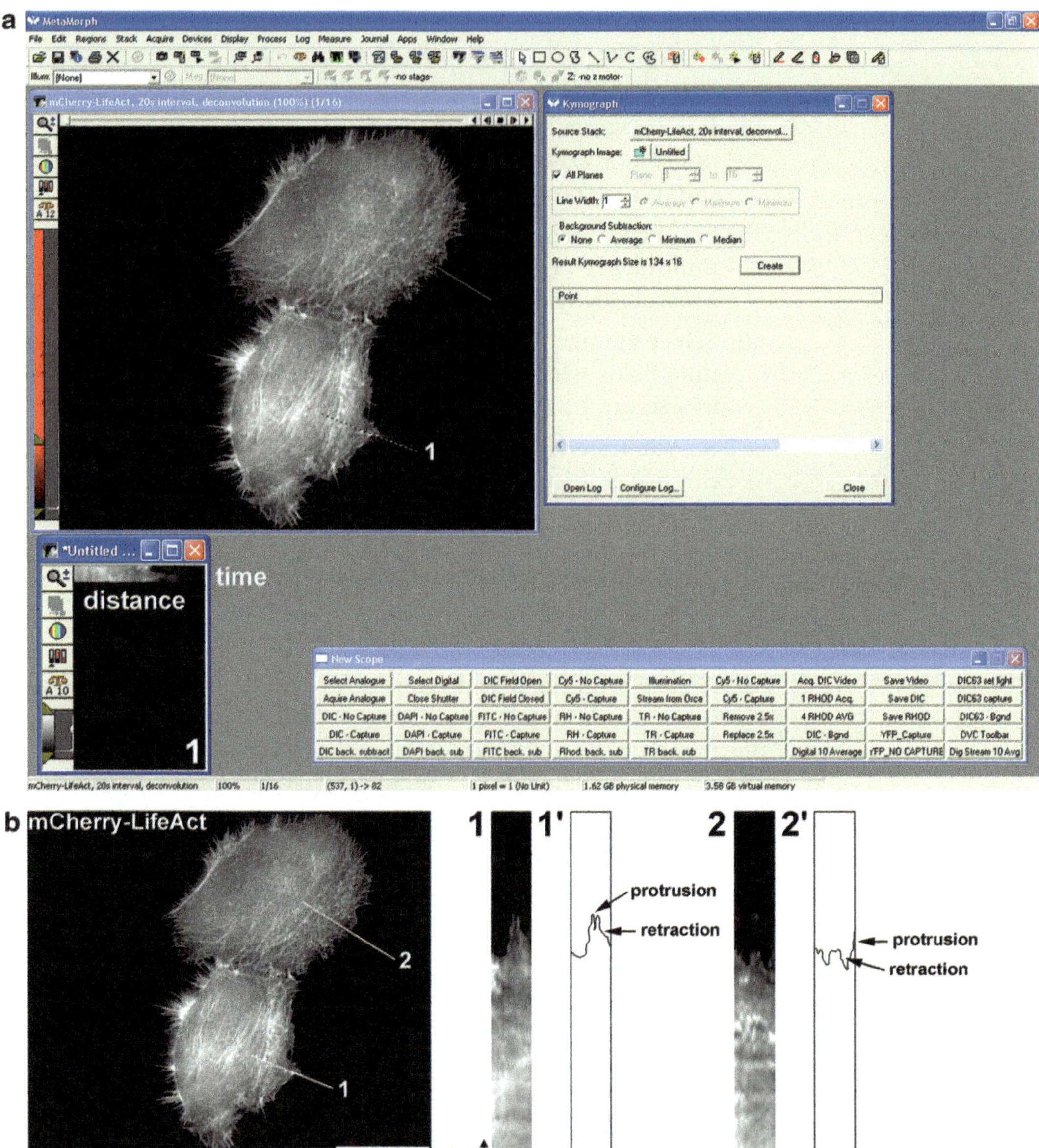

Fig. 4 Analysis of actin filament dynamics at the leading edge in migrating human oral keratinocytes. (**a**) Screen shot of the MetaMorph imaging and analysis software during an online measurement of untreated OKF6/TERT-2 cells migrating over the coverslip surface. The kymograph (time–distance plot below image) function enables the user to measure linear movement, such as distance, time, and velocity. A line or region of interest (ROI) was drawn along the leading edge and the grey values along this ROI of each frame from a time-series were read out with the MetaMorph software. The *y* axis of the kymograph becomes the time axis and the *x* axis is the distance along the line (ROI). (B) Single frame (*left*) taken from a fluorescence time-lapse series (0–5 min) obtained with the Zeiss Axio Observer Z1 using a 40× objective and corresponding kymographs (1 and 2, *right*) and protrusion maps (1′ and 2′, *right*) from two single cells transiently transfected with mCherry-LifeAct as shown in left image. *Lines* show ROIs used to create kymographs. Kymographs of leading edges that were created to analyze the actin filament dynamics at the cell periphery and cell migration velocity demonstrated that the free edges of these cells exhibit protrusive and retractive activity. *Bar* represents 20 μm

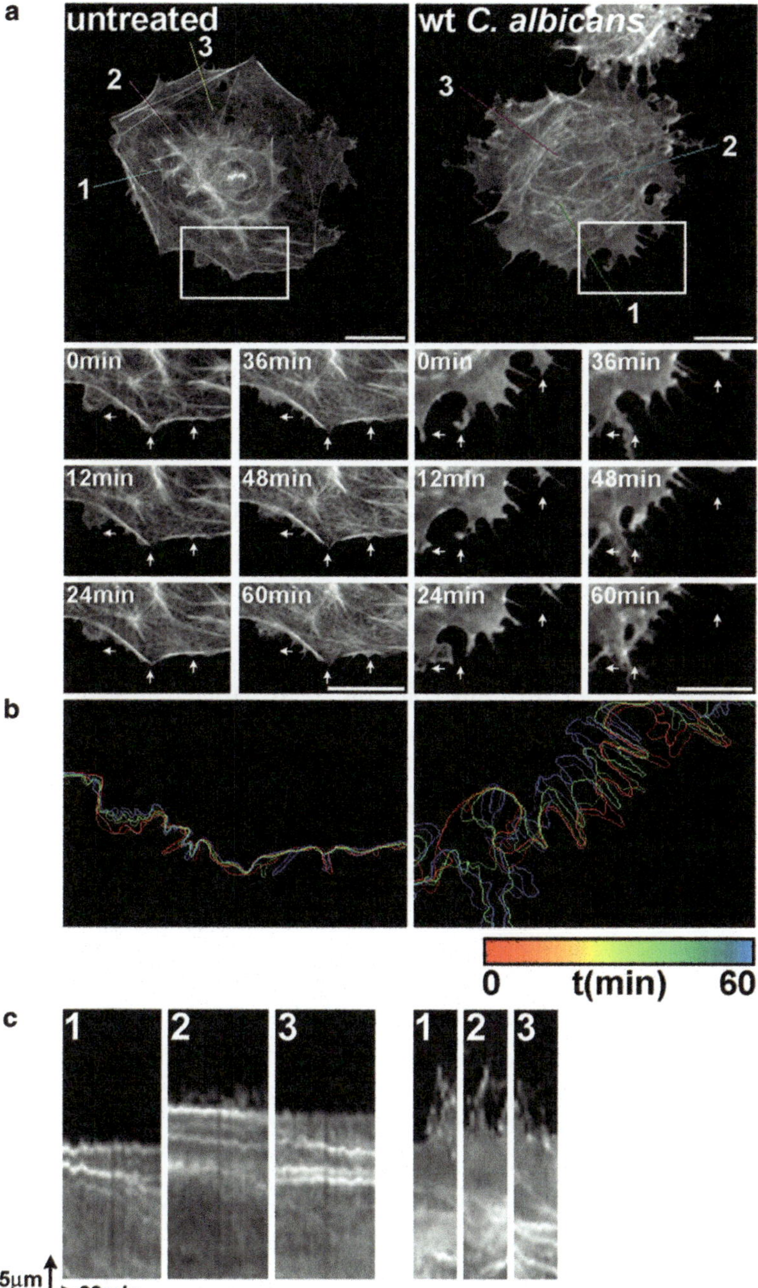

Fig. 5 Analysis of actin filament dynamics at the leading edge in migrating green monkey kidney cells (COS7). (**a**, *top row*) Single frames from fluorescence time-lapse movies of COS7 cells that were transiently transfected with mCherry-LifeAct and grown for 24 h. Untreated cultures of COS7 cells (*left panel*) and COS7 cells cocultured with biofilms formed by wild-type *C. albicans* (*right panel*) were then processed for live cell imaging. Images were taken with an Olympus CV1000 spinning disk confocal microscope using a 100× oil-immersion objective and 561 nm excitation wavelength laser as described in Subheading 2.4. (**a**, *bottom panel*) Micrographs (0–60 min) of the actin filament dynamics at the cell periphery (*see boxed regions* of the fluorescent micrographs shown in the *top row*). *Arrows* indicate regions of protrusion and retraction cycles in untreated cultures or during coculture with wild-type *C. albicans*. (**b**) Activity maps of the leading edge of migrating COS7 cells shown in (**a**). *Color* encodes time points of leading edge position. *Bars* represent 20 and 5 μm. (**c**) Lines in figures (**a**) (*top row*) show ROIs used to create kymographs of leading edges in untreated (*left panel*) or COS7 cells cocultured with *C. albicans* (*right panel*) to analyze the actin filament dynamics

4 Notes

1. Human oral keratinocytes (OKF6/TERT-2) resemble keratinocytes in the intermediate layers of the stratified squamous epithelium and can be obtained only from the Harvard Skin Disease Research Center and BWH/Partners.
2. The Optifect transfection reagent should be used for low-density cell cultures (10–70 % confluence) or cells that are sensitive to other commercial available transfection reagents because of it's reduced toxicity.
3. The Olympus CV1000 can hold up to a total of 6 objective lenses including dry (10× is standard), oil, water and long working distance objectives (20× and 40×).
4. Liquid *C. albicans* cultures can be prepared either from a glycerol stock stored at −80 °C or from a single colony that is picked from an YNB plate.
5. The cell culture medium can also be replaced with 2 mL of live cell imaging solution from Molecular Probes. This medium has been designed for different live cell imaging applications and provides better cell viability and signal-to-noise ratio.
6. The attachments can be removed from the stage incubator of the Olympus CV1000 and placed in a cell culture incubator to reach the desired temperature of 37 °C much faster.
7. Open source image analysis software such as ImageJ (NIH) or commercially available software packages including *Volocity* (Improvision), *IPLab* (BioVision) or *NIS elements* (Nikon Instruments Inc.) can be also used for motion analysis.
8. Kymographs combine pixel intensity values into a single image that displays distances (horizontal direction) versus time (vertical direction).

Acknowledgements

The authors would like to thank Tim Randall and Jim Foley from Olympus America Inc. and Caroline Nissan from Eicom USA for technical support.

References

1. Lauffenburger DA, Horwitz AF (1996) Cell migration: a physically integrated molecular process. Cell 84:359–369
2. Horwitz R, Webb DJ (2003) Cell migration. Curr Biol 13:R756–R759
3. Hall A, Nobes CD (2000) Rho GTPases: molecular switches that control the organization and dynamics of the actin cytoskeleton. Philos Trans R Soc Lond B Biol Sci 355: 965–970

4. Heasman SJ, Ridley AJ (2008) Mammalian Rho GTPases: new insights into their functions from in vivo studies. Nat Rev Mol Cell Biol 9:690–701
5. Etienne-Manneville S (2008) Polarity proteins in migration and invasion. Oncogene 27: 6970–6980
6. Joo HS, Otto M (2012) Molecular basis of *in vivo* biofilm formation by bacterial pathogens. Chem Biol 19:1503–1513
7. Zhu W, Filler SG (2010) Interactions of *Candida albicans* with epithelial cells. Cell Microbiol 12:273–282
8. Mayer FL, Wilson D, Hube B (2013) *Candida albicans* pathogenicity mechanisms. Virulence 4:119–128
9. Claesson D, Rozen DE, Kuipers OP, Søgaard-Andersen L, van Wezel GP (2014) Bacterial solutions to multicellularity: a tale of biofilms, filaments and fruiting bodies. Nat Rev Microbiol 12:115–124
10. Fox EP, Nobile CJ (2012) A sticky situation: untangling the transcriptional network controlling biofilm development in *Candida albicans*. Transcription 3:315–322
11. Douglas JL (2003) *Candida* biofilms and their role in infection. Trends Microbiol 11:30–36
12. Stephens DJ, Allan VA (2003) Light microscopy techniques for live cell imaging. Science 300:82–86
13. Dormann D, Weijer CJ (2006) Imaging of cell migration. EMBO J 25:3480–3493
14. Frigault MM, Lacoste J, Swift JL, Brown CM (2009) Live-cell microscopy—tips and tools. J Cell Sci 15:753–767
15. Dickson MA, Hahn WC, Ino Y, Ronfard V, Wu JY, Weinberg RA, Louis DN, Li FP, Rheinwald JG (2000) Human keratinocytes that express hTERT and also bypass a p16 (INK4a)-enforced mechanism that limits life span become immortal yet retain normal growth and differentiation characteristics. Mol Cell Biol 20:1436–1447
16. Spring KR (2013) Cameras for digital microscopy. Methods Cell Biol 114:163–178
17. Coates CG, Denvir DJ, McHale NG, Thurnbury KD, Hollywood MA (2004) Optimizing low-light microscopy with back-illuminated electron multiplying charge-couple device: enhanced sensitivity, speed and resolution. J Biomed Opt 9:1244–1252
18. Jung J, Weisenburger S, Albert S, Gilbert DF, Friedrich O, Eulenburg V, Kornhuber J, Groemer TW (2013) Performance of scientific cameras with different sensor types in measuring dynamic processes in fluorescence microscopy. Microsc Res Tech 76:835–843
19. Hing WA, Poole CA, Jensen CG, Watson M (2000) An integrated environmental perfusion chamber and heating system for long-term, high resolution imaging of living cells. J Microsc 199:90–95
20. Tam J, Cordier GA, Bálint Š, Sandoval Álvarez Á, Borbely JS, Lakadamyali M (2014) A microfluidic platform for correlative live-cell and super-resolution microscopy. PLoS One 29:e115512
21. Wöllert T, Langford GM (2009) High-resolution multimode light microscopy of cell migration: long-term imaging and analysis. Methods Mol Biol 586:3–21
22. Fonzi WA, Irwin MY (1993) Isogenic strain construction and gene mapping in *Candida albicans*. Genetics 134:717–728
23. Hinz B, Alt W, Johnen C, Herzog V, Kaiser HW (1999) Quantifying lamella dynamics of cultured cells by SACED, a new computer-assisted motion analysis. Exp Cell Res 251: 234–243

Chapter 2

Live-Cell Imaging of Mitochondria and the Actin Cytoskeleton in Budding Yeast

Ryo Higuchi-Sanabria, Theresa C. Swayne, Istvan R. Boldogh, and Liza A. Pon

Abstract

Maintenance and regulation of proper mitochondrial dynamics and functions are necessary for cellular homeostasis. Numerous diseases, including neurodegeneration and muscle myopathies, and overall cellular aging are marked by declining mitochondrial function and subsequent loss of multiple other cellular functions. For these reasons, optimized protocols are needed for visualization and quantification of mitochondria and their function and fitness. In budding yeast, mitochondria are intimately associated with the actin cytoskeleton and utilize actin for their movement and inheritance. This chapter describes optimal approaches for labeling mitochondria and the actin cytoskeleton in living budding yeast cells, for imaging the labeled cells, and for analyzing the resulting images.

Key words Yeast, Mitochondria, Fluorescent proteins, Vital staining, Microscopy, Live-cell imaging, Deconvolution, Ratio imaging, Actin, Cytoskeleton

1 Introduction

Live-cell imaging is the optimal approach to visualize dynamic structures, and several imaging techniques have been developed to efficiently visualize both actin and mitochondria in living yeast cells. These techniques enable visualization of events that cannot be seen in fixed cells, including mitochondrial motility, fusion, and fission, and retrograde actin cable flow. Some vital dyes can also be used to assess the function and quality of mitochondria, including mitochondrial membrane potential ($\Delta\psi$), mitochondrial DNA (mtDNA) content, and mitochondrial redox state.

Here, we describe methods to tag yeast genes at their chromosomal loci, determine functionality of proteins with fluorescent tags, visualize tagged gene products expressed either from endogenous or other promoters, and carry out quantitative analysis of mitochondrial and actin dynamics. We also describe methods to

Ray H. Gavin (ed.), *Cytoskeleton: Methods and Protocols*, Methods in Molecular Biology, vol. 1365,
DOI 10.1007/978-1-4939-3124-8_2, © Springer Science+Business Media New York 2016

measure mitochondrial quality using probes engineered to quantitatively measure mitochondrial Δψ, redox state, and mtDNA content.

1.1 Detection of Cellular Structures Using Targeted Fluorescent Proteins

Both mitochondria and the actin cytoskeleton can be visualized in living yeast cells using fluorescent proteins (FPs). Mitochondria can be visualized by tagging a protein native to mitochondria, or by adding a short mitochondrial targeting sequence to a fluorescent protein.

1.1.1 Choosing a Fluorescent Protein

Green fluorescent protein (GFP), discovered and cloned from the jellyfish *Aequorea victoria,* revolutionized live-cell imaging in cell biology. Now, GFP is one of a large and growing palette of FPs with different colors and molecular properties. Identification of novel FPs, such as mCherry derived from *Discosoma sp.* coral and Teal from *Clavularia sp.* coral, combined with laboratory mutagenesis, has yielded FPs with a variety of colors as well as improved brightness, faster folding, and decreased oligomerization [1].

To perform well in a live-cell experiment, an FP must be photostable enough to withstand repeated imaging. It also must be bright enough to allow the structure of interest to be seen over background fluorescence and detector noise, and spectrally distinct from any other labels being used. Finally, it must not be toxic or disrupt the behavior of the cell or the protein to which it is fused. The effective brightness of an FP is determined by its intrinsic brightness (the product of extinction coefficient and quantum yield), and the behavior of the FP upon ectopic expression (rate of folding and stability). The amount of this signal that is excited and detected depends on properties of the imaging system (light sources, lenses, filters and detectors). As an aid to selecting fluorescent proteins, Talley Lambert and Kurt Thorn have prepared an extensive database of physical properties of FPs, available at http://nic.ucsf.edu/FPvisualization/.

1.1.2 Tagging Endogenous Proteins

For chromosomal tagging of proteins in yeast, a double-stranded linear DNA that encodes the tag of interest and a selection marker is inserted into a target site in the genome by homologous recombination. This insertion cassette is most commonly produced by PCR using a tagging vector as a template. Tagging vectors have been developed for a variety of FPs (e.g., GFP, mCherry, mCitrine), epitopes (e.g., HA, myc), and affinity tags (e.g., GST, TAP, His) and have been linked to a variety of selection markers conferring drug resistance or rescue of auxotrophy. Families of tagging vectors have been constructed that share PCR-priming sequences. With these, a single set of primers can be used to insert any tag into a given target gene. Vectors are also available for expression of tagged genes from endogenous promoters, constitutively active promoters (e.g., *AHD1*, *GPD1*), and regulatable promoters (e.g., *GAL1*).

FP and epitope tags can also be used for biochemical techniques including affinity purification, immunoprecipitation, western blot analysis, and immunofluorescence.

Some tagging cassettes, including the pOM family, allow removal of the selectable marker after tagging [2]. In these cassettes, the selection marker is flanked by LoxP sites and thus can be removed by bacteriophage Cre recombinase that is conditionally expressed from a plasmid [3]. This technology is useful in several situations: (1) for N-terminal tagging without separation of the tagged protein from its endogenous promoter; (2) for inserting a tag internally within the coding region of the gene of interest; (3) for multiple rounds of tagging at the same locus; and (4) for use in yeast strains with a limited number of selectable markers. Some readily available tagging vectors are shown in Table 1.

When a tagging vector is used, primers should be designed with sequences to hybridize with both the tagging vector and the target site for homologous recombination within the yeast chromosome. An insertion cassette is then produced by PCR using the

Table 1
Yeast tagging cassette vectors

Plasmid family	Tag position	Promoter	Tags	Markers
pFA6a[a]	C terminal	Endogenous	GFP(S65T) 3×HA 13×Myc GST	*TRP1* *kanMX6* *HIS3MX6*
pFA6a-PGAL1[a]	N terminal or internal	*GAL1*	GFP(S65T) 3×HA GST	*TRP1* *kanMX6* *HIS3MX6*
pUR[b]	C terminal	Endogenous	DsRed	*HIS3* *URA3(K.l.)*
pYM[c]	C terminal	Endogenous	yEGFP EGFP EBFP ECFP EYFP DsRed, DsRedI RedStar, RedStar2 eqFP611 FlAsH 1×HA, 3×HA, 6×HA 3×Myc, 9×Myc 1×Myc + 7×His TAP Protein A	*kanMX4* *hphNT1* *natNT2* *HIS3MX6* *klTRP1*

(continued)

Table 1
(continued)

Plasmid family	Tag position	Promoter	Tags	Markers
pKT[d]	C terminal	Endogenous	yEGFP yECFP yEVenus yECitrine yESapphire yEmCFP[e] yEmCitrine tdimer2[f] yECitrine + 3×HA yECitrine + 13×Myc yECFP + 3×HA yECFP + 13×Myc	*KanMX* *SpHIS5* *CaURA3*
pOM[g]	N terminal or internal	Endogenous[h]	yEGFP 6×HA 9×Myc Protein A TEV-ProteinA TEV-GST-6×His TEV-ProteinA-7×His	*kanMX6* *URA3(K.l.)* *LEU2(K.l.)*
pCY[i]	C-terminal	Endogenous	Cerulean yECFP yEmCFP yEGFP Venus yEVenus yECitrine yEmCitrine mCherry mEos2 yEc-MYC yEHA yEFlAsH	*HygromycinB* *Zeocin*

[a][32]
[b][33]
[c][34]
[d][35]
[e]Monomeric version
[f]Tandem dimer of DsRed
[g][2]
[h]After Cre-mediated removal of auxotrophic marker
[i][36]

tagging vector as a template. The amplified DNA is transformed into yeast using a standard protocol [4]. The selection marker used in the insertion cassette is used to identify recombinants that carry the inserted tag, and PCR-based screening can be used to ensure that the homologous recombination occurred at the correct site.

1.1.3 Targeting FPs to Mitochondria Using a Signal Sequence

Mitochondria contain two membranes, the inner and outer membranes, and two soluble compartments, the intermembrane space and the matrix. Other submitochondrial compartments include the mtDNA nucleoids and contact sites between inner and outer membranes [5, 6]. Over 95 % of mitochondrial proteins are encoded in nuclear DNA, synthesized in the cytoplasm, and imported into the organelle. Targeted import depends on signal sequences, which can exist at the amino terminus or within nuclear-encoded mitochondrial proteins [7]. Therefore, nuclear-encoded mitochondrial proteins are typically tagged at their C-terminus to preserve signal sequence function. A full-length protein containing a mitochondrial targeting sequence, or simply the signal sequence alone, can be used to target FPs to specific compartments within mitochondria. It is also possible to localize the FPs to specific mitochondrial components by tagging endogenous proteins that are imported to sites of interest, such as Tom70p (*T*ranslocation through the *O*uter *M*itochondrial membrane) for outer membrane labeling and Cit1p (*Cit*rate synthase 1) for matrix targeting (Fig. 1a). Expression of FP-tagged Tom70p or Cit1p has no obvious effect on mitochondrial morphology, motility, or respiratory activity. They are excellent tools for investigating mitochondrial distribution and morphology (Fig. 1b). Plasmid-borne targeted FPs can also be used to label yeast mitochondria and produce a robust, fluorescent signal that is specific to the organelle (Table 2). However, as a result of cell-to-cell variability in plasmid copy number, fluorescence intensity is variable using plasmid-borne mitochondria-targeted FPs.

Mitochondria are highly dynamic structures that undergo constant anterograde movement (toward the bud) and retrograde movement (towards the tip of the mother cell distal to the bud). They also undergo cycles of fusion and fission [8–10]. Additionally, mitochondria are anchored at the bud tip and in the cortex of the mother cell by Mmr1p and Num1p, respectively [11–13]. Mitochondrial dynamics, including velocity of mitochondrial motility, total mitochondrial motility, anchorage, and fusion and fission can be monitored by time-lapse imaging of mitochondria-targeted FPs, such as GFP-tagged Cit1p (Fig. 2).

1.1.4 Detecting Mitochondrial DNA with DNA-Binding Dyes

In animal cells and fungi, mitochondria are the only extranuclear organelles containing DNA. Cytoplasmic DNA staining is therefore a reliable marker for mitochondria. In addition, the quantity and distribution of mitochondrial DNA (mtDNA) can illuminate mitochondrial structural integrity and fitness. In yeast and other eukaryotes, mtDNA associates with a complex of proteins involved in organization, replication, and expression of genes of mtDNA [6]. This complex of proteins and mtDNA assemble into punctate

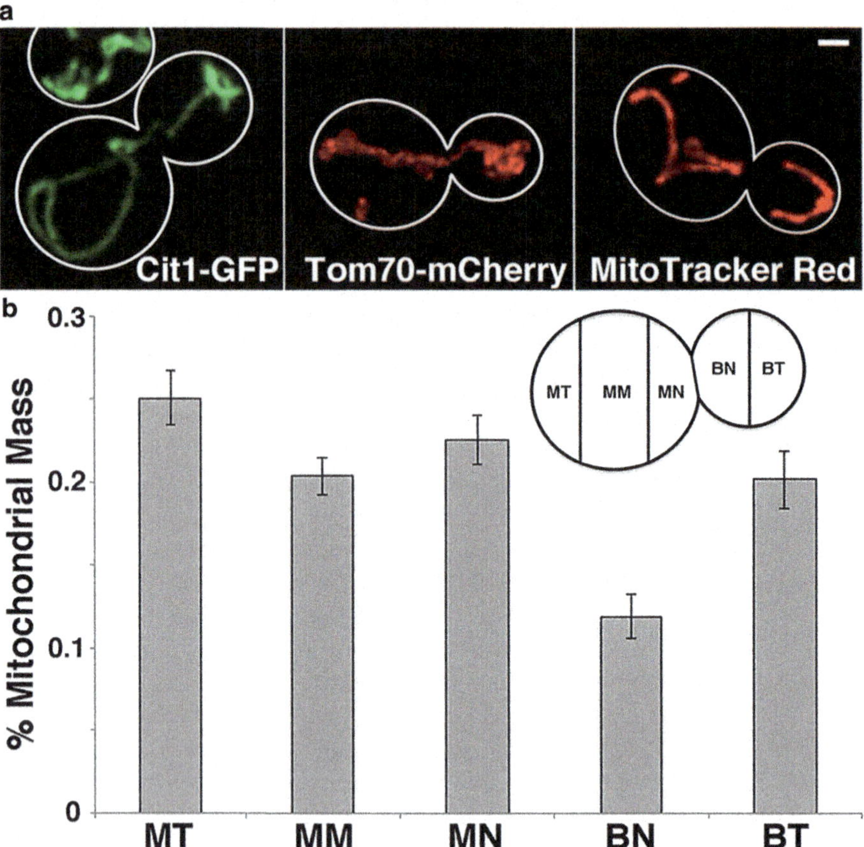

Fig. 1 Visualization of mitochondrial morphology and distribution. (**a**) Mitochondria were visualized by tagging Cit1p, a mitochondrial matrix protein, with GFP (*left panel*); Tom70p, a mitochondrial outer membrane protein, with mCherry (*middle panel*); and using the lipophilic dye, MitoTracker Red (*right panel*). Scale bar = 1 μM. Images were acquired using a standard GFP filter with 100 ms exposure time for GFP and a standard DsRed filter with 200 ms exposure time for mCherry and MitoTracker Red. (**b**) Mitochondrial distribution was measured by separating the yeast cell into five different compartments as shown in *inset*: tip of the bud distal to the mother (bud tip), tip of the bud adjacent to the mother (bud neck), tip of the mother cell adjacent to the bud (mother neck), middle of the mother cell (mother middle), and tip of the mother cell distal to the bud (mother tip). *MT* mother tip, *MM* mother middle, *MN* mother neck, *BN* bud neck, *BT* bud tip. Mitochondrial content in each region was assessed by measuring the integrated intensity of Cit1-GFP fluorescence in yeast cells bearing a large bud (0.60–0.80 ratio of bud to mother size). Error bars indicate SEM. $n = 40$. Data is representative of three experiments

structures localized to the inner leaflet of the inner mitochondrial membrane. Tagging of these proteins can be used to monitor nucleoid assembly and distribution. In yeast, a species in which mtDNA is dispensable for viability, DNA-binding dyes can also be used to determine if a strain lacks mtDNA (*rho*^{0}).

Many DNA binding dyes do not readily cross the yeast cell wall. However, DAPI (4′,6′-diamidino-2-phenylindole), which stains nuclear and mtDNA in *S. cerevisiae*, has become the most

Table 2
Mitochondria-targeted signal sequence-FP fusion proteins

Site	*Targeting*	Promoter	Vector	FP	Reference
Matrix	*OLI1*[2] signal sequence	*ADH1*	2 μ pRS426 derivative	HcRed	[22]
	CIT1[3] signal sequence	*CIT1*	CEN-*URA3*	bGFP (F99S, M153T, V163A)	[37]
	CIT1 signal sequence	*GAL1*	CEN-*URA3*	bGFP	[37]
	OLI1	*GAL1*	CEN-*URA3*	DsRed	[38]
OM	*TOM6* signal sequence	*GAL1*	CEN-*URA3*	bGFP	[37]
IM	*YTA10*	*GAL1*	CEN-*URA3*	bGFP	[37]
mtDNA	*ABF2*	*GAL1*	CEN-*URA3*	bGFP	[37]

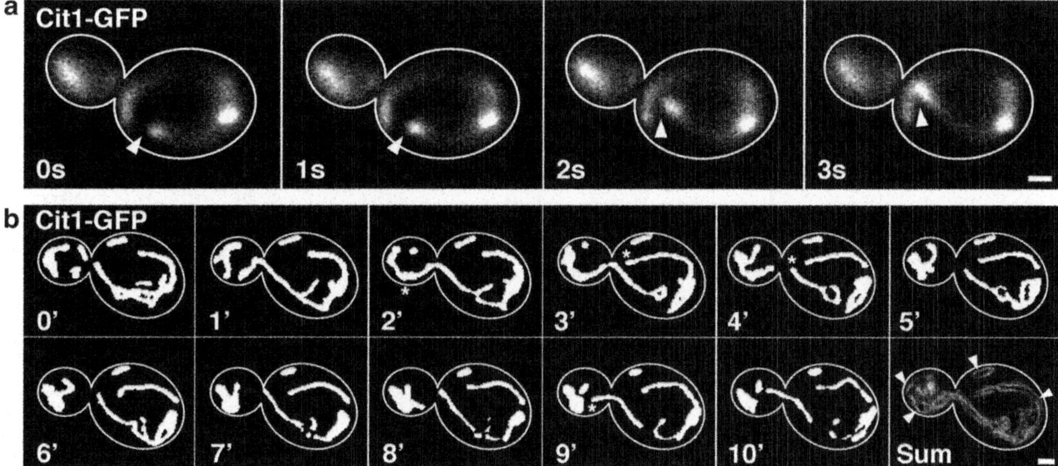

Fig. 2 Visualization of mitochondrial dynamics in budding yeast. (**a**) Velocity of mitochondrial motility was determined by taking single-plane images of Cit1-GFP every 1 s for 30 s. Images were captured using standard GFP filters and 100 ms exposure time. The optimal imaging plane is slightly above the center of the cell. *Arrowheads* follow the tip of a tubular mitochondrion moving in an anterograde direction. Scale bar = 1 μM. (**b**) Total motility of mitochondria was determined by taking z-stacks of Cit1-GFP every 30 s for 10 min. Images were captured using a standard GFP filter, a 50 % neutral density filters, and 100 ms exposure time. Thresholded maximum-intensity projections at 1-min intervals are shown. *Stars* indicate the largest movement events that were visualized during the 10 min period. The "Sum" image (*bottom-right corner*) is the sum of the difference between successive images calculated by the Total Motility plugin. This image depicts all motility events where *white* indicates movement and *indicates* indicate lack of movement. Edges of mitochondria commonly appear *gray* due to small non-directed movements. *Arrowheads* mark mitochondrial anchorage sites in the bud tip, mother tip, and mother cortex where mitochondrial movement is completely lacking. Scale bar = 1 μM

Table 3
Vital dyes for yeast mitochondria

Dye	IUPAC name	λ_{ex}	λ_{em}
$DiOC_6$ [39]	3,3′-dihexyloxacarbocyanine iodide	482	504
DASPMI	4-(4-(dimethylamino)styryl)-*N*-methylpyridinium iodide (4-Di-1-ASP)	475	605[a]
Rhodamine 123	2-(6-Amino-3-imino-3H-xanthen-9-yl)benzoic acid methyl ester	505	534[a]
MitoTracker	Various	Various	Various
DAPI	4′,6′diamidino-2-phenylindole	358	461
DHE[b]	Dihydroethidium	485	530

[a]Broad emission range; not recommended for green/red double-label studies

[b]DHE is not a mitochondrial vital dye, but has been included in this list as it can reliably be used for measurements of mitochondrial ROS levels (*see* Fig. 3)

commonly used DNA-binding dye in yeast (Table 3). Its fluorescence increases greatly when bound to DNA in comparison to the unbound form or DAPI bound with RNA, making DAPI a reliable and robust nuclear and mtDNA marker with little cytoplasmic background. Additionally, DAPI staining of mtDNA is not dependent upon the metabolic state of the organelle. Finally, DAPI can be used in both live and fixed cells, producing a robust and persistent fluorescent signal.

One disadvantage of DAPI is that due to its strong nuclear staining, mtDNA nucleoids close to the nucleus are not well resolved. In live cells, DAPI shows a preference for mtDNA over nuclear DNA, but nuclear staining does still occur, making accurate quantification of mtDNA using DAPI difficult. In addition, the UV illumination (350–400 nm) required to visualize DAPI is damaging. Therefore, it is not suitable for time-lapse or long-term live-cell imaging. In addition, because DAPI binds directly to DNA, it can reduce cell viability.

1.2 Visualization of Organelle Function: Mitochondrial Membrane Potential and Redox State

1.2.1 Measuring Mitochondrial Membrane Potential

Positively charged lipophilic fluorophores can be utilized to measure mitochondrial $\Delta\psi$ because functioning mitochondria have the most negative $\Delta\psi$ in the cell, causing the fluorophores to accumulate in the organelle. Unhealthy or damaged mitochondria having low $\Delta\psi$, such as those treated with the proton ionophore carbonyl cyanide-4-(trifluoromethoxy)phenylhydrazone (FCCP) or lacking mtDNA, fail to import these vital dyes into the organelle, making these dyes a useful tool in monitoring mitochondrial function in live cells.

$DiOC_6$, the styryl dye DASPMI, the cationic rhodamine derivative rhodamine 123, and the fixable stains of the MitoTracker family work particularly well in yeast. $DiOC_6$ and MitoTrackers are particularly useful for double-label experiments due to their narrow excitation and emission spectra, preventing bleed-through in more quantitative experiments. In addition, MitoTracker Orange, Red, and Deep Red are the only dyes that persist in mitochondria after aldehyde fixation and permeabilization by acetone or Triton X-100, making them the only membrane-potential dyes that can be used with immunofluorescence.

$DiOC_6$ can be used to monitor mitochondrial $\Delta\psi$ as a quantitative measurement of overall mitochondrial quality [14]. In this assay, cells are exposed to $DiOC_6$, and the total level of imported dye is normalized to total mitochondrial mass measured by a $\Delta\psi$-independent mitochondrial marker, such as Tom70p that is tagged at its chromosomal locus with mCherry (Fig. 3a). This normalization is crucial to correct for strains or conditions that may cause increased mitochondrial quantity, which may overestimate $DiOC_6$ measurements as an indicator for $\Delta\psi$. Treatment with a proton ionophore will cause collapse of mitochondrial $\Delta\psi$ by uncoupling the proton gradient from ATP synthesis, while growth in non-fermentable carbon sources, such as glycerol, will result in increased $\Delta\psi$ due to increased respiration. These treatments can be used to determine the dynamic range of this quantitative tool (Fig. 3b).

1.2.2 Measuring Mitochondrial Redox State

The components of the electron transport chain in mitochondria are key contributors to reactive oxygen species (ROS) production in cells. ROS can be detoxified by catalases and peroxidases both within the mitochondria and in the cytosol. However, accumulation of ROS and oxidative stress due to both age-dependent and age-independent factors contributes to decreased fitness and quality of the mitochondria overall. ROS and redox-sensing probes allow quantitative monitoring of these important aspects of mitochondrial function.

roGFP, a redox-sensitive GFP variant, reveals the mitochondrial redox state [15]. This ratiometric probe is constructed with cysteine molecules exposed on the surface of the GFP molecule. Oxidation of these cysteine residues results in a conformational change, shifting the optimal excitation wavelength to ~400 nm. Reduction of these cysteine residues favors excitation at ~480 nm. Thus, the ratio of GFP emission upon excitation of roGFP at 480 and 400 nm provides a relative readout of the redox state of the fluorophore's environment (Fig 3c) [16]. roGFP can be targeted to the mitochondrial matrix by fusion to a mitochondrial localization sequence and ultimately used to determine redox fitness of mitochondria. Because oxidizing ROS molecules are a primary source of mitochondrial damage, reducing environments indicate

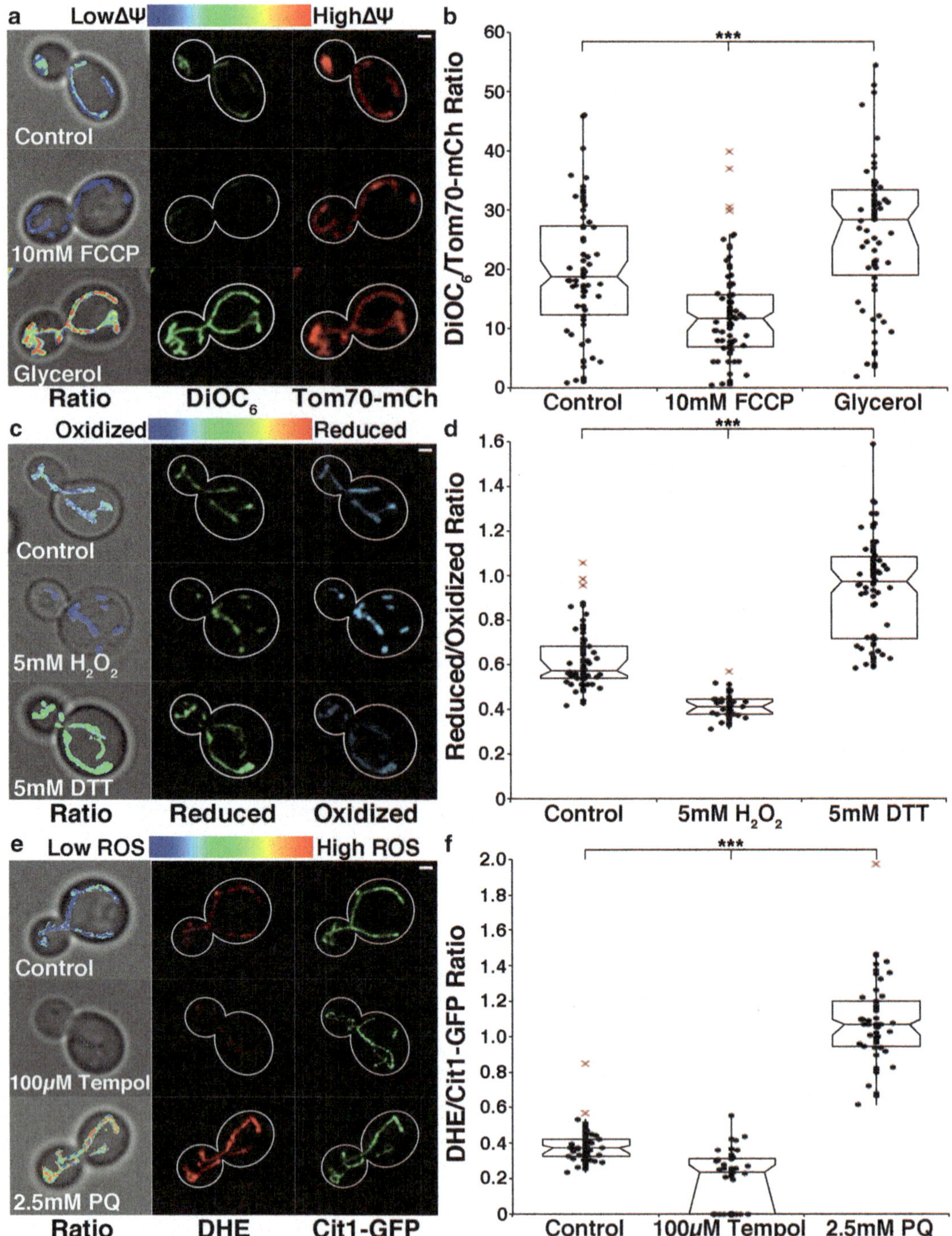

Fig. 3 Characterization of mitochondrial quality. A) $DiOC_6$ was used to visualize mitochondrial membrane potential in wild-type cells. Images were acquired using a 470 nm LED at 100 % power, 50 ms exposure time and standard GFP filters for $DiOC_6$ and 200 ms exposure time and metal-halide lamp with standard DsRed filters for Tom70-mCherry. Left panels: $DiOC_6$:Tom70-mCherry ratios overlaid on bright-field images. *Color scale* indicates ratio values; *higher numbers* and *warmer colors* indicate higher membrane potential. *Middle panels*:

healthier mitochondria and oxidizing environments indicate damaged or dysfunctional mitochondria. Treatment with H_2O_2 will result in a highly oxidizing environment, while treatment with dithiothreitol (DTT) will result in a highly reducing environment. These agents can be used to determine the dynamic range of roGFP (Fig 3d).

Dihydroethidium (DHE) is used to directly measure levels of mitochondrial superoxide. DHE fluoresces blue; however, when it is oxidized by superoxide anion, it emits red fluorescence. Since mitochondria are the main endogenous source of superoxide, DHE is an ideal probe to specifically measure mitochondrial ROS levels. MitoSOX Red (Life Technologies) is a cationic derivative of DHE that has been used to detect mitochondrial ROS. Mitochondrial ROS levels can be quantified with DHE by normalizing to a bulk mitochondrial label such as GFP-tagged Cit1p (Fig. 3e) [17]. Paraquat, a drug that specifically increases mitochondrial superoxide formation, and TEMPOL, an agent that promotes elimination of superoxide, can be used to determine the dynamic range of DHE as an indicator for ROS (Fig. 3f). DHE is best used for short-term measurements of acute ROS, because its oxidized form can intercalate with DNA, causing it to accumulate in the nucleus after ~15 min.

Fig 3 (continued) $DiOC_6$ images. *Right panels*: Tom70-mCherry images. Cell outlines were drawn from bright-field images. Scale bar = 1 μM. (**b**) Notched *dot box plot* of the average $DiOC_6$:Tom70-mCherry ratio in wild-type cells grown in glucose, 10 mM FCCP, and glycerol. Cells were treated with FCCP for 15 min prior to addition of $DiOC_6$ or grown in YPG overnight and allowed to recover in SC with glucose for 3 h. $n = 50$ cells for each strain. Data is representative of three experiments. *** = p-value < 0.001 using non–parametric Kruskal–Wallis testing with pairwise Bonferroni correction. (**c**) Mito-roGFP1 was used to visualize the redox environment of mitochondria in wild-type cells. *Left panels*: reduced–oxidized roGFP ratios overlaid on phase images. Images were acquired using a standard GFP filter with excitation filter removed, and LED excitation at 365 nm (25 % power) and 100 ms exposure for the oxidized channel and 470 nm (100 % power) and 100 ms exposure for the reduced channel. *Color scale* indicates ratio values; *higher numbers* and *warmer colors* indicate more reducing mitochondria. *Middle panels*: reduced roGFP. *Right panels*: oxidized roGFP. Cell outlines were drawn from bright-field images. Scale bar = 1 μM. (**d**) Notched dot box plot of the average reduced–oxidized mito-roGFP1 ratio in wild-type cells untreated and treated with 5 mM H_2O_2 and 5 mM DTT. Cells were treated with H_2O_2 or DTT for 30 min prior to imaging. $n = 50$ cells for each strain. Data is representative of 3 experiments. *** = p-value < 0.001 using nonparametric Kruskal–Wallis testing with pairwise Bonferroni correction. (**e**) DHE was used to visualize mitochondrial ROS levels in wild-type cells. *Left panels*: DHE:Cit1-GFP ratios overlaid on bright-field images. Cells were stained with DHE at 40 μM for 30 min and imaged using a standard rhodamine filter and 100 ms exposure for DHE and a standard GFP filter and 100 ms exposure for Cit1-GFP. *Color scale* indicates ratio values; *higher numbers* and *warmer colors* indicate lower ROS levels. *Middle panels*: DHE. *Right panels*: Cit1-GFP. Cell outlines were drawn from bright-field images. Scale bar = 1 μM. (**f**) Notched dot box plot of the average DHE:Cit1-GFP ratio in wild-type cells untreated and treated with 2.5 mM paraquat (PQ) and 100 μM Tempol. Cells were treated with PQ or Tempol for 30 min prior to addition of DHE. $n = 50$ cells for each strain. Data is representative of two experiments. *** = p-value < 0.001 using nonparametric Kruskal–Wallis testing with pairwise Bonferroni correction

1.3 Visualization of the Actin Cytoskeleton with Fluorescent Proteins

Actin patches and actin cables are two F-actin-containing structures that persist throughout the cell cycle in budding yeast. Actin patches are endosomes that are coated with F-actin filaments and localize to the bud and the mother–bud neck in polarized cells. Actin cables are dynamic bundles of F-actin that align along the mother–bud axis and continuously flow in a bud-to-mother direction via retrograde actin cable flow (RACF) [18]. These cables serve as tracks for the movement of cellular cargo including secretory vesicles, mRNA, spindle alignment elements, mitochondria, Golgi, and vacuoles. In addition, the actin cytoskeleton plays a major role in regulation of quality control of mitochondria [19].

Early studies revealed that tagging the actin-encoding *ACT1* gene of yeast with GFP compromised function of the protein. Specifically, *ACT1-GFP* expressed on a plasmid did not rescue loss of the endogenous *ACT1* gene [20]. Presumably, GFP tagging results in loss of function because every surface of the actin protein is involved in protein–protein interactions in microfilament nucleation and polymerization, interaction with motor proteins and other force generators, microfilament assembly, capping, cross-linking, and severing. As an alternative, actin-binding proteins are often more tolerant of tagging with FPs. GFP tagging of Abp140p and Abp1p produces a fluorescent signal that localizes to actin cables and actin patches, respectively (Fig. 4a) [18, 21].

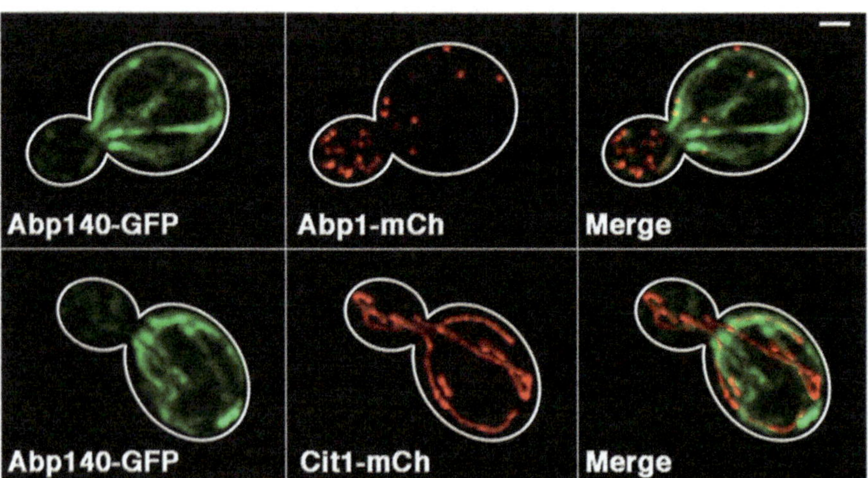

Fig. 4 Colocalization of the actin cytoskeleton with cargo. Yeast cells were grown in YPG and washed in SC medium prior to imaging. Actin cables were visualized with Abp140-GFP (left panels). Two cellular cargo structures, actin patches (*top*) and mitochondria (*bottom*), were visualized by tagging Abp1p and Cit1p with mCherry, respectively. Images were captured using standard GFP filters and 300 ms exposure time for GFP and standard DsRed filters and 200 ms exposure time for mCherry. Right panels: merged images indicating colocalization of actin patches and mitochondria along the actin cytoskeleton. Cell outlines were drawn from bright-field images. Scale bar = 1 μM

Table 4
GFP-tagged actin proteins in budding yeast

Actin-containing structure	GFP tagged proteins
Actin patch (early)	Las17p [40, 41],Sla1p [41, 42], Pan1p [41, 43], End3p [44], Sla2p [41, 45], Bzz1p [46], Vrp1 [47]
Actin patch (mid)	Myo3p [48], Myo5p [49], Bbc1p [46]
Actin patch (late)	Abp1p [20, 41, 50, 51], Arp2p [40], Arp3p [40], Arc15p [41, 52], Sac6p [51, 53], Cap1p [54], Cap2p [55], Scp1p [56], Cof1p [57]
Actin cable (vegetative)	Abp140p [18]
Actin cable (G_0)	Lifeact (Abp140 aa 1–17) [23]

Live-cell imaging of the actin cytoskeleton allows quantification of velocity of RACF and the visualization of cargo trafficking, such as the movement of mitochondria along actin cables (Fig. 4b) [22]. A drawback of Abp140-GFP is low signal, which can be boosted by culturing cells in non-fermentable carbon sources as actin cables are thicker in these culture conditions (*see* **Note 1**). As an alternative to tagging the full-length protein, the first 17 amino acids of Abp140p are sufficient to mediate actin localization of fluorescent proteins without altering actin cable dynamics. This protein, termed LifeAct, provides reliable and robust staining of the actin cytoskeleton [23]. Table 4 lists genes that have been tagged with FPs to serve as probes for actin patches and actin cables in budding yeast.

2 Materials

2.1 Yeast Growth Media

1. Amino acid supplements: For synthetic complete medium, supplement with all of the following. For dropout medium, omit one or more components to select for the desired strains: 10 mL adenine (2 mg/mL stock in 0.05 M HCl), 10 mL uracil (2 mg/mL stock in 0.5 % NaHCO3), 10 mL arginine 10 mL (1 mg/mL stock in H_2O), 10 mL histidine (1 mg/mL stock in H_2O), 10 mL leucine (10 mg/mL stock in H_2O), 10 mL lysine (3 mg/mL stock in H_2O), 10 mL methionine (2 mg/mL stock in H_2O), 10 mL phenylalanine (5 mg/mL stock in H_2O), 10 mL tryptophan (2 mg/mL stock in H_2O), 10 mL tyrosine (3 mg/mL stock in 0.05 M HCl).
2. Synthetic complete medium (SC): Dissolve 6.7 g yeast nitrogen base without amino acids, 20 g glucose, and amino acid supplements as needed in 800 mL distilled H_2O. Adjust pH to 5.5 with $NaHCO_3$ and bring volume to 1 L with distilled H_2O. Sterilize by autoclaving.

3. Yeast peptone-dextrose medium (YPD): For 1 L, dissolve 10 g yeast extract, 20 g Bacto peptone, and 20 g glucose in distilled H_2O and bring volume to 1 L with distilled H_2O. Sterilize by autoclaving.
4. Lactate medium: For 1 L, combine 3.0 g yeast extract, 0.5 g glucose, 0.5 g $CaCl_2$, 0.5 g NaCl, 0.5 g $MgCl_2$, 1 g KH_2PO_4, 22 mL 90 % lactic acid, and 7.5 g NaOH pellets. Dissolve ingredients in 800 mL distilled H_2O. Adjust pH to 5.5 with NaOH. Bring volume to 1 L with distilled H_2O.
5. Synthetic raffinose medium: For 1 L, combine 6.7 g of yeast nitrogen base without amino acids and amino acid supplements as needed. Dissolve ingredients in 600 mL distilled H_2O. Adjust pH to 5.5 with $NaHCO_3$. Bring volume to 900 mL with distilled H_2O. Supplement after autoclaving with 100 mL filter-sterilized 20 % raffinose.
6. Synthetic glycerol medium (SG): For 1 L, combine 6.7 g of yeast nitrogen base without amino acids, 30 mL of glycerol, 0.5 g glucose, and amino acid supplements as needed. Dissolve ingredients in 800 mL distilled H_2O. Adjust pH to 5.5 with $NaHCO_3$.
7. Synthetic galactose medium: For 1 L, combine 6.7 g of yeast nitrogen base without amino acids and amino acid supplements as needed. Dissolve ingredients in 600 mL distilled H_2O. Adjust pH to 5.5 with $NaHCO_3$. Bring volume to 950 mL with distilled H_2O. Supplement after autoclaving with 50 mL filter-sterilized 40 % galactose.

2.2 Reagents for PCR Amplification

1. PCR buffer: 100 mM Tris–HCl, pH 8.3 at 25 °C; 500 mM KCl; 15 mM $MgCl_2$; 0.01 % gelatin (Sigma, P2912).
2. $MgCl_2$ stock solution (25 mM).
3. dNTP stock solutions (10 mM each).
4. Forward and reverse primers (10 μM each).
5. Distilled H_2O.

2.3 Reagents for Yeast Transformation and Marker Excision

1. 50 % w/v polyethylene glycol 3350 in H_2O.
2. 1 M lithium acetate in H_2O.
3. 0.1 M lithium acetate in H_2O.
4. Carrier DNA: 2 mg/mL single-stranded calf thymus DNA or salmon sperm DNA in H_2O. Store at −20 °C. Boil for 5 min immediately before use. Do not subject to more than two freeze–thaw cycles.
5. DNA for transformation (one PCR reaction as described in protocol, or 1 μg).
6. Sterile distilled H_2O for resuspending cells before plating.
7. Selective plates for growth of cells containing plasmid or integrated DNA.

2.4 Materials for Imaging Living Cells

1. Agarose pad: Add 0.1 g low-melting agarose to a 50 mL conical-bottom tube and bring to a total volume of 5 mL with desired medium (final concentration will be 2 % agarose). Dissolve agarose in medium by boiling the tube in a water bath. Dispense into 200-μL or smaller aliquots in microfuge tubes and store at room temperature in the dark. Avoid repeated boiling.
2. VALAP: combine equal weights of petrolatum (Vaseline), lanolin, and paraffin (hard). Submerge mixture in a 70 °C water bath to melt and aliquot into 60 × 15 mm glass Petri dishes. Store at room temperature.

2.5 Materials for Measuring Mitochondrial Membrane Potential Using $DiOC_6$

1. HEPES: For 1 L, combine 2.5 g HEPES, 50 g glucose. Dissolve ingredients in 800 mL distilled H_2O. Adjust pH to 7.6 using 10 *N* NaOH. Bring volume to 1 L with distilled H_2O. Filter-sterilize; do not autoclave.
2. $DiOC_6$: Dissolve $DiOC_6$ in 100 % ethanol to a stock concentration of 17.5 mM. Store at −20 °C for up to 2 years. Dilute this stock to 17.5 μM in 100 % ethanol as a working solution (*see* **Note 2**). Store working solution at −20 °C for up to 2 weeks.
3. FCCP: Dissolve FCCP in 100 % DMSO to a stock concentration of 2 M. Store at −20 °C.

2.6 Materials for Measuring Redox Environment/ROS Levels

1. Tempol: Dissolve Tempol in distilled H_2O to a stock concentration of 100 mM. Store at −80 °C.
2. Paraquat: Dissolve Paraquat in distilled H_2O to a stock concentration of 100 mM. Store at −80 °C.
3. H_2O_2: Dilute H_2O_2 to a working concentration of 1 M in distilled H_2O_2 prior to use (*see* **Note 3**).

3 Methods

3.1 Modification of Yeast Genes at Their Chromosomal Locus

3.1.1 Primer Design

Primers will generally be 60 or more bases long. The 5′ end of each primer should contain at least 40 bases of perfect homology to the target site. For example, for C-terminal tagging with a FP, the 5′ end of the forward primer should contain the 40–45 bases directly upstream of the stop codon, while the 5′ end of the reverse primer should contain the reverse complement of the 40–45 bases directly downstream of the stop codon. The 3′ ends of the primers should contain 18–25 bases complementary to the sequences that will be inserted, which include the open reading frame of the FP, the transcription termination site, and the selectable marker (*see* **Note 4**).

The target gene and the FP tag are typically separated by a short linker, which is designed to allow proper folding and function of both the fluorescent tag and the protein of interest. For example, GFP fluorescence can often be optimized by increasing the length of the linker between the target gene and GFP molecule (we suggest starting with five alanines). The length and amino acid composition of this linker may need to be adjusted for different target proteins. Proteins with processive enzymatic functions, such as polymerases and telomerases, will benefit from poly-glycine linkers [24]. In C-terminal tagging, the linker is encoded in the in the forward primer by including DNA encoding the appropriate amino acids between the 40 bases of DNA homologous to the target gene and the 20 bases corresponding to the template. For N-terminal tagging, a similar strategy can be used in the reverse primer.

3.1.2 Amplification of Insertion Cassette from Tagging Vector

1. Prepare the PCR reaction: PCR buffer, 1.5–5 mM $MgCl_2$, dNTP mix (2.5 mM each dNTP), 0.1–1.0 μM forward and reverse primers, template DNA (*see* **Note 5**), 1.0–5.0 units of polymerase (*see* **Note 6**), distilled H_2O up to 50 μL. A single 50 μL PCR reaction will provide sufficient DNA for a single yeast transformation (minimally 1 μg of DNA) if a high-fidelity, high-yield polymerase is used. For lower-yield polymerases, multiple PCR reactions can be pooled.
2. Amplify the cassette using the following thermocycler conditions: Initial denaturation cycle at 95 °C for 2–5 min; 35 cycles of: 98 °C for 20 s, 60–75 °C (*see* **Note** 7) for 15 s, and 72 °C for 60 s/kb; final extension cycle at 72 °C for 1–5 min.
3. Use the PCR reaction directly in the lithium acetate transformation protocol described below. Alternatively, amplified DNA can be purified using any commercially available PCR purification kit (but *see* **Note 8**).

3.2 Lithium Acetate Transformation of Yeast

The lithium acetate transformation protocol is the most commonly used method for yeast transformation [4]. The following protocol is for one transformation reaction. A negative control containing no DNA should always be carried out in parallel.

1. Grow yeast to mid-log phase.
2. Transfer 10^7 cells from the culture to a 1.5 mL microfuge tube.
3. Pellet cells (30 s, 7000×*g*) and wash with 500 μL 0.1 M lithium acetate. Resuspend in 240 μL of 50 % w/v polyethylene glycol 3350.
4. Add 36 μL 1.0 M lithium acetate, 25 μL of carrier DNA, and 1 μg of DNA to be transformed (PCR-amplified insertion cassette or plasmid DNA) or 50 μL of H_2O for negative control.
5. Vortex vigorously and incubate in a water bath at 30 °C for 30 min.
6. Heat-shock in a water bath at 42 °C for a minimum of 20 min and a maximum of 3 h.

7. Concentrate cells by mild centrifugation (30 s, 7000 × *g*).
8. For transformations using auxotrophic markers, resuspend cell pellet gently in 100 μL of sterile distilled H_2O and plate on appropriate selective media.
9. For transformations using drug resistance markers (e.g., KanMX6), resuspend cell pellet gently in 500 μL of YPD or synthetic complete medium with dropouts as needed to maintain plasmids. Allow cells to recover at 30 °C for 2–4 h. Concentrate and plate cells as described in **steps** 7 and **8** on appropriate selective medium containing drug.

3.3 Marker Excision by Cre Recombinase

If a loxP-containing vector such as the pOM family has been used to tag the gene of interest, the selectable marker can be removed by the following method.

1. Verify tagging by PCR and sequencing, as described in Subheading 3.4.1.
2. Using methods described above, transform the tagged strain with a plasmid that encodes Cre recombinase under control of a galactose-inducible promoter [25]. Select transformants on appropriate medium selecting for the Cre plasmid.
3. Grow cells to mid-log phase in 5 mL liquid medium in a 50-mL conical-bottom tube. Use medium that selects for the Cre plasmid.
4. Pellet cells (5 min, 4000 × *g*) and resuspend in 5 mL galactose medium.
5. Incubate cells in galactose medium in a shaking incubator at 30 °C for 12–16 h.
6. Plate an aliquot of cells on non-selective (YPD or SC) medium.
7. Screen colonies for loss of the tagging cassette marker by replica plating on selective plates.
8. To induce dropout of the Cre plasmid, grow cells in non-selective liquid medium for 24–48 h, plate on non-selective medium, and screen for loss of the plasmid by replica plating on selective plates (*see* **Note 9**).

3.4 Validating and Characterizing FP-Tagged Cytoskeletal Proteins

3.4.1 PCR Screening and Sequencing

PCR screening should be used to validate the insertion of the tag into the target locus. This can be accomplished by isolating genomic DNA from transformants and using that as a template for a PCR reaction using a pair of primers that hybridize upstream and downstream of the tagging locus. Strains confirmed to have the correct tag at the target location should be validated by sequencing to confirm that the tagged genes do not carry any mutations.

3.4.2 Verification of Protein Expression and Function

1. The first step to verify successful expression of tagged proteins is visual inspection of transformed cells. Cells should be prepared for short-term imaging as described below.
2. Screen several colonies to find bright, consistent FP expression. Western blotting can also be used to monitor protein expression. This can reveal (1) expression below the threshold of detection of microscopy; (2) the presence of degradation products that may not localize or function properly; or (3) differences in expression level relative to the native protein.
3. Tagged constructs should also be evaluated for correct localization and function if knowledge of the gene of interest is available. FP tags can change the function and/or localization of proteins, and some are prone to oligomerization (which can produce artifacts due to protein–protein interactions).
4. If the localization of the target protein is known, correct localization of the FP-tagged protein suggests that this aspect of its function is preserved. For example, Cit1p is a mitochondrial matrix protein, and Cit1-GFP localization to the mitochondria will suggest that the tag does not perturb localization. Mitochondria can be counterstained by vital dyes or visualized using another tagged mitochondrial protein, such as Tom70p. If there is any question about the correct localization of the FP-tagged protein, it may be necessary to localize the target protein with a smaller tag such as myc or HA followed by fixation and immunostaining. After the correct localization is verified, the FP-tagged version can be used for live-cell analysis.
5. To check for proper function of a tagged protein, evaluate cell growth rate and any phenotypes characteristic of loss of function of the gene of interest. For example, if tagging of a nonessential gene compromises function of the protein, then the tagged strain may show a phenotype resembling that of cells mutated in that gene. Cell growth rate should be comparable to that of untagged strains. To demonstrate full wild-type function, a plasmid-borne version of the FP-tagged protein should rescue the wild-type phenotype in yeast cells with a deletion of the genomic copy of the gene of interest (*see* **Note 10**).

3.4.3 Characterization of Mitochondria-Targeted FPs

1. Mitochondrial targeting depends on signal sequences, which may be at the N terminus or internal to the protein, and the protein including the tag may need to traverse one or more membranes to be properly localized. FP-tagged proteins may mislocalize if targeting information is masked or if insertion of the fusion protein into the membrane is inhibited. For example, N-terminal tagging of Mdm33p inhibits mitochondrial localization, as the mitochondrial signal sequence of Mdm33p is at the N terminus. In contrast, GFP fusion to the C terminus

of Mdm33p impairs the function of the protein, but does not alter mitochondrial localization. This problem is circumvented by internal tagging of Mdm33p between the signal sequence and the mature protein product [26], yielding a protein that is fully functional and correctly localized.

2. Proper localization of mitochondria-targeted FPs can be assessed by visual inspection by counterstaining mitochondria with vital dyes or by biochemical methods for mitochondrial fractionation. Adverse effects of fusion proteins on mitochondrial morphology or respiratory function can be assessed by visual inspection of mitochondria and by analysis of growth rates on non-fermentable carbon sources, such as glycerol.

3.5 Staining Yeast Mitochondria with Vital Dyes

Vital dyes that are used to detect mitochondria in living cells, and conditions used for staining are listed in Table 5.

1. Grow yeast cells to mid-log phase in 5 mL liquid medium in a 50 mL conical tube (*see* **Note 11**).
2. Concentrate 10^7 cells by gentle centrifugation at ~8,000 × *g* for 30 s if necessary (*see* **Note 12**).
3. Add appropriate volume of dye directly to cell culture medium and mix thoroughly (*see* **Note 13**).
4. Incubate in the dark, shaking, at 30 °C for the desired amount of time.
5. Remove excess dye by three washes with growth medium.

Table 5
Conditions for staining yeast mitochondria with vital dyes

		Staining conditions	
Dye	**[Stock][a]**	**[Dye]**	**Incubation (min RT)**
$DiOC_6$[39][b]	17.5 mM	17.5 nM	15
DASPMI[c]	1 mg/mL in ethanol	10–100 μg/mL	30
Rhodamine 123[d]	10 mg/mL in DMSO[e]	5–10 μg/mL	15–30
MitoTracker[f]	1 mM in DMSO[5]	100 nM	30
DAPI	1 mg/mL in H_2O	0.1 μg/mL	15

[a]All stock solutions should be stored in the dark
[b][14]
[c][58]
[d][59]; Broad emission range; not recommended for green/red double-label studies
[e]DMSO can be toxic and inhibits partitioning of the dye into the aqueous environment of the cells. If DMSO is the solvent, use a stock concentration that is at least 100×
[f][60]

6. Concentrate cells by gentle centrifugation at ~8,000 × *g* for 30 s.
7. Mount cells for short-term or long-term observation, as described below.

3.6 Wide-Field Imaging of Living Cells

3.6.1 Equipment for Wide-Field Imaging

1. Budding yeast are among the smallest eukaryotic cells: buds are typically 2–3 μm and mother cells 4–6 μm in diameter. Resolving intracellular structures in such small cells requires a microscope equipped with a high-magnification objective lens for good spatial resolution, and a sensitive camera to capture the low signal levels emitted by small structures.
2. When imaging living cells, overall light throughput and sensitivity of the imaging system are especially important. High sensitivity supports lower-intensity illumination and shorter exposure times, which greatly reduce phototoxicity and photobleaching. In addition, shorter exposure time allows better time resolution when following rapid dynamics. Further improvements in time resolution are possible with a trigger connection between the computer and hardware components such as the camera and the focus drive, and by maximizing the speed and random access memory (RAM) of the computer.
3. An imaging system that works well for imaging yeast is based on a motorized inverted epifluorescence microscope (Zeiss AxioObserver.Z1) equipped with a metal halide lamp and Colibri LEDs for excitation; an EC Plan-NeoFluar 100×/1.3 NA or Plan-Apochromat 100×/1.4 NA objective lens; and an Orca ER cooled CCD camera with 1280 × 1024 pixel resolution. A software package such as ZEN (Zeiss) is used to control the camera and associated hardware, capture images at defined focus and time intervals, and export them for further analysis.

3.6.2 Equipment for 2-Color Imaging

1. Historically, multicolor imaging was first performed with dual-emission filters and color cameras. However, monochrome cameras offer superior dynamic range, resolution, and speed of detection. Therefore, for optimal multi-color imaging, each fluorophore should be imaged separately, and the images pseudocolored (e.g., green for EGFP and red for mCherry) and merged. Sequential imaging of multiple fluorophores is achieved by rapidly switching filters or light sources. Alternatively, a splitting device such as the Dual-View (Photometrics) can be used to image multiple colors simultaneously on subregions of the camera chip.
2. In our preferred imaging system there are two possible strategies for sequential imaging: (1) changing excitation and emission filters and (2) LED excitation switching. For example, for imaging of actin cables using Abp140-GFP and mitochondria using Cit1-mCherry, the sample is illuminated using a

broad-spectrum light source (e.g., a metal halide lamp) and an image is taken first with the green excitation and emission filter sets, then with the red. In a motorized microscope stand, this is typically accomplished by rotating the filter turret. In a non-motorized stand, or for somewhat faster performance, external motorized filter wheels can be inserted in the excitation and/or emission path while a multi-line dichroic beamsplitter is placed in the filter turret. Changing the entire filter cube provides the best signal, because the single-band filters have high throughput and can be tailored precisely to the fluorophore of interest. However, because filter switching requires large mechanical movements, the time interval between acquisition of channels, at least 35 ms, may be too long for highly dynamic events. In addition, the rotation of the filter wheel may generate vibrations, necessitating a pause in the imaging that slows the actual time interval between channels to ~200 ms.

3. A faster, more stable approach is to switch the excitation light through the light source. This can be accomplished with lasers or narrow-spectrum, high intensity light-emitting diodes (LEDs), such as those in the Colibri 2 system (Zeiss). This illuminator holds up to four interchangeable LEDs that match the spectrum of many commonly used FPs, and are controlled individually by electrical signals with virtually no time lag. This system used in conjunction with a multi-line dichroic/emission filter cube allows rapid changing of the excitation light without any mechanical motion. This method is preferred for dual-color imaging of dynamic structures and ratio imaging.

Table 6 provides suggested imaging conditions for methods described in this chapter.

3.6.3 Preparing Cells for Short-Term Imaging

For short-term visualization, concentrated cells in culture medium or staining buffers are added directly to a glass microscope slide and imaged immediately for maximally 10 min (*see* **Note 14**). After 10 min, cells will start to experience significant decrease in viability and may begin to exhibit signs of stress. The following protocol can be used for imaging actin cables labeled with Abp140-GFP together with mitochondria-targeted mCherry, as well as other cells expressing FPs and/or labeled with vital dyes as described above.

1. Inoculate cells from a colony into 5 mL of appropriate medium in a 50-mL conical-bottom tube and incubate overnight at 30 °C in a shaking incubator.
2. In the morning, dilute cultures to early log phase (OD_{600} = 0.01–0.1) and incubate under growth conditions at least 2 h, until cells are mid-log phase (OD_{600} = 0.2–0.6).
3. Transfer 1 mL of culture to a 1.5-mL microfuge tube and concentrate cells by centrifugation at ~8,000 × *g* for 30 s (*see* **Note 15**).

Table 6
Suggested imaging conditions for methods described in this chapter

FP/dye[a]	Excitation/emission	Light source	Exposure time (ms)[b]
Cit1-GFP	488/507	LED 470 nm (100 %) or metal-halide lamp + standard GFP filter	100
Tom70-mCherry	587/610	Metal-halide lamp + standard DsRed filter	200
MitoTracker Red (100 nM for 30 min)	581/644	Metal-halide lamp + standard DsRed filter	200
DiOC$_6$ (17.5 nM for 15 min)	482/504	LED 470 nm (100 %)	50
roGFP (Oxidized)	365/507	LED 365 nm (25 %)	100
roGFP (Reduced)	470/507	LED 470 nm (100 %)	100
DHE (40 μM for 30 min)	485/530	Metal-halide lamp + standard DsRed filter	100
Abp140-GFP	488/507	LED 470 nm (100 %) or metal-halide lamp + standard GFP filter	300
Abp1-mCherry	587/610	Metal-halide lamp + standard DsRed filter	200
Cit1-mCherry	587/610	Metal-halide lamp + standard DsRed filter	200

[a]These are recommended concentrations, and can be adjusted as needed
[b]These are recommended exposure times and can be adjusted as needed

4. Without disturbing the pellet, remove almost all of the supernatant, leaving a volume of supernatant in the tube approximately twice the volume of the pellet.
5. Resuspend pellet in residual medium and transfer ~1.5 μL of the cell suspension to a microscope slide that has been cleaned with 70 % ethanol and wiped dry with a Kimwipes.
6. Apply a #1.5 (170 μm thick) coverslip, taking care to avoid creating bubbles between slide and coverslip (*see* **Note 16**).
7. Acquire images. Prepare a fresh slide from the concentrated cell suspension after 10 min.

3.6.4 Preparing Cells for Long-Term Imaging

Cells can be immobilized on medium containing low-melting-point agarose, which will support cell growth at wild-type levels for up to 5 h. These agarose pads have minimal autofluorescence and remain transparent and thin enough for observation with oil-immersion lenses. Note that agarose pads take roughly 5 min to prepare and should not be left unused for extended periods of time as they may dry out. For time-sensitive studies, take care to prepare pads about 5 min before imaging.

1. Grow yeast cells to mid-log phase and concentrate in liquid medium or staining buffer as described above for short-term imaging (Subheading 3.6.3 **steps 1–3**).
2. Melt an aliquot of agarose bed material in a boiling water bath, approximately 2 min for a 200-μL aliquot.
3. Pipet 35 μL of agarose bed material onto a glass microscope slide that has been cleaned with 70 % ethanol and wiped dry with a Kimwipes (*see* **Note 17**), and cool for ~5 s (*see* **Note 18**).
4. Place a second cleaned microscope slide on top of the agarose and apply light pressure to spread the agarose bed to the diameter of a standard coverslip. Note that the agarose pad may adhere to either the top or bottom slide, so if using slides with charged surfaces, ensure that both slides have the charged side facing the agarose pad.
5. Allow pad to harden between microscope slides for ~10-60 sec.
6. Gently remove top slide by rotating and sliding it past the bottom slide. If the pad is wrinkled or torn at this step, discard it and make a new one.
7. Pipet 1.5 μL of concentrated cells onto the center of the agarose pad and cover with a 22 × 22 mm coverslip.
8. Seal coverslip edges with VALAP: Using a metal spatula, pick up a chunk of solid VALAP about the size of a grain of rice. Melt the VALAP on the spatula in a Bunsen burner flame and drag the spatula along one edge of the coverslip to dispense a thin line of VALAP at the interface with the slide. Repeat on all edges of the coverslip. For imaging longer than 2 h or at temperatures >25 °C, add a layer of nail polish over the VALAP for proper sealing.

3.7 Ratio Imaging

3.7.1 Ratio Imaging of $\Delta\psi$ with $DiOC_6$

1. Grow yeast cells containing a red mitochondrial marker (we recommend Tom70p-mCherry) to mid-log phase in 5 mL of liquid medium in a 50-mL conical-bottom tube.
2. Transfer 3×10^7 cells to a 1.5 mL microfuge tube and concentrate by centrifugation at ~8,000 × *g* for 30 s.
3. Resuspend cells in 1 mL HEPES and add $DiOC_6$ to a final concentration of 17.5 nM. For drug treatment (e.g., FCCP), add drug to cells in HEPES prior to addition of $DiOC_6$ (*see* **Note 19**).
4. Incubate cells in the dark at room temperature for 15 min.
5. Wash three times with HEPES.

6. Concentrate cells by centrifugation at ~8,000 × *g* for 30 s.
7. Mount 1.5 μL cells for short-term imaging (*see* **Note 20**).
8. Acquire images. Refer to Table 6 for suggested imaging conditions.

3.7.2 Ratio Imaging of Redox State Using roGFP

1. Grow yeast cells with mito-roGFP plasmid to mid-log phase in 5 mL of liquid medium in a 50-mL conical-bottom tube.
2. Transfer 3×10^7 cells to a 1.5 mL microfuge tube and concentrate by gentle centrifugation at ~8,000 × *g* for 30 s.
3. Mount 1.5 μL cells for short-term imaging.
4. Acquire images. Refer to Table 6 for suggested imaging conditions.

3.7.3 Ratio Imaging of ROS Using DHE

1. Grow yeast cells containing a green mitochondrial marker (we recommend Cit1-GFP) to mid-log phase in 5 mL of liquid medium in a 50-mL conical-bottom tube.
2. Transfer 3×10^7 cells to a 1.5 mL microfuge tube and concentrate by centrifugation at ~8,000 × *g* for 30 s.
3. Resuspend cells in 1 mL SC medium and add DHE to a final concentration of 40 μM. For drug treatment, add drug to cells in SC prior to addition of DHE.
4. Incubate cells in the dark at room temperature for 30 min.
5. Wash three times with SC.
6. Concentrate cells by centrifugation at ~8,000 × *g* for 30 s.
7. Mount 1.5 μL cells for short-term imaging.
8. Acquire images. Refer to Table 6 for suggested imaging conditions.

3.7.4 Optimizing Imaging Conditions and Preventing Toxicity

Checking for Photodamage

During live-cell imaging, illumination of cells for extended periods of time can lead to phototoxicity by two primary mechanisms: (1) photons at wavelengths close to the UV can cause direct cell damage; and (2) photons can react with cellular molecules to produce free radicals and reactive oxygen species. The following criteria can be used to check for photodamage to organelle and cytoskeletal function.

1. Cytoskeletal and mitochondrial structures: intense or long excitation can cause actin cable depolymerization and mitochondrial fragmentation.
2. Dynamics: While behavior of photodamaged mitochondria and cytoskeletal components has not been characterized, photodamage can be suspected if mitochondria or cytoskeletal elements change their velocity or other motility parameters in wild-type cells.

3. Staining with potential-sensitive dyes, such as $DiOC_6$ throughout the mitochondria in wild-type cells. Loss of membrane potential indicates mitochondria are not functioning properly.

Reducing Phototoxicity and Photobleaching

Even if cells remain healthy throughout imaging, excessive light exposure can cause photobleaching of fluorophores. To reduce phototoxicity and photobleaching, any or all of the following strategies can be used.

1. Reduce excitation intensity. With LEDs and many lasers, the intensity can be varied through the control software. With conventional light sources, neutral density filters can be used.
2. Reduce exposure time. In general, longer exposure time and lower intensity gives a comparable quality image with less photodamage.
3. Reduce spatial or time resolution. For example, increase the time interval and/or the *z*-section interval.

Increasing Signal-to-Noise Ratio Without Increasing Light Exposure

1. Apply binning in the camera. This will result in decreased spatial resolution, but this can be alleviated by adding a projection tube before the camera.
2. Increase camera gain.
3. Reduce camera readout speed, if possible.
4. Use a filter set with a broader spectral window or more efficient coatings to increase throughput.

3.7.5 Special Considerations for Quantitative Analysis

For quantitative fluorescence imaging, two criteria must be met: (1) emitted fluorescence must be proportional to the number of fluorescent molecules present; and (2) recorded pixel intensities must be proportional to the amount of light emitted by the sample. Under well-controlled conditions, these criteria are met at sufficient levels to measure many biological factors.

Absolute quantification of fluorescent molecules by imaging is a serious challenge due to many variables that complicate the linearity of fluorescence emission and pixel intensities [27]. However, proper controls and normalization can allow reliable measurements of relative changes in the volume and intensity of fluorescent structures, providing a powerful tool for hypothesis testing. At minimum, the following variables must be controlled in any wide-field fluorescence imaging experiment where intensities are compared:

1. Fluorescent probe concentration and age, if an exogenous probe, such as a dye, is used.
2. Sample age, unless photobleaching is known to be negligible.
3. Objective lens.

4. Illumination spectrum and intensity, including filters, field aperture setting, and light source alignment.
5. Camera model, binning, exposure time, gain, offset, readout speed, bit depth.
6. Imaging parameters must be adjusted to give pixel values significantly above the background level of the detector and sample, and below the saturation level of the detector. Enough cells should be imaged to ascertain the variability within the population. Whenever possible, experimental design should be validated by positive and negative controls. For dyes, specifically those that are quantitative, such as $DiOC_6$ for membrane potential or DHE for ROS, it is particularly useful to perform controls to determine the dynamic range of the dye (*see* Fig. 3). We also recommend normalizing to counterstains and producing ratio channels to compensate for local differences in mitochondrial mass or variability in thresholding (*see* **Note 21**).

3.8 Deconvolution of Wide-Field Fluorescence Images

Deconvolution is a computational method for removing out-of-focus light from wide-field fluorescence images. The result is a sharper image with a higher signal–noise ratio and enhanced three-dimensional information. Various algorithms are available, only some of which allow quantitation after processing [28]. The constrained iterative restoration algorithm (Volocity, PerkinElmer) preserves quantitative information.

For best results with deconvolution, *z*-series of images should be acquired at focus intervals of 0.2 μm. When studying fast dynamic processes such as mitochondrial motility, the z interval can be increased to 0.5 μm, resulting in some loss of spatial resolution but a doubled time resolution.

1. Acquire *z*-series images using short-term or long-term imaging conditions.
2. Load datasets in Volocity and verify that the *x*, *y*, and *z* scale are correctly set.
3. For each channel, generate a new calculated PSF. When prompted, enter the numerical aperture of the objective lens and the maximum wavelength of fluorescence emission (e.g., 507 nm for GFP and 620 nm emission wavelength for mCherry and MitoTracker Red).
4. Using the calculated PSF for each channel, perform iterative deconvolution with 60 iterations, 100 % confidence limit (*see* **Note 22**).

3.9 Quantification of Fluorescent Signals

To understand the mechanisms and consequences of actin-mediated mitochondrial trafficking and subsequent anchorage, it is helpful to quantify the amount of actin or mitochondria within a cell or subregions of a cell. There are two common measurements of fluorescence intensity: mean and integrated intensity.

Mean fluorescence intensity is the average pixel value. It approximates the concentration of fluorescent probes in a given area. It is also proportional to the total number of fluorescent molecules present when measuring areas of similar sizes. However, the mean is not appropriate for measurements of areas of different sizes, such as comparing the amount of mitochondria or F-actin in yeast buds of different sizes. A small bud and a large bud may have the same number of labeled actin structures, but the higher density of fluorescence in the small bud would produce a higher mean intensity than in the large bud where the molecules are more spread out. For scenarios where ROI sizes vary greatly, it is best to indicate the total amount of fluorescent probes via measurements of integrated density or integrated intensity. This is the sum of all pixel intensities in the area. For scenarios where structures may be superimposed and thus cannot be resolved by fluorescence imaging, as in the cases of actin cables or highly aggregated structures, the integrated intensity is also a more accurate assessment than area in indicating the total amount of a labeled structure. Here, we show how to obtain integrated intensity of mitochondria using Volocity and ImageJ.

3.9.1 Measurements Using Volocity

1. From the **Measurements** tab, create a protocol by selecting the following steps in order: **Find Objects Using Intensity, Exclude Objects by Size, Clip Objects to ROIs, Measure Objects**.
2. For images with multiple channels, make sure that the correct channel is selected under the **Find Objects Using Intensity** tab.
3. Under **Exclude Objects by Size,** select a size criterion that will exclude background pixels or other artifacts from image capture or deconvolution. This value will depend on imaging conditions and must be determined empirically.
4. Under the **Measure Objects** tab, click the gear icon to select which measurements you wish to use. Check **Intensity and Volume Measurements**.
5. Define the ROI you wish to measure. Volocity allows rectangular, circular, and freehand ROIs. Use the Zoom function to make it easier to outline the desired area.
6. Adjust the threshold under the **Find Objects Using Intensity** tab by manipulating the lower and upper limits. We suggest maximizing the upper limit to include all pixels that have high intensity values. The lower limit should be adjusted to remove enough background to only include signal that is found within the cellular structure of interest.
7. All measurements will appear in a tab below the image. Individual objects will be separated (for example, if mitochondrial fluorescence intensity is being measured and there are three

mitochondrial structures, three measurements will be provided, each color-coded to depict which structure they represent). These measurements can be copied into a spreadsheet application such as Microsoft Excel or saved within Volocity by clicking **Make Measurements** under the **Measurement** tab.

8. The Sum value is the integrated intensity (total fluorescence of all the pixels within the object).

3.9.2 Measurements Using ImageJ

1. From the **Analyze** menu, choose **Set Measurements**. Check the following: **Area, Mean Gray Value, Integrated Density, Display Label**.
2. Using the ROI tool, define the ROI you wish to measure.
3. From the **Analyze** menu, choose **Measure** (or press M).
4. Draw an ROI in the background (outside of cells) and measure background.
5. Copy or save the measurements that appear in the Results window.
6. Subtract the background integrated density from the integrated density of each measured ROI. The background integrated intensity for an ROI is the product of the background mean gray value and the area of the measured ROI.

3.10 Analysis of Intracellular Movement

Motility analysis can be used to test hypotheses about the mechanism and regulation of intracellular movements. The methods described here employ ImageJ [29] and Volocity (PerkinElmer) software as the tools of choice in our own laboratory. Similar functions are available in most software packages for image analysis and processing.

Two strategies are presented here for quantifying motility: tracking and total motility measurement. The most direct and comprehensive way to measure motility is tracking analysis. Tracking can provide several quantitative measurements including velocity, distance traveled, frequency of movement, persistence/processivity of movement, and direction of movement. Here, we show how to calculate velocity and distance traveled.

Tracking involves marking the position of an object at successive time points. For punctate or roughly circular structures, the position is usually defined as the center of the structure. For structures with elongated, irregular or dynamic shapes, a point can be defined on the object of interest and be used as a reference point for velocity measurements. For example, to track mitochondria, which are highly tubular and dynamic, the leading tip can be used. In the case of Abp140-GFP-labeled actin, heterogeneity of Abp140 binding produces bright dots along the actin cable, and these can serve as fiduciary marks to track movement [18].

The second method, measurement of total motility, finds differences between successive image frames to quantify movement. A structure that does not change position will disappear in the subtracted image; any portion that does move will appear in the difference image. Measuring the area that changes relative to the total provides a quantitative indicator of the degree of motility. This method is more easily automated than tracking, especially for irregular structures. In addition, the regions of the cell where movement occurs are visible. This is helpful for qualitatively assessing anchorage, e.g., of mitochondria.

3.10.1 Tracking Intracellular Movement

Measurements in Volocity

1. Import image into Volocity. Make sure spatial and time scales are correctly set.
2. View the movie to identify structures meeting motility criteria (*see* **Note 23**).
3. Go to the first timepoint in which the movement occurs. Use the point tool to mark the center or leading edge of the structure of interest. For example, place a point at the leading edge of a mitochondrial tip.
4. Go to the next time point and mark the new position of the structure of interest.
5. Export point measurements into a spreadsheet or statistical application such as Microsoft Excel. Use the Pythagorean theorem to determine the distance moved and divide this by the time elapsed to calculate velocity.

Measurements in ImageJ

1. From the **Analyze** menu, choose **Set Measurements**. Check the following boxes: **Display Label, Centroid**.
2. View the movie to identify structures meeting motility criteria (*see* **Note 23**).
3. Select the **Point Selection** tool from the ImageJ toolbar and double-click the tool icon to set parameters: Check **Auto-Measure** and **Auto-Next Slice**.
4. Go to the first time point in which the movement occurs and click the structure.
5. Continue marking the motile structure for the duration of observable movement.
6. Copy or save the measurements that appear in the results window.
7. Calculate distance traveled between each set of successive time points t1 and t2 using the Pythagorean Theorem, where d is distance and the positions are (x1,y1) and (x2, y2) respectively:

$$d_{(t_1,t_2)} = \sqrt{(x1 - x2)^2 + (y1 - y2)^2}$$

Calculate velocity as distance traveled divided by time elapsed. To find mean velocity, calculate the average of all incremental velocities.

3.10.2 Qualitative and Quantitative Measurement of Total Motility

In this protocol, 4D image capturing is crucial to include movement in the *z*-axis. This analysis is done in ImageJ using a plugin published by Kurt De Vos [30]. Total motility for mitochondria is described below to provide an example of how this measurement can be made (Fig. 2).

1. Collect time-lapse *z* series. For mitochondria, we recommend using the long-term imaging preparation to collect a z-stack every 30 s for 10 min, with each *z* stack covering 6 μm of depth at an interval of 0.5 μm. We recommend imaging mitochondria using Cit1-GFP due to the high intensity and reliability of this mitochondria-targeted FP.
2. Deconvolve images in Volocity to remove out-of-focus light.
3. Open the deconvolved dataset as a hyperstack in ImageJ.
4. Optional: Contrast can be automatically enhanced using **Process > Enhance Contrast**. Any cell movement or drift can be corrected using the **StackReg** plugin in ImageJ (**Note 24**) [31].
5. Generate a maximum-intensity projection for each timepoint: **Image > Stacks > Z project**; choose **Projection Type: Max Intensity**; check the box indicating **All Time Frames**.
6. Use **Image > Adjust > Threshold** to manually select a threshold for the projected stack. The thresholded areas should follow the outlines of mitochondria. Do not click Apply; rather, convert the stacks into binary images using **Process > Binary > Make Binary**.
7. Run the **Total Motility** plugin. The output will include a Results window and a new stack containing the difference images for each time point.
8. For quantitative analysis, each line of the Results window shows the percentage of mitochondrial area that was motile at a given time point.
9. For qualitative analysis, generate a sum projection of the new stack: **Image > Stacks > Z project**; choose **Projection Type: Sum Slices**. Here, the projection will summarize movement over the entire time course. This image depicts all motility events where white indicates movement and black indicates lack of movement.

3.10.3 Quantitative Ratiometric Measurements for Measuring Mitochondrial Quality

Many tools used to measure mitochondrial quality are dyes that are imported into the mitochondria, such as the membrane potential-dependent dye, $DiOC_6$. Measurements of total or mean intensity of these dyes are unreliable because they are skewed by the total mitochondrial mass. To correct for this, we suggest normalizing the dye intensities to a mitochondrial marker by calculating the ratio. The following protocol for quantitative ratiometric measurements using Volocity and ImageJ can be used for relative ratiometric measurements of any two fluorophores, including $DiOC_6$ for mitochondrial membrane potential, roGFP for mitochondrial redox state, and DHE for mitochondrial ROS levels.

Producing a Ratio Channel in Volocity

1. Draw an ROI around a field where there is zero fluorescence. This area will be used for background subtraction.
2. Under the **Tools** tab, select **Ratio…**
3. For Channel A (the numerator) select the dye to be quantified, such as $DiOC_6$ or DHE. For Channel B (the denominator) select what you are normalizing to, such as Tom70-mCherry or Cit1-GFP.
4. Under **Subtract**, click **Get From ROI** for both Channel A and Channel B to subtract the background from each channel.
5. Click **Calculate** to automatically threshold each channel. Threshold values can also be manually set (*see* **Note 25**).
6. Optional: A rainbow LUT (look-up table or color scale) can make differences more obvious in the ratio channel, with warmer, redder colors indicating higher dye concentration (e.g., more $DiOC_6$) and colder, bluer colors indicating lower dye concentration.
7. Click the **Ratio** button to produce the ratio channel.

Measurements of the Ratio Channel in Volocity

1. From the **Measurements** tab, create a protocol by selecting the following steps in order: **Clip Objects to ROIs, Measure Objects**.
2. Click on the gear icon in the **Measure Objects** tab and check **Intensity measurements ignoring zero values**. Ensure that the appropriate ratio channel is selected here.
3. Define the ROIs you wish to measure. You may select multiple ROIs for ratio measurements.
4. Under the **Measurements** tab, the measurements for each ROI are listed below the image and color-coded to match. These measurements can be copied into a spreadsheet application such as Microsoft Excel or saved within Volocity by clicking **Make Measurements** under the **Measurement** tab.
5. To compare values between strains, use the measurement item **Mean (ignoring zero values)**.

Creating and Analyzing Ratiometric Images in ImageJ

1. Open deconvolved images and change type to 32 bit: **Image > Type > 32 bit.**
2. Draw a region of interest (ROI) in an area where there are no cells. Calculate the mean intensity in this ROI: **Analyze > Measure.**

3. Subtract the calculated mean background from the stack: **Process > Math > Subtract.**
4. Using the subtracted *z*-stack, find the middle slice and threshold on mitochondria: **Image > Adjust > Threshold** and click **Apply** on the Threshold window. Apply to all slices in the stack. Check **Set background pixels to NaN.**
5. Create the ratio *z* stack: **Process > Image Calculator.** For the numerator, use the functional probe (DHE or $DiOC_6$) or the reduced roGFP signal. For the denominator, use the mitochondrial marker or the oxidized roGFP signal.
6. Draw an ROI around the area of interest. **Analyze > Tools > ROI Manager**, and click **Add** to record the ROI. Multiple regions may be stored in the manager. In ROI Manager, select all ROIs, then choose **More > Multi-Measure** to measure all regions in stack slices. Export data to a spreadsheet for analysis.

4 Notes

1. Abp140-GFP signal is significantly higher in non-fermentable carbon sources. Lactate medium is the preferred choice as it is not autofluorescent and it provides more nutrients than synthetic complete medium with glycerol (SG). However, there may be some mitochondrial fragmentation in lactate medium. If this is a problem, use glycerol-based media, such as SG.
2. $DiOC_6$ stock solutions are most stable at higher concentrations (~1000× working concentration). For best results, make working solutions fresh each day. Stock solutions may precipitate, so take care to resuspend prior to use.
3. H_2O_2 is unstable when diluted in water and diluted solutions should be prepared fresh and discarded after use for optimal results. DTT should be purchased in solution isolated in glass ampules for optimal results as reducing capacity can be lost upon contact with oxygen.
4. The sequences for the 3′ ends are often provided by the designers of the tagging plasmid. In many cases, the forward primer will include the sequence of the polylinker upstream of the FP tag while the reverse primer will include the reverse complement of the last few nucleotides of the selection marker, or some sequence downstream of the selection marker.
5. At least 250 ng of genomic DNA and 25 ng of plasmid DNA should be used.
6. We suggest using a high-fidelity polymerase for the amplification to prevent point mutations, especially when amplifying larger insertion cassettes.

7. The second step of the 35-cycle procedure is the primer annealing phase. The temperature here should be matched to the melting temperature of the DNA. For primers that have secondary structures, or when primer sequences cannot be manipulated to make the melting temperatures comparable, betaine, DMSO, or both can be added to the PCR reaction to a final concentration of 1 M and 5 %, respectively (10 μL and 5 μL, respectively for a 50 μL reaction).
8. For most PCR reactions, we find that PCR purification is unnecessary. While purification will eliminate some contaminants from the PCR reaction, the accompanying loss of DNA from the process often outweighs the benefits of performing the purification. We specifically find that when using the KAPA HiFi HotStart or Platinum Taq DNA Polymerase High Fidelity PCR reaction kits, purification to eliminate contaminants is unnecessary.
9. The galactose-inducible Cre recombinase plasmids are CEN plasmids, which can often be dropped out by continuous growth on non-selective medium. Alternatively, pSH47 can be used as this galactose-inducible Cre recombinase expression plasmid contains the *URA3* selectable marker and allows curing cells of the plasmid by counter-selection on 5-fluoroorotic acid (5-FOA) plates after marker selection.
10. When performing rescue/plasmid complementation assays, it is important to express the FP-tagged protein under the endogenous promoter of the gene. This will avoid unexpected errors, such as overexpression of a partially functional protein rescuing wild-type phenotypes.
11. For live-cell imaging, avoid using medium containing yeast extract (e.g., YPD, YPG) as the yeast extract is autofluorescent and will require repeated washing prior to imaging. Instead, use synthetic complete or lactate medium.
12. Generally, mid-log phase cultures (OD_{600} = 0.2–0.6) will have the optimal amount of cells for staining and do not require concentrating unless a higher number of cells is necessary for the experiment. A culture with OD_{600} = 1.0 contains ~10^7 cells/mL.
13. The dye concentrations suggested in Table 5 are a good starting point. The specificity of these dyes is dependent on mitochondrial membrane potential and dye concentration. For example, some dyes can accumulate in other organelles, such as ER and vacuoles, when concentrations are too high. Therefore, dye concentration and incubation times should be titrated when working with new strains. High concentrations of dyes should be avoided to prevent excessive accumulation of dye in mitochondria resulting in organelle swelling and respiratory defects.

14. Slides should be used for even less time for more sensitive applications. For example, $DiOC_6$ slides should only be kept for 2–3 min, and slides for visualization of actin cables should be used within 4 min due to fading of Abp140-GFP signal.
15. Excessive/extended centrifugation can cause actin cable depolymerization and subsequent mitochondrial fragmentation.
16. The volume is important since excess volume can cause cells to float and move during image acquisition, while insufficient volume can compress cells or cause uneven spreading. Slight pressure should be applied to the edges of the coverslip to stably trap cells between the slide and coverslip and avoid movement of cells during image acquisition. Take care not to put too much pressure or cells may burst.
17. The agarose pad material should be placed slightly off center as it is likely to move when removing the top slide in a subsequent step.
18. Excess cooling will cause the agarose pad to harden before it is properly spread. 5 s cooling of agarose pad is appropriate for rooms at 25 °C or higher. For colder rooms, no cooling is necessary.
19. It is not advisable to wash out the drug when staining with $DiOC_6$ unless it is known that the drug will react with $DiOC_6$ and make the dye unreliable. Because the incubation time is 15 min, the cells may recover or adapt during this time. If it is not desirable to keep cells in a specific drug for an additional 15 min, $DiOC_6$ can be added 15 min prior to the completion of drug treatment. For example, if a 30-min treatment is desired, the drug can be added to cells in HEPES, incubated for 15 min, then $DiOC_6$ can be added and incubation can continue for another 15 min.
20. $DiOC_6$ imaging can only be performed for a very short time (maximally 2–3 min). After 2–3 min of imaging, the dye begins to leak out of the cell and stain slides, making visualization and quantification unreliable.
21. For measurements of fluorescence intensity, thresholding can affect quantitative values greatly. When normalizing and contrasting to a stable and valid counterstain, thresholding can be more appropriately controlled. For example, when measuring fluorescence intensity of a membrane potential-dependent dye, such as $DiOC_6$, thresholding by normalizing to or contrasting signal intensities to Tom70-mCherry will help reduce subjective errors.
22. Artifacts occasionally appear in deconvolved images. These may include halo-like structures or the disappearance of structures that are visible in the raw data. If structures are lost, or halos appear, repeat the deconvolution with fewer iterations.

If artifacts do not appear, but the image is still blurry or faint, repeat with more iterations. If deconvolution fails to exceed the 95 % confidence limit, verify that optics are clean, light source is aligned, and cells are mounted under a 1.5 coverslip. Also make sure that images are acquired with pixel values at least in the middle of the detector range, and not saturating the detector.

23. For characterizing directed movement in mitochondria, we generally make velocity measurements on mitochondrial tubules that have made three or more consecutive movements in the same direction. For tracking of actin cables, we generally make measurements on fiduciary marks that have made four or more consecutive movements in the same direction. It is important to determine a selection criterion prior to measurement to decrease subjectivity of measurements.
24. The StackReg plugin is available from http://bigwww.epfl.ch/thevenaz/stackreg/.
25. Thresholding is very important in getting accurate measurements. The mitochondria-targeted FP should be used to threshold the ratio channel. Care must be taken to match the image of the ratio channel as close to the mitochondrial structure as possible [15]. In addition, it is important to keep the ratio between the two thresholds identical for each image to avoid introducing errors. For example, if the calculated thresholds are 220 and 200 for Channel A and Channel B respectively, they may be changed to 440 and 400.

Acknowledgments

This work was supported by awards from the Ellison Medical Foundation (AG-SS-2465) and the NIH (GM45735, GM45735S1, and GM096445) to L.A.P. GM45735S1 was issued from the NIH under the American Recovery and Reinvestment Act of 2009. The microscopes used for these studies were supported in part through a NIH/NCI grant (5 P30 CA13696) and obtained using funds from the NIH-NCRR (1S10OD014584) to L.A.P.

References

1. Bubnell J, Pfister P, Sapar ML, Rogers ME, Feinstein P (2013) beta2 adrenergic receptor fluorescent protein fusions traffic to the plasma membrane and retain functionality. *PLoS ONE* **8**: e74941.
2. Gauss R, Trautwein M, Sommer T, Spang A (2005) New modules for the repeated internal and N-terminal epitope tagging of genes in Saccharomyces cerevisiae. Yeast 22:1–12
3. Gueldener U, Heinisch J, Koehler GJ, Voss D, Hegemann JH (2002) A second set of loxP marker cassettes for Cre-mediated multiple gene knockouts in budding yeast. Nucleic Acids Res 30:e23
4. Gietz RD, Schiestl RH, Willems AR, Woods RA (1995) Studies on the transformation of intact yeast cells by the LiAc/SS-DNA/PEG procedure. Yeast 11:355–360

5. Simbeni R, Pon L, Zinser E, Paltauf F, Daum G (1991) Mitochondrial membrane contact sites of yeast. Characterization of lipid components and possible involvement in intramitochondrial translocation of phospholipids. J Biol Chem 266:10047–10049
6. Meeusen S, Nunnari J (2003) Evidence for a two membrane-spanning autonomous mitochondrial DNA replisome. J Cell Biol 163: 503–510
7. Koehler CM (2004) New developments in mitochondrial assembly. Annu Rev Cell Dev Biol 20:309–335
8. Westermann B (2010) Mitochondrial fusion and fission in cell life and death. Nat Rev Mol Cell Biol 11:872–884
9. Higuchi-Sanabria R, Pernice WM, Vevea JD, Alessi Wolken DM, Boldogh IR, Pon LA (2014) Role of asymmetric cell division in lifespan control in Saccharomyces cerevisiae. FEMS Yeast Research 14(8):1133–46
10. Vevea JD, Swayne TC, Boldogh IR, Pon LA (2014) Inheritance of the fittest mitochondria in yeast. Trends Cell Biol 24:53–60
11. Swayne TC, Zhou C, Boldogh IR et al (2011) Role for cER and Mmr1p in anchorage of mitochondria at sites of polarized surface growth in budding yeast. Curr Biol 21:1994–1999
12. Lackner LL, Ping H, Graef M, Murley A, Nunnari J (2013) Endoplasmic reticulum-associated mitochondria-cortex tether functions in the distribution and inheritance of mitochondria. Proc Natl Acad Sci U S A 110:E458–E467
13. Klecker T, Scholz D, Fortsch J, Westermann B (2013) The yeast cell cortical protein Num1 integrates mitochondrial dynamics into cellular architecture. J Cell Sci 126:2924–2930
14. Hughes AL, Gottschling DE (2012) An early age increase in vacuolar pH limits mitochondrial function and lifespan in yeast. Nature 492:261–265
15. Vevea JD, Wolken DM, Swayne TC, White AB, Pon LA (2013) Ratiometric biosensors that measure mitochondrial redox state and ATP in living yeast cells. J Vis Exp (77), e50633, doi:10.3791/50633
16. Hanson GT, Aggeler R, Oglesbee D, Cannon M, Capaldi RA, Tsien RY, Remington SJ (2004) Investigating mitochondrial redox potential with redox-sensitive green fluorescent protein indicators. J Biol Chem 279: 13044–13053
17. McFaline-Figueroa JR, Vevea J, Swayne TC, Zhou C, Liu C, Leung G, Boldogh IR, Pon LA (2011) Mitochondrial quality control during inheritance is associated with lifespan and mother-daughter age asymmetry in budding yeast. Aging Cell 10:885–895
18. Yang HC, Pon LA (2002) Actin cable dynamics in budding yeast. Proc Natl Acad Sci U S A 99:751–756
19. Higuchi R, Vevea JD, Swayne TC, Chojnowski R, Hill V, Boldogh IR, Pon LA (2013) Actin dynamics affect mitochondrial quality control and aging in budding yeast. Curr Biol 23: 2417–2422
20. Doyle T, Botstein D (1996) Movement of yeast cortical actin cytoskeleton visualized in vivo. Proc Natl Acad Sci U S A 93: 3886–3891
21. Huckaba TM, Lipkin T, Pon LA (2006) Roles of type II myosin and a tropomyosin isoform in retrograde actin flow in budding yeast. J Cell Biol 175:957–969
22. Fehrenbacher KL, Yang HC, Gay AC, Huckaba TM, Pon LA (2004) Live cell imaging of mitochondrial movement along actin cables in budding yeast. Curr Biol 14:1996–2004
23. Riedl J, Crevenna AH, Kessenbrock K et al (2008) Lifeact: a versatile marker to visualize F-actin. Nat Methods 5:605–607
24. Sabourin M, Tuzon CT, Fisher TS, Zakian VA (2007) A flexible protein linker improves the function of epitope-tagged proteins in Saccharomyces cerevisiae. Yeast 24:39–45
25. Cheng TH, Chang CR, Joy P, Yablok S, Gartenberg MR (2000) Controlling gene expression in yeast by inducible site-specific recombination. Nucleic Acids Res 28:E108
26. Messerschmitt M, Jakobs S, Vogel F, Fritz S, Dimmer KS, Neupert W, Westermann B (2003) The inner membrane protein Mdm33 controls mitochondrial morphology in yeast. J Cell Biol 160:553–564
27. Pawley J (2000) The 39 steps: a cautionary tale of quantitative 3-D fluorescence microscopy. Biotechniques 28(884–886):888
28. Wallace W, Schaefer LH, Swedlow JR (2001) A workingperson's guide to deconvolution in light microscopy. Biotechniques **31**: 1076–1078, 1080, 1082 passim
29. Schneider CA, Rasband WS, Eliceiri KW (2012) NIH Image to ImageJ: 25 years of image analysis. Nat Methods 9:671–675
30. De Vos KJ, Sheetz MP (2007) Visualization and quantification of mitochondrial dynamics in living animal cells. Methods Cell Biol 80: 627–682
31. Thevenaz P, Ruttimann UE, Unser M (1998) A pyramid approach to subpixel registration based on intensity. IEEE Trans Image Process 7:27–41

32. Longtine MS, McKenzie A 3rd, Demarini DJ, Shah NG, Wach A, Brachat A, Philippsen P, Pringle JR (1998) Additional modules for versatile and economical PCR-based gene deletion and modification in Saccharomyces cerevisiae. Yeast 14:953–961
33. Rodrigues F, van Hemert M, Steensma HY, Corte-Real M, Leao C (2001) Red fluorescent protein (DsRed) as a reporter in Saccharomyces cerevisiae. J Bacteriol 183:3791–3794
34. Janke C, Magiera MM, Rathfelder N et al (2004) A versatile toolbox for PCR-based tagging of yeast genes: new fluorescent proteins, more markers and promoter substitution cassettes. Yeast 21:947–962
35. Sheff MA, Thorn KS (2004) Optimized cassettes for fluorescent protein tagging in Saccharomyces cerevisiae. Yeast 21:661–670
36. Young CL, Raden DL, Caplan JL, Czymmek KJ, Robinson AS (2012) Cassette series designed for live-cell imaging of proteins and high-resolution techniques in yeast. Yeast 29: 119–136
37. Okamoto K, Perlman PS, Butow RA (1998) The sorting of mitochondrial DNA and mitochondrial proteins in zygotes: preferential transmission of mitochondrial DNA to the medial bud. J Cell Biol 142:613–623
38. Mozdy AD, McCaffery JM, Shaw JM (2000) Dnm1p GTPase-mediated mitochondrial fission is a multi-step process requiring the novel integral membrane component Fis1p. J Cell Biol 151:367–380
39. Riezman H, Hase T, van Loon AP, Grivell LA, Suda K, Schatz G (1983) Import of proteins into mitochondria: a 70 kilodalton outer membrane protein with a large carboxy-terminal deletion is still transported to the outer membrane. EMBO J 2:2161–2168
40. Madania A, Dumoulin P, Grava S, Kitamoto H, Scharer-Brodbeck C, Soulard A, Moreau V, Winsor B (1999) The Saccharomyces cerevisiae homologue of human Wiskott-Aldrich syndrome protein Las17p interacts with the Arp2/3 complex. Mol Biol Cell 10: 3521–3538
41. Kaksonen M, Sun Y, Drubin DG (2003) A pathway for association of receptors, adaptors, and actin during endocytic internalization. Cell 115:475–487
42. Warren DT, Andrews PD, Gourlay CW, Ayscough KR (2002) Sla1p couples the yeast endocytic machinery to proteins regulating actin dynamics. J Cell Sci 115:1703–1715
43. Miliaras NB, Park JH, Wendland B (2004) The function of the endocytic scaffold protein Pan1p depends on multiple domains. Traffic 5:963–978
44. Morishita M, Engebrecht J (2005) End3p-mediated endocytosis is required for spore wall formation in Saccharomyces cerevisiae. Genetics 170:1561–1574
45. Sun Y, Carroll S, Kaksonen M, Toshima JY, Drubin DG (2007) PtdIns(4,5)P2 turnover is required for multiple stages during clathrin- and actin-dependent endocytic internalization. J Cell Biol 177:355–367
46. Sun Y, Martin AC, Drubin DG (2006) Endocytic internalization in budding yeast requires coordinated actin nucleation and myosin motor activity. Dev Cell 11:33–46
47. Vaduva G, Martin NC, Hopper AK (1997) Actin-binding verprolin is a polarity development protein required for the morphogenesis and function of the yeast actin cytoskeleton. J Cell Biol 139:1821–1833
48. Evangelista M, Klebl BM, Tong AH, Webb BA, Leeuw T, Leberer E, Whiteway M, Thomas DY, Boone C (2000) A role for myosin-I in actin assembly through interactions with Vrp1p, Bee1p, and the Arp2/3 complex. J Cell Biol 148:353–362
49. Jonsdottir GA, Li R (2004) Dynamics of yeast Myosin I: evidence for a possible role in scission of endocytic vesicles. Curr Biol 14:1604–1609
50. Sekiya-Kawasaki M, Groen AC, Cope MJ et al (2003) Dynamic phosphoregulation of the cortical actin cytoskeleton and endocytic machinery revealed by real-time chemical genetic analysis. J Cell Biol 162:765–772
51. Huckaba TM, Gay AC, Pantalena LF, Yang HC, Pon LA (2004) Live cell imaging of the assembly, disassembly, and actin cable-dependent movement of endosomes and actin patches in the budding yeast, Saccharomyces cerevisiae. J Cell Biol 167:519–530
52. Boldogh IR, Yang HC, Nowakowski WD, Karmon SL, Hays LG, Yates JR 3rd, Pon LA (2001) Arp2/3 complex and actin dynamics are required for actin-based mitochondrial motility in yeast. Proc Natl Acad Sci U S A 98:3162–3167
53. Smith MG, Swamy SR, Pon LA (2001) The life cycle of actin patches in mating yeast. J Cell Sci 114:1505–1513
54. Karpova TS, Reck-Peterson SL, Elkind NB, Mooseker MS, Novick PJ, Cooper JA (2000) Role of actin and Myo2p in polarized secretion and growth of Saccharomyces cerevisiae. Mol Biol Cell 11:1727–1737
55. Waddle JA, Karpova TS, Waterston RH, Cooper JA (1996) Movement of cortical actin patches in yeast. J Cell Biol 132:861–870
56. Winder SJ, Jess T, Ayscough KR (2003) SCP1 encodes an actin-bundling protein in yeast. Biochem J 375:287–295

57. Okreglak V, Drubin DG (2007) Cofilin recruitment and function during actin-mediated endocytosis dictated by actin nucleotide state. J Cell Biol 178:1251–1264
58. McConnell SJ, Stewart LC, Talin A, Yaffe MP (1990) Temperature-sensitive yeast mutants defective in mitochondrial inheritance. J Cell Biol 111:967–976
59. Skowronek P, Krummeck G, Haferkamp O, Rodel G (1990) Flow cytometry as a tool to discriminate respiratory-competent and respiratory-deficient yeast cells. Curr Genet 18:265–267
60. Nunnari J, Marshall WF, Straight A, Murray A, Sedat JW, Walter P (1997) Mitochondrial transmission during mating in Saccharomyces cerevisiae is determined by mitochondrial fusion and fission and the intramitochondrial segregation of mitochondrial DNA. Mol Biol Cell 8:1233–1242

Chapter 3

Imaging of the Actin Cytoskeleton and Mitochondria in Fixed Budding Yeast Cells

Ryo Higuchi-Sanabria*, Theresa C. Swayne*, Istvan R. Boldogh, and Liza A. Pon

Abstract

The budding yeast *Saccharomyces cerevisiae* is widely used as a model system to study the organization and function of the cytoskeleton. In the past, its small size, rounded shape, and rigid cell wall created obstacles to explore the cell biology of this model eukaryote. It is now possible to acquire and analyze high-resolution and super-resolution multidimensional images of the yeast cell. As a result, imaging of yeast has emerged as an important tool in eukaryotic cell biology. This chapter describes labeling methods and optical approaches for visualizing the cytoskeleton and interactions of the actin cytoskeleton with mitochondria in fixed yeast cells using wide-field and super-resolution fluorescence microscopy.

Key words Yeast, Mitochondria, Fluorescent proteins, Vital staining, Microscopy, Live-cell imaging, Deconvolution, Ratio imaging, Actin, Cytoskeleton

1 Introduction

It was more than 30 years ago when the actin cytoskeleton was first visualized in fixed cells of the budding yeast *Saccharomyces cerevisiae* using fluorescent phallotoxins [1]. In the ensuing time fluorescence microscopy turned out to be a useful method to study actin-containing structures in yeast. Many of the actin-cytoskeletal proteins are similar to those found in other eukaryotes, and their organization and functions are often conserved. Therefore, budding yeast has emerged as a model system to study many aspects of actin-dependent cellular functions including establishment and maintenance of cell polarity, cell division, and movements of cellular organelles and other cargo.

In yeast there are two major actin-containing cytoskeletal structures that are maintained during the whole cell division cycle:

*Author contributed equally with all other contributors.

Ray H. Gavin (ed.), *Cytoskeleton: Methods and Protocols*, Methods in Molecular Biology, vol. 1365,
DOI 10.1007/978-1-4939-3124-8_3, © Springer Science+Business Media New York 2016

actin patches and actin cables. Actin patches are cortical structures that appear punctate by fluorescence microscopy. They consist of a coat of short actin filaments, nucleated by the Arp2/3 complex, around endosomal vesicles (reviewed in Ref. [2]). They contain conserved proteins including clathrin and its adaptors, Arp2/3 complex and its nucleation promoting factors (NPFs), kinases, actin bundling proteins (e.g., fimbrin), and membrane binding proteins. Actin patches localize to incipient bud sites, the bud tip during apical bud growth, the bud cortex during isotropic bud growth, and the bud neck immediately before and during cytokinesis. As a result, clustering of patches in the bud is used as an indicator of a functional cell polarization machinery.

Actin cables are bundles of actin filaments that align along the mother–bud axis, and can extend the whole length of a yeast cell. They contain conserved proteins including the actin-bundling proteins fimbrin (Sac6p) and Abp140p, and two tropomyosin isoforms (Tpm1p and Tpm2p) and are similar to actin bundles in microvilli, filopodia, and stereocilia. Through most of the cell cycle in *S. cerevisiae*, actin filaments within actin cables are oriented with their barbed (plus) ends toward the bud or bud neck. However, immediately before cytokinesis, actin cables rearrange such that their barbed ends of actin cables in mother cells and buds are directed toward the bud neck. In the budding yeast, actin cables are tracks for the movement of many intracellular elements including mRNA, spindle alignment elements, mitochondria, vacuoles, and secretory vesicles (reviewed in Ref. 3).

Actin structures can be visualized either by standard immunofluorescence or by binding the small peptide phalloidin. The main difference between these reagents is that phalloidin binds only polymerized, filamentous actin (F-actin), whereas antibodies will bind both F-actin and unpolymerized, globular actin (G-actin). Typically, the polymeric structures are of most interest, so phalloidin is the preferred reagent. It has the added advantage of a simpler staining protocol. However, in cases where visualizing G-actin is important (for example, to assess a change in equilibrium between F- and G-actin), the antibody may be a better choice.

As one important function of the cytoskeleton is to control intracellular movement, we also describe methods to visualize the nucleus and one of the cargos for cytoskeleton-driven movement, the mitochondrion. With appropriate antibodies, the immunofluorescence protocol can be used to visualize other cargo as well.

While observations in fixed cells cannot capture the dynamic nature of the actin cytoskeleton, they are very important in acquiring structural data. This is especially valuable in light of the recent introduction and increased availability of super-resolution structured illumination microscopy (SIM) techniques. SIM provides spatial resolution surpassing the diffraction limit of conventional fluorescence microscopy. As a result, structures and fine details that would otherwise be undetectable are made evident. Here, we describe methods for visualizing the actin cytoskeleton using both conventional wide-field fluorescence microscopy and SIM. We also describe how to quantify actin cables using freely available image analysis software.

2 Materials

2.1 Yeast Growth Media

Synthetic Complete medium (SC): 6.7 g yeast nitrogen base without amino acids, 20 g glucose. Dissolve ingredients in 800 mL distilled H_2O. Supplement with the following amino acids as needed: 10 mL adenine (2 mg/mL stock in 0.05 M HCl), 10 mL uracil (2 mg/mL stock in 0.5 % $NaHCO_3$), 10 mL arginine 10 mL (1 mg/mL stock in H_2O), 10 mL histidine (1 mg/mL stock in H_2O), 10 mL leucine (10 mg/mL stock in H_2O), 10 mL lysine (3 mg/mL stock in H_2O), 10 mL methionine (2 mg/mL stock in H_2O), 10 mL phenylalanine (5 mg/mL stock in H_2O), 10 mL tryptophan (2 mg/mL stock in H_2O), 10 mL tyrosine (3 mg/mL stock in 0.05 M HCl). Adjust pH to 5.5 with $NaHCO_3$ and bring volume to 1 L with distilled H_2O. Sterilize by autoclaving.

2.2 Materials for Fixation and Staining of Yeast Cells

1. Wash solution: Mix 12.5 mL 1 M potassium phosphate, pH 7.5 and 400 mL 1 M KCl. Bring volume up to 500 mL with distilled H_2O. Sterilize by autoclaving.
2. Tris/DTT: 10 mL 1 M Tris–SO_4, pH 9.4, 1 mL 1 M DTT. Bring volume to 100 mL with distilled H_2O. Make this solution on the day of use.
3. Zymolyase (Seikagaku Inc., Tokyo, Japan): Dissolve Zymolyase 20 T in wash solution to a concentration of 0.125 mg/mL. Make this solution on the day of use.
4. NS: 10 mL 1 M Tris–HCl, pH 7.5, 21.4 g sucrose, 1 mL 0.5 M EDTA, pH 8.0, 0.5 mL 1 M $MgCl_2$, 0.05 mL 1 M $ZnCl_2$, 0.05 mL 0.5 M $CaCl_2$. Adjust volume to 500 mL with distilled H_2O. Filter-sterilize and store in 40-mL aliquots at -20 °C.
5. NS+: On the day of use, supplement 10 mL NS with 50 μL of 200 mM phenylmethylsulfonylfluoride (PMSF) dissolved in ethanol, and 500 μL 10 % (w/v) NaN_3.
6. 10× PBS stock: 80 g NaCl, 2 g KCl, 14.4 g Na_2HPO_4, 2.4 g KH_2PO_4. Dissolve ingredients in 800 mL distilled H_2O. Adjust pH to 7.4 with HCl. Bring volume to 1 L with distilled H_2O.
7. PBS+: 5 mL 10× PBS, 0.5 g bovine serum albumin, 0.1 mL 10 % (w/v) NaN_3. Bring volume to 50 mL with distilled H_2O and filter-sterilize. Make the solution on the day of use (*see* **Note 1**).
8. PBT: 1 mL 10× PBS, 0.1 mL 10 % Triton X-100, 0.1 g bovine serum albumin, and 20 μL 10 % (w/v) NaN_3. Bring volume to

Table 1
Useful marker antigens for yeast mitochondria

Protein	Location	References
Porin	Mitochondrial outer membrane	[13, 14]
Cytochrome oxidase	Mitochondrial inner membrane	[15, 16]
Citrate synthase I	Mitochondrial matrix	[17]
OM14	Mitochondrial outer membrane	[18, 19]
Abf2p	Mitochondrial DNA	[20]

10 mL with distilled H_2O and filter-sterilize. Make the solution on the day of use (*see* **Note 1**).

9. Polylysine: Dissolve poly-l-lysine in distilled H_2O to 0.5 mg/mL. Filter-sterilize and store in aliquots at −20 °C.
10. ConA (Concanavalin A, Sigma C7275): Dissolve ConA in distilled H_2O to 2 mg/mL. Filter-sterilize and store in aliquots at −20 °C.
11. Fluorescent phalloidin (Rhodamine–phalloidin or Alexa Fluor–phalloidin, Invitrogen, Carlsbad, CA): Dissolve in 100 % methanol to 200 U/mL. Store in 50-μL aliquots at −20 °C.
12. Mounting solution: 10 mL 10× PBS, 100 mg p-phenylenediamine. Stir vigorously until dissolved. Adjust pH to 9 with NaOH. Add 90 mL glycerol and mix thoroughly. Store mounting solution in aliquots at −80 °C (*see* **Note 2**). Optional: For DAPI counterstaining of nuclear and mitochondrial DNA, add 20 μL of a stock solution of 5 mg/mL DAPI (4′,6-diamidino-2-phenylindole) in distilled H_2O (*see* **Note 3**).
13. Paraformaldehyde (*see* **Note 4**): 20 % solution, EM grade (Electron Microscopy Sciences 15713, Hatfield, PA).

2.3 Antibodies for Indirect Immunofluorescence

Several proteins can be used for immunofluorescence visualization of mitochondria and actin in *S. cerevisiae*. For actin, we suggest the C4 mouse monoclonal antibody, which recognizes a conserved region in actin that is common to yeast and many other cell types [4]. Antibodies to proteins of mitochondria and other cellular cargo are widely available. Table 1 lists antigens that enable visualization of different mitochondrial compartments.

3 Methods

The paraformaldehyde fixation detailed here (Subheading 3.1) preserves cellular structures by creating covalent hydroxymethylene bridges between spatially adjacent amino acid residues. Other fixatives, such methanol and acetone, are compatible with some immunofluorescence staining, including the staining of the actin cytoskeleton. However, we do not recommend methanol and acetone for visualizing mitochondria because they cause solubilization of many membranes [5].

After fixation, cells can be stained by various methods. For immunofluorescence staining, the cell wall must first be removed with zymolyase. The resulting spheroplasts are then exposed to pretreated primary and secondary antibodies in a staining chamber and mounted on a poly-lysine-coated coverslip (Subheading 3.2).

If antibody staining is not required, we recommend leaving the cell wall intact in order to preserve cell integrity and 3D shape. Intact cells can be stained directly with non-antibody agents such as fluorescent phalloidin (Subheading 3.3) or DAPI (Subheading 3.4), then mounted on a slide. For SIM imaging, intact cells can be further immobilized by binding to a coverslip using the lectin Concanavalin A (Subheading 3.5) (Fig. 2).

3.1 Fixation of Yeast Cells

1. Grow a 5 mL liquid culture of cells in a 50-mL conical-bottom tube to mid-log phase ($OD_{600} = 0.1–0.5$) (*see* **Note 5**).
2. Add paraformaldehyde to the culture medium to a final concentration of 3.7 % (*see* **Note 6**).
3. Incubate cells under normal growth conditions for 50–60 min.
4. Concentrate 1×10^7 cells by centrifugation (30 s at $10{,}000 \times g$).
5. Continue either to non-antibody staining (Subheadings 3.3 or 3.4), or to removal of the cell wall for immunofluorescence (Subheading 3.2).

3.2 Immunofluorescence Methods

3.2.1 Pretreatment of Antibodies with Yeast Cell Walls

Polyclonal antisera from rabbit and other species may contain antibodies that bind to yeast, because of the ubiquity of fungi in the environment. These antibodies usually recognize the cell wall, residues of which remain on spheroplasts. This may generate unwanted background staining. To avoid this, we recommend pretreating polyclonal antibodies with intact yeast cells to remove contaminating antibodies. Pretreated antibodies can be prepared in bulk and stored for later use.

1. Grow yeast cells to late log phase ($OD_{600} > 1.0$). A 5-mL culture is sufficient to treat 400 μL of undiluted antiserum.
2. Dilute antiserum to 1/25 in PBS+.
3. From the yeast cell culture, take a volume equal to ¼ the volume of diluted antibody (250 μL per mL of diluted antibody) and transfer to a microcentrifuge tube.
4. Concentrate cells by centrifugation (30 s at $10{,}000 \times g$) and wash three times with 1 mL of PBS+.
5. Resuspend cells in diluted antiserum and incubate with gentle mixing for 2 h at 4 °C.

6. Pellet cells by centrifugation and transfer antiserum into a fresh tube.
7. Repeat **steps 3–5** (exposing treated antiserum to a fresh batch of cells).
8. Pellet the cells by centrifugation and aliquot pretreated antiserum for storage at −20 °C.

3.3 Preparation of a Staining Chamber

Incubations are carried out in a dark, humid chamber to protect the samples from desiccation and unnecessary exposure to light.

1. Cut ten 10 × 15 cm sheets of Parafilm and press them together, crimping slightly at the edges, to make a flat, semi-rigid platform.
2. Place the Parafilm platform on a stack of damp paper towels. Ensure that the surface is flat to prevent unequal distribution of solution on coverslips.
3. Cover the platform and paper towels with an inverted opaque tray or box for the duration of the incubation.

3.3.1 Removal of the Cell Wall with Zymolyase

1. Resuspend fixed cell pellet (from Subheading 3.1) in 1 mL Tris/DTT and incubate for 20 min at 30 °C.
2. Concentrate cells by centrifugation (30 s at 10,000 × *g*). Resuspend pellet in 1 mL of zymolyase solution and incubate for 0.5–1.0 h at 30 °C. The actual incubation time varies from strain to strain (*see* **Note 7**).
3. Concentrate cells by centrifugation and wash cell pellet three times with 1 mL of NS+. Resuspend the final cell pellet in two volumes of NS+.
4. Store fixed spheroplasts at 4 °C for up to 1 week.

3.3.2 Attachment of Spheroplasts to Coverslip with Polylysine

Polylysine is a polycation that enhances the attachment of the negatively charged spheroplasts to a coverslip.

1. Place 20 μL of polylysine solution on the Parafilm platform.
2. Lay a 22 × 22 mm coverslip on the polylysine drop and apply some pressure to allow the entire coverslip to be covered by polylysine (*see* **Note 8**).
3. Incubate for ~1 min.
4. Remove the coverslip from the polylysine by slowly pipetting 200–300 μL distilled water under the edge of the coverslip. This will cause the coverslip to float off the Parafilm for easy removal.
5. Rinse off excess polylysine by pipetting water gently on the coverslip. Drain excess liquid by touching the edge of the coverslip to filter paper. Allow to air-dry.

6. Mix 10 μL of fixed spheroplast suspension (Subheading 3.2.3) with 100 μL of 1× PBS.
7. Place coated coverslip on Parafilm platform with coated side facing up, and apply the spheroplast mixture onto the coverslip, spreading it gently with the side of the pipet tip. Incubate 30 min at room temperature in the staining chamber.
8. Remove unbound spheroplasts by gently pipetting 200–500 μL PBT onto the coverslip and allowing it to run off. Drain excess liquid by touching the edge of the coverslip to filter paper. Proceed immediately to staining. Do not allow spheroplasts to remain dry for more than 1 min.

3.3.3 Indirect Immunofluorescence Staining of Attached Spheroplasts

For staining, a coverslip with attached spheroplasts is inverted on a small volume of diluted, pretreated antibody on a Parafilm platform. This method allows efficient use of antibodies.

1. Place 20–40 μL of diluted primary antibody on the Parafilm platform. Lay the spheroplast-coated side of the coverslip onto the drop, ensuring the solution is evenly spread over the coverslip. Incubate for 2 h at RT in the staining chamber (*see* **Note 9**).
2. Lift the coverslip by slowly pipetting 200–300 μL of PBT under the edge of the coverslip. Rinse coverslip by gently pipetting 200–500 μL PBT onto the coverslip and allowing it to run off. Drain excess liquid by touching the edge of the coverslip to filter paper. Do not allow spheroplasts to remain dry for more than a minute or two.
3. Place 20–40 μL of diluted secondary antibody on the Parafilm platform. Place the coverslip on the antibody as in **step 1**. Incubate 1 h at RT.
4. Lift coverslip, rinse, and drain as in **step 2**.
5. Place 1–2 μL of mounting solution on a microscope slide and gently place the coverslip, sample side down, onto the mounting solution. Dry any residual liquid from the edges of the coverslip. Seal the edges with clear nail polish.
6. After the nail polish is dry, rinse the coverslip surface with distilled water and dry gently with a Kimwipes. Visualize sample as soon as possible to maximize fluorescence signal. Most samples are stable for several days.

3.4 Visualization of the Actin Cytoskeleton in Fixed Cells Using Fluorescent Phalloidin

The following protocol is for staining of cells in suspension (*see* **Note 10**). Any fluorophore of choice conjugated to phalloidin can be used. We recommend using Alexa Fluor 488–phalloidin for SIM imaging due to its robust signal and photostability.

1. Wash ~1×10^7 cells from the fixed cell pellet (from Subheading 3.1) three times with 500 μL wash solution, concentrating each time by centrifugation (30 s at 10,000 × *g*).

2. Wash cells with 500 μL of PBT and resuspend in 30–50 μL of PBT.
3. Add fluorescent phalloidin to a concentration of 1.65 μM (*see* **Note 11**) and incubate for 35 min at RT.
4. Wash cells with 100 μL of 1× PBS.
5. Mount cells for imaging. For enhanced immobilization of intact cells, resuspend in 100 μL of PBS and mount on a ConA-coated coverslip (Subheading 3.5). For a simpler preparation, continue to the next step.
6. Resuspend cells in mounting solution (*see* **Note 12**). Place 1.5 μL of cell suspension onto a microscope slide. Place a 22×22 mm coverslip on top of the cells and seal the edges of the coverslip with nail polish.

3.5 Visualization of the Nucleus and mtDNANucleoids Using DAPI

DAPI is a common DNA-binding dye that can be used to visualize both the nucleus and mtDNA nucleoids (Fig. 1). Generally, DAPI is added to mounting solution to a final concentration of 1 μg/mL. However, if excessive background is seen with DAPI mounting solution, or if using SIM, use the following protocol to minimize over-staining.

1. Wash ~1×10^7 cells from fixed cell pellet (from Subheading 3.1) three times with 500 μL wash solution, concentrating each time by centrifugation (30 s at 10,000×*g*).

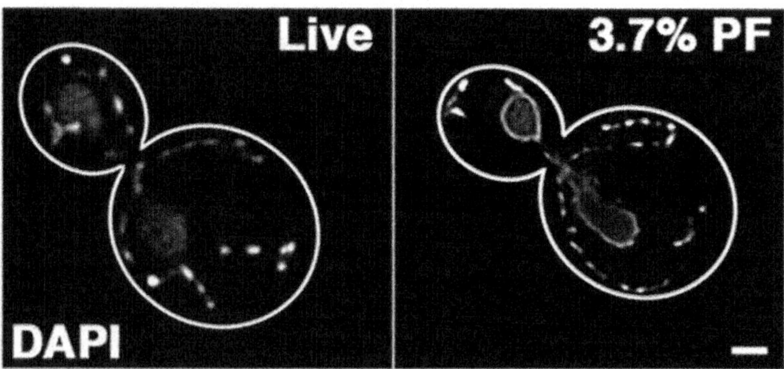

Fig. 1 Live-cell versus fixed-cell staining of DNA using DAPI. Wild-type cells were grown to mid-log and DAPI-stained in either live cells (*left panel*) or fixed cells (*right panel*). For live-cell imaging, cells were stained with 1 μg/mL DAPI directly in culture medium for 15 min, washed three times and imaged. For fixed cells, cells were stained with 1 μg/mL DAPI in PBS after fixation with 3.7 % paraformaldehyde. Both live- and fixed-cell imaging were performed using wide-field microscopy using 1 × 1 binning, 100 ms exposure time, and a metal halide lamp with standard DAPI filters. *Z*-series were collected through the entire cell at 0.5 μm intervals for 7 μm. Images were deconvolved using a constrained iterative restoration algorithm (Volocity, PerkinElmer) with the following parameters: 460 nm emission wavelength, 15 iterations, and 95 % confidence limit

2. Add DAPI to a concentration of 1 μg/mL and incubate for 15 min at room temperature.
3. Wash cells with 100 μL of 1× PBS.
4. Mount cells for imaging. For enhanced immobilization of intact cells, resuspend in 100 μL of PBS and mount on a ConA-coated coverslip (Subheading 3.5). For a simpler preparation, continue to the next step.
5. Resuspend cells in mounting solution (*see* **Note 12**). Place 1.5 μL of cell suspension onto a microscope slide. Place a 22 × 22 mm coverslip on top of the cells and seal the edges of the coverslip with nail polish.

3.6 Attachment of Intact Cells to a Coverslip with ConA

Cells can be attached to ConA-coated coverslips either before or after paraformaldehyde fixation (*see* **Note 13**) (Fig. 2).

1. Place a coverslip (18–22 mm square) on a Parafilm platform. For structured illumination microscopy (SIM), we recommend using a high-performance coverslip. Add 20 μL of 2 mg/mL ConA solution on top of the coverslip and spread using the side of a pipette tip. Incubate in staining chamber for 30 min at RT.
2. Remove unbound ConA by rinsing gently with distilled H_2O.
3. Allow coverslip to dry for at least 1 h at RT and store until needed.
4. Place 30 μL of fixed cell suspension on the coverslip and spread using the side of a pipette tip (*see* **Note 12**). Incubate in staining chamber for 15 min.
5. Wash off excess cells by gently pipetting PBS onto the coverslip.
6. Remove excess liquid by touching the edge of the coverslip to filter paper.
7. Place 5–10 μL of mounting solution (*see* **Note 14**) onto the center of a slide. For optimal signal preservation (which is essential for SIM), we recommend using Prolong Gold for organic dyes, such as Alexa 488–Phalloidin, and Prolong

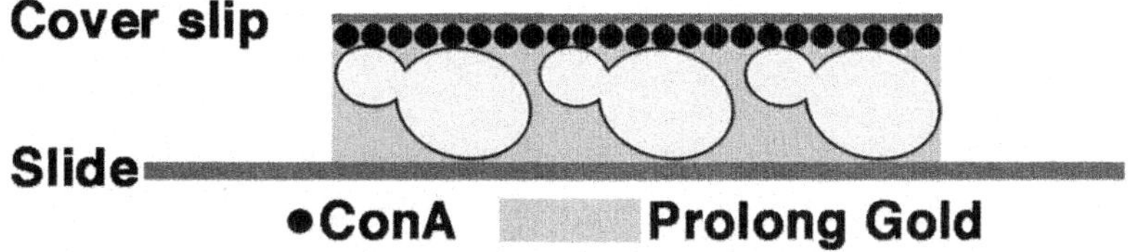

Fig. 2 Schematic for ConA immobilization of cells onto a coverslip. Coverslips are coated with ConA to immobilize cells. Prolong Gold or other mounting solution is placed onto the slide, and coverslips are placed directly on the mounting medium with cells facing down

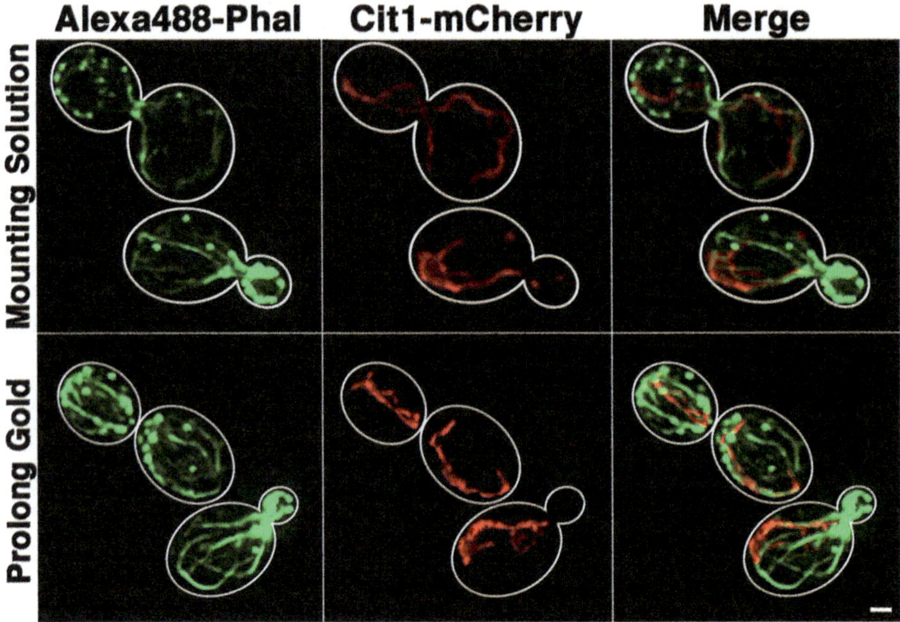

Fig. 3 Comparison of traditional mounting solution and Prolong Gold. Wild-type cells with mCherry-tagged Cit1p were fixed in grown to mid-log phase, fixed in 3.7 % paraformaldehyde, and stained with 1.65 μM Alexa 488–phalloidin. *Z*-series were collected through the entire cell at 0.3 μm intervals for 6 μm using wide-field microscopy using 1 × 1 binning, 200 ms exposure time, a metal halide lamp (with standard dsRed filter for mCherry and standard GFP filter for Alexa 488). Images were deconvolved using a constrained iterative restoration algorithm (Volocity, PerkinElmer) with the following parameters: 507 nm emission wavelength, 60 iterations, and 100 % confidence limit for Alexa 488 and 620 nm emission wavelength, 60 iterations, and 100 % confidence limit for mCherry. Scale bar represents 1 μm

Diamond for fluorescent proteins. Figure 3 shows a comparison of signal intensity between conventional mounting solutions and Prolong Gold.

8. Place the coverslip cell side down onto the mounting solution. Apply gentle pressure on the coverslip to push out any air pockets that may have formed.
9. Allow to cure at RT in the dark for 4 days.
10. Seal coverslip edges with nail polish. Image as soon as possible. Do not store for extended periods of time as this will diminish the fluorescent signal.

3.7 Imaging

Here we detail two methods for imaging fixed budding yeast cells: wide-field microscopy with deconvolution; and structured illumination microscopy. Deconvolution is a computational method for removing out-of-focus light from wide-field fluorescence images, revealing three-dimensional information [6]. The constrained iterative restoration algorithm used here (Volocity, PerkinElmer) allows quantitation of the processed images.

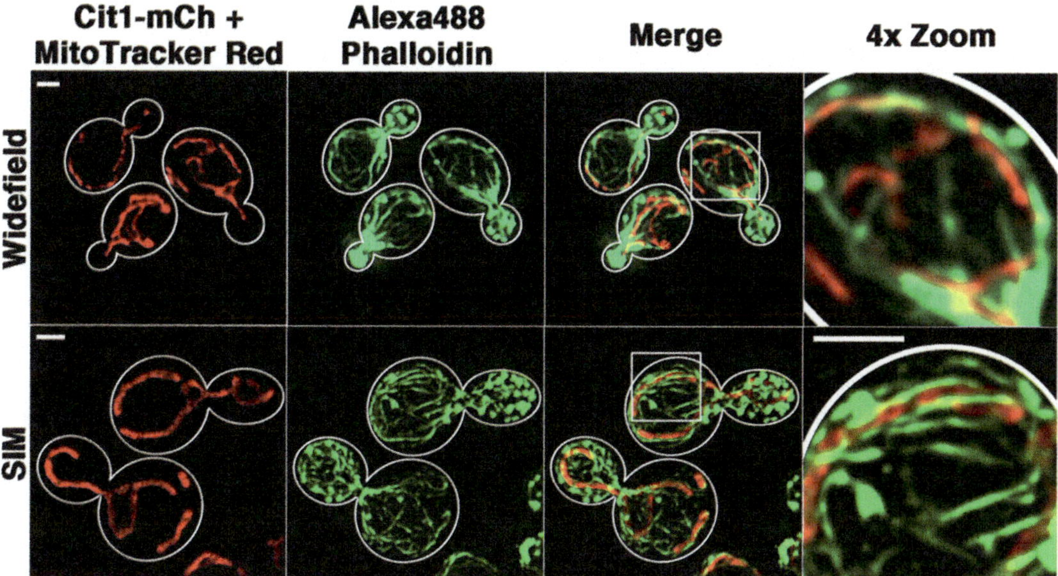

Fig. 4 Comparison of conventional wide-field microscopy with deconvolution and SIM. Wild-type cells containing mCherry-tagged Cit1p were grown to mid-log phase, stained with 100 nm MitoTracker Red, fixed in 3.7 % paraformaldehyde, and stained with 2.5 μM Alexa 488–phalloidin. Z-series were collected through 3 μm of the middle of the cell at 0.125 μm intervals using conventional wide-field microscopy (*top panel*) using 1 × 1 binning, 200 ms exposure time, and a metal halide lamp with standard GFP filter for Alexa 488, and 400 ms exposure time and a metal halide lamp with standard dsRed filters for mCherry/MitoTracker Red; and SIM (*bottom panel*) using 1 × 1 binning, 200 ms exposure, 460 nm laser at 40 % power and 150 gain for GFP and 400 ms exposure, 570 nm laser at 60 % power and 225 gain for mCherry/MitoTracker Red. Image reconstruction was performed using Nikon Elements software with the following parameters: illumination modulation contrast = 2, high-resolution noise suppression = 1, out of focus blur suppression = 0.2. *Right panels* additional 4× magnification of the *boxed area* for comparison of resolution. Scale bar, 1 μm. Quantification of actin cables using the two methods is shown in Table 2

Wide-field microscopy with deconvolution is an accessible method that provides better signal–noise ratios than point-scanning confocal imaging, and is more sensitive than spinning-disk confocal imaging.

Structured illumination microscopy (SIM) [7] allows resolution up to twice as good as conventional wide-field methods. SIM requires a specialized microscope system, and some changes to sample preparation. In addition, much more time is required to process images. However, for suitable samples SIM yields dramatically improved resolution (Fig. 4).

3.7.1 Visualization of Fixed Cells by Wide-Field Deconvolution

To image yeast cells, which average 4–6 μm in diameter, a high-magnification objective lens and high-resolution camera are required. Cameras that support low-intensity illumination and short exposure times are essential to minimize photobleaching.

Filter sets should be chosen to maximize light throughput, but should not allow bleed-through in multi-color imaging experiments.

An example of a wide-field microscopy system is the following, used for all wide-field images in this chapter: an inverted AxioObserver.Z1 microscope (Carl Zeiss Inc., Thornwood, NY) equipped with a metal halide lamp and Colibri LEDs for excitation; 100×/1.3 EC Plan-Neofluar oil-immersion objective; and an Orca ER cooled CCD camera with 1280×1024 pixel resolution (Hamamatsu, Bridgewater, NJ). ZEN software (Carl Zeiss) is used to control the camera and microscope, capture images at defined focus intervals, and export images for further analysis. Images are imported into Volocity software (PerkinElmer) for deconvolution and either Volocity or ImageJ [8] for quantification.

1. Optimize signal levels by adjusting fixation and staining conditions, and using as fresh a sample as possible.
2. Set exposure time and camera gain so that pixel values fill as much as possible of the dynamic range of the camera, without saturation.
3. Acquire *z*-series of images at a focus interval of ~0.2 μm.
4. Load datasets in Volocity and verify that the *x*, *y*, and *z* scale are correctly set.
5. For each channel, generate a new calculated PSF. When prompted, enter the numerical aperture of the objective lens and the maximum wavelength of fluorescence emission (e.g., 507 nm for Alexa Fluor 488–phalloidin).
6. Using the calculated PSF for each channel, perform iterative deconvolution with 60 iterations, 100 % confidence limit (*see* **Note 14**).

3.7.2 Visualization of Cells by Structured Illumination Microscopy (SIM)

SIM imaging can increase resolution by twofold compared to conventional wide-field microscopy [7]. This reveals complex details in fine structures, such as the actin cytoskeleton in yeast (Fig. 4, Table 2). The improvement in resolution with SIM comes from modulating the fluorescence excitation pattern by means of a grating. The interaction between the illumination pattern and fine sample details produces a Moiré effect. By varying the illumination pattern in a defined way, and analyzing the resulting set of Moiré patterns, an image can be reconstructed to restore details not captured by standard illumination. Importantly, current "super-resolution" or "high-resolution" SIM (SR-SIM) differs from previously marketed structured illumination systems such as Apotome. The Apotome and similar systems provide optical sectioning only. SR-SIM, in contrast, doubles resolution in the *x*, *y*, and *z* dimensions.

Table 2
Visualization of actin patches and cables using conventional wide-field microscopy and SIM

Method	Actin cables/cell (± SEM)	Actin patches/cell (± SEM)	n (cells)
Wide-field with deconvolution	4.69 ± 0.17	1.95 ± 0.29	39
SIM	9.13 ± 0.30	2.43 ± 0.29	51

The numbers of actin cables and patches were counted in the mother cell. Only actin cables parallel to the mother–bud axis and greater than half the length of the mother cell were counted

Sample Preparation and Imaging

There are several criteria that must be met for optimal SIM imaging:

1. Minimal movement in the sample. We recommend fixing and immobilizing cells, and using a hard-set mounting medium. In addition, the room and equipment should be stable in temperature and free from drafts.
2. Maximized signal intensity and photostability. Alexa 488–phalloidin is preferred for labeling the actin cytoskeleton. For labeling mitochondria, endogenous fluorescent proteins (mCherry-tagged Cit1p) can be enhanced with MitoTracker Red staining. Mounting medium should show excellent anti-bleaching performance; Prolong Gold or SlowFade Gold is recommended.
3. Predictable optical performance. A # 1.5 (0.170 mm) coverslip is recommended. Because standard coverslip thickness varies by tens of micrometers, which may induce spherical aberration, high-performance coverslips (manufactured with smaller variation in thickness) should be used. In addition, mounting medium should have a consistent, stable refractive index.
4. Adequate *Z* sampling. *Z*-series of images should be acquired at a focus interval of ~0.2 μm.
5. Appropriate pixel values in raw data. Exposure time and camera gain should be set so that pixel values fill no more than 1/4 of the dynamic range of the camera. EM gain for EMCCD cameras should not exceed 150.

3.7.3 Reconstruction

Parameters for SIM reconstruction vary with the software used. Here, we describe options for Nikon Elements software (Nikon Instruments, Melville, NY).

The following three parameters should be adjusted for each sample. Recommended values for images of actin and mitochondria are given.

1. Illumination modulation contrast. Higher values reveal more detail, but may induce artifacts. Recommended values: 0.5–2.
2. High-resolution noise suppression. Higher values suppress artifacts due to pixel noise, but also suppress fine details. Recommended values: <1.
3. Out-of-focus blur suppression. Higher values suppress out-of-focus signal, essential for densely labeled areas, but also reduce overall signal and resolution. Recommended values: 0.1–0.2.

3.8 Quantitative Analysis of Actin Cable Number, Thickness, and Intensity Using Line Profile Measurements

The quantity and structure of actin cables within a cell are affected by age, metabolic state, and other factors, through a host of actin-binding and regulatory molecules. To quantify these effects, we use a line profile analysis method. Because actin cables in the mother cell are organized in parallel along the length of the mother cell, a line profile drawn across the center of the mother cell effectively samples the cable population and can be used to determine the number, thickness, and actin content of actin cables (Fig. 5). When the pixel intensity along the line is plotted, cables appear as high-intensity peaks. The number of peaks represents the number of cables, and the width of each peak corresponds to the thickness of the actin cable (Table 3). The height of the peak or the area under the curve can be used to estimate relative actin content. The following protocol describes how to use ImageJ [8] to perform these measurements.

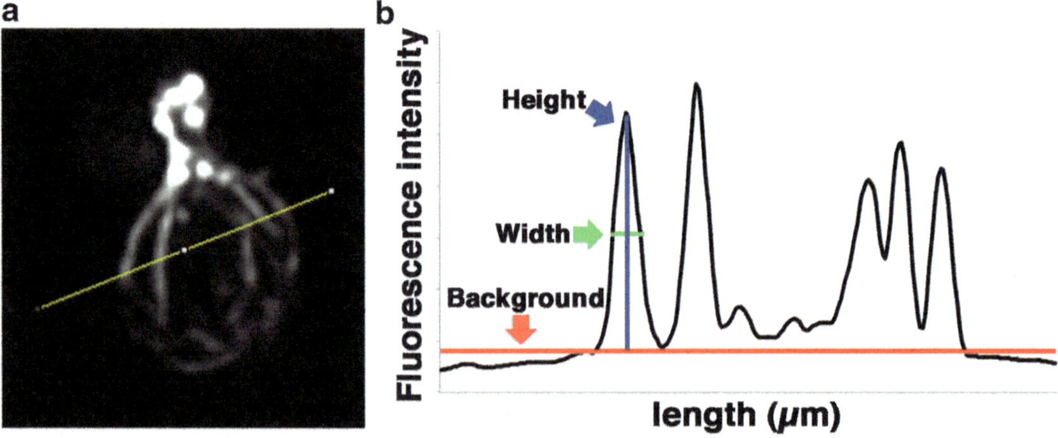

Fig. 5 Schematic for measurements of actin cable number, thickness, and content using line profiles. (**a**) Wild-type cells were imaged using Alexa 488–phalloidin as in (Fig. 3) using Prolong Gold. A line ROI (*yellow*) is drawn near the center of the mother cell perpendicular to the mother–bud axis. (**b**) Schematic of line profile measurement of actin cable thickness and actin content. The background level, determined by averaging the intensity of regions outside of the cell, is shown in *red*. A vertical line is then drawn from the maximum point of the peak to the background line as shown in *blue*. This represents the maximum intensity of the cable. A horizontal line (*green*) is then drawn across the peak at half of the maximum intensity. This represents the thickness of the actin cable

Table 3
Line profile measurements reveal carbon source-dependent changes in actin cable thickness

Medium	Actin cable thickness, µM (± SEM)	*n* (cells)
Glucose	0.241 ± 0.007	83
Lactate	0.288 ± 0.006	123

Wild-type cells were grown to mid-log phase in glucose or lactate medium. Cells were fixed in 3.7 % paraformaldehyde and stained with 1.65 µM rhodamine–phalloidin for 35 min. *Z*-series were collected throughout the cell at 0.3-µM intervals through 6 µM using wide-field microscopy using 1 × 1 binning, 200 ms exposure and a metal halide lamp and standard dsRed filters

3.8.1 Creating Line Profiles

1. Collect z-stack images of cells stained with fluorescent phalloidin. It is recommended that images be deconvolved using Volocity or comparable software to remove out-of-focus information and increase the signal–background ratio.
2. Open *z* stack in ImageJ.
3. Make a maximum-intensity projection of the *z* stack using **Image > Stacks > Z Project**; choose Projection Type: Max Intensity.
4. If the image has more than one channel, make sure the actin channel is selected.
5. Set up ImageJ with the following parameters:
   - **Edit > Options > Input-Output**. Under Results Table Options, check Copy Row Numbers and leave other options unchecked.
   - **Edit > Options > Profile Plot Options**. Set width and height equal to the width and height of your image. Set Minimum Y to 0 and Maximum Y to the Image Max, obtained from Analyze > Histogram.
   - **Analyze > Set Measurements**. Check Area, Display Label.
   - **Analyze > Set Scale**. Make sure that the spatial scale is set accurately.
6. Make sure the Line ROI tool is set to the Straight Line mode; right-click on the tool to change the mode if needed. Double-click on the Line ROI tool and set Line Width to 3. Do not check Spline Fit.
7. Draw a line ROI perpendicular to the mother–bud axis, near the middle of the mother and avoiding actin patches as shown in Fig. 5a. Both ends of the line should extend into the background outside the cell.
8. **Analyze > Plot Profile**. A Plot window should appear showing Gray Value (pixel value) versus Distance.

9. If desired, save the plot image (**File > Save**). Data can be saved (click **Save** on the Plot window) or copied and pasted into a spreadsheet application such as Excel (Click **Copy** on the Plot window, then paste into the other application). Proceed to the next subheading to quantify the profile data using ImageJ.

3.9 Measurements of Cable Number, Thickness, and Intensity

1. Count the number of peaks in the profile. This represents the number of actin cables within the mother cell (*see* **Note 15**).
2. Determine a background level by inspection or by using a spreadsheet to average the values from the parts of the profile that are outside the cell. In ImageJ, draw a horizontal line across the plot at that gray value, as shown in red in Fig. 5b.
3. Cable thickness is defined as the width of the peak at half-maximal height above background. To find the height, using the Line ROI tool, draw a vertical line from the maximum height of the peak to the background line, as shown in blue in Fig. 5b. Holding the shift key while drawing a line forces the line to be either vertical or horizontal.
4. Press T to add the vertical line to ROI Manager. **Analyze > Measure**. The Length value represents the maximum cable intensity.
5. In the ROI manager window, check Show All and Labels. A small number will appear in the middle of the vertical line, marking half the peak height.
6. Draw a horizontal line across the peak at half the peak height as shown in green in Fig. 5b. **Analyze > Measure.** The length of this horizontal line represents the thickness of the actin cable.
7. Click the Wand tool and select an area under the peak above the background line. **Analyze > Measure**. The **Area** measurement represents integrated density of the cable and can give a relative measurement of actin content for each cable.

4 Notes

1. Filter sterilization removes any contaminants and large aggregates of BSA.
2. Glycerol prevents freezing of samples at −20 °C, and p-phenylenediamine, an anti-photobleaching agent, protects fluorophores from destruction by oxygen radicals. Discard mounting solution if it becomes very dark. Slight discoloration, however, does not affect its function.
3. DAPI is a possible carcinogen. The solid compound should be handled with gloves, mask, and safety glasses.

4. Paraformaldehyde is toxic and carcinogenic. It is readily absorbed through the skin as a liquid and toxic to the respiratory tract as a gas. Gloves and safety goggles are recommended when handling paraformaldehyde and all handling should be performed in a chemical hood.
5. We recommend fixing cells in mid-log phase because internal structures, including mitochondria and the actin cytoskeleton, are better preserved, compared to cells picked from solid media. The choice of carbon source is also important when imaging actin and mitochondria [9, 10]. Glucose represses mitochondrial biogenesis, which decreases mitochondrial abundance and allows for higher-resolution imaging of individual mitochondrial tubules. Growth in a non-fermentable carbon source, such as lactate or glycerol, selects for cells with mtDNA and higher mitochondrial metabolic potential.
6. Common manipulations, including centrifugation and temperature changes, can alter internal structures, especially cytoskeletal elements [11]. Adding fixative directly to a liquid culture under growth conditions will minimize disruptions. Other important variables are the concentration of the fixative and pH. High paraformaldehyde concentration and low pH increases the number of cross-links formed and improves structural preservation, but excessive cross-linking can result in artifacts or background fluorescence, owing to cross-linking of fluorescent substances in the media to cellular proteins. Therefore, it is necessary to optimize fixation protocols.
7. To check for cell wall removal, examine 1–2 μL of cells under a microscope using transmitted light. Cell edges become less refractile when the wall is removed. Over-treatment with zymolyase will cause cell lysis.
8. Marking one corner of the coverslip with a solution-resistant marker will help differentiate between the coated and uncoated sides.
9. Two antigens can be simultaneously used by mixing the two primary antibodies in the first incubation and mixing the two secondary antibodies in the second step. Primary antibodies must be raised in different species and fluorophores used for visualization must have sufficiently separated excitation and emission spectra.
10. This protocol can be adapted for spheroplasts that are attached to coverslips. For spheroplasts, add ¼ volume of fluorescent phalloidin solution (200 U/mL) to the secondary antibody solution and proceed as described in the immunofluorescence protocol.

11. For SIM, signal must be as bright as possible, so we recommend using a higher concentration of fluorescent phalloidin, such as 2.5 μM Alexa 488–Phalloidin.
12. Cell density should be around 3×10^8/mL but can be adjusted as needed to produce good distribution of cells on the coverslip.
13. Fixation prior to attaching cells to coverslips may impair rigid immobilization [12]. However, it also leads to loss of cells from the coverslip during staining manipulations. We find that using a hard-set mounting medium (such as Prolong Gold) effectively immobilizes the cells, even if they are fixed before binding. A protocol for fixation of cells after binding to a coverslip can be found in [12].
14. Artifacts occasionally appear in deconvolved images. These may include halo-like structures or the disappearance of structures that are visible in the raw data. If structures are lost, or halos appear, repeat the deconvolution with fewer iterations. If artifacts do not appear, but the image is still blurry or faint, repeat with more iterations. If deconvolution fails to exceed the 95 % confidence limit, verify that optics are clean, light source is aligned, and cells are mounted under a 1.5 coverslip. Also make sure that images are acquired with pixel values at least in the middle of the detector range, and not saturating the detector.
15. The line plot profile method for counting cables is reliable and more objective than counting by eye. However, it is important to keep in mind that the measurements are only performed in the center of the mother cell. This may result in underrepresentation of the true cable count due to missing cables that do not traverse the region where the line is drawn or due to cables that are too close for the software to pick up as two distinguishable cables.

Acknowledgments

This work was supported by awards from the Ellison Medical Foundation (AG-SS-2465) and the NIH (GM45735, GM45735S1, and GM096445) to L.A.P. GM45735S1 was issued from the NIH under the American Recovery and Reinvestment Act of 2009. The microscopes used for these studies were supported in part through a NIH / NCI grant (5 P30 CA13696) and obtained using funds from the NIH-NCRR (1S10OD014584) to L.A.P.

References

1. Adams AE, Pringle JR (1984) Relationship of actin and tubulin distribution to bud growth in wild-type and morphogenetic-mutant Saccharomyces cerevisiae. J Cell Biol 98:934–945
2. Girao H, Geli MI, Idrissi FZ (2008) Actin in the endocytic pathway: from yeast to mammals. FEBS Lett 582:2112–2119
3. Pruyne D, Legesse-Miller A, Gao L, Dong Y, Bretscher A (2004) Mechanisms of polarized growth and organelle segregation in yeast. Annu Rev Cell Dev Biol 20:559–591
4. Lessard JL (1988) Two monoclonal antibodies to actin: one muscle selective and one generally reactive. Cell Motil Cytoskeleton 10:349–362
5. Boldogh IR, Yang HC, Nowakowski WD, Karmon SL, Hays LG, Yates JR 3rd, Pon LA (2001) Arp2/3 complex and actin dynamics are required for actin-based mitochondrial motility in yeast. Proc Natl Acad Sci U S A 98:3162–3167
6. Wallace W, Schaefer LH, Swedlow JR (2001) A workingperson's guide to deconvolution in light microscopy. BioTechniques 31: 1076–1078, 1080, 1082 passim
7. Gustafsson MG (2000) Surpassing the lateral resolution limit by a factor of two using structured illumination microscopy. J Microsc 198:82–87
8. Schneider CA, Rasband WS, Eliceiri KW (2012) NIH Image to ImageJ: 25 years of image analysis. Nat Methods 9:671–675
9. Damsky CH (1976) Environmentally induced changes in mitochondria and endoplasmic reticulum of Saccharomyces carlsbergensis yeast. J Cell Biol 71:123–135
10. Visser W, van Spronsen EA, Nanninga N, Pronk JT, Gijs Kuenen J, van Dijken JP (1995) Effects of growth conditions on mitochondrial morphology in Saccharomyces cerevisiae. Antonie Van Leeuwenhoek 67:243–253
11. Lillie SH, Brown SS (1994) Immunofluorescence localization of the unconventional myosin, Myo2p, and the putative kinesin-related protein, Smy1p, to the same regions of polarized growth in Saccharomyces cerevisiae. J Cell Biol 125:825–842
12. Mund M, Kaplan C, Ries J (2014) Localization microscopy in yeast. Methods Cell Biol 123:253–271
13. Mihara K, Sato R (1985) Molecular cloning and sequencing of cDNA for yeast porin, an outer mitochondrial membrane protein: a search for targeting signal in the primary structure. EMBO J 4:769–774
14. Roeder AD, Hermann GJ, Keegan BR, Thatcher SA, Shaw JM (1998) Mitochondrial inheritance is delayed in Saccharomyces cerevisiae cells lacking the serine/threonine phosphatase PTC1. Mol Biol Cell 9:917–930
15. Taanman JW, Capaldi RA (1993) Subunit VIa of yeast cytochrome c oxidase is not necessary for assembly of the enzyme complex but modulates the enzyme activity. Isolation and characterization of the nuclear-coded gene. J Biol Chem 268:18754–18761
16. Poot M, Zhang YZ, Kramer JA, Wells KS, Jones LJ, Hanzel DK, Lugade AG, Singer VL, Haugland RP (1996) Analysis of mitochondrial morphology and function with novel fixable fluorescent stains. J Histochem Cytochem 44:1363–1372
17. Ruprich-Robert G, Zickler D, Berteaux-Lecellier V, Velot C, Picard M (2002) Lack of mitochondrial citrate synthase discloses a new meiotic checkpoint in a strict aerobe. EMBO J 21:6440–6451
18. Riezman H, Hay R, Gasser S, Daum G, Schneider G, Witte C, Schatz G (1983) The outer membrane of yeast mitochondria: isolation of outside-out sealed vesicles. EMBO J 2:1105–1111
19. McConnell SJ, Stewart LC, Talin A, Yaffe MP (1990) Temperature-sensitive yeast mutants defective in mitochondrial inheritance. J Cell Biol 111:967–976
20. Diffley JF, Stillman B (1991) A close relative of the nuclear, chromosomal high-mobility group protein HMG1 in yeast mitochondria. Proc Natl Acad Sci U S A 88:7864–7868

Chapter 4

Imaging of the Cytoskeleton Using Live and Fixed *Drosophila* Tissue Culture Cells

Derek A. Applewhite, Christine A. Davis, Eric R. Griffis, and Omar A. Quintero

Abstract

In recent years, the convergence of multiple technologies and experimental approaches has led to the expanded use of cultured *Drosophila* cells as a model system. Their ease of culture and maintenance, susceptibility to RNA interference, and imaging characteristics have led to extensive use in both traditional experimental approaches as well as high-throughput RNAi screens. Here we describe *Drosophila* S2 cell culture and preparation for live-cell and fixed-cell fluorescence microscopy and scanning electron microscopy.

Key words Fluorescence microscopy, S2 cells, Transfection, Microtubule, Actin, Electron microscopy

1 Introduction

The marriage of Drosophila tissue culture and cytoskeletal research has been a fruitful one. Numerous cytoskeletal proteins have been discovered using S2 cells [1], which is a testament to their practicality. Major advances in our understanding of cytoskeletal dynamics and function have been achieved using S2 cells, an indication of their versatility and strength as a model system [2–10]. Drosophila cells are cultured at room temperature in a medium that does not require buffering with CO_2. RNAi is both effective and easily administered [11]. Drosophila cells do not produce an interferon response to dsRNAs. Therefore, dsRNAs can be produced in vitro, eliminating the need to buy siRNAs. S2 cells readily incorporate dsRNAs when added to tissue culture medium. dsRNAs can also be generated against the 5′ and 3′ untranslated regions (UTRs) allowing for the depletion of endogenous target genes, and testing of exogenous genes for their ability to rescue RNAi phenotypes. RNAi protocols can involve as few as 3 days, depending upon the turnover of a given protein and frequently lead to substantial protein knockdown [12]. The Drosophila genome represents a pared

Ray H. Gavin (ed.), *Cytoskeleton: Methods and Protocols*, Methods in Molecular Biology, vol. 1365,
DOI 10.1007/978-1-4939-3124-8_4, © Springer Science+Business Media New York 2016

down form of the mammalian genome and while approximately 75 % of human disease causing genes are retained in the Drosophila genome, there is far less functional redundancy [13–15]. Whereas the human or mouse genome may have three genes that are functionally interchangeable, the Drosophila genome will often have one. Collectively, these qualities have led to hundreds of RNAi screens of the full Drosophila genome at a fraction of the cost of screens preformed in mammalian tissue culture systems.

The usefulness of Drosophila S2 cells as tools for cytoskeletal research goes beyond their use in RNAi screens. S2 cells in particular adopt a flat, fried-egg-like morphology when plated on a concanavalin A (Con A)-coated surface [16]. This conformation makes them highly amenable to high-resolution microscopy [17], and their sessile nature eliminates the problems produced by motile cells. S2 cells form a circumferential actin-rich lamellipodium with characteristic fast actin dynamics where proteins such as the Arp2/3 complex, Capping Protein, and Ena/VASP proteins localize [5]. This lamellipodium is followed by a contractile network highly reminiscent of a lamella with its own characteristic slower actin dynamics and signature proteins such as tropomyosin, non-muscle myosin II, and alpha-actinin (Fig. 1) [18, 19]. This is even more evident as these actin-rich zones of S2 cells display similar phenotypes to those of migrating cells following RNAi depletion of these key actin-binding proteins [5, 20–22]. Thus, despite being non-motile, the behavior and dynamics of the actin-rich cell periphery of S2 cells mimics that of migrating cells. S2 cells have also been quite useful in studying microtubule dynamics. Our current

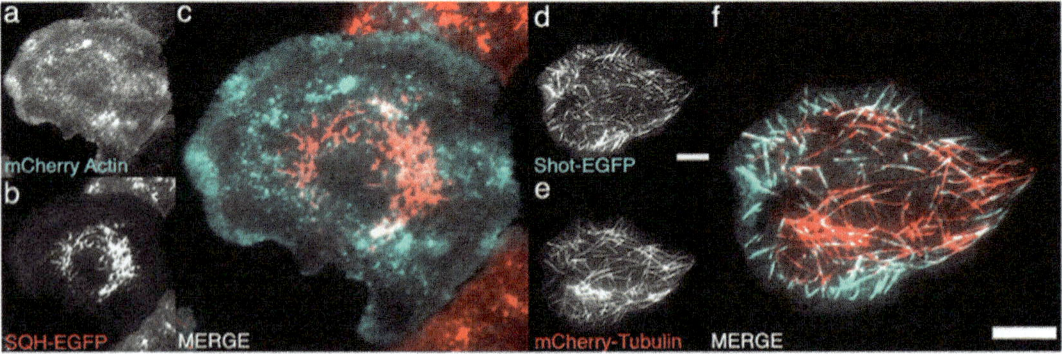

Fig. 1 Live imaging of cytoskeletal probes in *Drosophila* S2 cells. (**a**–**f**) Shown is a single time point from a live-cell image sequence of S2 cells imaged by TIRF-M. (**a**) An S2 cell co-expressing mCherry-Actin and (**b**) the regulatory chain of non-muscle myosin II (Spaghetti-Squash, Sqh), tagged with EGFP. (**c**) Shown at higher magnification, the merged image of (**a**) and (**b**) where mCherry actin pictured in (**a**) is shown in *cyan* and EGFP-Sqh pictured in (**b**) is shown in *red*. (**d**) An S2 cell co-expressing EGFP-Short stop (Shot) and (**e**) mCherry-α-Tubulin. (**f**) The merged image of (**d**) and (**e**) where Shot-EGFP pictured in (**d**) is shown in *cyan* and mCherry-α-Tubulin pictured in (**e**) is shown in *red*. Scale bars in both low and high magnification images are 10 μm

understanding of spindle formation, in particular poleward flux, kinetochore attachments, and the role of molecular motors during mitosis owe much to experiments performed in S2 cells [16, 23–30]. In addition, the acentriolar interphase microtubule array offers a unique system to study with great detail the dynamics of the microtubule plus-end and minus-end as well as interactions between the actin and microtubule cytoskeletons [16, 31–36].

S2 cells are easily transfected with tagged protein (fluorescent or otherwise), but are naturally non-adherent and must be induced to attach to a substrate. If the cells have not been attached to a substrate, they will be washed away during fixation, permeabilization, and staining. There are two main coatings that will induce cell attachment, concanavalin A (Con A) and polylysine. These two coatings differ in how the cell responds to them. Con A is a lectin (sugar binding protein) that is commonly purified from the jackbean (*Canavalia ensiformis*), and it presumably induces cell spreading in a Rac and Arp2/3-dependent manner as cells try to engulf the Con A coated substrate [5]. Due to the strong inducement of spreading as well as the engagement of the phagocytic machinery to the basal cortex of the cell, many membrane trafficking events (including cytokinesis) are disrupted in Con A plated cells, and these side effects should be considered when deciding whether to use Con A to induce adherence. Polylysine is a positively charged polymer of lysine that electrostatically interacts with negatively charged cell membranes and induces cell attachment and mild amounts of cell spreading. Polylysine is available as both L and D forms and available in a range of molecular weights. We have found that Drosophila cells need higher concentrations of polylysine than mammalian cells to attach to the substrate, and they tend to be highly sensitive to the oxidizing effects of thimerosal. Therefore, it is not recommended to purchase premade polylysine that uses this chemical as a preservative.

2 Materials

2.1 Culturing and Transient Transfection of Drosophila S2 Cells

1. Schneider's Drosophila Medium (Gibco/Life Technologies).
2. Fetal bovine serum, heat inactivated.
3. 100× antibiotic/antimycotic (Gibco/Life Technologies).
4. pMT/V5 His, pIZ/V5, and/or pAc5.1 vectors (Life Technologies).
5. $CuSO_4$: 100 mM aqueous solution.
6. FugeneHD (Promega).
7. Amaxa Kit V (Lonza) (optional).
8. Nucleofector Machine (Lonza) (optional).

2.2 Fixation and Staining of Drosophila S2 Cells for Immuno-fluorescence

1. 1 mg/mL Polylysine: Dissolve Poly-D-Lysine or Poly-L-lysine (70,000–150,000 average molecular weight) in 50 mM Tris–HCl, pH 8.0. Aliquot and freeze at −20 °C.
2. 0.5 mg/mL concanavalin A(Con A): Dissolve Con A (Type IV-S, tissue culture grade) at 0.5 mg/mL in distilled water. Filter the Con A solution through a 0.22 μm filter into a fresh tube or centrifuge at 18,000 × *g* (14,000 RPM) for 5 min to pellet aggregates. Retain the supernatant, and store it in 0.5 mL aliquots in 1.5 mL tubes at −20 °C.
3. 1 M Tris–HCl, pH 8.0.
4. #1.5 glass coverslips, glass-bottom or optical plastic multi-well plates.
5. Humidified chambers.
6. Paraformaldehyde: EM Grade packed in ampoules or made fresh.
7. Glutaraldehyde: EM Grade.
8. PHEM Buffer pH 6.9: 60 mM PIPES, 25 mM HEPES, 10 mM EGTA, 2 mM $MgCl_2$. This buffer can be made as a 2× stock solution, autoclaved, and stored until use.
9. Sodium borohydride: aqueous solution 1 mg/mL.
10. Saponin: 10 mg/mL aqueous stock solution.
11. Donkey or goat serum.
12. Bovine serum albumin.
13. Triton X-100.
14. Mounting medium: (Dako, ProLong (Life Technologies), Fluoromount (Sigma), VECTASHIELD (Vector Laboratories), or a solution of 90 % glycerol, 20 mM sodium bicarbonate (from 0.5 M aqueous solution, pH 9), 3 % *n*-propyl gallate.
15. CoverGrip (Biotium) or nail polish.

2.3 Live Cell Imaging

1. #1.5 glass-bottom dishes.
2. Titanium Step Drill Bit—1/4″ to 1-3/8″ Increments, ten steps (Neiko) (optional).
3. 120 V 10-in. Drill Press (optional).
4. Norland Optical Adhesive 81 (Norland) (optional).
5. UV transilluminator (optional).
6. Cell-Tak (Corning) (optional).
7. Poly-lysine (optional).

2.4 Fixation and Preparation of Drosophila S2 Cells for Scanning Electron Microscopy

1. Silver paint.
2. Double-sided conductive tape.
3. Critical point dryer.
4. Coverslip holder for critical point dryer.
5. 200 proof ethanol.
6. Sputter coater.

3 Methods

3.1 Culture of Drosophila S2 Cell (*See* Note 1)

1. Maintain *Drosophila* S2 cells at room temperature (preferably between 19 and 27 °C) and in semi-suspension in 25-cm or 75-cm flasks with plug-end caps.
2. Passage cells by pipetting them from the culture medium, making sure that the cells are adequately resuspended and that any cells lightly adhered to the tissue culture plastic are removed. Generally, a 1:4 split of the cells into fresh cell culture medium will yield a cell density that will require passage every 3–4 days. It is important to note that S2 cell viability decreases when the cell density falls below 5×10^5 cells/mL.

3.2 Transient Transfection of Drosophila S2 Cells (*See* Notes 2 *and* 3)

1. Plate S2 cells at 40–80 % confluency in a 6 or 12-well tissue culture plate 15–30 min prior to transfection. Alternatively, plate cells a day in advance of transfection. This step is needed as the transfection protocol calls for the exchange of medium, and the cells will need this time to loosely adhere to the tissue culture plastic.
2. If using the Fugene HD transfection reagent (Promega) (*see* **Note 3**), prepare the DNA–Fugene HD complex by diluting the desired amount of DNA in water or serum free Schneider's medium to a final volume of 100 μL. Next, add 6–8 μL of Fugene HD reagent directly to the diluted DNA with special care taken to avoid pipetting the reagent down sides of the microfuge tube. Incubate the DNA–Fugene HD mixture for 15–20 min at room temperature.
3. While the incubation is in progress, gently remove the cell culture medium from the tissue culture plate, taking precaution not to disturb the weakly adherent cells. Replace the cell culture medium with 900 μL of fresh cell culture medium. Add the 100 μL DNA–Fugene HD complex dropwise to the 900 μL of medium containing the cells. The transfection is carried out overnight, and the medium is replaced with fresh medium the following day.
4. Alternatively, transiently transfect S2 cells by using electroporation systems such as Amaxa nucleofector from Lonza. The Amaxa nucleofector kits are tailored to specific cell lines. The manufacturer suggests Kit V for Drosophila S2.

5. The Amaxa kit provides a nucleofection solution and a supplement solution. Mix these two solutions at a ratio of 4.5:1 prior to use. The mixed nucleofector reagent can be stored at 4 °C for up to 3 months.
6. Prepare DNA for electroporation by diluting the desired amount of DNA, typically between 0.1 and 2 μg of DNA, in the nucleofector reagent to a final volume of 100 μL in a microfuge tube. This can be prepared and set aside until the cells are ready.
7. Determine the appropriate volume of cells and pellet them by gentle centrifugation at 300 × *g* at room temperature for 5–10 min.
8. Next, remove the supernatant (tissue culture medium) (*see* **Note 4**).
9. Resuspend the pellet of cells in the diluted DNA–nucleofector mixture, and transfer the mixture to the electroporation cuvette (*see* **Note 5**).
10. After electroporation, transfer cells to fresh medium, and allow them to recover prior to $CuSO_4$ induction if using pMT/His-5 vectors (*see* **Note 6**).

3.3 Fixation and Staining of Drosophila S2 Cells for Immunofluorescence (*See* Note 7)

3.3.1 Preparing the Surface of Coverslips (*See* Notes 8 *and* 9)

1. For flame cleaning, use fine tipped forceps to grab an individual coverslip, and dip it into absolute ethanol. Allow most of the ethanol to run off the edge of the coverslip, and then wave the coverslip through an open flame. As the ethanol burns, continue to tilt and move the coverslip to prevent the ethanol from pooling in a single corner and superheating the glass.
2. Place flamed coverslips in a rack and rinse thoroughly in distilled water. Then place the coverslip rack over a heat block to allow them to dry thoroughly. Cover and store to block dust accumulation.

3.3.2 Coating of Substrates

Con A Coating of Coverslips

1. Cover the bottom of a broad shallow chamber with Parafilm, and then place the cleaned coverslips onto the Parafilm.
2. For 22 × 22 mm coverslips, put a 50 μL drop of Con A on the center of the coverslip and then evenly smear the drop over the surface of the glass.
3. For 18 × 18 mm coverslips use a 35 μL drop.
4. Place the coverslips in an out-of-the-way location where they will not accumulate dust, and allow to dry overnight.

Con A Coating of 96-Well Plates

1. Dilute Con A to 0.05 mg/mL and then:
2. Add 80 μL of Con A to each well and allow to dry over at least 2 days.

Polylysine Coating of Coverslips

1. Place coverslips into the Parafilm lined chamber.
2. Spread 50 μL of polylysine onto coverslips and let sit at room temperature for 3 h.
3. Rinse coverslips three times with distilled deionized water, and then aspirate excess water.
4. Allow the coverslips to dry, and then store at 4 °C until needed.

Coating of Coverslip-Bottom Imaging Chambers

1. Add a 100 μL drop of Con A or polylysine to the center of the coverslip, and spread it evenly over the surface.
2. Use the tip of a p100 pipette to remove any excessive fluid accumulation that might pool in the interface between the glass bottom and plastic side.
3. Allow Con A coated chambers to dry overnight. Allow lysine-coated chambers to sit for 3 h before rinsing them 3× with distilled deionized water. Dry them by aspirating any leftover fluid and then store them at 4 °C.

3.3.3 Plating of Cells (See **Note 10***)*

1. Resuspend cells that have been growing in a 96-well plate or 6-well dish.
2. Place prepared coverslips into a 6-well plate or 35 mm dish.
3. Place a drop of 100–200 μL of fresh medium onto the center of the coverslip.
4. Place a drop of the cell suspension into the fresh medium and pipette to mix. This step is very empirical. Allow the cells to settle to the surface and observe their density. Initially, add fewer cells to the coverslip, plate, or chamber than you think you will need. After 20 min the cells can be checked for density, and more cells can be added to achieve the desired density.
5. Add additional medium to the wells after the cells have attained the desired density and have begun to attach to the coverslip.
6. Initiate fixation, observation, or further manipulation after 1 h.

3.3.4 Fixation and Staining (See **Note 11***)*

1. Remove the culture medium from each well of the six-well plate or 35 mm dish, and fix cells for 10 min by adding in a solution of 0.3 % glutaraldehyde, 4 % formaldehyde in 1× PHEM supplemented with saponin (1 mg/mL).
2. Remove fixative and incubate cells with PHEM-T (1× PHEM + 0.5 % Triton X-100) for 5 min.
3. Rinse cells in PHEM.
4. Reduce unreacted aldehydes in the sample by three treatments for 10 min with a solution of 1 mg/mL sodium borohydride dissolved in PHEM (Make immediately before use).

5. Wash the coverslips three times in PHEM-Wash (1× PHEM + 0.1 % Triton X-100)—5 min per wash.
6. Block nonspecific antibody binding sites and reduce background by incubating cells in a blocking solution (PHEM-Wash + 5 % Boiled donkey serum) for 30 min to 1 h at room temperature (*see* **Note 12**).
7. Incubate cells in primary antibody for 1 h at room temperature or overnight at 4 °C if antibody staining is desired. Dilute antibodies to the appropriate concentration with blocking solution; determine concentrations empirically for each antibody. Dot 50 μL drops of antibody solution onto a Parafilm sheet in a humidified chamber, and place a coverslip on top of each drop. Ensure that the cell-adhering side of the coverslip makes contact with the drop of antibody solution.
8. Wash cells 4×5 min in PHEM-Wash.
9. Perform secondary antibody incubation as described for primary antibody using fluorescent secondary antibodies and additions such as DAPI or fluorescent-labeled phalloidin diluted in blocking buffer.
10. Wash coverslips 4×5 min in PHEM-Wash.
11. Rinse cells 1× in PHEM and 1× in distilled water to remove salts.
12. Wick away excess fluid from coverslips and then mount them (cell side down) on slides using one drop of a mounting media.
13. Use vacuum aspiration to remove any excess mounting media from around the edges of the coverslips and then seal them with a hardening material such as nail polish, VALAP, or CoverGrip (Biotium).

3.4 Live Cell Imaging (*See* Note 13)

1. Plate S2 cells on glass-bottom dishes (*see* **Note 14**) that have been coated with a lectin solution, concanavalin A (con A). Con A should be applied to the glass surface and allowed to incubate for 1–2 min.
2. Remove con A, carefully ensuring that the reagent is completely removed, and allow the dishes to dry before the addition of cell culture medium containing cells. Allow 30–60 min for S2 cells to fully attach and spread on the con A coated surface.
3. Alternatively, plate S2 plate on Polylysine or Cell-tak (Corning Cat. No. 354240) coated glass. See manufacturer's instructions for coating glass-bottom dishes with Cell-tak.
4. Perform imaging in Schneider's Insect Medium supplemented with 10 % heat-inactivated FBS and antibiotic. Imaging is conveniently carried out at room temperature without CO_2.

3.5 Fixation and Preparation of Drosophila S2 Cells for Scanning Electron Microscopy

3.5.1 Fixation (See Note 15)

1. Equilibrate fixative to the same temperature as the cell culture. Remove the cell medium from the coverslips, and replace it with a solution of 4 % glutaraldehyde with 4 % paraformaldehyde in 0.1 M Phosphate Buffered Saline.
2. Incubate the sample in the fixative for at least 30 min at room temperature.
3. Store samples for a maximum of 2 weeks at 4 °C if desired.

3.5.2 Dehydration and Critical Point Drying (See Note 16)

1. Wash the sample three times for 5 min in distilled water to remove fixative.
2. Exchange distilled water for 30 % ethanol for 5 min, and continue through dehydration with 5 min steps in 50, 70, 90, 100 % ethanol, using three washes at each ethanol concentration.
3. Under 100 % ethanol, stack coverslips in a CPD holder with a washer between each.
4. Fill the CPD chamber with enough dehydrated ethanol to cover the holder. Place the holder in the chamber. Operate the CPD as directed by the manufacturer's instructions.

3.5.3 Mounting and Sputter Coating (See Note 17)

1. Remove dry coverslips from the CPD chamber and adhere onto SEM stubs with double-sided conductive tape.
2. To further facilitate conductivity, apply silver paint to create a bridge from the surface of the coverslip to the conductive tape or SEM stub.
3. After the silver paint has dried, place the samples in a sputter coater chamber.
4. Operate the sputter coater according to manufacturer's instructions.
5. Image the coverslips in a SEM (Fig. 2).

4 Notes

1. S2 cells can be maintained in either Schneider's Drosophila medium (available from many suppliers) supplemented with 10 % FBS and antibiotics, or serum free media that are designed for SF9 cell culture. However, live cell imaging should be performed in Schneider's medium as serum free media both produces background fluorescence and contains trace amounts of divalent ions, which tend to induce expression from pMT/V5-His vectors. S2 do not require CO_2, and they are tolerant of and perform well at high cell densities. Cells should be passaged when they become confluent. S2 cells are loosely adherent on tissue culture plastic that has been treated for cell culture and thus do not require trypsin/EDTA treatments.

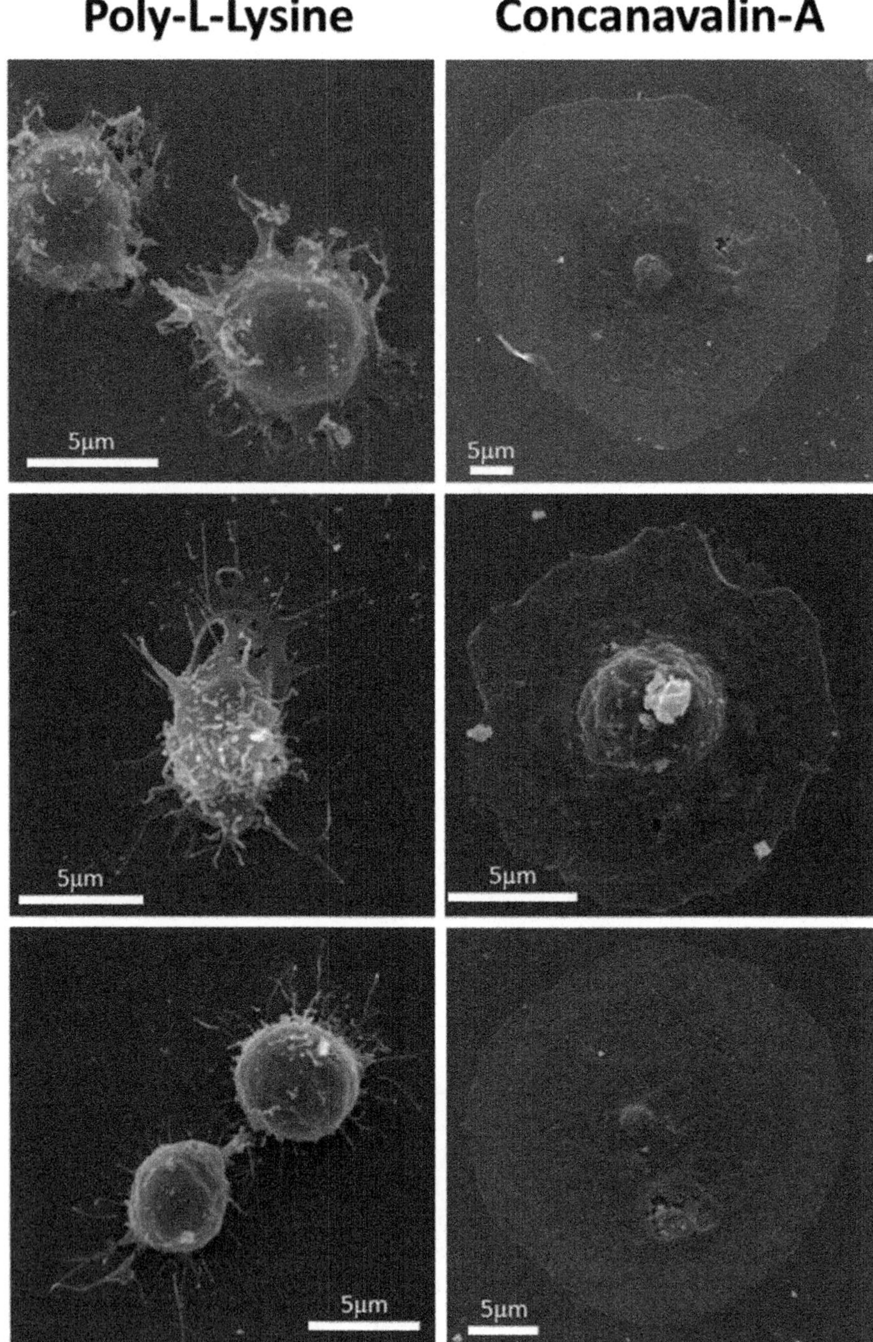

Fig. 2 Scanning electron micrography of *Drosophila* S2 cells. Shown are representative images of S2 cells allowed to adhere to glass coverslips coated with either poly-L-lysine or concanavalin-A. Cells were allowed to adhere for 1 h prior to fixation and processing for SEM. While the size and surface morphology for cells was somewhat variable, nearly all cells grown on concanavalin-A coated coverslips displayed a smooth dorsal surface morphology with few projections, a smooth-edge morphology and were well-spread. Scale bars are 5 μm

2. The amount of plasmid DNA transfected depends on the cytoskeletal protein being imaged and the *Drosophila* expression vector used. pMT/V5 His vectors that are copper inducible are typically transfected at a range of 0.1–2 μg DNA. An advantage of using pMT/V5-His vectors is that the amount of protein expressed can be titrated with the amount of $CuSO_4$ supplemented in the tissue culture medium prior to imaging. Successful imaging experiments have been carried out using vectors that have constitutively active promoters such as the pIZ/V5 vector, which uses the OpIE2 promoter and the pAc5.1/V5 which uses the *Drosophila* actin 5C promoter. However, it is often best to titrate the amount of DNA transfected in order to achieve optimal protein expression levels with these vectors. It is also possible to transfect in vectors containing the UAS promoter, which are usually generated for producing transgenic flies. To work with UAS vectors, one can co-transfect a pMT-Gal4 to allow for inducible expression of the UAS promoter. For live cell imaging of fluorescently tagged cytoskeletal proteins good results have been achieved by an overnight induction of 25–600 μM $CuSO_4$. It is particularly important to use the low end of this range of $CuSO_4$ when imaging microtubule plus-end tracking proteins (+TIPS) that are sensitive to mis-localization as a consequence of overexpression. Expression of other cytoskeletal proteins such as tubulin, actin, or actin-binding proteins that are less sensitive to the artifacts generated by overexpression, can be achieved by overnight induction at the higher end of this range.

   Transient transfection of cytoskeletal proteins is generally carried out 1–2 days prior to live-cell imaging. The exact timing is highly dependent on the vector and the expression of the protein of interest and an effort should be made to optimize this timing. For visualizing mitosis, it is usually best to wait 3 or more days before imaging. Mitosis is a rare event in a cells lifetime, and only 5 % of cells will be in mitosis at any point in time. If the transfection efficiency is 50 %, a culture will only have 2.5 % of cells transfected and in mitosis at any point during imaging. For this reason, stably transfected cells are usually preferred for studies of mitosis.

3. We have achieved good results using Fugene HD transfection reagent (Promega). It is always best to follow the manufacturer's instructions while preparing both the plasmid DNA and cells for electroporation. Lonza recommends 2×10^6 cells per transfection and from 0.1 to 2 μg of DNA per transfection.

4. It is important to remove as much tissue culture medium as possible without disturbing the pellet as residual media can interfere with the electroporation.

5. The electroporation machine provided by the Lonza system contains a preset S2 cell protocol, G-030 (or G-30 for Nucleofector I device).
6. Electroporation does lead to cell death but in general if the protocol is adhered to, the amount of cell death does not interfere with the progress of the experiment. Very high transfection efficiencies can be achieved with electroporation, however, transfection efficiency is highly dependent on plasmid DNA quality and the health of the cells prior to electroporation.
7. Drosophila S2 cells cannot be fixed and stained like mammalian cells. They are non-adherent and must be induced to adhere to coverslips or multiwell plates; they do not tolerate low-density cultures and therefore must be seeded carefully in order to function optimally. And they do not tolerate hypertonic buffers and therefore require specific buffers for washing and fixation.
8. In order to plate cells on glass, the hydrophobic coating found on most coverslips must be removed. There are several ways to do this: plasma cleaning, hydrogen peroxide + sulfuric acid soaking, and flame cleaning. Flame cleaning does not produce any chemical waste that requires storage and disposal or require a special piece of equipment.
9. S2 cells are only weakly adherent to glass surfaces, which must be functionalized in order to keep cells immobilized for fixation and staining.
10. S2 cells are extremely sensitive to plating density. If plated at too low of a concentration, they will not form extensive microtubule networks. Additionally, if one wishes to visualize mitotic cells, a too low density will make it very difficult to observe enough cells. However, plating at a too high density will inhibit their spreading and lead to cells stacking upon each other, resulting in out of focus noise in images. It is advisable to empirically determine how many cells to plate for any given application.
11. Many existing protocols for cell fixation work well for S2 cells. Several methods may need to be tested to find one that is compatible with both fixing the structure you wish to observe and that preserves the antigenicity of the molecules you wish to localize. Fixatives that work for S2 cells include −80 °C methanol, a mixture of methanol and formaldehyde (90 % methanol, 3.2 % formaldehyde, 5 mM sodium bicarbonate, pH 9 (from a 0.5 M stock) chilled to −80 °C prior to fixation), 4–10 % formaldehyde, and ice-cold acetone. The protocol in Subheading 3.3.5 uses a common fixation method that contains a mild detergent to give simultaneous fixation and permeabilization. Simultaneous fixation/permeabilization works well for many

antigens and structures as the detergent speeds up the access of the fixative while allowing some loosely bound protein to leave the cells, thereby reducing background. Alternatively, fixation followed by permeabilization or rapid fixation (10–60 s) followed by permeabilization and then 10 min of a second fixation can be utilized as well.

12. We use donkey serum and the highly cross-subtracted donkey secondary antibodies from Jackson ImmunoResearch diluted in 1× TBS + 0.1 % Triton X-100 + 2 % BSA.

13. Imaging is performed on an inverted microscope. Both the actin and microtubule cytoskeletons are sensitive to phototoxicity. Exposure to excess fluorescent light can quickly lead to a slowing or altering of the natural filament dynamics. Imaging should be performed to minimize the fluorescent light exposure. Both spinning-disk confocal and total internal reflection (TIRF) microscopy lend themselves to limiting the amount of light exposure while providing a good signal-to-noise ratio ideal for imaging proteins that are expressed at low levels. The speed of image acquisition by both these types of microscopy is also ideal for capturing the dynamics of the actin and microtubule cytoskeletons of live cells.

14. Glass bottom dishes can be made using a step drill bit (Neiko Titanium Step Drill Bit—1/4″ to 1-3/8″ Increments, 10 steps) installed in an upright drill press (SKIL 3320-01 120 V 10-in. Drill Press); however, commercial dishes with their superior flatness are highly recommended for TIRF. The bit is used to bore a circular opening in the bottom of a 35 mm plastic dish 1.5–2 cm in diameter. Cover-slips can then be glued to the dish, using any number of quick drying plastic compatible adhesives. We prefer an ultraviolet curing adhesive available from Norland (Norland Optical Adhesive 81) which quickly cures following a brief exposure (5–10 min) to ultraviolet light from a transilluminator.

15. Chemical fixation halts deterioration of the sample by crosslinking proteins. This crosslinking also provides some support for dehydration and critical point drying. Chemical fixatives are hazardous. Gloves and a fume hood should be used during fixation.

16. Samples must be dry to withstand the high vacuum of a scanning electron microscope. Air-drying creates enormous surface tension that causes samples to crumple losing their original surface shapes. To preserve a sample's structure, dehydration in ethanol and critical point drying (CPD) in carbon dioxide are used to dry the sample. During CDP 100 % ethanol, which is miscible with water and carbon dioxide, is exchanged for liquid carbon dioxide. Heat and pressure are applied to the

sample to reach the point where liquid carbon dioxide changes into gaseous carbon dioxide (the critical point). The sample is now dry. Having never been exposed to surface tension the original structures are preserved.

17. Scanning Electron Microscopy (SEM) requires conductivity throughout the system. Biological samples lose their conductivity once water is removed. Mounting samples on aluminum stubs facilitates conductivity from the SEM stage to the sample and provides a stable surface. Conductivity must continue across the surface of the sample. This requires the deposition of a thin layer of metal. Sputter coating uses a heavy inert gas (typically Argon) to create a plasma field around a metal target. The plasma field erodes the target so that metal atoms are ejected onto the surface of the sample.

References

1. Schneider I (1972) Cell lines derived from late embryonic stages of *Drosophila melanogaster*. J Embryol Exp Morphol 27:353–365
2. Somma MP, Fasulo B, Cenci G, Cundari E, Gatti M (2002) Molecular dissection of cytokinesis by RNAi interference in Drosophila tissue culture cells. Mol Biol Cell 13:2448–2460
3. Pearson AM, Baksa K, Rämet M, Protas M, McKee M, Brown D, Ezekowitz RA (2003) Identification of cytoskeletal regulatory proteins required for efficient phagocytosis in Drosophila. Microbes Infect 10:815–824
4. Kiger AA, Baum B, Jones S, Jones MR, Coulson A, Echeverri C, Perrimon N (2003) A functional genomic analysis of cell morphology using RNA interference. J Biol 2:27
5. Rogers SL, Wiedemann U, Stuurman N, Vale RD (2003) Molecular requirements for actin-based lamella formation in Drosophila S2 cells. J Cell Biol 162:1079–1088
6. Eggert US, Kiger AA, Richter C, Perlman ZE, Perrimon N, Mitchison TJ, Field CM (2004) Parallel chemical genetic and genome-wide RNAi screens identify cytokinesis inhibitors and targets. PLoS Biol 12:e379
7. Goshima G, Wollman R, Goodwin SS, Zhang N, Scholey JM, Vale RD, Stuurman N (2007) Genes required for mitotic spindle assembly in Drosophila S2 cells. Science 316:417–421
8. D'Ambrosio MV, Vale RD (2010) A whole genome RNAi screen of Drosophila S2 cell spreading performed using automated computational image analysis. J Cell Biol 191:471–479
9. Moutinho-Pereira S, Stuurman N, Afonso O, Hornsveld M, Aguiar P, Goshima G, Vale RD, Maiato H (2013) Genes involved in centrosome-independent mitotic spindle assembly in Drosophila S2 cells. Proc Natl Acad Sci 110:19808–19813
10. Toret CP, D'Ambrosio MV, Vale RD, Simon MA, Nelson WJ (2014) A genome-wide screen identifies conserved protein hubs required for cadherin-mediated cell-cell adhesion. J Cell Biol 201:265–279
11. Caplen NJ, Fleenor J, Fire A, Morgan RA (2000) dsRNA-mediated gene silencing in cultured Drosophila cells: a tissue culture model for the analysis of RNA interference. Gene 252:95–105
12. Rogers SL, Rogers GC (2008) Culture of Drosophila S2 cells and their use for RNAi-mediated loss-of-function studies and immunofluorescence microscopy. Nat Protoc 3: 606–611
13. Adams MD et al (2000) The genome sequence of *Drosophila melanogaster*. Science 287: 2185–2195
14. Reiter LT, Potocki L, Chien S, Gribskov M, Bier E (2001) A systematic analysis of human disease-associated gene sequences in *Drosophila melanogaster*. Genome Res 11:1114–1125
15. Beir E (2005) Drosophila, the golden bug, emerges as a tool for human genetics. Nat Rev Genet 39:715–720
16. Rogers SL, Rogers GC, Sharp DJ, Vale RD (2002) Drosophila EB1 is important for proper assembly, dynamics, and positioning of the mitotic spindle. J Cell Biol 158: 873–884
17. Kner P, Chhun BB, Griffis ER, Winoto L, Gustafsson MG (2009) Super-resolution video microscopy of live cells by structured illumination. Nat Methods 6:339–342

18. Iwasa JH, Mullins RD (2007) Spatial and temporal relationships between actin-filament nucleation, capping, and disassembly. Curr Biol 17:395–406
19. Uehara R, Goshima G, Mabuchi I, Vale RD, Spudich JA, Griffis ER (2010) Determinants of myosin II cortical localization during cytokinesis. Curr Biol 20:1080–1085
20. Biyasheva A, Svitkina T, Kunda P, Baum B, Borisy G (2004) Cascade pathway of filopodia formation downstream of SCAR. J Cell Sci 117:837–848
21. Kim JH, Cho A, Yin H, Schafer DA, Mouneimne G, Simpson KJ, Nguyen KV, Brugge JS, Montell DJ (2011) Psidin, a conserved protein that regulates protrusion dynamics and cell migration. Genens Dev 25:730–741
22. Bai SW, Herrera-Abreu MT, Rohn JL, Racine V, Tajadura V, Suryavanshi N, Bechtel S, Wiemann S, Baum B, Ridley AJ (2011) Identification and characterization of a set of conserved and new regulators of cytoskeletal organization, cell morphology and migration. BMC Biol 9:54
23. Maiato H, Sampaio P, Lemos CL, Findlay J, Carmena M, Earnshaw WC, Sunkel CE (2002) MAST/Orbit has a role in microtubule-kinetochore attachment and is essential for chromosome alignment and maintenance of spindle bipolarity. J Cell Biol 157:749–760
24. Logarinho E, Bousbaa H, Dias JM, Lopes C, Amorim I, Antunes-Martins A, Sunkel CE (2004) Different spindle checkpoint proteins monitor microtubule attachment and tension at kinetochores in Drosophila cells. J Cell Sci 117:1757–1771
25. Maiato H, Rieder CL, Khodjakov A (2004) Kinetochore-driven formation of kinetochore fibers contributes to spindle assembly during animal mitosis. J Cell Biol 167:831–840
26. Goshima G, Nédélec F, Vale RD (2005) Mechanisms for focusing mitotic spindle poles by minus end-directed motor proteins. J Cell Biol 171:229–240
27. Maiato H, Hergert PJ, Moutinho-Pereira S, Dong Y, Vandenbeldt KJ, Rieder CL, McEwen BF (2006) The ultrastructure of the kinetochore and kinetochore fiber in Drosophila somatic cells. Chromosoma 115:469–480
28. Griffis ER, Stuurman N, Vale RD (2007) Spindly, a novel protein essential for silencing the spindle assembly checkpoint, recruits dynein to the kinetochore. J Cell Biol 117: 1005–1015
29. Zhang D, Rogers GC, Buster DW, Sharp DJ (2007) Three microtubule severing enzymes contribute to the "Pacman-flux" machinery that moves chromosomes. J Cell Biol 177:231–242
30. Maresca TJ, Salmon ED (2009) Intrakinetochore stretch is associated with changes in kinetochore phosphorylation and spindle assembly checkpoint activity. J Cell Biol 184:373–381
31. Rogers GC, Rusan NM, Peifer M, Rogers SL (2008) A multicomponent assembly pathway contributes to the formation of acentrosomal microtubule arrays in interphase Drosophila cells. Mol Biol Cell 19:3163–3178
32. Rogers SL, Wiedemann U, Häcker U, Turck C, Vale RD (2004) Drosophila RhoGEF2 associates with microtubule plus ends in an EB1-dependent manner. Curr Biol 14: 1827–1833
33. Mennella V, Rogers GC, Rogers SL, Buster DW, Vale RD, Sharp DJ (2005) Functionally distinct kinesin-13 family members cooperate to regulate microtubule dynamics during interphase. Nat Cell Biol 7:235–45
34. Goodwin SS, Vale RD (2010) Patronin regulates the microtubule network by protecting microtubule minus ends. Cell 143: 263–274
35. Rothenberg ME, Rogers SL, Vale RD, Jan LY, Jan YN (2003) Drosophila pod-1 crosslinks both actin and microtubules and controls the targeting of axons. Neuron 39:779–791
36. Applewhite DA, Grode KD, Keller D, Zadeh AD, Slep KC, Rogers SL (2010) The spectraplakin Short stop is an actin-microtubule cross-linker that contributes to organization of the microtubule network. Mol Biol Cell 21:1714–1724

Chapter 5

Imaging Cytoskeleton Components by Electron Microscopy

Tatyana Svitkina

Abstract

The cytoskeleton is a complex of detergent-insoluble components of the cytoplasm playing critical roles in cell motility, shape generation, and mechanical properties of a cell. Fibrillar polymers—actin filaments, microtubules, and intermediate filaments—are major constituents of the cytoskeleton, which constantly change their organization during cellular activities. The actin cytoskeleton is especially polymorphic, as actin filaments can form multiple higher order assemblies performing different functions. Structural information about cytoskeleton organization is critical for understanding its functions and mechanisms underlying various forms of cellular activity. Because of the nanometer-scale thickness of cytoskeletal fibers, electron microscopy (EM) is a key tool to determine the structure of the cytoskeleton.

This article describes application of rotary shadowing (or metal replica) EM for visualization of the cytoskeleton. The procedure is applicable to thin cultured cells growing on glass coverslips and consists of detergent extraction of cells to expose their cytoskeleton, chemical fixation to provide stability, ethanol dehydration and critical point drying to preserve three-dimensionality, rotary shadowing with platinum to create contrast, and carbon coating to stabilize replicas. This technique provides easily interpretable three-dimensional images, in which individual cytoskeletal fibers are clearly resolved, and individual proteins can be identified by immunogold labeling. More importantly, replica EM is easily compatible with live cell imaging, so that one can correlate the dynamics of a cell or its components, e.g., expressed fluorescent proteins, with high resolution structural organization of the cytoskeleton in the same cell.

Key words Electron microscopy, Cytoskeleton, Critical point drying, Rotary shadowing, Actin, Microtubules, Immunogold, Correlative microscopy

1 Introduction

Electron microscopy (EM) has been instrumental in discovering the cytoskeleton in the first place, and also in investigating its structural organization in different cells and conditions. The initial progress in the cytoskeletal studies closely paralleled the development of EM techniques. Thus, the introduction of heavy metal fixation led to the discovery of actin filaments in non-muscle cells [1], while the discovery of microtubules [2] was made possible after the introduction of aldehyde fixation [3].

Ray H. Gavin (ed.), *Cytoskeleton: Methods and Protocols*, Methods in Molecular Biology, vol. 1365,
DOI 10.1007/978-1-4939-3124-8_5, © Springer Science+Business Media New York 2016

A great value of EM is its ability to obtain structural information at a high resolution level, which for biological samples is limited by a sample preparation procedure rather than by the power of a transmission electron microscope (TEM). Vacuum in the TEM column and electron beam irradiation impose strict restrictions on how samples should be prepared, which in turn greatly affect the quality of images and the rate of success. A large number of different EM protocols have been developed over the years to improve the quality of samples and the amount of collected information, and to avoid artifacts. Each technique has its pluses and minuses, making it more suitable for some applications than for others.

The thin sectioning technique was initially a dominant way to visualize the cytoskeleton [4, 5]. It involves the embedding of chemically fixed specimens into a resin followed by thin sectioning to allow for beam penetration. Contrast is generated by positive staining of the sections with heavy metal salts, and the limited ability of stains to bind bioorganic material reduces the resolution of this technique. Thin sections provide a 2D view of the sample at a single plane, and a series of sections is required to retrieve the 3D information. Such reconstruction works well with relatively large and simple objects, but is not efficient in revealing the details of complex and delicately organized cytoskeletal structures, such as, for example, the actin filament networks in lamellipodia of locomoting cells.

Different versions of whole mount EM have been used to investigate the structural organization of the cytoskeleton in its entirety. Thus, the structural arrangement of actin filaments in lamellipodia was first visualized by the negative staining EM of cultured cells [6]. In this technique, partially permeabilized cells growing on EM grids are immersed into a heavy metal stain solution, which is blotted off shortly after, and the samples are dried in open air. The dried stain generates a dark amorphous background on which the structures appear as translucent shapes. Negative staining EM provides high resolution and allows one to see the thin regions of a cell all the way through. Weaknesses of the procedure include sample flattening during air drying, relatively low contrast, and low stability of the samples.

In cryo EM technique, the samples are quickly frozen (to prevent ice crystal formation) and viewed while still embedded in amorphous ice, either as whole mounts or after cryosectioning, so that they remain hydrated and the proteins retain their natural conformation [7, 8]. To view frozen samples, a TEM should be equipped with a chilled sample holder and electron beam power and the observation time should be minimized to keep the specimen frozen. No contrasting procedures are used in this technique except for the specimen's own contrast, which is quite low. Therefore, significant image processing is required for the presentation and analysis of images and many structures still remain undetected. The major limitation of the technique is the significant difficulty in obtaining successful samples.

In metal replica EM, heavy metals are evaporated onto a 3D sample at an angle, which reveals its surface topography [9]. The quality of the samples is greatly enhanced if rotary, and not unilateral, coating is used, as it helps to avoid deep featureless shadows. As metal coating is not cohesive, it is subsequently stabilized by a layer of carbon, which keeps metal grains together and is fairly transparent for the electron beam. The coated sample, or just a metal–carbon replica, is subsequently removed from its original support and placed onto EM grids. The resolution of replica EM is quite high, but it depends on the metal grain size, the thickness of the coating, and the angle of shadowing. Platinum is the most popular metal, as it provides a good compromise between the grain size and an ease of evaporation. The replica technique was initially introduced to study freeze-fractured samples [10], but it is applicable for a large range of samples, such as single molecules [11, 12], cells [13–15], and tissues [16, 17]. This approach can reveal the 3D structure in great detail, but it is limited by the depth of shadowing penetration.

For our studies of the cytoskeleton organization in cultured cells, we chose platinum replica EM, in which detergent extraction is used to expose the cytoskeleton; chemical fixation helps to preserve the sample structure; ethanol dehydration followed by critical point drying (CPD) preserves the cell's 3D organization; and rotary shadowing with platinum creates contrast. Over the years, we have found a good combination of individual steps to develop a reliable and relatively simple protocol that consistently produces highly informative images with excellent yield that can be combined with immunochemistry [18–20]. However, this approach is not universal, but is limited to relatively thin samples attached to glass surfaces. Also, because of extensive fixation and dehydration, it can achieve the molecular level of resolution only for very large molecules [21, 22], but is optimal for analyses of the fine cytoskeletal architecture with a single filament resolution at the scale of a whole cell.

As EM, in general, cannot work with live samples, investigators can only guess the kind of activity the cell was involved in at the moment of fixation, and what it would do next. A partial solution for this problem is provided by correlative light and EM, in which the dynamics of a living cell is followed by time-lapse optical imaging, and the same sample is subsequently analyzed by EM. Our EM protocol made it possible to perform correlative light and EM routinely, as it allowed us to obtain high quality structural information for a cell of interest with high probability [15, 18–20]. Several other EM techniques have also been used in a correlative approach, including thin sections of resin-embedded samples [23–25], cryoEM [26–29], and negatively contrasted cells [30].

2 Materials

2.1 Cell Culture and Extraction

1. Small (6–12 mm) coverslips that can be made by cutting regular coverslips with a diamond pencil. Trapezoidal (square with one oblique side) coverslips lacking mirror symmetry are helpful to easily determine the cell-containing side. Commercially available 12 mm round coverslips are also acceptable (*see* **Note 1**).
2. Phosphate-buffered saline (PBS) with Ca^{2+} and Mg^{2+}.
3. PEM buffer: 100 mM PIPES (free acid), pH 6.9 (adjust with KOH), 1 mM $MgCl_2$, and 1 mM EGTA. Working buffer is prepared from 2× stock solution, which can be stored up to 1 month at 4 °C (*see* **Note 2**).
4. Extraction solution: 1 % Triton X-100 in PEM buffer supplemented (optionally) with 1–4 % polyethyleneglycol (PEG) (MW 20,000–40,000), 2 μM phalloidin, and/or 2 μM taxol (paclitaxel) (*see* **Note 3**). Use a stirrer and allow 15–20 min to dissolve PEG. Extraction solutions can be stored for up to 3 days at 4 °C, but phalloidin and taxol should be added before use. Stock solutions (1000×) of phalloidin and taxol are made in dimethylsulfoxide (DMSO) and stored at −20 °C in aliquots.

2.2 Fixation

1. Glutaraldehyde solution: 2 % EM grade glutaraldehyde in 0.1 M sodium cacodylate, pH 7.3. The working solution can be stored at 4 °C for up to 1 week, 2× stock solution of sodium cacodylate is stable at 4 °C. *Caution*: Glutaraldehyde is toxic and volatile, so a fume hood should be used when working; sodium cacodylate is toxic.
2. Tannic acid solution: 0.1 % tannic acid in distilled water. Use within a day.
3. Aqueous uranyl acetate solution: 0.2 % uranyl acetate in distilled water. Use a stirrer to dissolve. Store at room temperature. *Caution*: Uranyl acetate is toxic.

2.3 Dehydration and Critical Point Drying

1. Graded ethanol solutions: 10, 20, 40, 60, 80 and 100 % ethanol in distilled water.
2. Alcohol uranyl acetate solution: 0.2 % uranyl acetate in 100 % ethanol. Use a stirrer to dissolve. Use within a day. *Caution*: Uranyl acetate is toxic.
3. Dehydrated ethanol: Wash molecular sieves (4 Å, 8–12 mesh) with several changes of water to remove dust. Dry in the air, bake at 160 °C overnight, cool down, and add to a bottle of 100 % ethanol (~50 g per 500 mL). Seal with Parafilm and store at room temperature. Do not shake, as the beads are fragile and easily generate dust.

4. CPD sample holder and scaffolds: A holder with a lid and two scaffolds are homemade with a stainless-steel wire mesh. The holder should fit into the chamber of the CPD apparatus. A scaffold is required to maintain the holder above the stirring bar during stirring. It should fit a beaker in which dehydration will be processed, e.g., a 50 mL glass beaker.
5. CPD device: We use Samdri PVT-3D (Tousimis) CPD with manual operation, but other devices are also appropriate.
6. Carbon dioxide: Use liquid dehydrated CO_2 (bone dry grade) in a tank with a siphon (deep tube) and a water and oil-absorbing filter (Tousimis). Siphonized tanks make it possible to take the liquid phase of CO_2 from the bottom of the tank.

2.4 Platinum and Carbon Coating

1. Vacuum evaporator: We use Auto 306 coater (Boc Edwards) equipped with a water-cooled diffusion pumping system, carbon and metal evaporation sources, a rotary stage, and a thickness monitor.
2. Metals for evaporation (Ted Pella): Tungsten wire (0.76 mm), platinum wire (0.2 mm), carbon rods (3 mm).

2.5 Preparation of Replicas

1. Hydrofluoric acid (HF): 5–10 % HF in distilled water is prepared from the concentrated acid (49 %). Do not use glassware to handle acid containing solutions. *Caution*: Extremely volatile and toxic, it causes severe skin burns. Use a fume hood and gloves.
2. Platinum loop: The optimal loop diameters are 3–5 mm.
3. EM Grids: Formvar-coated EM grids with low mesh size (e.g., 50) to provide a large viewing area. Other options are acceptable.

2.6 Immunogold EM

1. Immunogold buffer: 20 mM Tris–HCl, pH 8.0, 0.5 M NaCl, and 0.05 % Tween 20. For dilution of antibodies, the buffer is supplemented with 1 % bovine serum albumin (BSA); for washing, 0.1 % BSA is added. Stock solutions (5×) are stable at 4 °C for several months. Sodium azide can be added to the stock solution to prevent microbial contamination. *Caution*: Sodium azide is toxic.
2. Quenching solution: 2 mg/mL sodium borohydrate ($NaBH_4$) in PBS. Use immediately.
3. Blocking solution: 1 mg/mL glycine (or lysine) in PBS. Stable at 4 °C.

2.7 Correlative EM

1. Marked coverslips: Homemade coverslips with reference marks are prepared by evaporating gold through a finder grid placed in the middle of a 22 × 22 square or 25 mm round coverslip. A variety of finder grids are available commercially. Baking

(160 °C overnight) of gold-coated coverslips is necessary for the firm adhesion of gold to glass (*see* **Note 4**). For light microscopy, choose cells on the clear footprint of the finder grid.

2. Light microscopy: The light microscopic system should be equipped with an environmental chamber to maintain normal cell behavior; it should allow for fast exchange of culture medium to extraction solution to quickly stop cellular activity during imaging. A simple option is to use open dishes on a heated stage with the cells growing in a bicarbonate-free medium.

3 Methods

In a basic form, replica EM can be used to study the cytoskeleton architecture in a cell population. In an advanced form, it can be combined with immunogold staining to detect specific proteins in the cytoskeleton (Fig. 1), and with light microscopy to correlate the cytoskeleton organization with cell behavior or with the distribution and dynamics of fluorescent probes (Figs. 1 and 2). The basic procedure consisting of extraction, fixation, dehydration, CPD, metal shadowing, and preparation of replicas is described first, and it is followed by the description of the advanced applications.

A major source of artifacts in this technique is the failure to perform a genuine CPD, which may occur if wet samples are transiently exposed to air, or water is not fully exchanged to ethanol or ethanol to CO_2, or if the dried samples absorb ambient humidity. In order to get a high quality preparation, it is critical not to allow a liquid–gas interface to touch the samples at any point during the procedure. Practically, it means keeping the cells away from air while they are wet, and away from water, while they are dry. Changes of solutions need to be done quickly, with a layer of liquid always being retained above the cells. After drying, the cells should be kept at low humidity until they are coated with carbon.

3.1 Cell Culture and Extraction

Detergent extraction is used to expose the internal cytoskeletal structures, while the carrier buffer is designed to maximally preserve them until fixation. Additional preservation may be achieved using specific and nonspecific stabilizers.

1. Put small glass coverslips into a culture dish and plate the cells. Cell culture conditions are specific for each system and are not discussed here. If several coverslips are placed into the same dish, make sure that they do not overlap and that there are no bubbles underneath.
2. When the cells are ready, remove the culture medium from the dish using a pipette or by pouring out; quickly rinse with the pre-warmed to 37 °C PBS (*see* **Note 5**).

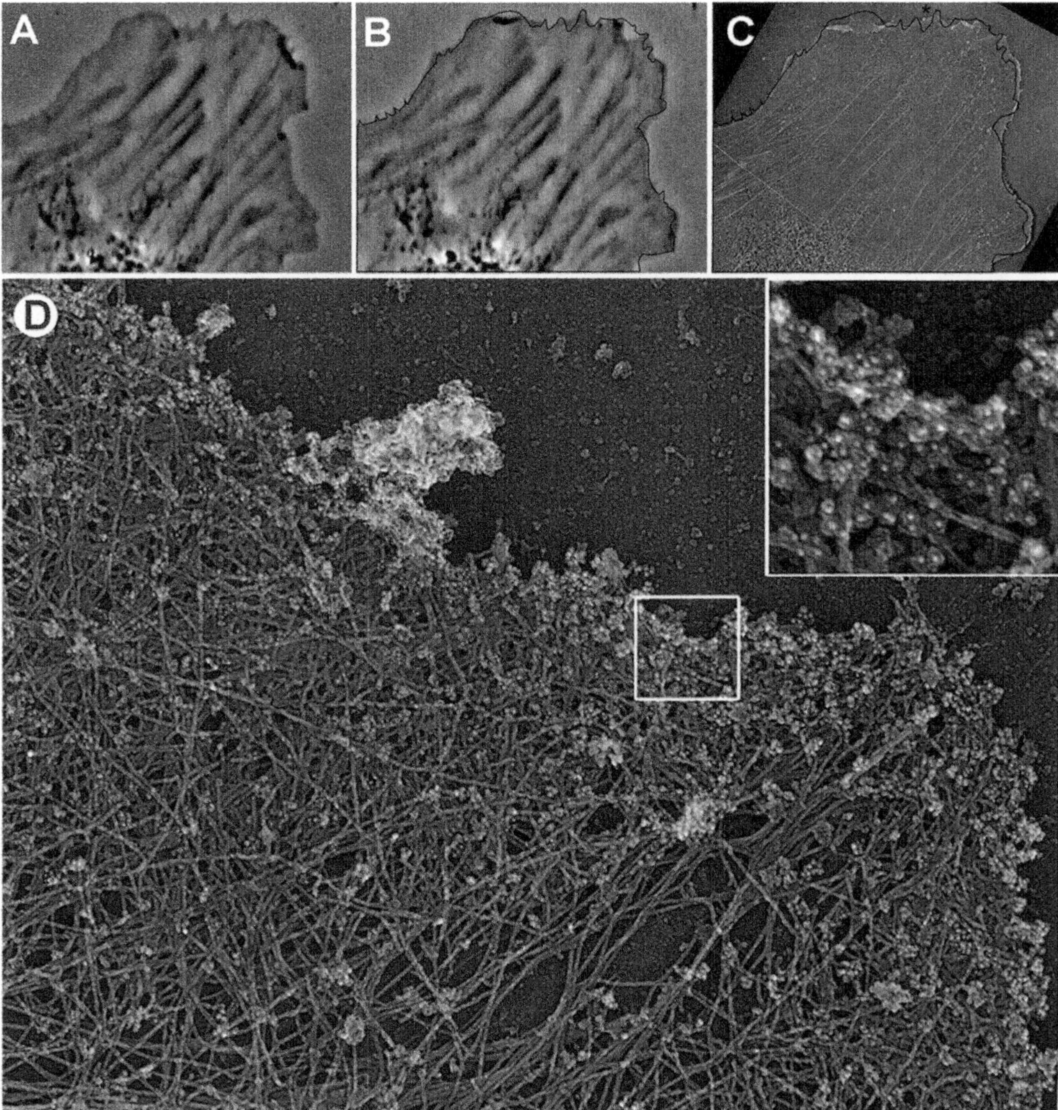

Fig. 1 Correlative phase-contrast and EM of cultured Rat-2 fibroblast combined with immunogold staining of ADF/cofilin. (**a** and **b**) Frames from time-lapse sequence showing the last live cell image (**b**) and an image 12 s earlier (**a**). *Black line* in (**b**) shows the cell edge outline from (**a**). (**c**) Low magnification EM image of the cell overlaid with the cell outline, as in (**b**). (**d**) EM of the protruding edge (*asterisk* in (**c**)) comprising a lamellipodium filled with dense actin network. Before the EM processing, the sample was immunogold labeled by cofilin antibody; *inset* in (**d**) shows gold particles as white dots

3. Immediately, but gently, add extraction solution equilibrated to room temperature; gently stir the dish to ensure that extraction solution instantly reaches all the cells (*see* **Note 6**). Incubate for 3–5 min at room temperature.
4. Rinse cells with PEM buffer 2–3 times, for a few seconds each time, at room temperature (*see* **Note 7**).

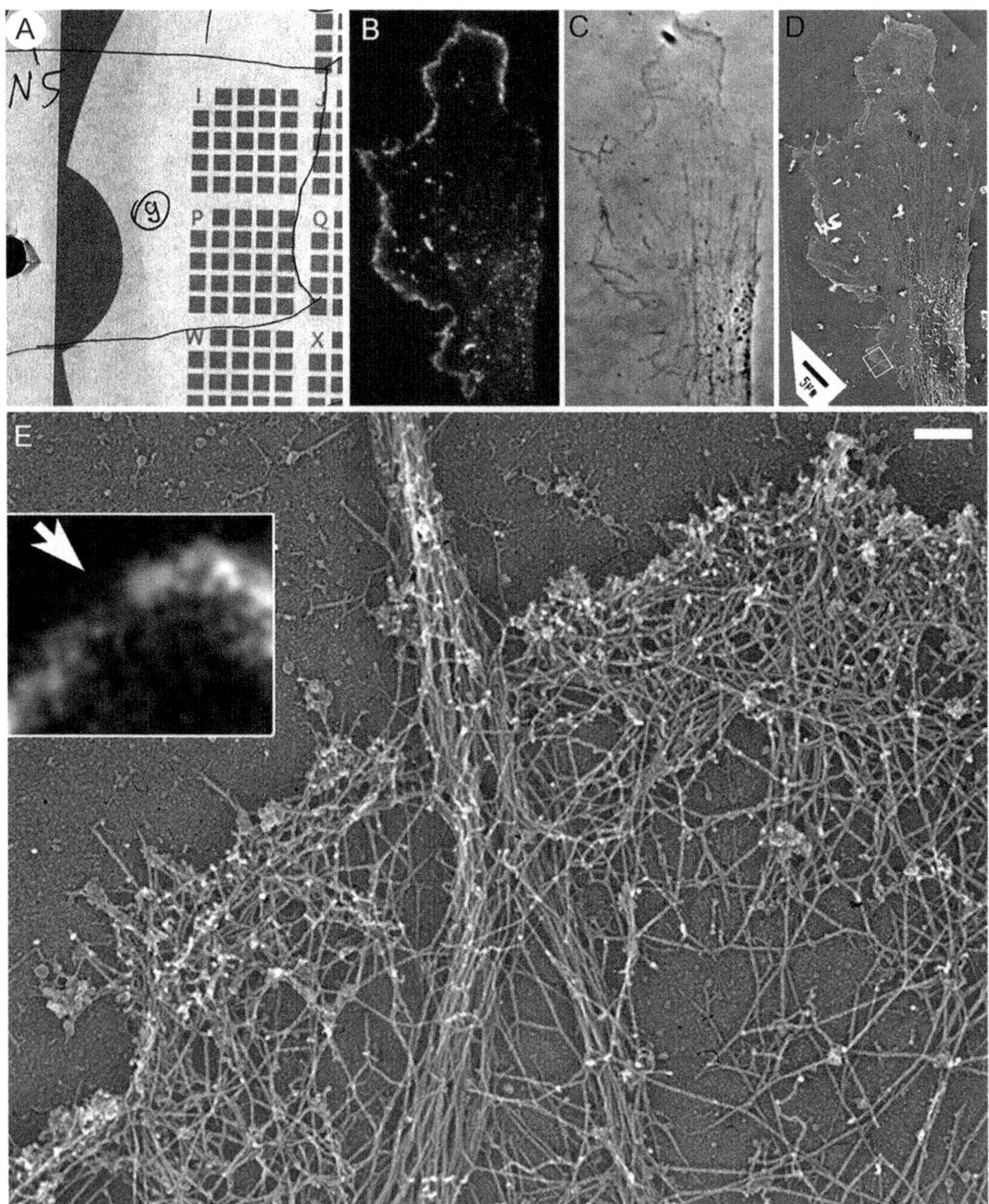

Fig. 2 Correlative fluorescence and EM of cultured B16F1 mouse melanoma cell expressing EGFP-capping protein. (**a**) Map showing position of the cell (*number 9 in a circle*) relative to the reference marks on the coverslip. (**b**) Fluorescence image the cell showing localization of EGFP-capping protein to the edge of lamellipodia and puncta in lamella. (**c**) Phase contrast image of the same cell. (**d**) Low magnification EM image of the same cell. *Box* indicates a region enlarged in (**e**). (**e**) High magnification EM of the boxed region from (**d**) showing actin filament bundle in a filopodium in the center and dense branched network of actin filaments in lamellipodia. *Inset* shows the same region by fluorescence microscopy. Bright fluorescence corresponds to lamellipodia, while the dim region (*arrow*) corresponds to the filopodium

3.2 Fixation

Chemical fixation provides cytoskeletal structures with physical resistance against subsequent procedures, especially dehydration and CPD. It is a three-step procedure using different fixatives: glutaraldehyde, tannic acid and uranyl acetate.

1. After the last PEM rinse, add glutaraldehyde solution and incubate for at least 20 min at room temperature. If necessary, the samples can be stored at this stage at 4 °C for up to 3 weeks. Take care to prevent evaporation during storage.
2. Transfer coverslips to another container with tannic acid solution, or change solutions in the same dish (*see* **Note 8**). Make sure that the cells remain covered with liquid during transfer. No washing is necessary before this step. Incubate for 20 min at room temperature.
3. Take the coverslips out of the tannic acid solution one by one, rinse by dipping sequentially into two water-filled beakers, and place in a new plate with distilled water. Do not keep the coverslips out of the solution longer than necessary. Incubate for 5 min (*see* **Note 9**).
4. Take the coverslips out of the water one by one, rinse twice again by dipping into water, and place in a new plate with aqueous uranyl acetate solution. Avoid drying during transfer. Incubate for 20 min at room temperature.
5. Wash off uranyl acetate solution with distilled water by transferring the samples or exchanging the solutions.

3.3 Dehydration and CPD

Drying of the samples is necessary to expose the surfaces for metal coating in a vacuum. However, plain drying in the open air generates major structural distortions. When the liquid–gas interface passes through the samples, the forces of the surface tension that are enormous at the cellular scale flatten the samples. During CPD, the temperature and pressure of a liquid are raised above its critical point, at which the phase boundary and surface tension do not exist. In this state, the liquid can well be considered as compressed gas. When the pressure is released, the samples remain dry with the 3D organization intact, because they never experienced the surface tension. Carbon dioxide has reasonably low values of critical point pressure and temperature that can be tolerated by biological samples. However, a direct transfer of the samples from water to CO_2 is not possible and ethanol, which is freely miscible with both water and CO_2, is used as an intermediate. For dehydration and CPD, the coverslips are stacked in the sample holder with pieces of lens tissue as spacers, and processed simultaneously.

1. Cut lens tissue with loosely arranged fibers into pieces fitting the size of the CPD sample holder or a little larger (*see* **Note 10**).
2. Place the CPD sample holder into a wide container with distilled water. Put a piece of lens tissue at the bottom of the holder,

place a coverslip with the cell side up on the lens tissue, and cover with another piece of lens tissue. Continue loading the coverslips into the holder, alternating them with pieces of lens tissue. All loading should be done under water. Make sure that the coverslips are minimally exposed to air during loading. Do not overload the holder, as it will interfere with the exchange of solutions (*see* **Note 11**). Keep track of sample identity based on its position in the holder.

3. Put a stirrer bar into a 50 mL glass beaker, place a scaffold over it, and add 10 % ethanol in the amount sufficient to cover the CPD holder. Quickly transfer the CPD holder with samples from the water-filled container into the beaker. Place the beaker on a magnetic stirrer, and stir for 5 min.
4. Prepare another beaker with a stirrer bar and a scaffold and add 20 % ethanol. Transfer the CPD holder from the first beaker into the second, and stir for 5 min.
5. Continue dehydration by transferring the CPD holder sequentially through the remaining graded ethanols, 5 min in each: 40, 60, 80, and 100 % (twice). Alternate the two beakers with their scaffolds and stirrer bars (*see* **Note 12**).
6. Place the CPD holder into a beaker with 0.2 % uranyl acetate in ethanol to fully cover it. Incubate for 20 min. Stirring is not necessary.
7. Transfer the CPD holder through beakers with 100 % ethanol (twice) and dehydrated 100 % ethanol dried over a molecular sieve (twice), as in **steps 3–5**. Stir for 5 min in each.
8. Fill the chamber of the CPD apparatus with dehydrated ethanol, just sufficient to cover the CPD holder. Place the holder in the ethanol and close the lid. Operate the CPD according to the manufacturer's instructions or following the procedure below (*see* **Note 13**).
9. Open the CO_2 tank. Cool down the CPD chamber to 10–15 °C to keep the CO_2 in a liquid state. Maintain this temperature until the heating step (*see* **Note 14**).
10. Open the inlet valve on the CPD to allow the CO_2 to fill the chamber. If the device is equipped with a magnetic stirrer, turn it on and keep it running till the heating step. Wait for 5 min. This is a mixing step, during which the ethanol in the chamber, the sample holder, and the samples, are equilibrated with the liquid CO_2 from the tank.
11. With the inlet valve still open, open the outlet valve slightly until you clearly see that the liquid exchange is happening in the chamber, but do not allow the level of the liquid to go below the top of the sample holder. Wait for 30 s. This is a washing step, when the ethanol–CO_2 mixture is released from

the chamber and replaced with pure CO_2 from the tank. If the CPD is not equipped with a magnetic stirrer, shake the CPD manually during the washing to mix the chamber contents better (*see* **Note 15**).

12. Close the outlet valve. Wait for 5 min.
13. Repeat **steps 11** and **12** nine more times (total ten washes). After the last wash, wait only until the chamber is completely filled with CO_2; that may take less than 5 min.
14. Close both the inlet and outlet valves and turn on the heat. As the chamber is isolated, heating will raise both the temperature and the pressure. Wait until both parameters exceed the critical values for CO_2 (critical pressure = 1072 psi or 73 atm, critical temperature = 31 °C) and reach values of ~1250 psi (85 atm) and ~40 °C. If one value is reached sooner than the other, maintain the former at a steady-state level by turning the heater or the outlet valve on and off until both the values are reached.
15. Open the outlet valve slightly to slowly release the pressure until it reaches the atmospheric pressure. It should take about 10 min (*see* **Note 16**).
16. Open the CPD chamber, remove the sample holder and immediately put it into a desiccator.

3.4 Platinum and Carbon Coating

Platinum shadowing generates the contrast of the samples. The angle and the thickness of the coating are critical parameters influencing the quality of the image. Lower angles provide higher contrast, but do not penetrate deep into the sample. Thinner coats provide higher resolution, but lower contrast. For cellular studies, we shadow platinum at a ~45° angle with rotation to achieve a ~2 nm thickness of the coat, which is controlled by the thickness monitor. Carbon is applied from the top of the samples with a thickness of 3.5–5 nm. The basic steps of coating are listed below. Use the equipment manual for detailed operation.

1. Open the coating chamber of the vacuum evaporator. Load the evaporation materials, platinum and carbon. Adjust the angles of coating.
2. Mount the samples onto the rotary stage of the vacuum evaporator using a double-sided tape. This will prevent the dislodging of the samples during rotation (*see* **Note 17**). To prevent damage of the samples by ambient humidity, perform mounting as quickly as possible (*see* **Note 18**).
3. Pump down the coating chamber till $\sim 5 \times 10^{-6}$ atm.
4. Turn on the stage rotation, shadow with platinum (2 nm) and then with carbon (3.5–5 nm) (*see* **Note 19**).
5. Vent the coating chamber. Remove the samples together with the mounting tape and place in a Petri dish. The samples can be safely stored in a room at this stage.

3.5 Preparation of Replicas

The release of the replicas from the coverslips is achieved by floating the coverslips onto the surface of hydrofluoric acid solution, which dissolves glass. After that, the replicas are washed and mounted on EM grids.

1. While the coverslips are still attached to the sticky tape, scratch the coated surface of each coverslip with a needle or a razor blade to make regions fitting the size of an EM grid.
2. Fill the wells of a 12-well plate with ~10 % HF almost to the top, which makes replica handling easier.
3. Detach a coverslip from the sticky tape and place it on the surface of the HF solution, so that the coverslip remains floating. Wait until the glass sinks, leaving the platinum replica floating. The replica will fall apart along the scratches made in **step 1** (*see* **Note 20**).
4. Fill the wells of another 12-well plate with distilled water, and add a trace amount of a detergent to decrease the surface tension of water. Stock solution of the detergent is prepared by dissolving ~1 drop of detergent in ~20 mL of water. To make a working solution, dip a platinum loop into the stock solution, take it out (this will create a film on the loop), and dip the loop into a water-filled well. The final concentration of detergent is ~10^{-6} % (*see* **Note 21**).
5. Using a platinum loop, transfer the replica pieces onto the surface of the detergent-containing water. Wait for 1 min or more.
6. Fill the wells of another 12-well plate with distilled water without detergent. Transfer the replica pieces onto the surface of the water. Wait for 1 min or more.
7. Pick up the replica pieces onto Formvar-coated EM grids with the lower side of the replica attached to the Formar film. The technique of replica mounting is similar to the way thin sections are picked up onto grids. Fasten a grid in a pair of forceps, partially submerge the grid into water at a ~45° angle, bring the grid close to a piece of replica and allow them to make contact; then, gently pull the grid out of the water making sure that the replica piece remains attached to, and spreads over the grid (*see* **Note 22**).
8. Examine the samples in TEM. Present the images in inverse contrast (as negatives) because it gives a more natural view of the structure, as if illuminated with scattered light.

3.6 Immunogold EM

Structural information has much greater value if the identity of the structures is known. Immunostaining is a conventional way to identify cellular components. For EM purposes, the antibodies are labeled with electron-dense markers. A popular marker, colloidal gold, has a higher electron density than platinum and thus is appropriate for platinum replica EM. For successful immunogold replica EM, a primary antibody should work after glutaraldehyde fixation, which optimally preserves the structure (*see* **Note 23**).

1. After glutaraldehyde fixation (Subheading 3.2, **step 1**), wash the samples with three changes of PBS. Incubate for at least 5 min in the last change.
2. Quench samples with $NaBH_4$ for 10 min at room temperature. Shake off the bubbles occasionally. Rinse with PBS as in **step 1**.
3. Incubate with blocking solution for 20 min at room temperature.
4. Apply primary antibody at concentration giving bright staining by light microscopy. Incubate for 30–45 min at room temperature. Wash with PBS as in **step 1**.
5. Rinse once in immunogold buffer with 0.1 % BSA. Apply gold-conjugated secondary antibody diluted ~1:10 in immunogold buffer with 1 % BSA. Incubate overnight at room temperature in a sealed container in moist conditions (*see* **Note 24**).
6. Rinse in immunogold buffer with 0.1 % BSA as in **step 1**, and perform the remaining steps starting from Subheading 3.2, **step 1**.

3.7 Correlative EM

The correlative light and EM combines the advantages of both the microscopic techniques, namely, the high spatial resolution of EM and the high temporal resolution of live imaging. In this procedure, the cell dynamics is recorded by light microscopy, and then the same cell is analyzed by EM. The correlative analysis is extremely important from at least two points of view: to control for potential artifacts and to establish functional connections between the cytoskeletal organization and the cell's motile behavior or the dynamics of cytoskeletal components [31, 32]. Modifications of the basic procedure as required for correlative EM are described below.

1. Cell Culture: Grow cells in dishes with marked coverslips mounted over a hole in the bottom of the dish. To make a dish, drill a 18-mm hole in the bottom of a 35-mm dish; polish the edges to remove any burrs; apply a thin layer of vacuum grease just outside the edges of the hole; place a marked coverslip symmetrically over the hole with the coated side facing up; press firmly to spread the grease until it forms a clear circle and all the air bubbles are gone (*see* **Note 25**).
2. Light microscopy: While imaging a region of interest, mark its position on a map with a pattern of reference marks. Because of the large difference in the resolution of light and EM, perform light microscopy at the highest possible resolution. EM is able to reveal even minor photodamage, not recognizable at

the light microscopic level; therefore, keep the illumination of the samples to a minimum.

3. Extraction: Change the culture medium to extraction solution as soon as possible after the acquisition of the last live images, and take another image after extraction. It will serve as a reference to correlate the light and EM images.
4. Fixation: Perform all the fixation and washing steps with the marked coverslips still attached to the dishes, by exchanging solutions.
5. Dehydration: Before loading the coverslips into the CPD sample holder, excise the central marked area of the coverslip as described in **steps 6–10**.
6. Wipe away the immersion oil from the bottom of the dish using dry cotton swabs first, followed by ethanol-soaked swabs.
7. Detach the marked coverslip from the bottom of the dish and immediately place it into a water-filled 100 mm Petri dish.
8. Lightly press down the coverslip to allow the vacuum grease on the underside to secure the coverslip in the dish. Make sure that grease does not contaminate the region of interest.
9. Use a diamond pencil as a cutter and a razor blade as a ruler to cut off the greasy margins of the coverslip. Do not press the pencil hard, but instead make several light cuts along the same line, guided by the razor blade, until the margin detaches. Move it out of the way and cut off another margin. This procedure should be done with the coverslip completely submerged in the water (*see* **Note 26**). When done, transfer the excised central region to another water-filled container. Use a new dish for the next coverslip, as the remaining glass crumbs may cause shattering of the coverslip.
10. All subsequent processing, including coating, is performed as in the basic procedure.
11. Preparation of replicas: After the samples are coated, the region of interest needs to be specifically recovered for EM analysis as described in **steps 12–16**.
12. Immobilize a coated coverslip in the middle of a 100 mm Petri dish, with two pieces double-sided tape positioned under opposite corners of the coverslip, so that the region of interest is not obstructed and the coverslip is not attached too strongly.
13. Under a dissection microscope, localize the cells of interest using locator marks. Dried and shadowed cells have sufficient contrast for their shape to be seen even at low magnification. If necessary, use a regular light microscope with a low power lens.
14. Using a razor blade (or a needle), make cuts in the platinum–carbon layer around the region of interest (*see* **Note 27**). To

facilitate the separation of this region from the rest of the replica after HF treatment, make additional cuts connecting the region of interest with the edges of the coverslip.

15. Make a drawing on the map to depict the exact shape of the outlined region of interest, which will help to identify it from among the other replica pieces during replica preparation.
16. Perform the washing and mounting on grids, as described in the basic protocol, handling only the replica piece with the region of interest. While mounting on a grid, use a dissection microscope to make sure that the cell of interest does not go to a grid bar; or use single-hole grids.

4 Notes

1. Small coverslips allow for better exchange of solutions during dehydration and CPD, and thus for better quality of samples at the end.
2. Stock solutions with a concentration of more than 2× change pH significantly after dilution to the working concentration. Free acid PIPES is not soluble in water and forms a milky suspension, but becomes soluble upon neutralization. KOH granules can be used for neutralization initially, until the solution almost clears. However, remember to allow enough time for the granules to dissolve before adding more. Finish the pH adjustment with 1 N KOH. KOH is preferable over NaOH, because K^+-containing buffer, more faithfully imitates the cytoplasm composition.
3. PEG is a nonspecific stabilizer of the cytoskeleton; phalloidin and taxol are specific stabilizers of actin filaments and microtubules, respectively.
4. Commercially available etched coverslips are not suitable for replica EM, as the marks are not visible in TEM.
5. Rinsing with PBS is optional, but if omitted, the extraction solution at the next step should be added in sufficient quantity to overcome the potentially harmful effects from the remaining medium and serum.
6. The choice of the extraction solution depends on a cell type and a goal. For a new experimental system, try different options in the preliminary experiments. Basic extraction solution (Triton X-100 in PEM) gives a better clarity of the cytoskeleton, but it is easier to damage the cells during extraction. If using this protocol, handle the samples extremely gently, and use phalloidin and taxol to better preserve the actin filaments and microtubules, respectively. The addition of PEG to the extraction solution provides for better preservation of the cells,

but it also retains many cytoskeleton-associated components, which may partially obscure the filament arrangement. Such an effect is increased with PEG concentration and molecular weight, but PEGs in the range of 20,000–40,000 act similarly. We typically use 2 % PEG (35,000). Phalloidin and taxol are not as necessary in this case. For extremely fragile and poorly attached cells, low concentrations of glutaraldehyde can be used as stabilizing supplements for the extraction solution. In this case, the detergent and fixative compete with each other, and the results depend on their ratio. The extraction solution containing 0.5 % Triton X-100 and 0.25 % glutaraldehyde in PEM buffer worked well in our experiments.

7. For PEG-containing extraction solutions, use a longer washing time, at least 1 min in each change. If drugs are used during extraction, add them also to the rinsing buffer in a fourfold to fivefold lower concentration.
8. It is convenient to use a multiwell plate with numbered wells (24-well for 6–8 mm coverslips or 12-well for 9–12 mm coverslips) to transfer the samples. This makes it possible to combine samples from different experiments for EM processing while keeping parallel samples in the original container as a backup.
9. Uranyl acetate and tannic acid react with each other and form a precipitate. Extensive washing is important to avoid the formation of debris on the samples.
10. Pieces of lens tissue slightly larger, than the holder's bottom area, will make minor wrinkles which promote looser packing of the coverslips in the holder and facilitate the liquid exchange.
11. The acceptable number of samples for a load depends on the sizes of the holder and the coverslips. For an 18 × 12 mm holder and ~7 × 7 mm coverslips, the maximum load is 12. For larger coverslips, the load should be decreased. Larger holders may accept more samples, especially if the coverslips are staggered.
12. It is not necessary to dry the beakers before the next incubation, as the ethanol concentration may not be exact, except for 100 % ethanol, when it is better to dry the beakers and scaffolds with tissue. Incubation for 5 min is minimal. For larger coverslips or greater loads, increase the incubation time.
13. The process of CPD is most commonly used for scanning EM and production of microelectronics. Consequently, the protocols suggested by the manufacturers or incorporated into automated procedures of CPDs are designed for those applications. Replica TEM, however, is more demanding in terms of sample quality. We adjusted the CPD processing to fully remove all traces of ethanol from the samples before bringing the CO_2 to the critical point; this helps to eliminate minor artifacts that

appear as a fusion of closely positioned filaments in the cytoskeleton. The CPD operation described here is applicable to manual CPDs, such as Samdri PVD-3D (Tousimis), which we use in the lab, or to semi-automatic CPDs switched to a manual mode of operation, e.g., Samdri-795 (Tousimis).

14. Lower temperatures are acceptable, but the diffusion of ethanol from the samples will be slower, so that longer washing time is needed. Warming up the chamber till the ambient temperature is allowed if the outlet valve of the CPD is closed, and the CO_2 remains pressurized and in liquid form. However, it is important to cool down the chamber back to 15 °C before opening the outlet valve for purging out the ethanol–CO_2 mixture.
15. Letting the liquid level go below the samples will irreversibly damage them. On the other hand, too low a rate of liquid exchange is also a mistake. Adjust the outlet valve to get a steady-state liquid level, about halfway from the top of the holder to the top of the chamber. This will also make the liquid mixing more efficient. Although the shaking step sounds a bit amusing, it does make a difference by helping to remove the ethanol from the samples.
16. Fast release of pressure may cause condensation of CO_2 back to liquid state and ruin the dried samples.
17. Conventional Scotch double-sided tape becomes too sticky in a vacuum, preventing the safe detachment of the samples after coating. To avoid this problem, sandwich the double-sided tape between the glued parts of two Post-It notes, with the sticky sides exposed.
18. Humidity in the room should be below 35 %; the 35–50 % humidity level may be acceptable, but much caution and the speedy mounting of the samples is required; humidity >50 % is not acceptable. Try to run a powerful dehumidifier in the latter case.
19. If the evaporator is not equipped with a thickness monitor, the thickness of the coating may be adjusted in the preliminary experiments based on the amount of coating material loaded (for platinum) or used (carbon) for evaporation.
20. To safely float a coverslip, grab it with the forceps from the top for parallel edges, lift the coverslip, and carefully place it onto the liquid surface, keeping it in a horizontal position. Practice first by placing a coverslip onto a clean solid surface. Alternatively, a coverslip can be placed first on a platinum loop bent at 90° angle and then loaded horizontally onto the liquid surface. If the replica does not fall apart along the scratches, use the platinum loop to reach the replica from below, lightly touch it and pull or shake it to detach it from other pieces.

Extreme care should be used not to ruin the replicas with these manipulations.

21. Water has much greater surface tension than HF, which may cause severe replica breakage, if detergent is not added. Test the detergent concentration before applying it to the samples. An overdose of detergent causes shrinkage and drowning of the replicas. Stock solutions should be changed at least every 2 weeks. Old detergents leave contamination on the samples, looking like semi-transparent films between filaments. Household non-colored detergent, such as Ivory, works fine. Triton X-100 can also be used, but it should be prepared fresh every time.
22. Sometimes, replicas appear to be repelled by the grid, making it difficult to establish the initial contact between a replica piece and a grid. Try to gently guide a piece of replica to the wall of the well to restrict its motility, and then pick it up. However, there is a danger of smashing the replica against the wall with this approach. In severe cases, use glow discharge to treat grids.
23. The efficiency of staining may be improved if the cells are fixed with a lower (e.g., 0.2 %) glutaraldehyde concentration before staining. For some antibodies that do not work after glutaraldehyde fixation, it may be possible to stain unfixed samples by incubating them with primary antibodies diluted in PEM for 10–15 min, then fixing with glutaraldehyde, and quenching and staining with a secondary antibody.
24. Gold size of 10–20 nm is optimal for this technique, as smaller particles are poorly visible, and larger particles are too disruptive for an image.
25. The coverslips can be mounted either inside or outside the dish, but inside mounting is more convenient at later stages, when the centerpiece of the coverslip needs to be cut out. For mounting, use a minimal amount of grease, just sufficient to seal the dish; excessive grease causes complications at later stages. Commercially available glass-bottom dishes have coverslips permanently glued to the bottom, which makes it difficult to remove them for EM processing.
26. Cutting under water is more difficult than in the air; therefore, use a sharp diamond pencil and avoid glass crumbs.
27. To reduce the effect of shaky hands, hold a razor blade with one hand with the sharp blade corner pointing down; stabilize the blade by putting the index finger of the other hand onto the blunt blade corner pointing up; rest the forearms on the table and the other fingers of both hands on the microscope stage and/or dish edges; keep the blade above the sample and find its unfocused image in the microscope; slowly bring down the sharp corner of the blade until it almost comes to focus;

bring the blade corner to a region where a cut is to be made; under microscope control, bring it down to the sample and make a scratch.

Acknowledgments

The author acknowledges the current support from NIH grant R01 GM 095977.

References

1. Wohlfarth-Bottermann KE (1962) Weitreichende fibrillare Protoplasmadifferenzierungen und ihre Bedeutung für die Protoplasmastromung. I. Elektronenmikroskopischer Nachweis und Feinstruktur. Protoplasma 54:514–539
2. Ledbetter MC, Porter KR (1963) A "microtubule" in plant cell fine structure. J Cell Biol 19(1):239–250
3. Sabatini DD, Bensch K, Barrnett RJ (1963) Cytochemistry and electron microscopy. The preservation of cellular ultrastructure and enzymatic activity by aldehyde fixation. J Cell Biol 17:19–58
4. Abercrombie M, Heaysman JE, Pegrum SM (1971) The locomotion of fibroblasts in culture. IV. Electron microscopy of the leading lamella. Exp Cell Res 67(2):359–367
5. Wohlfarth-Bottermann KE (1964) Differentiations of the ground cytoplasm and their significance for the generation of the motive force of amoeboid movement. In: Allen RD, Kamiya N (eds) Primitive motile systems in cell biology. Academic, New York, pp 79–109
6. Small JV, Isenberg G, Celis JE (1978) Polarity of actin at the leading edge of cultured cells. Nature 272(5654):638–639
7. Lucic V, Rigort A, Baumeister W (2013) Cryo-electron tomography: the challenge of doing structural biology in situ. J Cell Biol 202(3):407–419. doi:10.1083/jcb.201304193
8. Ben-Harush K, Maimon T, Patla I, Villa E, Medalia O (2010) Visualizing cellular processes at the molecular level by cryo-electron tomography. J Cell Sci 123(Pt 1):7–12
9. Heuser J (1981) Preparing biological samples for stereomicroscopy by the quick-freeze, deep-etch, rotary-replication technique. Methods Cell Biol 22:97–122
10. Steere RL (1957) Electron microscopy of structural detail in frozen biological specimens. J Cell Biol 3(1):45–60
11. Heuser J (1989) Protocol for 3-D visualization of molecules on mica via the quick-freeze, deep-etch technique. J Electron Microsc Tech 13(3):244–263
12. Loesser KE, Franzini-Armstrong C (1990) A simple method for freeze-drying of macromolecules and macromolecular complexes. J Struct Biol 103(1):48–56
13. Heuser JE, Kirschner MW (1980) Filament organization revealed in platinum replicas of freeze-dried cytoskeletons. J Cell Biol 86(1): 212–234
14. Svitkina TM, Shevelev AA, Bershadsky AD, Gelfand VI (1984) Cytoskeleton of mouse embryo fibroblasts. Electron microscopy of platinum replicas. Eur J Cell Biol 34(1):64–74
15. Svitkina TM, Verkhovsky AB, Borisy GG (1995) Improved procedures for electron microscopic visualization of the cytoskeleton of cultured cells. J Struct Biol 115(3):290–303
16. Hirokawa N (1989) Quick-freeze, deep-etch electron microscopy. J Electron Microsc (Tokyo) 38(Suppl):S123–S128
17. Meyer HW, Richter W (2001) Freeze-fracture studies on lipids and membranes. Micron 32(6):615–644
18. Svitkina T (2007) Electron microscopic analysis of the leading edge in migrating cells. Methods Cell Biol 79:295–319
19. Svitkina TM, Borisy GG (2006) Correlative light and electron microscopy studies of cytoskeletal dynamics. In: Celis J (ed) Cell biology: a laboratory handbook, vol 3, 3rd edn. Elsevier, Amsterdam, pp 277–285
20. Svitkina TM, Borisy GG (1998) Correlative light and electron microscopy of the cytoskeleton of cultured cells. Methods Enzymol 298:570–592
21. Shutova MS, Spessott WA, Giraudo CG, Svitkina T (2014) Endogenous species of mammalian nonmuscle myosin IIA and IIB include activated monomers and heteropolymers.

Curr Biol 24(17):1958–1968. doi:10.1016/j.cub.2014.07.070

22. Svitkina TM, Verkhovsky AB, Borisy GG (1996) Plectin sidearms mediate interaction of intermediate filaments with microtubules and other components of the cytoskeleton. J Cell Biol 135(4):991–1007
23. Kukulski W, Schorb M, Welsch S, Picco A, Kaksonen M, Briggs JA (2011) Correlated fluorescence and 3D electron microscopy with high sensitivity and spatial precision. J Cell Biol 192(1):111–119
24. Perkovic M, Kunz M, Endesfelder U, Bunse S, Wigge C, Yu Z, Hodirnau VV, Scheffer MP, Seybert A, Malkusch S, Schuman EM, Heilemann M, Frangakis AS (2014) Correlative light- and electron microscopy with chemical tags. J Struct Biol 186(2):205–213. doi:10.1016/j.jsb.2014.03.018
25. Watanabe S, Punge A, Hollopeter G, Willig KI, Hobson RJ, Davis MW, Hell SW, Jorgensen EM (2011) Protein localization in electron micrographs using fluorescence nanoscopy. Nat Methods 8(1):80–84
26. Fernandez-Busnadiego R, Schrod N, Kochovski Z, Asano S, Vanhecke D, Baumeister W, Lucic V (2011) Insights into the molecular organization of the neuron by cryo-electron tomography. J Electron Microsc (Tokyo) 60(Suppl 1):S137–S148. doi:10.1093/jmicro/dfr018
27. Kandela IK, Bleher R, Albrecht RM (2008) Immunolabeling for correlative light and electron microscopy on ultrathin cryosections. Microsc Microanal 14(2):159–165
28. Sartori A, Gatz R, Beck F, Rigort A, Baumeister W, Plitzko JM (2007) Correlative microscopy: bridging the gap between fluorescence light microscopy and cryo-electron tomography. J Struct Biol 160(2):135–145
29. Verkade P (2008) Moving EM: the rapid transfer system as a new tool for correlative light and electron microscopy and high throughput for high-pressure freezing. J Microsc 230(Pt 2):317–328
30. Nemethova M, Auinger S, Small JV (2008) Building the actin cytoskeleton: filopodia contribute to the construction of contractile bundles in the lamella. J Cell Biol 180(6):1233–1244
31. Svitkina TM, Bulanova EA, Chaga OY, Vignjevic DM, Kojima S, Vasiliev JM, Borisy GG (2003) Mechanism of filopodia initiation by reorganization of a dendritic network. J Cell Biol 160(3):409–421
32. Yang C, Czech L, Gerboth S, Kojima S, Scita G, Svitkina T (2007) Novel roles of formin mDia2 in lamellipodia and filopodia formation in motile cells. PLoS Biol 5(11):e317

Chapter 6

Purification and Localization of Intraflagellar Transport Particles and Polypeptides

Roger D. Sloboda

Abstract

The growth and maintenance of almost all cilia and flagella are dependent on the proper functioning of the process of intraflagellar transport (IFT). This includes the primary cilia of most human cells that are in the G_0 phase of the cell cycle. The model system for the study of IFT is the flagella of the biflagellate green alga *Chlamydomonas*. It is in this organism that IFT was first discovered, and genetic data from a *Chlamydomonas* mutant first linked the process of IFT to polycystic kidney disease in humans. The information provided in this chapter addresses procedures to purify IFT particles from flagella and localize these particles, and their associated motor proteins, in flagella using light and electron microscopic approaches.

Key words Flagella, Cilia, Intraflagellar transport, IFT, Motility, Immunofluorescence, Immunogold EM

1 Introduction

Intraflagellar transport (IFT) refers to the bidirectional movement of protein complexes from the base to the tip and back again in flagella and cilia.[1] The process was initially discovered by Kozminski et al. [2] in the flagella of the green alga *Chlamydomonas*, but IFT is now known to exist in almost all cilia and flagella that have been studied to date. IFT is required for the assembly and maintenance of flagella, and this was demonstrated using cells defective in one of the motor proteins that power the process of IFT [3]. IFT can be visualized using high resolution differential interference contrast microscopy [4]; this and related microscopy techniques are detailed elsewhere in this volume. Fluorescence microscopy was first used to visualize IFT in the sensory cilia of chemosensory

[1] The term "flagella" is used throughout this chapter, but readers should realize that the process of IFT occurs—and is just as important—in cilia as well. Indeed, many developmental problems and diseases of humans are related to defects in primary cilia, some of which derive from defects in IFT (*see* ref. [1] for a recent overview of this field).

Ray H. Gavin (ed.), *Cytoskeleton: Methods and Protocols*, Methods in Molecular Biology, vol. 1365,
DOI 10.1007/978-1-4939-3124-8_6, © Springer Science+Business Media New York 2016

neurons in *C. elegans* [5]; a *Chlamydomonas* cell line is also available in which the non-motor subunit (KAP) of kinesin-2, the anterograde IFT motor, is fused to GFP [6]. IFT in these cells can also be visualized with standard epifluorescence microscopy or via total internal reflection interference fluorescence microscopy (TIRF, [7]).

The flagella of *Chlamydomonas* represent the only microtubule-based motility system in which *all* of the following have been established: (a) the microtubule-based motors powering the movements of IFT particles in both directions along flagella are known [3, 8–10]; (b) individual strains carrying ts mutations in one or the other motors are available [3, 8, 11]; (c) IFT particles can be isolated, their constituent polypeptides have been cloned and sequenced, and mutations have been identified in a number of these polypeptides [12, 13]; (d) IFT particle motility occurs in clearly defined directions, i.e., out and back along the flagellum, essentially in only one dimension; IFT particle motility occurs between the outer doublet microtubules and the membrane [3, 14]; (e) motility occurs in a biochemically defined, easily isolated, membrane-bound compartment; (f) microtubule assembly and disassembly—indeed, assembly and disassembly of the entire organelle—occurs only at the flagellar tip [15]; (g) because IFT particles move from the base to the tip or from the tip to the base without pauses [2], the enzyme activity regulating the IFT motors must also be restricted to the flagellar tip; and (h) structures unique to the flagellar tip complex (FTC) have been observed [16–18], and specific protein components are beginning to be identified [19, 20]. Thus, it is clear that *Chlamydomonas* offers many advantages for the study of the mechanism of IFT, organelle assembly, control of motor enzyme activity, cargo unloading and loading, etc. One can readily observe IFT in *Chlamydomonas* flagella with video enhanced differential interference contrast (DIC) microscopy or fluorescence microscopy (the latter with GFP-tagged proteins), isolate IFT particles in quantities sufficient for biochemical characterization, and localize their constituent proteins by light and electron microscopic immunocytochemistry.

2 Materials

All solutions are prepared with deionized (18 MΩ) water.

2.1 Cells and Culture Conditions

1. *Chlamydomonas* cells are available from the stock center http://chlamycollection.org/. Unless a specific genotype or mutation is required for a specific analysis, the standard wild type strain is CC125 (wild type, mt+137c), which was originally isolated in 1945 near Amherst, MA, by G.M. Smith.

2. TAP medium: Stock solutions for TAP (Tris–Acetate–Phosphate, *see* Gorman and Levine [21]) are: 100× Tris–acetate: 242 g Tris base dissolved in 750 ml of water; pH to 7.0 with glacial acetic acid and bring to 1 l with water.

   TAP salts: 40 g NH_4Cl, 10 g $MgSO_4{\cdot}7H_2O$, 5 g $CaCl_2{\cdot}2H_2O$, water to 1 l.

   1 M Phosphate solution: 28.8 g K_2HPO_4, 14.4 g KH_2PO_4, water to 100 ml.

   A working solution of TAP medium is then made by combining 10 ml of 100× Tris–acetate, 10 ml TAP salts, 1 ml phosphate, 1 ml Hutner trace metals (*see* Subheading 2.1, **item 3** below) and 1 ml glacial acetic acid. Bring to 1 l with water.

3. Hutner's trace metals: This solution is based on the work of Hutner et al. [22] and the procedure below is adapted from Surzycki [23]. Preparation of this stock solution is complex and time consuming (*see* **Note 1**). Begin by dissolving the number of grams indicated in Table 1 for each trace metal salt in the volume of water listed. Combine these seven solutions and bring to a boil. Boil a separate volume of 250 ml of water and add 50 g of the disodium salt of EDTA to the boiling water. Then combine this with the seven other ingredients. When the solution is clear and greenish in tint, indicating all components have dissolved, cool to 70 °C. Adjust the pH to 6.7 *while the solution is at 70 °C* using hot 20 % KOH (*see* **Note 2**). Bring to the final volume of 1 l with water. Place the 1 l solution in a 2 l flask, stopper with a cotton plug to allow gas exchange, and let the solution rest at RT for 1 week or more. Swirl the solution briefly each day. Eventually the solution will turn purple and a brown precipitate will form (*see* **Note 3**). Filter with Whatman #1 paper and repeat the filtration, if necessary, until the solution is

Table 1
Components of Hutner trace metals stock solution

Trace metal salt	g	ml
$ZnSO_4{\cdot}7H_2O$	22.0	100
H_3BO3	11.4	200
$MnCl_2{\cdot}4H_2O$	5.1	50
$CoCl_2{\cdot}6H_2O$	1.6	50
$CuSO_4{\cdot}5H_2O$	1.6	50
$(NH_4)_6MO_7O_{24}{\cdot}4H_2O$	1.1	50
$FeSO_4{\cdot}7H_2O$	5.0	50

clear; two or three sheets of #1 filter paper can be used at the same time, if needed. Store the trace metals stock at 4 °C.

4. TAP culture plates/slants: sterile 1 % agarose in TAP, poured into petri dishes or slants in screw cap tubes.

2.2 Flagellar Isolation and Extraction

1. HMDS buffer [24]: 10 mM HEPES, pH 7.5, 5 mM $MgSO_4$, 1 mM DTT, 5 % sucrose.
2. Sucrose cushion: HMDS in which the sucrose concentration is increased to 20 %.
3. Extraction solution: HMDS containing 10 mM Mg-ATP and 0.1 % NP-40.
4. Pefabloc stock: 200 mM Pefabloc (Roche) in water (*see* **Note 4**).
5. HMDEK buffer: 10 mM HEPES, pH 7.5, 5 mM $MgSO_4$, 1 mM DTT, 0.5 mM EDTA, 25 mM KCl, 2 mM Pefabloc.
6. 100 mM Mg-ATP.
7. 10 % NP-40.
8. Sucrose gradient solutions: 10 % sucrose in HMDEK; 30 % sucrose in HMDEK.

2.3 Gel Electrophoresis

The gel solutions in this section are based on the buffer formulations of LaemmLi [25].

1. Acrylamide–bis (30:0.8). This can be made from powdered acrylamide and bis-acrylamide; however, acrylamide in the unpolymerized state is a neurotoxin and thus should be handled with appropriate precautions (fume hood and mask for weighing, protective gloves always). Alternatively, premade solutions can be purchased at reasonable cost from Sigma-Aldrich and other vendors.
2. 3 M Tris base, brought to pH 8.8 with HCl.
3. 0.5 M Tris base, brought to pH 6.8 with HCl.
4. 20 % SDS. The powder is a bronchial irritant, and so care (fume hood and mask) should be taken in weighing and handling this reagent. Very pure (expensive) SDS is not recommended, as the tubulin subunits (α- and β-) will not separate well in the presence of purified SDS [26, 27].
5. TEMED (*N,N,N′,N′*-tetramethylethylenediamine).
6. Ammonium persulfate, 10 % in water, made fresh and stored at RT for up to 1 week, or in small aliquots at −20 °C for up to several months.
7. 2× Sample buffer: 0.0625 M Tris–HCl, pH 6.8 (i.e., an eightfold dilution of the 0.5 M Tris stock from Subheading 2.3, **item 3** above, 4 % SDS (fivefold dilution from the 20 % stock of Subheading 2.3, **item 4** above), 20 % glycerol, 10 % β-mercaptoethanol, and 10 μg/ Pyronin-Y, the latter to serve

as a tracking dye. In cases where sample protein concentrations are dilute, it is often preferable to use a more concentrated sample buffer. To prepare, for example, 10 ml of 5× sample buffer: In a fume hood mix 1.25 ml of 0.5 M Tris–HCl, pH 6.8, 2.5 ml β-mercaptoethanol, 1 g SDS, 5 ml glycerol, and 10 μg/ml Pyronin-Y. Stir slowly to dissolve the SDS, and bring to 10 ml with water.

8. 10× Stock electrode buffer: 0.25 M Tris and 1.92 M glycine, weighed very accurately, and combined. These amounts will generate the appropriate pH of 8.3. For a working solution, the stock is diluted tenfold and adjusted to a final concentration of 0.1 % SDS using 5 ml of the 20 % SDS stock (Subheading 2.3, **item 4** above) per liter of solution.
9. Staining and destaining solutions [28]: Fix and stain the gel in 25 % isopropanol, 10 % acetic acid, and 0.05 % Coomassie blue; destain in 10 % isopropanol, 10 % acetic acid [28].

2.4 Immunoblotting

1. Electrode buffer: tenfold dilution of the 10× electrode buffer stock noted in Subheading 2.3, **item 8**, with no further additions (*see* **Note 5**).
2. Nitrocellulose (NC) membrane (nylon supported NC is easier to handle than plain NC).
3. Sheets of Whatman 3 MM filter paper, cut to the size of the blotting apparatus/gel.
4. Tris buffered saline (TBS), 10× stock: 250 mM Tris base, 1.5 M NaCl, adjusted to pH 7.4 with HCl.
5. TBS-T: TBS adjusted to 0.05 % Tween 20 (500 μl Tween 20 per liter of TBS).
6. Blocking buffer: TBS plus 5 % nonfat dry milk and 0.02 % NaN_3.
7. Primary and secondary antibodies suitably diluted in blocking buffer. Secondary antibodies conjugated with peroxidase (*see* **Note 6**). For this purpose we use horseradish peroxidase (HRP)-conjugated goat anti-rabbit or anti-mouse antibodies from Thermo Fisher Scientific.
8. ECL reagents from GE Healthcare; X-ray film.

2.5 Immunofluorescence Microscopy

1. For processing several antibodies on a single slide, we use precleaned, PTFE printed ten well slides (Electron Microscopy Sciences catalog number 63424-06). These have wettable surfaces and minimize the use of expensive antibodies.
2. 0.1 % polyethyleneimine (PEI) in water.
3. 10 mM HEPES, pH 7.4.
4. 10× PBS: To make 1 l of 10× phosphate buffered saline (PBS), dissolve 80 g NaCl, 2 g KCl, 14.4 g $Na_2HPO_4{\cdot}2H_2O$, and

2.4 g KH_2PO_4 in 800 ml water and bring to 1 l final volume (*see* **Note 7**).

5. Blocking buffer: 1× PBS containing 5 % nonfat dry milk, 1 mg/ml BSA, and 10 % goat serum.
6. 100 % methanol, prechilled to −20 °C.

2.6 Scanning Electron Microscopy

1. HMDEK, described above in Subheading 2.2, **item 5**.
2. Round glass coverslips (#1 thickness, 12 mm diameter, e.g., Bellco Glass catalog number 1943-10012).
3. Secondary antibodies labeled with colloidal gold (e.g., 12 nm gold from Jackson ImmunoResearch or 25 nm gold from Electron Microscopy Sciences).
4. Electron microscopy grade glutaraldehyde, purchased as an 8 % stock in ampoules sealed under nitrogen, diluted to 1 % with HMDEK just before use.

3 Methods

3.1 Growth and Maintenance of Cell Cultures

1. The main cell line stocks are maintained on agar slants (1 % agarose in TAP) in screw cap tubes in an incubator at 18 °C, provided with 14 h of light and ten of dark. Working plates are made from the stocks in petri dishes containing 1 % agarose in TAP and are maintained in a walk in incubator at 23 °C. The incubator is also rigged to provide a light–dark cycle of 14:10. The fluorescent lamps for illumination are fixed to the underside of the shelf above (Fig. 1a). The working plates are made fresh every few weeks, while the stocks on slants are made fresh every 4–6 weeks (*see* **Note 8**).
2. Cells are grown in liquid culture for a given experiment. To do this, starter flasks of 150 ml sterile TAP in 250 ml flasks are prepared. The flasks are closed with a foam plug through which is inserted a cotton plugged 5.75 in. Pasteur pipette (e.g., Fisher catalog # 13-678-8A), and the entire assembly (pipette, foam plug, and neck of the flask) is wrapped in aluminum foil and autoclaved. Using sterile technique a loop of cells from a single colony growing on a TAP plate is suspended in the TAP of a starter flask, and then the flask is placed in the culture room and aerated with 0.5 μm filtered air (Fig. 1b). The cells will begin growing and are ready for use in a few days, at late log phase, when they are at a density of 10^6–10^7 cells/ml. Before proceeding further, it is a good idea to check an aliquot of cells to ensure they are flagellated (the flagella—two per cell—should be 10–12 μm in length) and that the culture is free of microbial contamination.

Fig. 1 Growing cultures of *Chlamydomonas*. (**a**) Cells growing on 1 % agarose in TAP medium in large and small petri dishes and in liquid TAP in 24 well microtiter plates (the latter used for analyzing RNAi strains). (**b**) Cells growing in starter flasks (125 ml). (**c**) Cells growing in two large carboys (8 l each) and a 3 l flask. The samples are illuminated from above in panel (**a**) and from the side in panels (**b**) and (**c**)

3. For immunofluorescence a single starter flask of cells is sufficient. However, for biochemical experiments more flagella, and hence a greater volume of cells, are often required. In such cases, the starter flask is decanted into either (a) a 3-l flask containing 1.5 l of sterile TAP and a 9″ cotton-plugged Pasteur pipette, or (b) a 10-l Nalgene jug (e.g., United States Plastic Corporation, item # Item No.: 72021) containing 8 l of sterile TAP (Fig. 1c). For the latter, aeration is provided by a sintered glass [29] dispersion tube (Ace Glass cat. # 7202-18). The fluorescent lights for illumination of the liquid cultures (starter flasks and larger volumes) are fixed to the wall of the incubator (Fig. 1b, c).

3.2 Flagellar Isolation and Fractionation

1. Collect cells from the culture vessel by centrifugation at 1500 × *g* for 5 min and resuspend, using a large bore measuring pipet, in 1/5 the original volume of HMDS; aerate the cells in the incubator or on the bench top in a water bath at 23 °C for an hour, with illumination provided by a low watt halogen desk

lamp. Make one last check of the cells by phase microscopy to see that they are healthy: two full length flagella, actively swimming, and no contamination with microorganisms. Deflagellate the cells by pH shock [30]. To do this, stir the cells while monitoring pH, and rapidly drop the pH to 4.8 using 0.5 N acetic acid. Hold the cells at pH 4.8 for 60 s, and then rapidly return the pH to 7.5 using 0.5 N KOH. Be careful not to overshoot the pH in either direction. Check that the cells are deflagellated using a phase microscope.

2. Separate the cell bodies from the released flagella by centrifugation at 1500 × *g* for 5 min, and remove the supernatant being careful to minimize contamination of it by the cell bodies in the relatively soft pellet. Centrifuge the supernatant at 12,000 × *g* for 10 min at 4 °C, decant the supernatant, and resuspend the flagellar pellet in HMDS. Underlay the solution of flagella with HMDS in which the sucrose concentration is increased to 20 %, and centrifuge in a swinging bucket rotor (Sorvall HB-4 or equivalent) at 1500 × *g* for 5 min at 4 °C. This will sediment any contaminating cell bodies into the sucrose cushion. Remove the supernatant containing the flagella, and collect the flagella again as above. Resuspend in HMDS. Check the flagella for cell body and/or debris contamination by phase or differential interference contrast microscopy and repeat the sucrose cushion centrifugation as needed to remove contaminating cell bodies and cell fragments. Two such steps are usually sufficient for small quantities of flagella (i.e., from 1 to 2 starter flasks), but more will be needed for larger volumes (i.e., from a 10 l jug). Adjust the final flagellar sample to 2 mM Pefabloc to block protease activity during subsequent steps.

3. For immunofluorescence of intact, isolated flagella, proceed from this point to the immunofluorescence Subheading 3.5 below. To continue with fractionation of the flagella into membrane-matrix (MM) and axoneme (Axo) fractions, and to purify IFT particle polypeptides, collect the isolated flagella as above and resuspend in a small volume (~1 ml) of HMDEK. Adjust the sample to 10 mM Mg-ATP and 0.1 % NP40, and incubate at RT for 20–30 min with occasional gentle agitation. Centrifuge the sample at 12,000 × *g* for 15 min. The supernatant is the MM fraction and contains solubilized membrane proteins, soluble components of the flagellar matrix, IFT particles, and IFT motor proteins. The pellet contains the axoneme (9 + 2 structure) comprised of outer doublet microtubules, dynein arms and associated regulatory proteins, radial spokes, and the central pair apparatus.

4. The paradigm for IFT particle isolation is the procedure developed by Cole et al. [12] and on which the following method is based. To purify IFT particle polypeptides, prepare a

10–30 % sucrose gradient in HMDEK. We use tubes that fit a Beckman-Coulter SW41Ti rotor for this purpose. The centrifuge tube has a volume of 12 ml; we prepare an 11 ml gradient and load ~1 ml of the MM fraction, described in item #3 above, isolated from 16 to 32 l of cells. Centrifuge the gradient at 150,000 × g_{ave} for 18 h at 4 °C. We then use an Isco Model 185 density gradient fractionator (Teledyne Isco) to fractionate the gradient into 0.6–0.7 ml fractions. The IFT particle polypeptides will be found toward the bottom of the gradient, sedimenting at about 16S (*see* **Note 9**). Two complexes of IFT particles have been identified [12] when the gradient is run at low ionic strength (10 mM HEPES, pH 7.4). These are IFT Complex A composed of five polypeptides sedimenting at 16.1–16.4S and IFT Complex B, sedimenting at 15–16S and composed of ~11–12 polypeptides [13]. As the ionic strength increases, Complex A and B will fractionate more closely together on the gradient [12]. Note that flagella with an elevated level of IFT particle polypeptides can be obtained from *fla2* cells, which, as reported initially by Cole et al. [12], contain 2–3× as much Complex A and B as wild type flagella.

3.3 SDS Polyacrylamide Gel Electrophoresis (SDS-PAGE)

1. These instructions are designed for pouring and running two mini-gels in a Bio-Rad Mini Protean 3 gel apparatus. Component ratios can be easily adapted for gels of slightly different dimensions. To begin, scrub the glass plates (two per gel) with a good laboratory detergent, and then rinse well with tap water, distilled water, and finally 70 % ethanol. Our gel plates have affixed, 1.5 mm spacers; mount the plates in the pouring stand of the apparatus.
2. Prepare the separating gel (8 %) by combining the following components in the order listed: 2.2 ml acrylamide stock, 1.0 ml 3 M Tris–HCl, pH 8.8, 5.0 ml water, 5 μl TEMED, and 50 μl 10% ammonium persulfate (*see* **Note 10**). Swirl to mix after each addition, and immediately pour the solution between the gel plates to a height roughly 1 cm below the top surface of the plates. The separating gel solution must fill the plates to a level that leaves room to cast a stacking gel of adequate length. For example, when the comb is inserted to make the wells in the stacking gel (*see* Subheading 3.3, **step 4**) the height of the stacking gel (i.e., the space between the bottom of the sample wells and the top of the separating gel) is at least equal to the depth of the sample well. This is required to provide the samples enough time in the stacking gel so that they stack effectively before entering the separating gel. Carefully overlay the separating gel solution with 0.1 % SDS (*see* **Note 11**). The interface between gel and overlay solution will gradually "disappear" and then reappear as a sharply defined interface when polymerization is complete (~45 min at RT).

3. Prepare the stacking gel solution as follows: 0.5 ml acrylamide stock, 1.25 ml 0.5 M Tris–HCl, pH 6.8, 3.2 ml water, 5 μl TEMED, and 25 μl 10 % ammonium persulfate. When the separating gel has polymerized, pour off the overlay solution, and wick out the remainder by touching the edge of the gel plate sandwich with a paper towel. Rinse the top of the gel with a small amount of stacking gel solution, pour and wick this off, and then fill the space with fresh stacking gel solution. Insert a comb with the desired number of wells (ten or 15 for the Bio-Rad apparatus) into the stacking gel solution, and allow it to polymerize (~45 min at RT).
4. Meanwhile, prepare the samples as follows: Dilute each sample 1:1 if using 2× reducing solution, and 1:4 if using 5× reducing solution. The latter is preferable, especially for sucrose gradient samples, as these tend to be dilute. Mix well, and immediately boil the sample for 3 min in a boiling water bath (*see* **Note 12**).
5. To run the gel prepare an adequate volume of 1× electrode buffer containing 0.1 % SDS. Gently remove the combs from the gels, fill the spaces with electrode buffer, and shake out the buffer. Mount the gels in the apparatus, and add electrode buffer to the bottom and top compartments. Load the samples into the wells (*see* **Note 13**). Run the gels at a constant current; this is important to produce an effective and reproducible stack, and hence migration, using the LaemmLi buffer system (*see* **Note 14**). We run two gels at 20 mA for about 2.5–3 h or until the tracking dye is near the bottom of the gel. The ion front runs behind (i.e., more slowly) than the Pyronin Y tracking dye (not true of other dyes, such as bromophenol blue); the ion front will appear as a faint refractile line. Thus, the gel can be run until the dye has just run off the bottom of the gel to maximize the separation distance of the polypeptides in the sample. However, if the gel is to be immunoblotted, note that the Pyronin Y, if left in the gel, will transfer to the blot and mark the individual lanes of the gel, which can be helpful when analyzing the results of an immunoblot experiment (*see* Subheading 3.4).
6. If the gel is to be stained, remove the gel plates from the apparatus and carefully pry them apart with the plastic instrument provided. The gel will stick to one glass plate or the other. It can be carefully picked up by the bottom corners to remove it from the plate. Alternatively, place the plate with the gel in a clean container (we use plastic storage boxes from K-Mart) and cover with stain. Slight rocking will dislodge the gel from the plate, and the latter can be removed. Stain the gel overnight with gentle rocking at RT, and in the morning replace the gel stain with destain solution (several changes). The destained gel is then photographed for documentation; it can also be stored almost indefinitely in destain or distilled water.

3.4 Immunoblotting and Immunodetection

1. The Bio-Rad Mini Protean 3 apparatus used above for SDS-PAGE can also be rigged for immunoblotting, and this is the apparatus described here. Near the end of the gel run, set up the blotting sandwich. Use a clean Pyrex baking dish in which the plastic sandwich can be opened and lay flat. Prepare 1 l of 1× electrode buffer (*no* SDS). Cut a piece of nitrocellulose (NC) paper slightly larger than the gel area to be blotted. Wet this in water with shaking for 10 min, and then in electrode buffer for 10 min. Place one of the ScotchBrite pads (provided with the apparatus) on one side of the plastic sandwich, followed by three sheets of Whatman 3 MM paper. Pour enough electrode buffer into the dish to cover the pads and paper.
2. Remove the gel and separate the plates as described above. The easiest way to place the gel against the nitrocellulose is as follows: As noted above (*see* Subheading 3.3, **step 6**) the gel will adhere to one or the other of the glass plates. Lay this down on the bench, gel side up. Using a razor blade, carefully scrape away the stacking gel. Then lay one of the wetted pieces of 3 MM paper on the gel, turn the plate upside down, and carefully, using the tip of a weighing spatula, tease a corner of the gel away from plate. Gravity will eventually take hold and the gel with the 3 MM paper adhering to it will peel off onto your gloved hand. Place the 3 MM paper, gel side up, on top of the other two pieces of 3 MM paper on the plastic sandwich lying in the Pyrex dish. Cover the gel with the wetted NC paper, and use a gloved finger or a 5.75″ Pasteur pipette to roll out any bubbles that are between the NC and the gel. It is essential that there be no bubbles between the NC and the gel otherwise the transfer will not occur in the region contaminated by bubbles. Cover the NC with three pieces of 3 MM paper previously soaked in electrode buffer, followed by the second piece of ScotchBrite pad. Close the gel sandwich and, if one is provided, slide the plastic lock closed to hold it together. Fill the reservoir of the apparatus with electrode buffer, and insert the sandwich, being careful to place it in the apparatus with the NC side facing the positive electrode (red, anode).
3. It is important to mark the NC relative to the gel so that the position of the lanes, and hence the samples, can be determined. One way to do this is to cut off a small piece of the bottom corner of the gel where lane #1 is, and then make a corresponding cut in the corner of the NC. If pre-stained gel standards are used and always loaded on the gel asymmetrically (i.e., to one side or another), their position on the NC will serve as a reference point for the identification of the relative positions of the other sample lanes of the gel.
4. Run the blot either in the cold room or, if available, insert into the buffer chamber a block of ice cast in the plastic holder that

comes with the apparatus. Run the blot at 340 mA (or at 12 V/cm, measuring the distance between the electrodes of the apparatus) for 60 min. After 60 min, disassemble the sandwich, peel off the NC, and place it, protein side up, in blocking buffer and rock for an hour at RT or overnight at 4 °C.

5. Prepare a suitable dilution of primary antibody in blocking buffer. Remove the blocking buffer from the gel and add the primary antibody. Incubate with rocking at RT, 1–4 h, depending on the relative affinity of the antibody being used. Remove the antibody solution and save it at 4 °C. Depending on the initial dilution it can be used many times over the next few weeks.
6. Wash the blot by immersion in TBS-T, with rapid agitation on a shaker table, three times, 5 min each. Discard the third wash and add secondary antibody diluted in blocking buffer (usually 1:15,000 or 1:20,000, depending on the vendor). Incubate with rocking 30–60 min.
7. Wash the blot by immersion in TBS-T, with rapid agitation, four times, 10–15 min each. Discard the last wash and drain the adhering wash buffer off of the blot. Add ECL reagent according to the manufacturer's instructions; the surface area of the blot being treated determines the amount of ECL reagent required. The volume used will be relatively small, so distribute the ECL reagent by hand over the blot using a Pasteur pipet every 5–10 sec for 5 min to ensure that the blot is equally covered and saturated with ECL solution.
8. Remove the blot and press it by hand between two paper towels to remove excess ECL reagent. Mount the blot either between the leaves of an acetate sheet protector, or wrap it in one layer of clear plastic food wrap, avoiding wrinkles in the plastic wrap on the side of the NC to which the proteins were blotted and to which the film will be exposed.
9. Working in the darkroom with 8 × 10 X-Ray film and an X-ray film cassette to hold the blot and film, expose different regions of a single sheet of the film to the blot for varying times (e.g., 5 s, 10 s 30 s, 1 min). This allows one film to contain multiple exposures and decreases film and processing costs. Based on these initial test exposures, re-expose the blot a second time, if necessary, to generate a blot of adequate exposure and background.
10. Finally, be sure to place the exposed film over the blot and, using a Sharpie pen, mark the positions of the corners of the gel, the standards, gel lanes, etc.

3.5 Immunofluorescence

1. The following procedure (based on the work of Pedersen et al. [31]) can be used with whole cells, isolated flagella, or axonemes. Prepare a ten well slide fresh for each experiment as

follows. We use Teflon printed 10 well (6 mm diameter) slides, cat. no. 63424-06 (Electron Microscopy Sciences). Fill each well with 50 μl of 0.1 % polyethyleneimine (PEI) and allow to sit for 5 min at RT. Using an aspirator or other suitable device (e.g., a Pasteur pipette) withdraw the PEI solution, and wash the wells three times with water by alternately adding water to the wells and removing it with the aspirator. Air-dry the slide in a covered petri dish to prevent dust etc. from settling on the slide. The PEI produces a surface to which the cells adhere tightly. Alternatively, 1 mM poly-L-lysine can also be used.

2. Place the sample of cells, flagella, or axonemes in a microfuge tube and sediment at top speed for 1 min. Using an autopipettor whose plastic tip has been cut off with a razor to widen the bore, gently resuspend the sample in 100 μl 10 mM HEPES, pH 7.4. Add 1.1 ml of 100 % methanol prechilled to −20 °C, and incubate the sample at −20 °C for 20 min.
3. Collect the sample as above, and gently resuspend in PBS to a suitable dilution. Place the resuspended sample(s) in the wells of the PEI coated slide and allow to adhere for ten min at RT.
4. Without removing the samples, immerse the slide in a Coplin jar filled with methanol at −20 °C. Incubate at −20 °C for 10 min.
5. Remove the slide from the methanol and air-dry. Rehydrate the wells with PBS, 10 min, and repeat once with fresh PBS.
6. Block the wells with immunofluorescence blocking buffer for 1 h at RT.
7. Add 20–40 μl of primary antibody at an appropriate dilution, in blocking buffer, to each sample well and incubate the slide overnight at 4 °C in a covered petri dish. Line the inside of the dish with moist filter paper or paper towel to maintain a humid environment to prevent the sample from drying out.
8. Wash the wells 3× with PBS, allowing a few minutes between each change of PBS.
9. Add second antibody diluted in blocking buffer, and incubate 1–2 h at RT.
10. Wash the wells 3× with PBS, allowing a few minutes between each change. Remove the last wash solution and add a drop of an antifade reagent to each well to stabilize the fluorescent signal to photobleaching during observation. We use Prolong Gold (Invitrogen/Molecular Probes, cat. no. P36935) as an antifade reagent, supplied premixed in 2 ml aliquots in dropper bottles. Cover the slide with a 24 × 50 coverslip.
11. Allow the slide to sit in the dark overnight, and then seal the edges of the coverslip to prevent drying prior to observation. We seal the edges with clear nail polish. For observation, a high

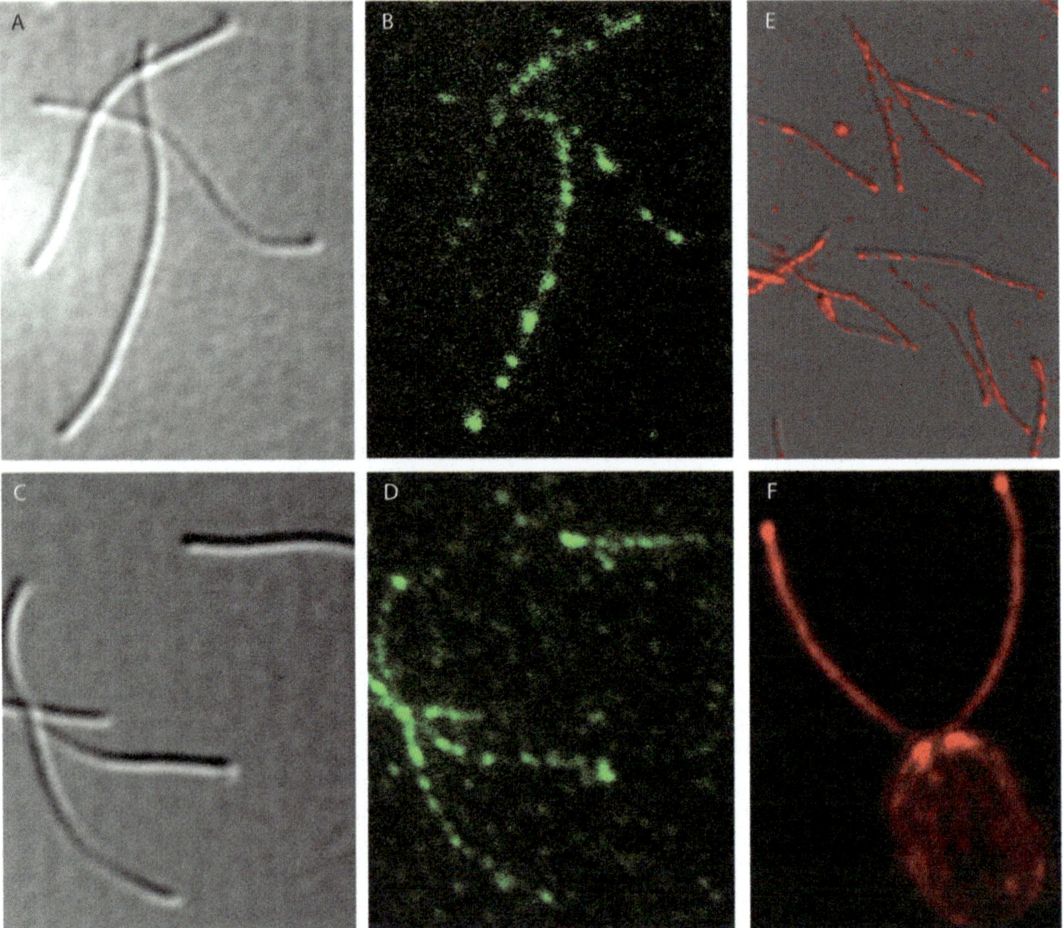

Fig. 2 Comparison of the distribution of IFT particle polypeptides and tubulin in *Chlamydomonas* flagella. (**a**) Differential interference contrast (DIC), and (**b**) fluorescent image of isolated flagella labeled with antibodies to IFT 139, an IFT complex A polypeptide. (**c**, **d**) Flagella labeled with IFT172, a complex B polypeptide. (**e**) Overlay of DIC and fluorescent images of isolated flagella labeled with antibodies to IFT88, an IFT complex B polypeptide (the magnification in **e** is less than that of **a**–**d**). The staining along the flagella in these images is punctate, indicating the particles were dispersed along the length of the flagella in discrete units (IFT particles, also called IFT trains) at the time of fixation. (**f**) *Chlamydomonas* cell stained with antibodies to α-tubulin, which label the flagella continuously along their entire length. This image is a maximum projection of a deconvolved *Z*-stack of 35 individual images

numerical aperture low magnification objective is required. This is because the fluorescence intensity projected to the image plane of the detector varies directly with numerical aperture and inversely with total magnification. Suitable objectives are 63×/1.4 NA (Zeiss, Leica) or 60×/1.4NA (Nikon, Olympus). IFT particles will appear punctate when detected via immunofluorescence in this manner (*see* Fig. 2a, b), while tubulin antibodies will stain the flagellar along their entire length (Fig. 2c).

3.6 Immunogold Scanning Electron Microscopy

1. These procedures describe immunolocalization performed on isolated axonemes at the limit of resolution afforded by the scanning electron microscope (SEM) [24]. The instrument we use is an FEI XL-30 field emission gun scanning electron microscope. It is operated at 15 kV with the spot size set at three; this generates a scan probe having a beam diameter of 1.7 nm [24]. For biological specimens of the type described here, the resolution is ~5 nm.
2. Collect the flagellar axonemes at 12,000 × *g* for 15 min and resuspend in a small volume of HMDEK. Divide into 50–100 μl aliquots in microfuge tubes, one aliquot for each primary antibody being assayed.
3. Add primary antibody to the aliquots of sample to achieve an appropriate final dilution of antibody and incubate at RT for 2 h.
4. Collect the axonemes in a swinging bucket rotor (e.g., Sorvall HB-4) at 12,000 × *g* for 15 min, resuspend in HMDEK. Repeat this step once to remove any antibodies nonspecifically trapped in the pellet.
5. Add diluted secondary antibodies labeled with colloidal gold (10 and/or 25 nm) and incubate at RT for 2 h.
6. Collect the axonemes in a swinging bucket rotor (e.g., Sorvall HB-4) at 12,000 × *g* for 15 min, resuspend in HMDEK, and repeat twice for a total of three washes.
7. Place the samples on round, 12 mm diameter glass coverslips previously coated with 0.1 % PEI as described above in Subheading 3.5, **step 1**. Allow the samples to adhere for 10 min at RT.
8. Fix the samples by immersing the coverslips in 1 % glutaraldehyde in HMDEK in a small petri dish. Incubate at 4 °C overnight.
9. Critical point dry the samples (we use a Samdri 795 for this purpose) and coat them with 2–3 nm of osmium using an SPI plasma coater. During the coating process the osmium atoms form a plasma resulting in a metal coat that is almost perfectly amorphous. The grain size of the osmium coating is about the same size as the osmium atoms (the calculated atomic radius is 0.185 nm), and thus is well below the 5 nm resolution limit noted above. The result is a stabilizing coat of metal that does not preclude detection of the gold label on the antibodies.
10. For clear detection of the gold particles, and hence the distribution of the antigen(s) in the sample, operate the SEM to detect backscattered electrons (Fig. 3a). For a conventional three dimensional view of the sample, operate the scanning electron microscope to detect secondary electrons (i.e., conventional SEM, Fig. 3b).

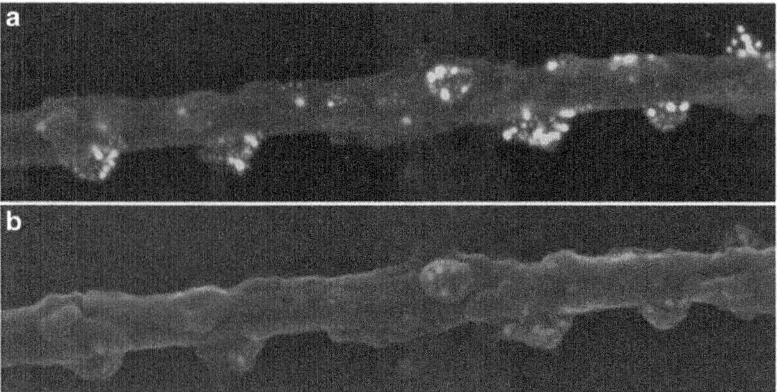

Fig. 3 Localization of IFT particle polypeptides by double label immunogold scanning electron microscopy. Twelve nanometer gold identifies the IFT complex A protein IFT139, and 25 nm gold identifies the motor enzyme kinesin-2, which powers anterograde (i.e., base to tip) IFT. (**a**) Image generated by backscattered electrons clearly showing the two sizes of gold particles (*white*, 12 and 25 nm average diameter) on a relatively *dark* background. The gold particles are localized to the globular structures (IFT particles) while the underlying microtubule-containing axoneme is free of label. (**b**) Conventional scanning electron microscope image (generated by secondary electrons) of the same field shown in panel (**a**). These images are reproduced, with permission, from Fig. 4 in ref. [24]

4 Notes

1. Readers should estimate how much TAP they might expect to use over the course of a few months, and prepare the trace metal solution accordingly (1 l of stock trace metal solution will yield 1000 l of TAP). One web site (which is no longer active http://149.152.32.229/~mikeadams/growthmedium.html) suggested simply borrowing some trace elements solution from a colleague!
2. Do *not* use NaOH to adjust the pH of the trace metals stock solution, and be sure to standardize the pH meter at 70 °C as well.
3. It has been reported that the time it takes for the solution to turn purple can be shortened by bubbling with filtered air, but I have never done this.
4. Pefabloc is 4-(2-Aminoethyl) benzenesulfonyl fluoride hydrochloride.
5. We use neither SDS nor methanol in the electrode buffer for immunoblotting with no negative consequences. A positive consequence of the lack of methanol is that special disposal procedures and the associated costs are avoided.

6. NaN_3 (sodium azide) is used to inhibit bacterial growth in the protein laden blocking buffer. NaN_3 may be present in many commercial antibody preparations for the same reason. Note that NaN_3 will also inhibit the action of peroxidase, and this enzyme is often linked to the secondary antibody whose activity is required for detection by ECL. Normally, the TBS-T washes of the blot prior to ECL detection remove the residual NaN_3 from the blocking buffer and antibody solutions. Thus, the TBS and TBS-T solutions should never contain NaN_3.
7. The pH of 10× PBS will be ~6.8–6.9, but will adjust to the working pH of 7.4 upon tenfold dilution.
8. The reason to keep the working cultures of cells separate from the cell stocks on slants is that only lab folks *highly* experienced with sterile technique are allowed to handle the slant cultures, thus protecting them from contamination by less experienced lab personnel.
9. In a gradient producing 17 fractions (~0.6–0.7 ml each) IFT particles will be found in fractions 14–16 (*see* [19]).
10. SDS is not required in either the separating gel or stacking gel solutions [27], but SDS is required in the electrode buffer (Subheading 3.3, **step 5**) for electrophoresis (but not in the buffer used for immunoblotting, *see* **Note 5**). Adding SDS to the gel solutions merely increases the chances that bubbles will form when the solutions are poured between the glass plates.
11. Other procedures recommend overlaying the separating gel with water-saturated isobutanol. Isobutanol, however, etches the plastic components of many gel apparatuses. We have found 0.1 % SDS works just as well and easily and does not react with the plastic of the gel apparatus.
12. Never leave a gel sample on the bench top without boiling. SDS denatures most proteins to the random coil configuration exposing the peptide bonds to contaminating protease activity (some proteases are active in SDS concentrations even higher than those in the sample buffer). Boiling, however, in the presence of SDS and reducing agent denatures all protein in the sample.
13. Depending on your preference, samples can also be loaded into the wells before the electrode buffer is added to the upper chamber. In either case, the glycerol in the sample buffer increases sample density sufficiently to prevent the mixing of the sample and the electrode buffer during either sample application approach.
14. The opposite is true of those pre-cast gels that do not contain a separate stacking gel and thus do not have a discontinuous pH. Run these gels at constant voltage according to the manufacturer's instructions.

Acknowledgements

Work in the author's lab is supported by the NSF (MCB 0950402). This support is greatly appreciated.

References

1. Brown JM, Witman GB (2014) Cilia and diseases. Bioscience 64:1126–1137
2. Kozminski KG, Johnson KA, Forscher P, Rosenbaum JL (1993) A motility in the eukaryotic flagellum unrelated to flagellar beating. Proc Natl Acad Sci U S A 90:5519–5523
3. Kozminski KG, Beech PL, Rosenbaum JL (1995) The Chlamydomonas kinesin-like protein FLA10 is involved in motility associated with flagellar membrane. J Cell Biol 131: 1517–1527
4. Kozminski KG (1995) High-resolution imaging of flagella. Methods Cell Biol 47:263–271
5. Orozco JT, Wedaman KP, Signor D, Brown H, Rose L, Scholey JM (1999) Movement of motor and cargo along cilia. Nature 398:674
6. Mueller J, Perrone CA, Bower R, Cole DG, Porter ME (2005) The FLA3 KAP subunit is required for localization of kinesin-2 to the site of flagellar assembly and processive anterograde intraflagellar transport. Mol Biol Cell 16:1341–1354
7. Engel BD, Ludington WB, Marshall WF (2009) Intraflagellar transport particle size scales inversely with flagellar length: revisiting the balance-point length control model. J Cell Biol 187:81–89
8. Pazour GJ, Dickert BL, Witman GB (1999) The DHC1b (DHC2) isoform of cytoplasmic dynein is required for flagellar assembly. J Cell Biol 144:473–481
9. Porter ME, Bower R, Knott JA, Byrd P, Dentler W (1999) Cytoplasmic dynein heavy chain 1b is required for flagellar assembly in Chlamydomonas. Mol Biol Cell 10:693–712
10. Signor D, Wedaman KP, Orozco JT, Dwyer ND, Bargmann CI, Rose LS, Scholey JM (1999) Role of a class DHC1b dynein in retrograde transport of IFT motors and IFT raft particles along cilia, but not dendrites, in chemosensory neurons of living Caenorhabditis elegans. J Cell Biol 147:519–530
11. Walther Z, Vashishtha M, Hall JL (1994) The Chlamydomonas FLA10 gene encodes a novel kinesin-homologous protein. J Cell Biol 126: 175–188
12. Cole DG, Diener DR, Himelblau AL, Beech PL, Fuster JC, Rosenbaum JL (1998) Chlamydomonas kinesin-II-dependent intraflagellar transport (IFT): IFT particles contain proteins required for ciliary assembly in Caenorhabditis elegans sensory neurons. J Cell Biol 141: 993–1008
13. Cole DG (2003) The intraflagellar transport machinery of Chlamydomonas reinhardtii. Traffic 4:435–442
14. Pigino G, Geimer S, Lanzavecchia S, Paccagnini E, Cantele F, Diener DR, Rosenbaum JL, Lupetti P (2009) Electron-tomographic analysis of intraflagellar transport particle trains in situ. J Cell Biol 187:135–148
15. Johnson KA, Rosenbaum JL (1992) Polarity of flagellar assembly in Chlamydomonas. J Cell Biol 119:1605–1611
16. Dentler WL, Rosenbaum JL (1977) Flagellar elongation and shortening in Chlamydomonas. III. structures attached to the tips of flagellar microtubules and their relationship to the directionality of flagellar microtubule assembly. J Cell Biol 74:747–759
17. Dentler WL (1980) Structures linking the tips of ciliary and flagellar microtubules to the membrane. J Cell Sci 42:207–220
18. Sale WS, Satir P (1977) The termination of the central microtubules from the cilia of Tetrahymena pyriformis. Cell Biol Int Rep 1: 56–63
19. Schneider MJ, Ulland M, Sloboda RD (2008) A protein methylation pathway in Chlamydomonas flagella is active during flagellar resorption. Mol Biol Cell 19:4319–4327
20. Satish Tammana TV, Tammana D, Diener DR, Rosenbaum J (2013) Centrosomal protein CEP104 (Chlamydomonas FAP256) moves to the ciliary tip during ciliary assembly. J Cell Sci 126:5018–5029
21. Gorman DS, Levine RP (1965) Cytochrome f and plastocyanin: their sequence in the photosynthetic electron transport chain of Chlamydomonas reinhardi. Proc Natl Acad Sci U S A 54:1665–1669
22. Hutner SH, Provasoli L, Schatz A, Haskins CP (1950) Some approaches to the study of the role of metals in the metabolism of microoganisms. Proc Am Phil Soc 94:152–170
23. Surzycki S (1971) Synchronously grown cultures of Chlamydomonas reinhardi. Meth Enzymol 23:67–84

24. Sloboda RD, Howard L (2007) Localization of EB1, IFT polypeptides, and kinesin-2 in Chlamydomonas flagellar axonemes via immunogold scanning electron microscopy. Cell Motil Cytoskeleton 64:446–460
25. LaemmLi UK (1970) Cleavage of structural proteins during the assembly of the head of bacteriophage T4. Nature 227:680–685
26. Best D, Warr PJ, Gull K (1981) Influence of the composition of commercial sodium dodecyl sulfate preparations on the separation of alpha- and beta-tubulin during polyacrylamide gel electrophoresis. Anal Biochem 114:281–284
27. Stephens RE (1998) Electrophoretic resolution of tubulin and tektin subunits by differential interaction with long-chain alkyl sulfates. Anal Biochem 265:356–360
28. Fairbanks G, Steck TL, Wallach DF (1971) Electrophoretic analysis of the major polypeptides of the human erythrocyte membrane. Biochemistry 10:2606–2617
29. Hunt GE (1947) A technique for aeration of sterile liquid culture medium. Science 105:184
30. Witman GB, Carlson K, Berliner J, Rosenbaum JL (1972) Chlamydomonas flagella. I. Isolation and electrophoretic analysis of microtubules, matrix, membranes, and mastigonemes. J Cell Biol 54:507–539
31. Pedersen LB, Geimer S, Sloboda RD, Rosenbaum JL (2003) The Microtubule plus end-tracking protein EB1 is localized to the flagellar tip and basal bodies in Chlamydomonas reinhardtii. Curr Biol 13:1969–1974

Chapter 7

Fluorescence Imaging of the Cytoskeleton in Plant Roots

Julia Dyachok, Ana Paez-Garcia, Cheol-Min Yoo, Karuppaiah Palanichelvam, and Elison B. Blancaflor

Abstract

During the past two decades the use of live cytoskeletal probes has increased dramatically due to the introduction of the green fluorescent protein. However, to make full use of these live cell reporters it is necessary to implement simple methods to maintain plant specimens in optimal growing conditions during imaging. To image the cytoskeleton in living Arabidopsis roots, we rely on a system involving coverslips coated with nutrient supplemented agar where the seeds are directly germinated. This coverslip system can be conveniently transferred to the stage of a confocal microscope with minimal disturbance to the growth of the seedling. For roots with a larger diameter such as *Medicago truncatula*, seeds are first germinated in moist paper, grown vertically in between plastic trays, and roots mounted on glass slides for confocal imaging. Parallel with our live cell imaging approaches, we routinely process fixed plant material via indirect immunofluorescence. For these methods we typically use non-embedded vibratome-sectioned and whole mount permeabilized root tissue. The clearly defined developmental regions of the root provide us with an elegant system to further understand the cytoskeletal basis of plant development.

Key words Actin, Arabidopsis, *Medicago truncatula*, Microtubules, Green fluorescent protein, Living cells, Roots, Immunofluorescence, Fixed plant material, Sectioning

1 Introduction

An important requirement for research on the plant cytoskeleton is the ability to routinely and reliably image its organization in the cell. This goal has been facilitated in large part by advances in fluorescence microscopy not only with regard to the development of advanced microscope systems but also through the introduction of reagents that allow for in vivo labeling of cytoskeletal elements [1–3]. Microtubules and actin filaments (F-actin), the two major components of the cytoskeleton, can now be visualized readily in living plant cells using an array of green fluorescent protein (GFP) constructs (e.g., [4–6]). Despite the rapid implementation of live cell imaging tools, methods for optimal fixation of plant tissues for cytoskeletal studies will remain an essential tool for plant biologists. This is because these fixation approaches not only verify results

Ray H. Gavin (ed.), *Cytoskeleton: Methods and Protocols*, Methods in Molecular Biology, vol. 1365, DOI 10.1007/978-1-4939-3124-8_7, from live cell probes, they also have led to the discovery of novel cytoskeletal structures not always revealed by in vivo cellular reporters (e.g., [7–9]). Furthermore, live imaging of the cytoskeleton has for the most part been limited to cells located on the plant surface such as epidermal cells, trichomes, and root hairs (e.g., [10–12]). In cells located within the plant interior especially those from plant species with thicker organ systems, it is necessary to utilize conventional methods of sectioning and permeabilization to allow access of the cytoskeletal label (e.g., [12–14]). In this chapter we describe procedures for fluorescent labeling of the cytoskeleton in plant roots for observation with confocal microscopy. We focus on plant roots because they provide an elegant system for studying the cytoskeletal basis of plant development [15]. We first describe procedures for fluorescent antibody labeling of non-embedded root sectioǹs and permeabilized whole roots. We then outline methods for preparing Arabidopsis and larger roots of the model legume *Medicago truncatula* for live cell imaging of fluorescently labeled cytoskeletal structures.

2 Materials

2.1 Vibratome Sectioning and Whole Mount Immunolabeling of Fixed Roots

1. 37 % formaldehyde.
2. Paraformaldehyde (16 % solution).
3. Dimethyl sulfoxide (DMSO).
4. PME buffer: 50 mM *P*IPES (piperazine-*N,N'* bis [2-ethanse-sulfonic acid]), 2 mM *M*gCl_2 and 10 mM *E*GTA. Adjust pH of PME to 7.0 using 10 N NaOH.
5. Phosphate buffered saline (PBS), pH 7.2 for diluting antibodies and pH 8.5 for preparing the mounting medium. PBS buffer: 135 mM NaCl, 25 mM KCl, 5 mM Na_2HPO_4, and 2 mM KH_2PO_4. To prepare PBS with a pH of 8.5, omit the KH_2PO_4 and adjust the pH with 10 N NaOH.
6. Cellulase YC and Pectolyase Y23.
7. 1 % (v/v) Triton X-100 in PME buffer 8. Mowoil 4-88 or any mounting reagent (e.g., Citifluor and VECTASHIELD).
8. Vibratome.
9. Superglue.
10. Double edged razor blades and fine forceps.
11. Bovine serum albumin and normal goat serum.
12. Maleimidobenzoyl-*N*-hydroxysuccinimide ester (MBS). To prepare 0.16 M MBS stock solution, dissolve 50 mg MBS in 1 DMSO. MBS stock solution can be stored at −20 °C for several months.
13. Methanol prechilled to −20 °C.

14. PEMT: 0.05 % (v/v) Triton X-100 in PME buffer.
15. PBST: 0.05 % (v/v) Triton X-100 in PBS buffer, pH 7.0.
16. Tissue culture inserts (TCIs, Corning Netwell Plate, 12-Well Cluster, 500 μm pore size).
17. Rotating shaker table.
18. Actin and tubulin antibodies.
19. Fluorescently labeled phalloidin.

2.2 Preparing Arabidopsis Seedlings for Live Cell Imaging of Roots

1. Agar or Agargel (*see* **Note 1**).
2. Murashige & Skoog (MS) basal salt mixture.
3. 48 × 64 × 0.13–0.17 mm coverslips.
4. Sterilized pointed end toothpicks.
5. Polystyrene sterile petri dishes.
6. Sterilized filter paper.

2.3 Preparing Medicago truncatula Seedlings for Live Cell Imaging of Roots

1. Sandpaper.
2. Whatman filter paper circles 70 mm diameter.
3. Polystyrene sterile petri dishes.
4. Germination paper 18 × 13 in.
5. Plastic trays.

3 Methods

3.1 Immunolabeling of Fixed Roots

In both sectioning and whole mount permeabilization of roots, the general methods for fixation to preserve cytoskeletal structures are roughly similar. The only major difference is that the latter requires specialized sectioning equipment while the former relies on extensive digestion of the root with cell wall degrading enzymes.

3.1.1 Immunolabeling the Cytoskeleton in Vibratome Sectioned Roots

1. For plant species with large roots (ca. 300 μm to 1 mm in diameter) such as *Zea mays* or *Medicago truncatula*, excise the terminal 5 mm of the primary root with a razor blade and immerse root segments in 3.7 % formaldehyde and 5 % DMSO in PME buffer (v/v) (*see* **Note 2**). We typically use clear 20 mL scintillation glass vials with rubber-lined caps to hold the fixative containing the root segments. Place uncapped vials containing fixative and the collected root tissues in a small vacuum dessicator jar equipped with a T connector. Apply a light vacuum for 10–20 min. After releasing the vacuum most of the root samples should sink to the bottom of the vial.
2. After incubating samples in fixative for 2 h, slowly aspirate the fixative with a pipette and dispose excess fixative in clearly marked waste bottles in a fume hood. Wash the samples by

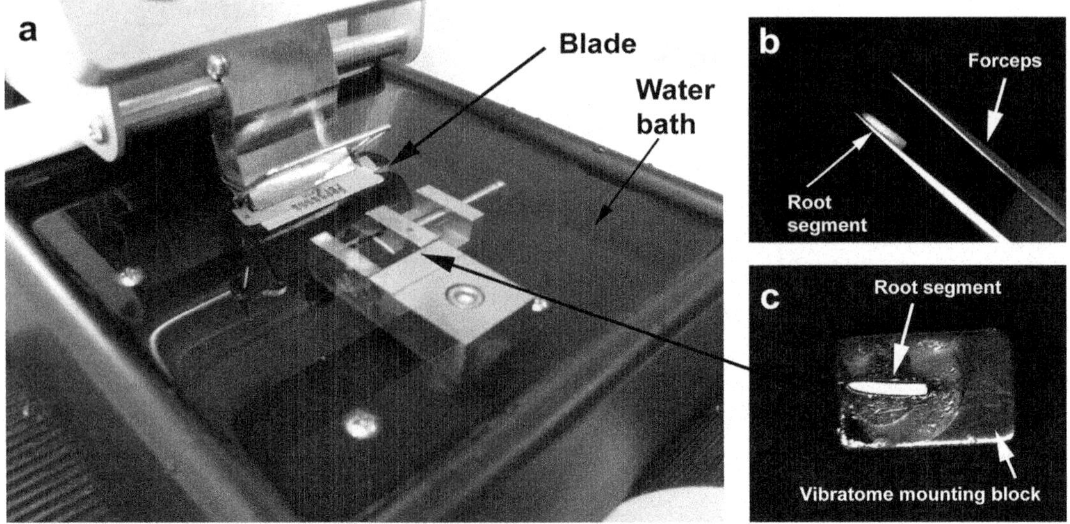

Fig. 1 Handling root tissues for vibratome sectioning. (**a**) The specimen holder of a vibratome 1000 showing the blade and block holder. (**b**) A segment of a fixed maize primary root on the tip of a pair of fine forceps. (**c**) Handling fixed root tissue as shown in (**b**) facilitates mounting the root on superglue coated vibratome block

immersing them in PME buffer for 3–5 min. Repeat this process three times to fully remove residual fixative.

3. For larger roots (i.e., >300 μm in diameter), longitudinal or cross sections can be easily obtained using a vibratome 1000 (Fig. 1a). To prepare root segments for vibratome sectioning, gently lift the excised root segment with a fine pair of forceps and blot excess liquid using filter paper or kimwipes (Fig. 1b). Spread a thin layer of superglue onto the surface of the vibratome mounting block and carefully position the root segment onto the block (Fig. 1c). To obtain good longitudinal sections, the long axis of the root must be perpendicular to the cutting surface of the blade (Fig. 1a). When the root contacts the thin layer of superglue, it takes about 60 s or less for the glue to polymerize allowing the root segment to firmly attach to the block (Fig. 1c; *see* **Note 3**).
4. Break a double edged razor blade in half and fasten half of the broken blade to the vibratome. Fill the vibratome specimen bath with deionized water and secure the block with the adhered root segment onto the block holder making sure that the root sample is fully immersed in water (Fig. 1a). Set the vibratome to section at 50–100 μm thickness following the manufacturers' instructions (*see* **Note 4**). Depending on the size of the root, one can obtain about 3–8 good quality longitudinal sections for cytoskeletal labeling.
5. Collect sections from the vibratome water bath using the wide end of a Pasteur pipette or a pair of fine forceps and transfer

sections to polystyrene petri dishes containing PME buffer. Alternatively, sections can be directly placed onto 22×22 coverslips or glass slides. Use a small piece of filter paper to remove excess PME buffer by touching the liquid adjacent to the root sections. Allow sections to partially dry onto the surface of the slide or coverslip (*see* **Note 5**).

6. When sections are secured onto the coverslip, apply a cocktail of 1 % cellulose YC in PME buffer for 10 min. The cellulose solution should cover the root sections during incubation. Slowly decant the cellulose solution and wash sections three times with PME buffer. During the washing steps, PME buffer can simply be added onto the surface of the coverslips containing the root sections and decanted (*see* **Note 6**). After cell wall treatment, incubate samples in 1.0 % Tritron X-100 in PME for 15–20 min and wash three times with PME buffer.
7. Apply primary antibody diluted in PBS, pH 7.2 onto the surface of the coverslips containing the root sections and place in a humid chamber for 2–3 h. Wash three times with PME buffer and apply secondary antibody conjugated to a fluorescent dye that specifically recognizes the primary antibody. Incubate sections in a humid chamber for another 2 h (*see* **Notes 7** and **8**).
8. Prepare the mounting medium while samples are incubating in the secondary antibody. This is accomplished by preparing 20 % Mowiol 4-88 in PBS, pH 8.5 (v/v). A 2 mL volume of mounting media is sufficient for mounting about 10–20 samples. The Mowiol-PBS solution can be prepared in 20 mL scintillation vials similar to those used for tissue fixation (*see* **Step 1** above). To dissolve the Mowiol crystals, stir the solution for 2 h using a small magnetic stir bar (*see* **Note 9**).
9. After secondary antibody treatment, wash samples with PME buffer three times and allow the sections to partially dry. Blot any excess liquid with filter paper and use the wide end of a Pasteur pipette to collect mounting medium (*see* **Note 10**). Carefully drop Mowiol onto the surface of the coverslip and mount sections on clean glass slides. After allowing the Mowiol to polymerize overnight, root sections can then be observed with a confocal microscope. Examples of the quality of cytoskeletal labeling from vibratome-sectioned roots is shown in Fig. 2 (*see* **Note 11**).

3.1.2 Immunolabeling of the Cytoskeleton in Whole Mount Arabidopsis Roots

1. Gently pick 3–4-day-old *Arabidopsis* seedlings growing on nutrient supplemented gel media (*see* Subheading 3.2 below for tips on growing Arabidopsis seedlings) with a fine pair of forceps and directly immerse seedlings into the fixative. For preserving F-actin in Arabidopsis roots we use 2 % (v/v) paraformaldehyde, 0.1 % (v/v) Triton X-100, 400 μM MBS in PME buffer, while for microtubules we use 3.7 % (v/v) paraformaldehyde.

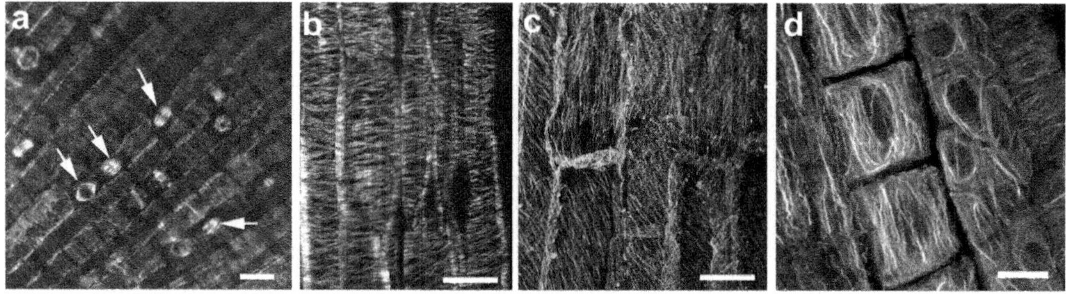

Fig. 2 Confocal microscopy of the cytoskeleton in root vibratome sections. (**a**) Microtubules in the meristematic region of a *Medicago truncatula* root. Arrows indicate mitotic figures. Transverse and oblique cortical microtubules in the elongation (**b**) and maturation zone (**c**) of a maize primary root. (**d**) F-actin organization in the vascular region of maize roots. Bars = 20 μm (**a**, **d**); 10 μm (**b**, **c**)

We perform fixation in 20 mL glass scintillation vials, adding 1–5 mL fixative per vial. Incubate samples in fixative for 1 h, applying a light vacuum for the initial 30 min. Place uncapped vials containing fixative and collected seedlings in a small vacuum dessicator jar equipped with a T connector. Open the vacuum vent slowly and keep it open until the hissing sound subsides. Close the vent and continue incubation. Release the vacuum slowly after 30 min and continue incubation so that the total time of fixation is about 1 h.

2. Subsequent steps, including washes, permeabilization and immunolabeling, are carried out in tissue culture inserts (TCIs), on a rotating shaker table. Upon fixation carefully transfer seedlings with fine forceps into a TCI inserted in a well of a 12 well plate containing 2 mL PEMT (*see* **Note 12**). Rinse seedlings in PEMT three to four times for 10 min each.
3. Permeabilize samples for 1 h in 1 % Triton X-100 in PME, followed by rinsing three times in PEMT, 5 min each.
4. For cell wall digestion, incubate samples for 20 min in 0.05 % (w/v) Pectolyase Y-23, 0.1 % (v/v) Triton, 1 % Bovine serum albumin, 0.4 M mannitol in PME. Wash by incubating three times in PEMT for 10 min each.
5. Permeabilize samples for 10 min in cold (−20 °C) methanol (*see* **Note 13**). When labeling microtubules, the cold methanol treatment should be omitted. Wash three times in PBS, 5 min each.
6. Incubate samples in blocking solution (5 % normal goat serum in PBS) for 1 h.
7. Primary and secondary antibodies are diluted in blocking solution (*see* **Note 14**). Primary antibody, mouse monoclonal anti-chicken gizzard actin ("C4," Chemicon, Temecula, CA) is used at a dilution of 1:1000. For microtubules, we use the

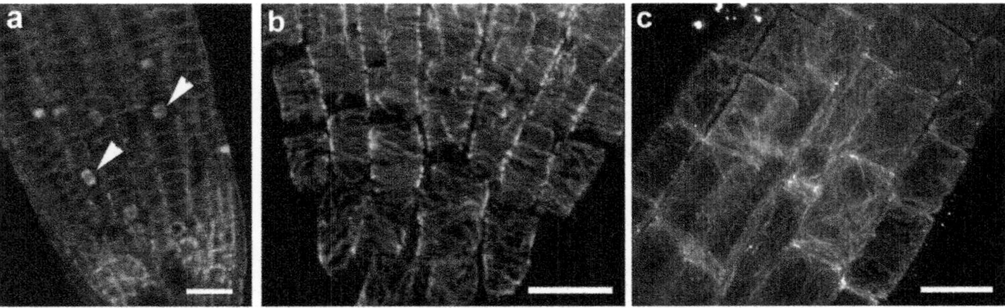

Fig. 3 Confocal microscopy of the cytoskeleton in whole mount Arabidopsis roots. (**a**) Low magnification image of microtubule labeling in the root meristem. *Arrowheads* indicate mitotic figures. (**b**) Transverse cortical microtubules in the root cap. (**c**) F-actin organization in the root distal elongation zone. Bars = 20 μm

monoclonal rat anti-yeast tubulin antibody (YOL1/34, Accurate Chemicals, Westbury, NY; *see* Subheading 3.1.1). Incubate samples with the primary antibodies for 16 h (usually overnight) at room temperature on a rotating shaker table.

8. Wash samples with PBST three to four times for a total duration of 1 h.
9. Add secondary antibodies (fluorescently labeled monoclonal anti-mouse for actin or anti-rat for microtubules) to the samples. We use Alexa Fluor 488 goat anti-mouse or goat anti-rat as the secondary antibody, diluted 1:200 in blocking solution (*see* **Note 15**). Incubate samples with the secondary antibodies for 5–6 h at room temperature on a rotating shaker table.
10. Wash samples with PBST three to four times for the total duration of 1 h and once with PBS for at least 10 min. Seedlings can be mounted for immediate observation or stored in PBS at 4 °C for up to 1 week (*see* **Note 16**).
11. For observation, mount seedlings in VECTASHIELD and image with a confocal microscope. Examples of the quality of cytoskeletal labeling from whole mount Arabidopsis roots is shown in Fig. 3.

3.2 Preparing Arabidopsis Roots for Live Cell Imaging

Most of the live cell imaging work on the plant cytoskeleton has been conducted using the model plant Arabidopsis. This is because Arabidopsis is readily transformed with various GFP cytoskeletal reporters and the smaller roots of Arabidopsis make sample preparation for live cell imaging very convenient. However, a number of precautions have to be taken to ensure that growth of the root is not compromised by excessive physical handling of the seedling. If at all possible, imaging of live Arabidopsis roots should be done directly on the media where the seeds are germinated.

1. Surface sterilize the seeds by immersing them for 3 min each in 95 % ethanol, 20 % bleach and sterilized deionized water.

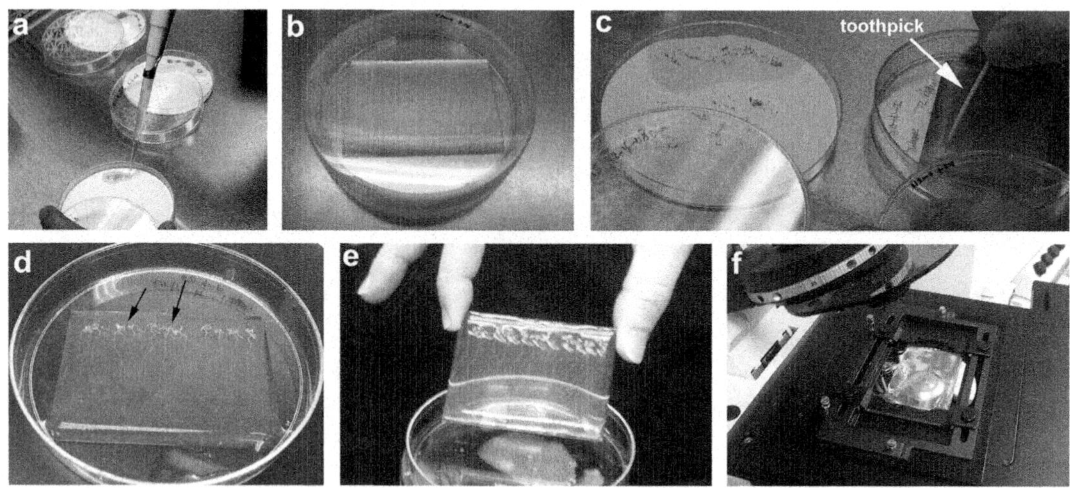

Fig. 4 Arabidopsis seed sterilization and planting for live cell confocal imaging of the cytoskeleton. (**a**) Spreading sterilized seeds on filter paper. (**b**) Agar–coverslip setup for planting Arabidopsis seeds. (**c**) Sterilized seeds can be picked individually using pointed toothpicks and planted directly on the agar–coverslip system. (**d**) Seven-day-old Arabidopsis seedlings (*arrows*) growing on the agar–coverslip system. The agar–coverslip system with the growing seedlings can be easily picked up by gently pressing the bottom of the polystyrene petri dish (**e**) and directly transferred to the stage of the confocal microscope for imaging (**f**)

Rinse the seeds in deionized water at least three times after bleach treatment. Transfer seeds to sterilized 90 mm filter paper in 100 × 15 mm polystyrene petri dishes, and spread the seeds evenly on the filter paper using a 1 mL pipette (Fig. 4a). Let the seeds dry in a laminar flow bench overnight. The seeds can be planted immediately after spreading onto the filter paper-lined polystyrene petri dishes but for long term storage they must be dried thoroughly prior to sealing with Parafilm.

2. Prior to seed sterilization a coverslip–agar medium for germinating seeds must be prepared in advance. This system consists of sterilized coverslips with about a 1 mm layer of nutrient supplemented polymerized 0.5–1 % agar. The coverslip agar system is placed inside 100 × 15 mm polystyrene petri dishes, wrapped with Parafilm, and kept at 4 °C for future use (Fig. 4b).
3. To prepare the coverslip-agar system, autoclave 0.5× MS salts in 0.5–1 % agar (v/v) supplemented with 0.5 mg/mL pyridoxine-HCl, 0.5 mg/mL nicotinic acid and 1 mg/mL thiamine. The vitamins can be dissolved in water as 1000× stock, filter-sterilized and stored at 4 °C prior to use. Add 0.10 g/L myo-inositol, 0.5 g/L MES, and 1 % sucrose to the MS vitamin solution. Adjust pH to 5.7 with 10 M KOH, add agar and autoclave. After autoclaving, carefully add the agar-MS solution on top of the coverslip using a pipette until the entire surface of the coverslip is covered with agar (Fig. 4b; *see* **Note 17**).

4. When the agar has fully polymerized and cooled, pick Arabidopsis seeds individually with a sterile toothpick and gently push the seed into the agar media so that the seed touches the bottom of the coverslip (Fig. 4c). This helps the roots grow straight inside the medium facilitating microscopic observation. Dried seeds can be easily picked up with a moist toothpick. This is done by immersing the tip of the toothpick into the agar and gently touching the seed.
5. After planting, seal the petri dish with Parafilm and transfer to a growth chamber. The petri dish should be positioned at a 60° angle from the horizontal to allow the roots to direct their growth toward the surface of the coverslip. It takes about 3–4 days for the roots to be ready for imaging (Fig. 4d).
6. When the roots have reached the desired age for imaging, remove the coverslip containing the seedlings by gently pressing the bottom of the polystyrene petri dish and lifting the coverslip making sure that the polymerized agar does not slide off (Fig. 4e).
7. Mount the coverslip directly on the stage of a confocal microscope for observation (Fig. 4f).

3.3 Preparing *Medicago truncatula* Roots for Live Cell Imaging

1. To induce germination, *Medicago truncatula* seed have to be scarified either mechanically using sandpaper or chemically with sulfuric acid [19]. For imaging *Medicago truncatula* roots we typically use mechanical scarification since not a lot of seedlings are needed. Place about 20 dry seed on the surface of a square piece of sandpaper (Fig. 5a). Rub the seed with a second piece of sand paper to abrade the seed coat (Fig. 5b).
2. Transfer abraded seeds to a petri dish lined with moist filter paper (Fig. 5c) and leave overnight at room temperature in darkness. The following day, the imbibed seeds are ready for transplanting (Fig. 5d).
3. Live imaging of roots is more convenient if they are straight. To obtain seedlings with straight roots, place a layer of germination paper in a plastic tray. Standard cafeteria trays can be used for this purpose. Moisten the germination paper with tap water and carefully arrange seeds side by side along a straight line (Fig. 5e).
4. Cover seeds with another layer of germination paper and add additional water if needed. Use a second plastic tray to the cover the seed and position the entire set-up vertically on a plastic reservoir with water about 2 in. deep. Capillary action will keep the germination paper moist for a few days.
5. After 2–3 days, roots that are about 3–4 cm in length should be available for imaging (Fig. 5e). Select seedlings with straight roots and mount in a glass slide (Fig. 5f).

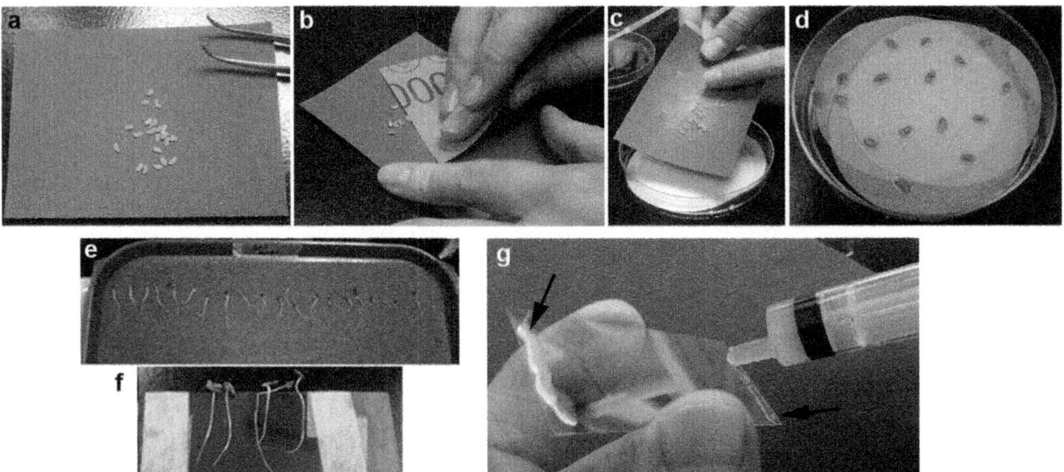

Fig. 5 Processing *Medicago truncatula* seeds for live cell imaging of roots. Mechanical scarification of dry seed with sandpaper treatment (**a**, **b**). Transfer of scarified seed to a petri dish lined with two or three layers of moist filter paper (**c**). Imbibed seeds ready for planting (**d**). Three-day-old seedlings growing on moist germination paper with primary roots fully emerged (**e**). Seedlings selected for imaging are transferred to a glass slide (**f**) and secured with a coverslip lined with vacuum grease (*arrows*, **g**)

6. Add a small amount of water to the roots and select a cover glass with the appropriate size. For *Medicago truncatula* roots, a 50 mm × 22 mm cover glass is typically used (*see* **Note 18**).
7. Apply vacuum grease evenly to opposite sides of the cover glass and gently cover the roots (Fig. 5g; *see* **Note 19**). Examples of cytoskeletal structures in living Arabidopsis and *Medicago truncatula* roots are shown in Fig. 6.

4 Notes

1. There are several companies that sell agar for plant culture, which can be used for preparing the agar–coverslip system. For imaging live Arabidopsis roots, however, we found that Agargel (A3301: Sigma-Aldrich) gives the most reliable images when overlayed on the coverslip.
2. In our experience, root microtubules have been preserved adequately with fixative solution containing 3–4 % (v/v) formaldehyde or paraformaldehyde [16]. On the other hand we have resorted to lower concentrations of formaldehyde (1.5–2 %) to optimally preserve the delicate F-actin structures in roots [14, 17, 18].
3. Make sure that superglue is spread evenly and thinly on the vibratome block surface so that only the outer cells of the root are in contact with the superglue. Typically a touch of superglue

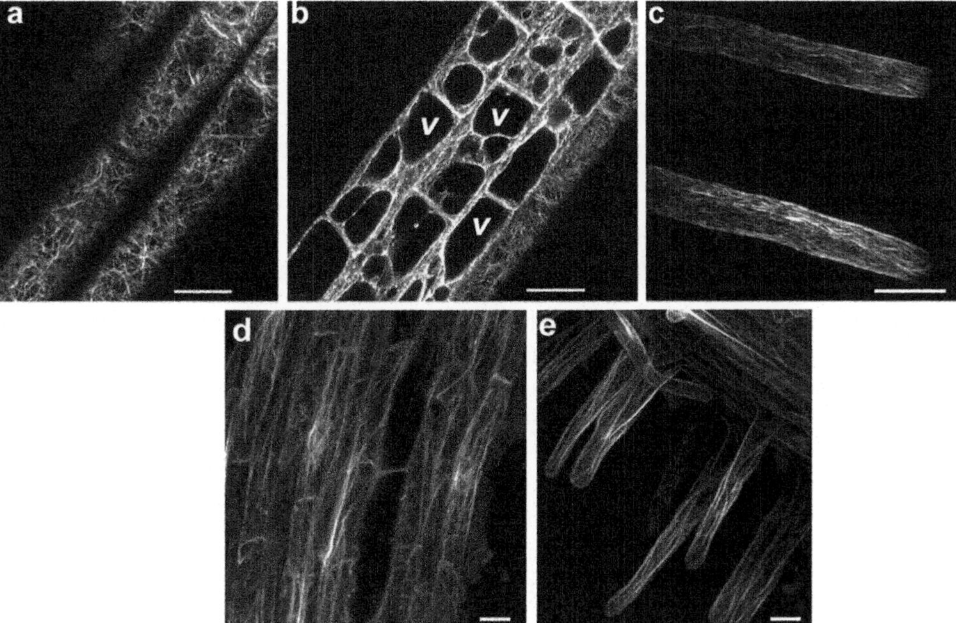

Fig. 6 Live cell imaging of Arabidopsis and *Medicago truncatula* roots expressing F-actin and microtubule GFP reporters. Cortical (**a**) and cytoplasmic (**b**) F-actin networks in epidermal cells from the root elongation zone of Arabidopsis. Images in (**a**) and (**b**) were taken at two different focal planes. Vacuoles (*V*). (**c**) Longitudinal cortical microtoubules in growing Arabidopsis root hairs imaged with a spinning disk confocal microscope. Confocal *Z*-stacks of epidermal cells in the root maturation zone (**d**) and root hairs (**e**) of *Medicago truncatula* stably expressing a UBQ10::GFP-ABD2-GFP construct [6]. Bars = 20 μm

is sufficient for the root tissue to adhere to the block surface. If too much glue is used, sectioning and subsequent labeling steps can be compromised.

4. Without any embedding support, vibratome sections of plant tissues typically break apart at a thickness of less than 40 μm.
5. The sections can be left on the lab bench to allow remaining PME buffer to evaporate. However, make sure that sections remain moist. Do not allow sections to dry completely as this could lead to distortion of cells during imaging. One way to secure the sections onto coverslips for subsequent antibody labeling is to use a thin film of bactoagar as originally proposed in Brown and Lemmon [13]. Dissolve 0.75 g bactoagar in 100 mL deionized water in a 250 mL beaker. Microwave the solution until it comes to a boil. When contents of the beaker start to rise, turn off the microwave and briefly swirl the contents of the beaker. Microwave the solution for a second time to completely dissolve the bactoagar granules. A loop can be constructed using copper or chromium-nickel wire. Material for making such loops can easily be found in the lab. For example, we have successfully used wire from the base of test tube

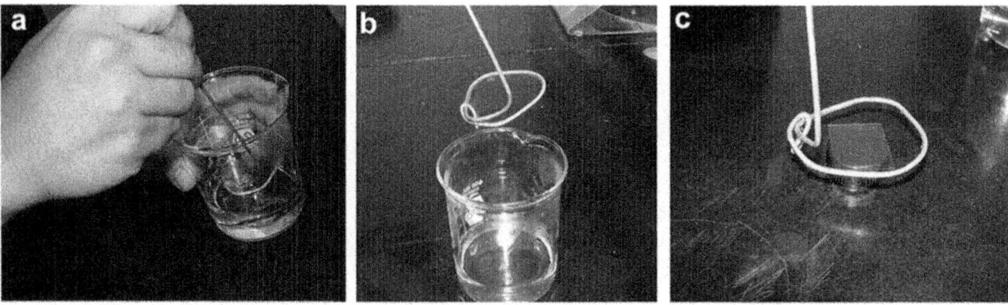

Fig. 7 Procedure for securing root vibratome sections on coverslips for indirect immunofluorescence labeling of the cytoskeleton. A thin film of agar is collected on a loop from molten 0.75 % agar solution (**a**, **b**). The coverslip containing the root section is placed on top of a cap of a scintillation vial and the loop with the thin agar film is passed over the coverslip to hold the root section in place (**c**)

brushes to make agar casting loops. To secure the sections, allow the agar film to solidify on the loop and slowly cover the sections with the agar film. Placing the coverslip on the caps of vials facilitates layering of the agar film onto the sections (Fig. 7). Layering of the agar film requires patience since the thin agar film can rupture easily. All solutions for labeling the sections can easily penetrate the agar film.

6. During subsequent treatments and washing steps one has to be careful that sections do not detach from the coverslip. Although the thin agar film described in **Note 5** typically prevents sections from floating away during sample treatment, there are cases when the bactoagar film will slide off because of constant exposure to the liquid solutions. When this happens, one has to take extra care that samples are not lost during the various washing steps.
7. Overnight incubation in primary antibody has sometimes led to improvements in the quality of cytoskeletal labeling and also allows flexibility with the labeling schedule if time becomes an issue. When overnight incubation in the primary antibody becomes necessary, the samples can be kept at 4 °C. A humid chamber can be constructed using 9 mm round polystyrene petri dishes lined with moist filter paper. Coverslips or glass slides containing the root sections can be laid on top of rubber lined caps of scintillation vials to prevent moisture from spreading onto the sample (Fig. 7c). Laying coverslips on top of vial caps also facilitates casting of the thin bactoagar film used for securing sections (*see* **Note 5**).
8. The choice of primary antibodies is critical for the success of labeling. There are a number of commercially available anti-tubulin and actin antibodies. For microtubules, we have been successful with the Rat anti-yeast tubulin (monoclonal rat and yeast tubulin antibody YOL1/34, Accurate Chemicals,

Westbury, NY) and anti-actin antibodies (C4, Chemicon, Temecula, CA). As an alternative to antibodies, fluorescently conjugated phalloidin can be used to label F-actin in roots [14, 17]. Commercial primary and secondary antibodies and fluorescently conjugated phalloidin typically come with recommendations for dilutions. However, it is advisable to test the dilution that is most effective for the plant tissue being studied. Antibodies and fluorescently conjugated phalloidin should be aliquoted into small volumes and stored at −20 °C prior to use to avoid repeated freezing and thawing.

9. Do not worry if residual Mowiol crystals remain after 2 h of stirring. Just be sure that you do not include any of the crystals when mounting your samples.
10. The Mowiol mounting medium will be viscous after 2 h of stirring and will occasionally contain air bubbles. Only a small drop (ca. 100–200 μL) of the Mowiol solution is needed to mount one coverslip. Be sure to avoid air bubbles when mounting your samples as this could affect the quality of imaging.
11. Because vibratome sections are thick the best images are obtained using a confocal microscope.
12. Solutions (2 mL per well) are added and removed directly to and from the well using a pipette. TCIs with sample seedlings are transferred from one solution to another by moving between wells. Tapping the TCI gently on a paper towel before inserting it into the well facilitates solution exchange. This arrangement allows for better preservation of fragile seedlings during the procedure.
13. Place a 12-well plate with methanol (2 mL per well) at −20 °C 1 h in advance.
14. To minimize the volume of antibody solution, TCI bottoms can be sealed with Parafilm before inserting into the well. Antibody solution (350–400 μL) is then added directly to the TCI with the sample. Seedlings should be fully submerged into the antibody solution. After the incubation with the antibodies, remove Parafilm from the bottoms of TCIs with samples and tap TCIs gently on a paper towel to remove the antibody solution. Proceed with washes.
15. Sealing TCI bottoms with Parafilm can also be used at this stage to minimize the volume of antibody solution required to cover samples fully.
16. During storage, keep plates with samples covered in aluminum foil to prevent loss of fluorescent signal.
17. Surface tension of the molten agar will form a dome on the surface of the coverslip. Avoid pouring excess agar on the coverslip as this will break the dome and cause the agar to flow out of the coverslip.

18. Observation of the cytoskeleton in living *Medicago* roots is more challenging because of the thicker roots. Imaging is typically limited to the epidermis and outer cortical cells. It is important to select a region of the root that is pressed closely to the surface of the coverslip. To overcome difficulties in imaging thick roots, some investigators have resorted to tissue-specific promoters and sectioning of live roots [20] or have used non-destructive clearing methods [21].
19. The vacuum grease serves two purposes. First, it keeps the cover glass securely attached to the glass slide and as a result, a confocal microscope with an inverted platform can be used. Second, it prevents roots from being crushed as the vacuum grease acts as a spacer between the glass slide and cover glass.

Acknowledgements

Work on root biology and the cytoskeleton in the Blancaflor laboratory is supported by the Samuel Roberts Noble Foundation Forage 365 initiative and National Aeronautics and Space Administration (NASA grant NNX12 AM94G).

References

1. Li J, Blanchoin L, Staiger CJ (2015) Signaling to actin stochastic dynamics. Annu Rev Plant Biol 66:415–440. doi: 10.1146/annurev-arplant-050213-040327
2. Eng RC, Wasteneys GO (2014) The microtubule plus-end tracking protein ARMADILLO-REPEAT KINESIN1 promotes microtubule catastrophe in Arabidopsis. Plant Cell 26: 3372–3386
3. Rocchetti A, Hawes C, Kriechbaumer V (2014) Fluorescent labelling of the actin *cytoskeleton* in *plants* using a cameloid antibody. Plant Methods 10:12
4. Marc J, Granger CL, Brincat J, Fisher DD, Kao T-H, McCubbin AG, Cyr RJ (1998) A GFP-MAP4 reporter gene for visualizing cortical microtubule rearrangements in living epidermal cells. Plant Cell 10:1927–1940
5. Vidali L, Rounds CM, Hepler PK, Bezanilla M (2009) Lifeact-mEGFP reveals a dynamic apical F-actin network in tip growing plant cells. PLoS One 4:e5744
6. Dyachok J, Sparks JA, Liao F, Wang Y-S, Blancaflor EB (2014) Fluorescent protein-based reporters of the actin cytoskeleton in living plant cells: fluorophore variant, actin binding domain and promoter considerations. Cytoskeleton 71:311–327
7. Lovy-Wheeler A, Wilsen KL, Baskin TI, Hepler PK (2005) Enhanced fixation reveals the apical cortical fringe of actin filaments as a consistent feature of the pollen tube. Planta 221:95–104
8. Collings DA, Wasteneys GO (2005) Actin microfilament and microtubule distribution patterns in the expanding root of Arabidopsis thaliana. Can J Bot 83:579–590
9. Wilsen KL, Lovy-Wheeler A, Voigt B, Menzel D, Kunkel JG, Hepler PK (2006) Imaging the actin cytoskeleton in growing pollen tubes. Sex Plant Reprod 19:51–62
10. Yoo CM, Quan L, Cannon AE, Wen J, Blancaflor EB (2012) AGD1, a class 1 ARF-GAP, acts in common signaling pathways with phosphoinositide metabolism and the actin cytoskeleton in controlling Arabidopsis root hair polarity. Plant J 69:1064–1076
11. Yoo CM, Blancaflor EB (2013) Overlapping and divergent signaling pathways for ARK1 and AGD1 in the control of root hair polarity in *Arabidopsis thaliana*. Front Plant Sci 4:528
12. Dyachok J, Shao MR, Vaughn K, Bowling A, Facette M, Djakovic S, Clark L, Smith L (2008) Plasma membrane-associated SCAR complex subunits promote cortical F-actin accumulation and normal growth characteristics in Arabidopsis roots. Mol Plant 1:990–1006

13. Brown RC, Lemmon BE (1995) Methods in plant immunolight microscopy. In: Galbraith DW, Bohnert HJ, Bourque DP (eds) Methods in cell biology. Academic Press, San Diego, CA, pp 85–107
14. Blancaflor EB, Hasenstein KH (2000) Methods for detection and identification of F-actin organization in plant tissues. In: Staiger C, Baluška F, Volkmann D, Barlow PW (eds) Actin: a dynamic framework for multiple plant cell functions. Kluwer Academic Publishers, Dordrecht, The Netherlands, pp 601–618
15. Blancaflor EB, Wang Y-S, Motes CM (2006) Organization and function of the actin cytoskeleton in developing root cells. Int Rev Cyto 252:153–198
16. Blancaflor EB, Zhao L, Harrison MJ (2001) Microtubule organization in root cells of *Medicago truncatula* during development of an arbuscular mycorrhizal symbiosis with *Glomus versiforme*. Protoplasma 217:154–165
17. Blancaflor EB, Hasenstein KH (1997) The organization of the actin cytoskeleton in vertical and gravíresponding primary roots of maize. Plant Physiol 113:1447–1455
18. Hou G, Mohamalawari DR, Blancaflor EB (2003) Enhanced gravitropism of roots with a disrupted cap actin cytoskeleton. Plant Physiol 131:1360–1373
19. Garcia J, Barker DG, Journet E-P (2006) Seed storage and germination. In: Mathesius U, Journet EP, Sumner LW (eds) The *Medicago truncatula* handbook. ISBN 0-9754303-1-9. http://www.noble.org/MedicagoHandbook/
20. Ivanov S, Harrison MJ (2014) A set of fluorescent protein-based markers expressed from constitutive and arbuscular mycorrhiza-inducible promoters to label organelles, membranes and cytoskeletal elements in *Medicago truncatula*. Plant J 80:1151–1163
21. Warner CA, Biedrzycki ML, Jacobs SS, Wisser RJ, Caplan JL, Sherrier DJ (2014) An optical clearing technique for plant tissues allowing deep imaging and compatible with fluorescence microscopy. Plant Physiol 166: 1684–1687

Chapter 8

Microtubules in Plant Cells: Strategies and Methods for Immunofluorescence, Transmission Electron Microscopy, and Live Cell Imaging

Katherine Celler, Miki Fujita, Eiko Kawamura, Chris Ambrose, Klaus Herburger, Andreas Holzinger, and Geoffrey O. Wasteneys

Abstract

Microtubules (MTs) are required throughout plant development for a wide variety of processes, and different strategies have evolved to visualize and analyze them. This chapter provides specific methods that can be used to analyze microtubule organization and dynamic properties in plant systems and summarizes the advantages and limitations for each technique. We outline basic methods for preparing samples for immunofluorescence labeling, including an enzyme-based permeabilization method, and a freeze-shattering method, which generates microfractures in the cell wall to provide antibodies access to cells in cuticle-laden aerial organs such as leaves. We discuss current options for live cell imaging of MTs with fluorescently tagged proteins (FPs), and provide chemical fixation, high-pressure freezing/freeze substitution, and post-fixation staining protocols for preserving MTs for transmission electron microscopy and tomography.

Key words Microtubules, EB1, GFP, Kinesin, MAP4, MBD, MOR1, ARK1, Electron tomography, Live cell imaging, Correlative light and electron microscopy, Immunofluorescence

1 Introduction

Microtubules (MTs) are a unifying feature of eukaryotic cells. Studying them in plants is not only important for understanding mechanisms of plant growth and development but also of broader interest for understanding the mechanisms that generate MT spatial organization, including the role of accessory proteins in MT dynamics. The current literature describes the role of MTs in spindle formation and division plane organization [1–3], transition from division to expansion [4, 5], cell wall formation and morphogenesis in diffusely expanding organs [6–13], tip-growing root hairs and pollen tubes [14–19], movement of stomata [20], endomembranes [21–23] or nuclei [24–27], chloroplast organization and positioning [28–30], or in association with amyloplasts [31].

Ray H. Gavin (ed.), *Cytoskeleton: Methods and Protocols*, Methods in Molecular Biology, vol. 1365,
DOI 10.1007/978-1-4939-3124-8_8, © Springer Science+Business Media New York 2016

This research requires a variety of experimental strategies, ranging from basic description of orientation patterns, MT dynamics and associations with other cellular structures, plant hormones, and mechanical stress, to pharmacological and genetic perturbations. With cytoskeleton research increasing in relevance to different plant research fields, it is especially important for researchers to be provided with a variety of standard techniques and tools for imaging MTs in plant cells. In this chapter, we outline preparation methods for light and electron microscopy. Methods for imaging root MT and actin filament arrays are provided in Chapter 7 of this book.

1.1 Immuno-fluorescence Options

Immunofluorescence microscopy remains a useful approach for MT imaging at the level of resolution of visible light [32]. Despite the convenience of live cell imaging with transgene-introduced fluorescent reporter proteins (e.g., [33, 34]), immunofluorescence remains essential under many circumstances. Immunofluorescence can be utilized immediately on any interesting plant material because it does not require the laborious and time-consuming cloning, transformation and selection procedures required for introducing transgenic reporter proteins. This is especially relevant for non-model systems for which genetic transformation is not yet feasible or when wild material is collected in its natural habitat. We demonstrate this first point with images of MTs in leaf cells of the high alpine plant *Oxyria digyna* (Fig. 1a–c) for which the "freeze shattering" method [35] was adapted to examine MTs in relation

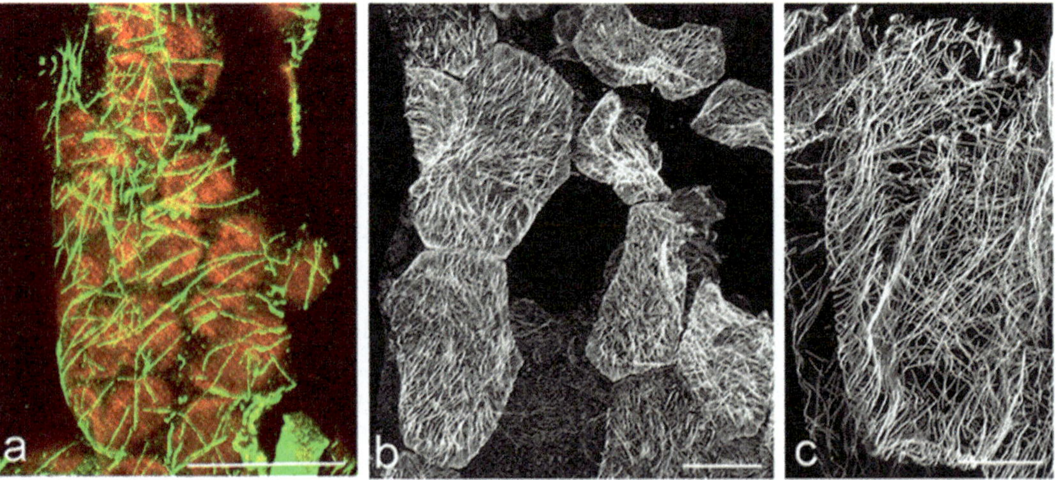

Fig. 1 Immunolocalization of MTs in the alpine vascular plant *Oxyria digyna*: (**a**) merge of MT staining (*green*) and chloroplast autofluorescence showing the close vicinity of MTs with chloroplasts. (**b**) MTs in epidermal cells, (**c**) MTs in parenchyma cells; note the very dense network in z-stack projections. The image is composed of approximately 100 images, comprising a depth of about 20 μm. Bars, 20 μm. (**b**) and (**c**) reprinted from [29] with permission from Thieme Verlag KG

to stress-induced chloroplast protuberances [29]. Previous work had demonstrated that MTs were a critical factor in the establishment of chloroplast stromules in the model system *Arabidopsis thaliana* [36]. Immunofluorescence enabled MTs in *Oxyria* leaves to be documented [29] without the need to develop protocols for transformation of this non-model system.

Immunofluorescence microscopy is also a useful adjunct technology for corroborating evidence from live cell strategies, where, as discussed below, artifact needs to be ruled out (Fig. 2). It is the best option for detecting low-abundance proteins, which require high-intensity light and long dwell times or when large series of

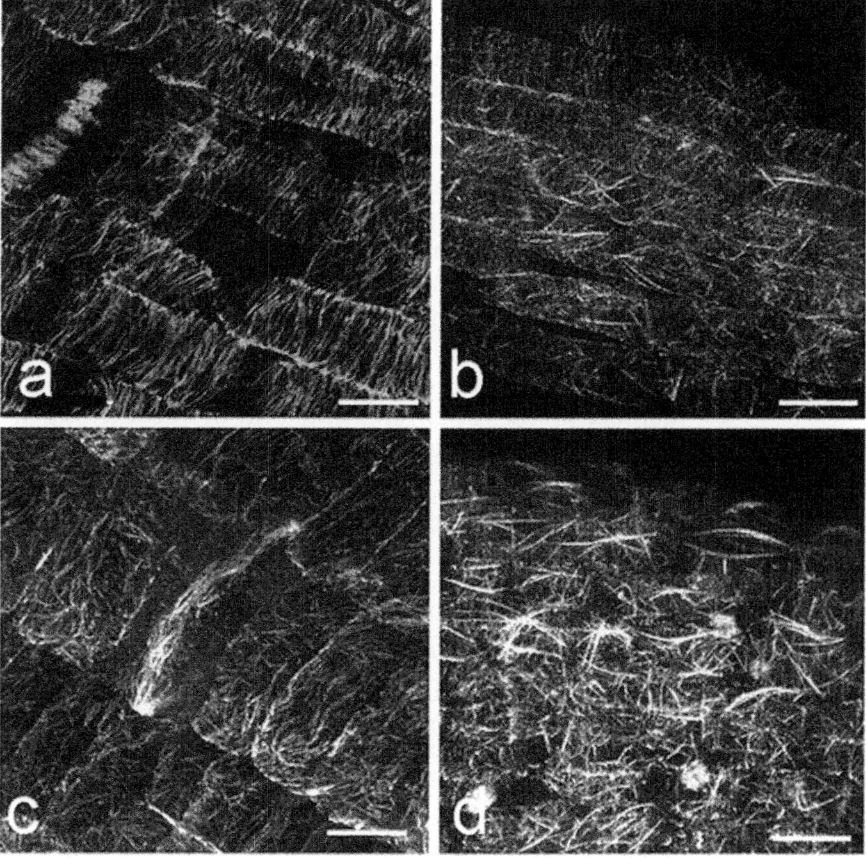

Fig. 2 Comparison of wild-type, wt (**a**, **b**) and *mor1-1* (**c**, **d**) MT bundles via immunolocalization (**a**, **c**) and GFP-MBD live cell imaging (**b**, **d**) in *Arabidopsis thaliana* at 31 °C. Plants were grown at 21 °C for 5 days and then transferred to 31 °C. (**a**) wt immunolabeled (fixed in a fixative that was preheated to 31 °C and labeled with anti-tubulin) and (**b**) wt GFP-MBD expressing root. Images were taken using a temperature-controlled stage that kept the specimens at 31 °C, 2–3.5 h after temperature shifts; (**c**) immunolabeled *mor 1-1*, (**d**) *mor1-1* at 31 °C showing apparent MT bundles in the GFP-MBD fusion protein expressing line. GFP-MBD labeling identifies bright, bundle-like structures at 31 °C. Although some evidence for this is observed in wild-type cells, MT remodeling is much more extensive in *mor1-1*. These thickened MT structures are not observed with immunolabeling, suggesting this putative bundling of MTs is dependent on the GFP-MBD fusion protein. Bars, 10 μm

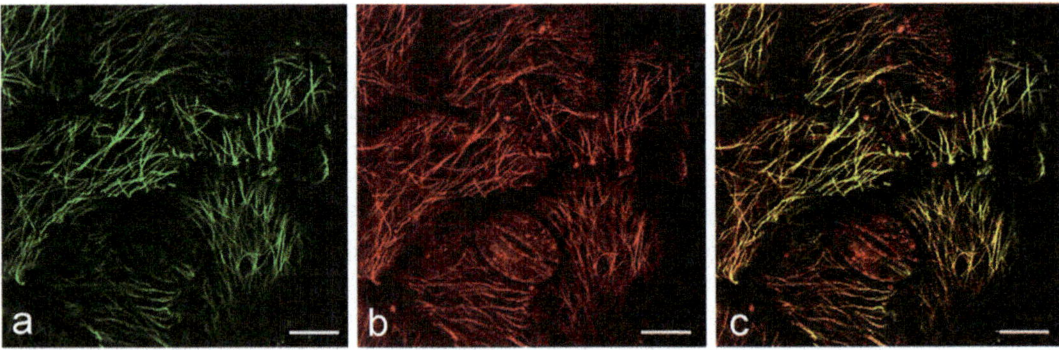

Fig. 3 Immunolabeling of tubulin (**a**) and MOR1 (**b**) in wild-type background at 31 °C. For immunolabeling, 11-day-old wild-type plants grown at 21 °C were cultured at 31 °C for 1 day and fixed in fixatives that were preheated to 31 °C and then double immunolabeled. (**a**) Tubulin, (**b**) MOR1, (**c**) merge of (**a**) and (**b**). The abaxial sides of first leaves are shown in all images. Bars, 10 μm

optical sections need to be collected because these procedures are toxic to living cells. Although immuno-procedures kill the samples, after fixation and appropriate fade protection, immunolabeled material can be scanned repeatedly and stored for months with little loss of image quality. Immunofluorescence is also an excellent method for describing the distribution of MT-associated proteins (MAPs) along MTs using double-labeling strategies (Fig. 3). Finally, at present, immunofluorescence is the only practical method to use when applying one of the exciting new super-resolution technologies, which approach transmission electron microscopy in terms of resolution but generally require extensive dwell times.

1.2 Live Cell Imaging with Microtubule Reporter Proteins

Ever since the successful exploitation of green fluorescent protein (GFP) in the early 1990s, a vast body of literature has been generated on the use of this and other intrinsically fluorescent proteins as reporters of structural proteins [33]. Many constructs, cell and plant lines are now available for observing MTs in living plant cells including the MT-binding domain (MBD) of mammalian MAP4, various tubulins (TUB), EB1, and other MAPs (Table 1). There are clear advantages to being able to follow MTs in living cells. Time-lapse imaging enables the growth and shrinkage of MTs to be followed in near real-time [37–39]. Due to the difficulties in obtaining clear MT images from deep tissue, most MT dynamics data come from observing cortical MTs in epidermal cells. Spinning disk confocal microscopy is especially suitable for study of MT dynamics because of the quick image acquisition possible compared to conventional laser scanning confocal microscopy. Another option would be to use total internal reflection fluorescence (TIRF) microscopy or more precisely near-TIRF (also referred to as variable-angle epifluorescence) microscopy. TIRF and near-TIRF

Table 1
Transgenic reporter lines for observing microtubules in live cells of *Arabidopsis thaliana*

	Promoter	Note	Reference
Tubulin marker line			
CFP-TUA1	35S		[81]
YFP-TUA5	35S		[6]
mCherry-TUA5	35S		[89]
GFP-TUA6	35S	Right-handed helical growth, no expression in roots	[65]
GFP-TUB6	35S	No expression in roots	[58]
mRFP-TUB6	35S		[7]
mRFP-TUB6	UBQ1		[60]
Microtubule-binding protein marker line			
GFP-MBD (from mammalian MAP4)	35S	MT bundling	[68, 80]
mRFP-MBD	35S		[90]
mCherry-MAP4-MBD	35S		[89]
GFP-EB1	35S	Overexpressing EB1 extensively binds MTs including minus end of MTs and endomembrane	[78, 79]
EB1b-GFP	EB1	MT plus-end marker	[70]
ARK1-GFP ARK1 ΔARM-GFP	ARK1	Expressing all cell types and label MT plus end	[91]

allow fluorescence to be captured close to the cover slip, such as in the cell cortex of epidermal cells. By adjusting the incident illumination angle, the excitation depth can be changed, eliminating or decreasing the cytoplasmic background fluorescence caused by free tubulin (Fig. 4). However, due to the limitation in depth of excitation, endoplasmic MTs or MTs in deep tissues cannot be acquired with TIRF and near-TIRF microscopy. Multiphoton fluorescence microscopy is ideal for such cases, and can also be used to acquire images from deep tissue such as spongy mesophyll cells in the leaf (Fig. 5). Conventional laser scanning confocal microscopy also remains a powerful tool for obtaining optical sections in thick samples and for following processes such as mitosis in the root [40, 41] as well as quantitative analysis for co-localization of MT-associated proteins with MTs [42]. There are, however, several limitations to live cell imaging of microtubules and great care needs to be exercised to avoid acquisition of erroneous information. Some of these problems are outlined in detail by Shaw [43]. In Subheading 4, we

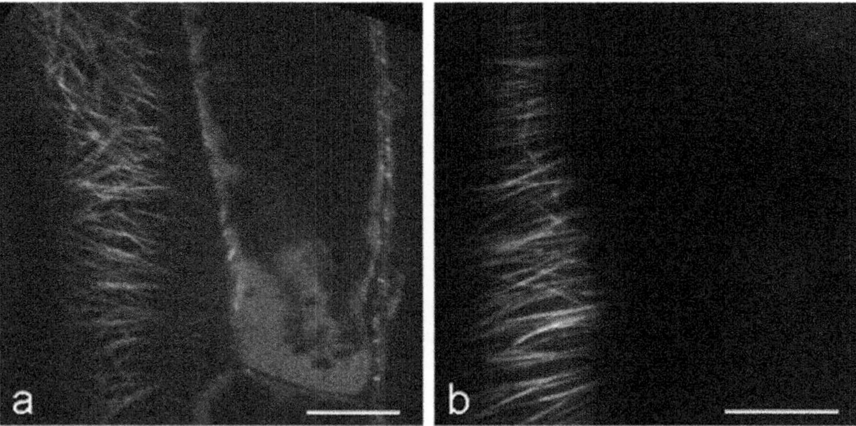

Fig. 4 Images of RFP-TUB6 labeled MTs in *Arabidopsis* hypocotyl cells acquired using spinning disk confocal microscopy (**a**) and near-TIRF microscopy (**b**). (**a**) Single-frame image was acquired with a spinning disk confocal microscope (Perkin Elmer UltraView Vox) with a 63× NA 1.3 glycerol immersion lens. In the image, MTs are in focus in the left cell, and free tubulin in the cytoplasm is shown in the focal plane in the right cell. Occasionally we observe free tubulin as background signal. (**b**) This single-frame image was acquired with a near-TIRF microscope (Zeiss Laser TIRFIII) with a 63× NA 1.46 oil immersion lens. MTs are in focus in only one cell. Due to the thin excitation depth, MTs in neighboring cells are not captured. Bars, 10 μm

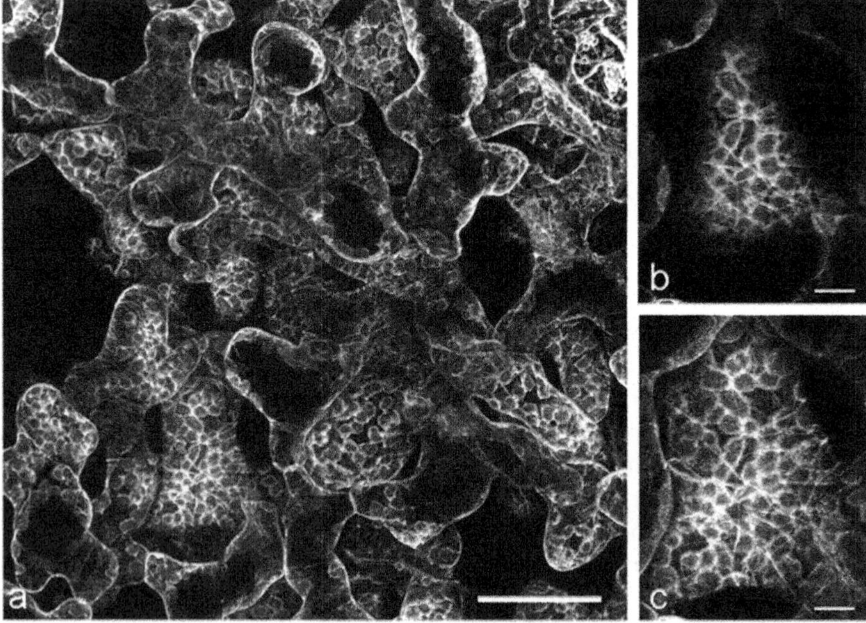

Fig. 5 MTs observed in *Arabidopsis* inner leaf tissues using two-photon imaging. (**a**) Spongy mesophyll cells expressing GFP-TUB6. Leaf was mounted in perfluoroperhydrophenanthrene (PP11) and the image was acquired with a two-photon fluorescence microscope (Olympus FV1000 MPE) with a 25× NA 1.05 water immersion lens. GFP-TUB6 is localized around chloroplasts and MTs are observed in some mesophyll cells. (**b**) MTs in spongy mesophyll cell from (**a**), focal plane in the cell cortex, (**c**) same cell as in (**b**), maximum *Z*-projection of 5 slices (5 μm total thickness), encompassing the outer periclinal cortex of a spongy mesophyll cell. Bars, 50 μm (**a**), 10 μm (**b**, **c**)

outline several important issues to consider when live cell imaging MTs, such as phototoxicity, choice of promoter, fusion protein reporter construction and culture temperature.

1.3 Super-Resolution Microscopy

Standard diffraction-limited optical microscopy techniques cannot resolve cellular details that are closer together than ~200 nm in the focal plane or ~450 nm along the optical axis. This is far from resolving MTs, which have a 25 nm diameter. Scanning tunnelling, atomic force, or electron microscopes provide much higher resolutions but are restricted to cell surfaces or fixed specimens. In contrast, super-resolution light microscopy techniques increase resolution to below the diffraction limit with the possibility to explore a whole cell in 3D [44]. Compared to traditional confocal microscopy, a dramatically increased lateral resolution is achieved by stimulated emission depletion (STED) and ground-state depletion (GSD) microscopy. These techniques are particularly suitable to visualize MTs after standard immunofluorescence preparation [45–47]. Stochastic optical reconstruction microscopy (STORM) or photoactivation localization microscopy (PALM), combined with single-particle tracking, allow mapping transport trajectories of, for example, lysosomes [48] or other cargo on individual MTs by using photoactivatable fluorescent proteins. Interferometric photoactivated localization microscopy (iPALM) has even been applied successfully to measure the 25 nm MT diameter by visualizing a fluorescent tagged α-tubulin (Fig. 6a–c; [49]). Until now, these methods have mainly been restricted to mammalian cells, as super resolution microscopes are very sensitive to auto-fluorescent molecules such as chlorophylls and accessory pigments. Their excitation with class 4 lasers at 592 or 660 nm (e.g., as applied by STED microscopy) causes excessive heating and structural damage. However, in the future, these limitations could be bypassed by irradiating samples at a higher wavelength (775 nm) to avoid excessive autofluorescence. In plants, structured illumination microscopy (SIM) of GFP-fused MT-associated proteins was used to visualize MT dynamics in hypocotyl epidermal cells of *Arabidopsis thaliana* at the subdiffraction level [50]. Although still mainly practical for fixed material, super resolution techniques are of great value for MT imaging, as they bridge the gap between cellular ultrastructure and superstructure.

1.4 Transmission Electron Microscopy

Despite the emergence of super-resolution imaging, transmission electron microscopy (TEM) remains a valuable tool when high-resolution analysis of individual MTs is required, especially when other cellular structures need to also be visualized. Since the 1960s, when MTs were first observed in plant electron micrographs [51], there has been debate over the best fixation protocol. Chemical fixation does a reasonable job, though after glutaraldehyde-osmium fixation (Fig. 7a) or potassium hexacyanoferrate

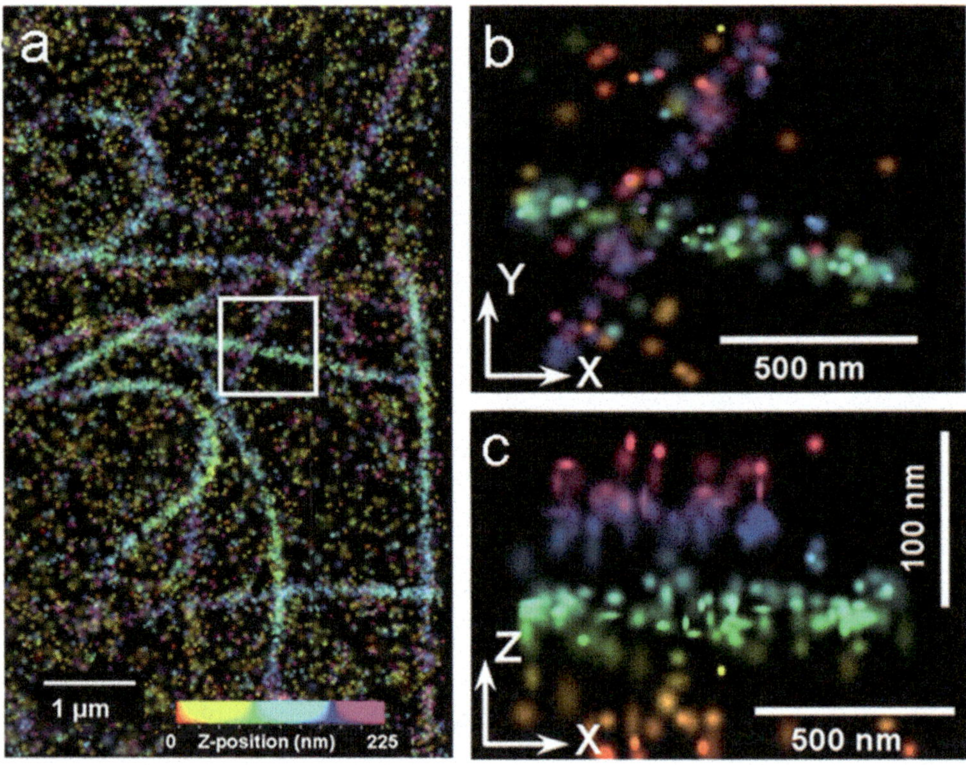

Fig. 6 Super-resolution iPALM image of fluorescently labeled microtubules (m-KikGR fused to α-tubulin), rendered with *z*-axis color-coding. (**a**) Large area overview. (**b**) *X–Y* projection and corresponding (**c**) *Z–Y* projection of the area bound by the *white box* in (**a**). Reprinted from [49], National Academy of Sciences, with permission

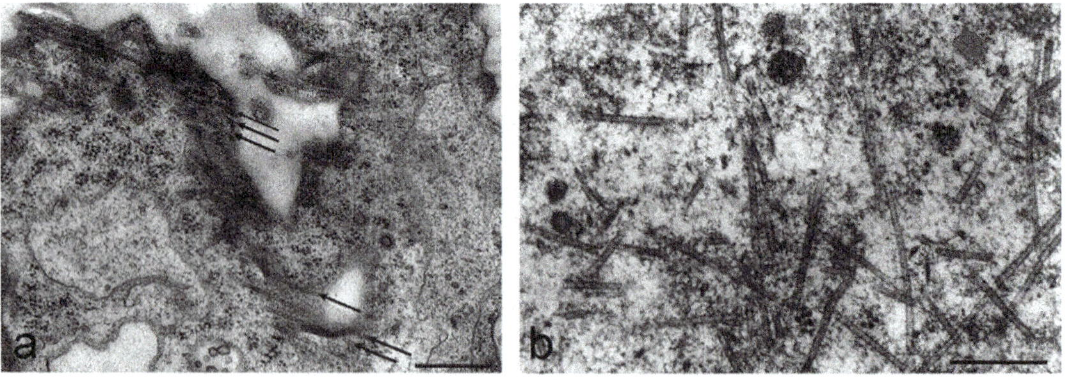

Fig. 7 TEM visualization of MTs in different algal cells: (**a**) chemical fixation protocol in *Klebsormidium crenulatum* visualizing cortical MTs in an adult cell. The *arrows* indicate the parallel orientation of the MTs. (**b**) plunge-frozen and freeze-substituted MTs in *Micrasterias muricata*. MTs in association with a migrating nucleus show various orientations in a MT organizing centre. Note the clear and sharp outer surfaces of the MTs. Bars, 0.2 µm

fixation, MTs appear slightly wrinkled (though well-preserved). Presently, high-pressure freezing followed by freeze substitution is considered the best method for preserving plant cellular ultrastructure (e.g. [52–55]) but involves specialized apparatuses and cannot be performed in many research facilities. We therefore also describe plunge freezing as an alternative, which can preserve cortical regions of cells up to thicknesses of ~20 μm very well. This method is acceptable for examining the ultrastructure of cortical MTs in single cells (Fig. 7b).

Despite the static picture of MT organization provided, TEM remains essential when resolution beyond the limit of conventional fluorescence microscopy is required, such as when examining the higher order structure of MT bundles, the presence of cross-linking proteins and the interaction of MTs with endomembrane components and other structures. Using gold-coupled antibodies, it is also possible to analyze how proteins colocalize [56], identify motor proteins of the kinesin family [24], and investigate how cross-linking MAP65s interact with MTs [57]. It is important to note, however, that immunolocalization by TEM requires specialized preparation to preserve antigenicity of the protein epitopes. Best results are obtained after high pressure freeze fixation, freeze substitution and embedding in LR white resin [24]. Within the context of TEM, we also discuss electron tomography for plant MT imaging. Electron tomography is a three-dimensional TEM imaging technique in which the internal ultrastructure of a sample is captured including, for example, MT orientation, bundling, and association with MAPs, within a thin resin-embedded or frozen section. Electron tomography has the potential to provide a complete picture of MT organization within the cortical cytoplasm.

1.5 Correlative Microscopy

Finally, a chapter on imaging techniques would not be complete without mention of correlative microscopy. Correlative microscopy combines, in the broadest sense, two or more different microscopy techniques in order to obtain complementary information about a sample. In practice, it often involves the use of light and electron microscopy (and is therefore termed 'correlative light and electron microscopy' or CLEM). Given proper preparation, a sample can be imaged with fluorescence microscopy to identify a location of interest (for example, a fluorescently tagged MT-associated protein) and the same sample subsequently imaged with TEM or tomography to obtain high-resolution 2D or 3D information at the location of interest. CLEM has tremendous potential to combine information gained by specific tagging, localization, and dynamics studies with 3D cellular ultrastructure. The reader is referred to Barton et al. [45] for more information.

2 Materials

2.1 Plant Material

1. *Arabidopsis thaliana* seeds, constructs: Seed stocks, including transgenic and mutant lines, as well as some DNA stocks, can generally be obtained from the Arabidopsis Biological Resource Centre. The Arabidopsis Information Resource (TAIR) website provides details and ordering information (http://www.arabidopsis.org/abrc/index.jsp).
2. Plant leaf material (this can vary according to the researchers' demands). Immunofluorescence and TEM protocols generally work with every type of plant leaf.
3. Plant cell cultures (e.g., suspension cell cultures, unicellular green algae).
4. Specific transgenic reporter lines for observing MTs in live cells of *Arabidopsis thaliana* are listed in Table 1 (*see* **Note 1**):

2.2 Reporter Lines (*See* Note 1 and Table 1)

1. *35Spro-GFP-TUB6*: This construct from the Hashimoto lab (NAIST, Japan; contact Dr. T. Hashimoto: hasimoto@bs.naist.jp) fuses GFP to the N-terminus of the β-tubulin 6 (At5g12550) isoform of *Arabidopsis thaliana* [58] (Fig. 8a–c) (*see* **Note 2**).
2. *35Spro:GFP-MBD*: This construct was developed in Dr. Richard Cyr's lab (Pennsylvania State University; rjc8@psu.edu) and is a heterologous MT reporter that works in plants, constructed from the MT binding domain (amino acid residues 935–1084) of the mammalian MAP4 [59] (Figs. 2b, d and 8a) (*see* **Note 3**).
3. *EB1bpro:EB1b-GFP*: This reporter generates comet-like fluorescent patterns at the growing plus ends of MTs and can be used to measure MT growth and to analyze MT polarity within an array (Fig. 9) (*see* **Note 4**).

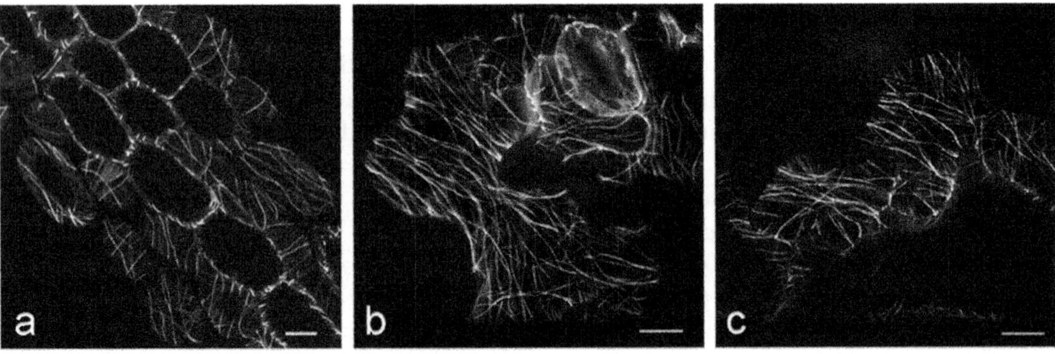

Fig. 8 Comparison of different GFP-transgenic lines in wild-type background of *Arabidopsis thaliana*. (**a**) GFP-MBD hypocotyl of a 10-day-old plant grown at 21 °C, (**b**) 11-day-old GFP-TUA, (**c**) 12-day-old GFP-TUB expressing plants grown at 21 °C were imaged within a few hours after the temperature was increased to 31 °C. Bars, 10 μm

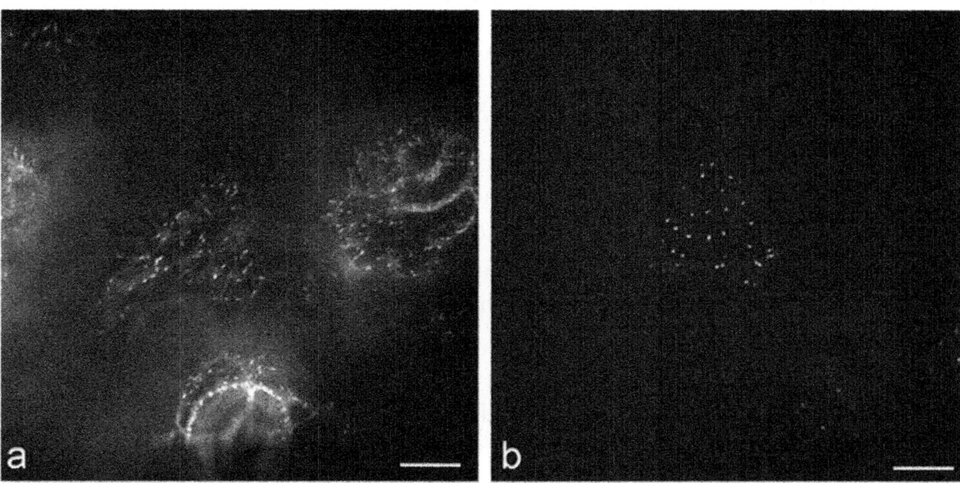

Fig. 9 Images of *35Spro:GFP-EB1* (**a**) and *EB1pro:EB1-GFP* (**b**) expressing *Arabidopsis thaliana* cells at 21 °C. (**a**) *35Spro:GFP-EB1*, GFP-EB1 forms a comet-like shape and weak associates with the side walls of MTs, and the background fluorescence is high. (**b**) *EB1pro:EB1-GFP*, plus-end labeling by EB1-GFP is more concentrated showing dot-like forms and the background signal is very low. Abaxial surface of first leaves from 12-day-old plants were used. Plants were grown at 21 °C and imaged with a Quorum Wave FX Spinning Disc Confocal System (Quorum, Guelph, Ontario, Canada) with a 63× NA 1.3 glycerol-immersion lens. A Bionomic Controller BC-110 stage together with a Heat Exchanger HEC-400 kept the specimens at 21 °C. Bars, 10 μm

4. *UBQ1pro:mRFP-TUB6*: Engineered for and first described in Ambrose et al. [60] by the Wasteneys Lab (Department of Botany, University of British Columbia; contact geoffrey.wasteneys@ubc.ca), Also available through the Arabidopsis Biological Resource Centre. This reporter is useful when the localization of a GFP- or YFP-tagged protein needs to be assessed in relation to MTs. It has several advantages over previous constructs (*see* **Note 5**).
5. *ARK1pro:ARK1-GFP* and *ARK1pro:ARK1ΔARM-GFP*: This construct is from the Wasteneys lab [91]. ARMADILLO-REPEAT KINESIN 1 (ARK1) is an MT plus-end tracking protein. These reporter lines show asymmetric plus-end labeling of MTs in many cell types in seedlings.

2.3 Technical Equipment

1. Temperature-controlled stage: Bionomic Controller BC-110 together with a Heat Exchanger HEC-400, a Bionomic Controller BC-100 (20-20 Technology Inc., Wilmington, NC, USA).
2. Objective lens heater (BIOPTECHS, Butler, PS, USA).
3. Culture chambers for live cell imaging (obtained from Electron Microscopy Sciences, Hatfield, USA).

4. Thermocouple device (FLUKE52; John Fluke MFG. Co., Inc., Everett, Washington, USA).
5. High-pressure freezer (formerly Balzers HPM 010, later taken over by BAL-TEC; then produced by ABRA Fluid AG, Widnau, Switzerland; currently available as "High Pressure Freezing Machine HPM 010" by Boeckeler Instruments Inc., Tucson/AZ, USA).
6. Freeze substitution device (LEICA EM AFS, Leica Microsystems GmbH, Vienna, Austria).
7. Plunge-freezing device (LEICA EM GP)

2.4 Chemicals and Reagents

All buffers are stored at 4 °C but should be at room temperature when in use, as MTs depolymerize at cold temperatures.

2.4.1 Chemicals and Reagents for Fluorescence Microscopy

1. 2× PME buffer: 50 mM PIPES, 5 mM EGTA, 1 mM manganese sulfate, pH 7.2
2. Fixative: 0.5 % glutaraldehyde, 1.5 % formaldehyde in 1× PME buffer, pH 7.2. Prepare before use. Use of 25 % glutaraldehyde in ampules (EM grade) is preferred. Due to its tendency to polymerize easily, make aliquots of 25 % glutaraldehyde and store at −20 °C. Freeze-thawing of glutaraldehyde will induce polymer formation, which affects the fixation efficiency. Use of 16 % formaldehyde in ampules (EM grade) is preferred. Store at 4 °C or room temperature.
3. 1× PMET buffer: 0.05 % Triton X-100 in 1× PME buffer, pH 7.2.
4. 1× PBS buffer: 137 mM sodium chloride, 2.7 mM potassium chloride, 4.3 mM sodium hydrogen phosphate, 1.4 mM potassium dihydrogen phosphate, pH 7.5.
5. Cell wall digestion enzyme solution: 0.05 % pectolyase (Pectolyase Y-23, MP Biomedicals, USA), 0.4 M mannitol, 1 % BSA in 1× PBS. Solution can be stored at −20 °C.
6. Permeabilization buffer: PBS, 1 % Triton X-100 in 1× PBS, pH 7.5.
7. Sodium borohydride solution: 1 mg/mL sodium borohydride in 1× PBS (needs to be fresh, prepare before use).
8. Incubation buffer: 50 mM glycine in 1× PBS.
9. Blocking buffer: 1 % BSA, 50 mM glycine in 1× PBS.
10. Primary antibody solution: primary antibody in blocking buffer. For example, mouse anti-α-tubulin, clone B 512 (Sigma) works well at a 1:1000 dilution.
11. Secondary antibody solution: secondary antibody in blocking buffer. For mouse anti-α-tubulin antibody B 512, Alexa 488-conjugated goat anti-mouse IgG (GE Healthcare) could be used at a 1:100 dilution.
12. Citifluor AF1 antifading agent (Citifluor Ltd., UK).

2.4.2 Chemicals and Reagents for Live Cell Imaging

1. Agar (plant cell culture tested), 1.5 %.
2. Hoagland's growth media or MS media.

   Hoagland's growth media: 2 mM KNO_3, 5 mM $Ca(NO_3)_2$, 2 mM $MgSO_4$, 1 mM KH_2PO_4, 90 μM EDTA, 46 μM H_3BO_4, 9.2 μM $MnCl_2$, 0.77 μM $ZnSO_4$, 0.32 μM $CuSO_4$, 0.11 μM MoO_3, autoclave the media.

   MS media: Dissolve 4.31 g of Murashige and Skoog (MS) basal salt mixture, 0.5 g of 2-(*N*-Morpholino) ethanesulfonic acid (MES) in 1 L water. Adjust pH to 5.7 using KOH, autoclave the media.
3. Agarose (low gelling temperature) 2 %.
4. Perfluoroperhydrophenanthrene (56919 FLUKA, Sigma) as a mounting media for leaves (*see* **Note 6**).

2.4.3 Chemicals and Reagents for Transmission Electron Microscopy

1. Cacodylic buffer: 50 mM cacodylic acid, sodium salt, pH 7.2.
2. Cacodylate-glutaraldehyde: 50 mM cacodylate buffer, 1 % glutaraldehyde (EM-grade), pH 6.8.
3. Cacodylate-glutaraldehyde-calcium chloride: 50 mM cacodylate buffer, 1 % glutaraldehyde, 5 mM calcium chloride, pH 7.2.
4. Cacodylate-osmium: 50 mM cacodylate buffer, 1 % osmium tetroxide.
5. Cacodylate-osmium-ferricyanide: 50 mM cacodylate buffer, 1 % osmium tetroxide, 0.8 % potassium hexacyanoferrate III (ferricyanide), pH 7.2.
6. 2 % aqueous uranyl acetate dihydrate.
7. Ethanol solutions: 15, 30, 40, 50, 60, 70, 80, 90, 95, 100 %.
8. 1,2-Propylene oxide reagent.
9. Propylene oxide-ethanol mixture: 1,2-propylene oxide and ethanol (1:1 v/v).
10. A suitable embedding resin: Embed 812, Araldite 502, Spurr's resin, Agar Low Viscosity Resin (LV) (Agar Scientific Ltd. Essex CM24 8DA, England) (*see* **Note 7**).
11. Formvar 1595 E.
12. Chloroform.
13. Reynold's lead citrate.

2.4.4 Chemicals and Reagents for High-Pressure Freezing and Freeze Substitution

1. Lecithin: L-α-Phosphatidyl-choline; 100 mg/mL dissolved in chloroform.
2. Sucrose: 100 mM sucrose.
3. Acetone-osmium-tannic acid: 2 % osmium tetroxide and 0.1 % tannic acid in acetone.
4. Acetone-osmium-uranyl acetate: 2 % osmium tetroxide and 0.05 % uranyl acetate in acetone.

2.4.5 Chemicals and Reagents for Immunogold Labeling TEM

1. TBS: 50 mM Tris-buffered saline, Trizma pre-set crystals; pH 7.5.
2. TBS-blocking solution: TBS, 1 % BSA (fraction V), 1 % acetylated BSA (BSAac), 0.2 % Tween-20.
3. TBS-incubation solution: TBS, 1 % BSA, 0.1 % BSAac, 0.01 % Tween-20.
4. TBS-BSA: TBS, 1 % BSA.
5. Primary antisera/antibodies (e.g., monoclonal antibody directed against bovine brain kinesin, K-1005; anti-α tubulin, cloneB-5-1-2, T 6074, Sigma).
6. 10 nm—collodial gold-conjugated corresponding secondary antibody (e.g., goat anti-rabbit IgG, G-7402; goat anti-mouse IgM, G-5652, Sigma).
7. Formvar-coated nickel grids.

3 Methods

3.1 Microtubule Staining of Fixed Material

3.1.1 Standard MT Labeling Method Using Wall-Degrading Enzymes: Suitable for Plant Organs Lacking Thick Waxy Cuticles Such as Roots

Ensure that samples are always in solution. Dried areas increase the chance of getting non-specific binding of antibodies. This whole mount protocol is modified from [61]. The root-tip squash method is available in [40].

1. Fix samples for 40 min in 0.5 % glutaraldehyde, and 1.5 % formaldehyde fixative solution. Make sure that the sample is immersed in the fixative.
2. Wash three times for 10 min in PMET buffer.
3. Incubate the sample with cell wall digestion enzyme solution for 20 min.
4. Rinse three times for 10 min in PMET on the shaker.
5. Incubate the sample with permeabilization buffer for 3 h at room temperature on the shaker.
6. Rinse three times for 10 min in PBS, and make sodium borohydride solution.
7. Incubate the sample in sodium borohydride solution for 20 min. This additional step helps to reduce aldehyde-derived autofluorescence by reducing free aldehydes groups.
8. Discard sodium borohydride solution and add blocking buffer. Incubate for 30 min.
9. Place the primary antibody solution in a microscope well slide, which allows use of less antibody solution (50–100 μL). Place the sample in the antibody solution and incubate in a humid box (place wet kimwipes in the box) at 4 °C overnight.
10. Transfer samples to a small Petri dish using a brush and rinse five times for 10 min each in incubation buffer on the shaker.

11. Incubate in blocking buffer for 30 min. Another option for blocking buffer would be to use the non-immune serum from the animal species used to raise the secondary antibody. For example, 5 % normal goat serum (Pierce Biotechnology) in incubation buffer.
12. Place secondary antibody solution in the well slide and samples. Incubate in the humid box wrapped with aluminum foil for 3 h at 37 °C.
13. Transfer samples to small Petri dish using a brush and rinse three times for 10 min in 1× PBS on the shaker. Cover the Petri dish to exclude light.
14. Mount sample in CitiFluor AF1 antifading mounting reagent.
15. Seal the cover slip with non-fluorescent nail polish.
16. The slides can be stored at −20 °C for long-time storage.

3.1.2 MT Staining Using the Freeze Shattering Method [35] with Useful Modifications According to [29]: Suitable for Aerial Organs

1. Fix samples for 40 min in 0.5 % glutaraldehyde, and 1.5 % formaldehyde fixative solution. Large organs should be cut into smaller pieces to facilitate infiltration of fixative.
2. Wash in 1× PMET buffer.
3. Blot samples to remove excess buffer, place between two glass slides and dip into liquid nitrogen. Remove and quickly apply pressure by compressing the slides with a pair of pliers before the tissues thaw (*see* **Note 8**).
4. Transfer samples to permeabilization buffer containing PBS for 1 h.
5. Transfer samples to PBS (pH 7.4) for 10 min, followed by 20-min incubation in PBS containing 1 mg/mL sodium borohydride.
6. Incubate in primary antibody overnight at 4 °C.
7. Wash extensively in buffer to remove any unbound antibody (e.g., three times 10 min in PBS).
8. Incubate in secondary antibody (Alexa488-conjugated goat anti-mouse IgG, 1:200) for 1 h at 37 °C.
9. Mount samples in Citifluor AF1 antifade agent.
10. Examine with a Confocal Laser Scanning Microscope (CLSM), using a high Numerical Aperture (NA) objective lens (e.g., Zeiss Axiovert, 63× magnification, 1.4 NA).
11. Generate excitation with an Argon laser at 488 nm.
12. Collect long pass (LP) 560 nm filtered emission and band pass (BP) 505–530 nm filtered emission simultaneously in two channels; if possible collect a DIC image with another channel.
13. Capture Z-stacks according to requirements.
14. Generate projections, e.g., using ImageJ software (freeware; http://www.imageJ).

3.2 Fluorescent Proteins to Visualize the MT Cytoskeleton

This section describes a standard method of preparing aerial organs such as leaves or cotyledons of *Arabidopsis thaliana* for live imaging on an inverted microscope. It can be used for any organs including roots, stems and floral organs and can of course be modified for any plant expressing a fluorescent reporter.

1. Plant surface-sterilized seeds on Petri plates containing nutrient medium in 1.5 % agar. Wrap plates with porous surgical tape.
2. Store plates at 4 °C for 3–5 days to synchronize germination, and then transfer to a growth cabinet, keeping the plants growing under constant light (80 μmol/m^2/s) or in an 16-h light, 8-h dark cycle at 21 °C for 11–12 days.
3. Excise whole leaves or pieces of them and place them on the cover slip that forms the bottom of microscope culture dishes (*see* **Note 9**). We obtain dishes suitable for this purpose from Electron Microscopy Sciences, Hatfield, USA. If you are using an upright microscope, the leaves can be placed directly onto a glass slide and covered with cover slips held in place at corners with small amounts of vacuum grease.
4. Mount leaf cuttings in water or perfluoroperhydrophenanthrene (*see* **Note 6**), taking care to ensure that the surface of interest is facing the cover slip. A small square of 2 % agarose can be placed on top of the leaf to stabilize it from drifting during imaging.
5. Allow samples to equilibrate under standard conditions for at least 1 h prior to observation.
6. Place culture chambers on microscope stage and ensure that images are recorded according to the spectral properties of the fluorescent protein. Images can be taken every 8 s over 3–5 min for GFP-TUB and every 5 or 8 s over 40–60 s for EB1b-GFP (*see* **Notes 10** and **11**).
7. To keep the temperature stable around the mounted samples, use a temperature-controlled stage: Bionomic controller BC-110 equipped with Heat exchanger HEC-400 (20/20 Technology Inc.) (*see* **Note 12**). An objective lens heater (Bioptechs) is sometime also required, especially when using immersion lenses. Monitor the temperature of the sample immediately after imaging by measuring the temperature of glycerol on the cover slip using a thermocouple device.
8. Avoid recording images over periods longer than 4 h.

3.3 TEM Preparations to Visualize MTs in Plant Cells

3.3.1 Glutaraldehyde-Osmium Fixation Protocol for Ultrastructure Studies (with Modifications After [29, 59])

1. Cut small blocks of tissue or cells and place in glass vials. Fix tissue or cells in 10–50 mM cacodylate buffer (*see* **Note 13**), containing 1 % glutaraldehyde for 30–90 min (*see* **Note 14** and Fig. 7a, *Klebsormidium crenulatum*).
2. Remove fixative and wash cells three times 5 min in cacodylate buffer.
3. Postfix cells in cacodylate-osmium for 2 h at room temperature or 12 h at 4 °C.
4. Wash cells three times 5 min with distilled water.
5. Dehydrate the cells in increasing concentrations of cold ethanol, 15 min per concentration: 15, 30, 40, 50, 70, 80, 90, 95, and 100 %. Keep in 100 % for 30 min.
6. Transfer the cells to a 1:1 mixture of propylene oxide:ethanol and allow to equilibrate for 10 min.
7. Transfer to propylene oxide for 10 min.
8. Transfer the cells to a 1:1 mixture ofpropylene oxide:embedding resin.
9. Rotate the cells in this mixture for 24 h in order to allow the propylene oxide to evaporate and for sufficient penetration of the cells with the resin.
10. Transfer the cells to freshly prepared resin in a polymerization chamber or aluminum dish.
11. Incubate in a desiccator for 4 days.
12. Polymerize the resin for 16 h at 60 °C.
13. Select samples and section using an ultramicrotome. Section thickness should generally be in the range of 50–200 nm. The ideal section thickness for transmission electron microscopy with accelerating voltages between 50 and 120 kV is about 30–100 nm. Thicker sections (up to 200 nm) can be visualized using a 200 kV accelerating voltage and may be beneficial when considering tomography, in which case a larger reconstructed volume can provide more information about MT configuration within the sample.
14. Collect the sections on formvar-coated copper grids (*see* **Notes 15** and **16**).
15. Counter stain the sections by floating them, section-side down, on small 10 μL droplets of 2 % aqueous uranyl acetate for 20 min (*see* **Note 17**).
16. Wash for 2 min with distilled water and allow to dry fully.
17. Next, counterstain on droplets of with Reynold's lead citrate for 10 min (*see* **Note 18**).
18. Wash for 2 min with distilled water and allow to dry fully.
19. Investigate and image with a transmission electron microscope. Some considerations for electron tomography are included below.

3.3.2 Considerations for Electron Tomography

When preparing grids for tilt series acquisition, one must take the following points into consideration:

1. Copper grids, which are non-ferromagnetic, should be used (and not nickel grids, *see* **Note 15**). Nickel grids distort the magnetic field of the objective lens and reduce the lens' ability to reproduce the object plane.
2. Generally, slot grids are preferred. Plant tissue sections are often larger than the mesh size on a mesh grid; using a mesh grid may result in some of the sample being obstructed by the grid bar. When using a slot grid, tilt series acquisition can be performed along the axis of the slot.
3. Thicker (200 nm) sections are advised in order to maximize the reconstructed volume in the final tomogram.
4. Electron dose must be taken into consideration during tilt series acquisition, especially when performing CEMOVIS (*see* **Note 21**). For vitreous samples, a "low-dose" imaging protocol is necessary to avoid excessive electron irradiation of the sample.

   For further information, including operating protocols, the reader is referred to [62].

3.3.3 High-Pressure Freeze Fixation Protocols to Preserve Microtubules

The protocol described here uses a commercially available hyperbaric freezing device. The methods basically follow the methods of [52] (Fig. 10), with the addition of notes from own experience and information gained from several other works as stated in **Notes 19–26**.

1. Dip specimen cups in lecithin (L-α-phosphatidylcholine; 100 mg/mL dissolved in chloroform) and allow to dry until specimen holders are covered with lecithin.
2. Prepare samples in their regular culture media; for certain plant tissues it may be useful to transfer samples to 100 mM sucrose (*see* **Note 19**).
3. Transfer the samples into gold or aluminum specimen cups.
4. Ensure that no air bubbles are present in the preparation as they will damage the samples with the high pressures involved in this process.
5. Close two cups face to face or, depending on the sample, orient the face toward the bottom.
6. Mount the cups in the specimen holder and transfer to the High Pressure Freezing Machine HPM 010 (*see* **Note 20**).
7. Freeze samples in liquid nitrogen according to the manufacturer's instructions.
8. Remove samples immediately after the freezing procedure and transfer them to liquid nitrogen, where they can be stored until further processing, either by freeze substitution followed

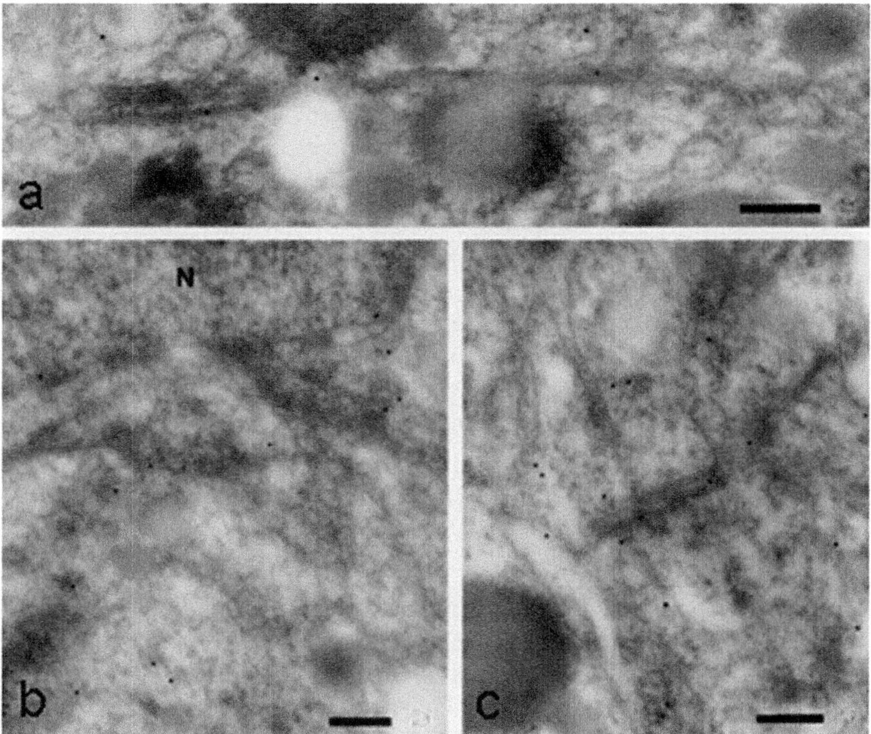

Fig. 10 High-pressure frozen MTs in the vicinity of a migrating nucleus in the green alga *Micrasterias denticulata* immuno-gold (10 nm) stained with an antibody to brain kinesin (a-c). Bars, 0.2 μm. Reprinted from [24], Elsevier, with permission

by sectioning (as outlined below), or by cryo-sectioning (CEMOVIS) (*see* **Note 21**).

9. Freeze substitute samples in 2 % OsO_4 in acetone or 0.1 % tannic acid in acetone at −80 °C for 24 h and a mixture of 2 % OsO_4 and 0.05 % uranyl acetate in acetone at −80 °C for 28 h (*see* **Note 22**).
10. Raise the temperature to −30 °C in the same medium over the course of 10 h. Bring to room temperature and rinse in acetone.
11. Embed and further process the samples as described in Subheading 3.3.1, **step 7**

3.3.4 Plunge Freezing of Thin Samples

As an alternative to high-pressure freezing, we describe plunge freezing [63]. Protocol is based on [53, 64].

1. Prepare the cryo-fixation device: switch on the plunge-freezing device (Leica EM GP) and fill the water container for the humidifier with 60 mL distilled water. Mark a cryo-storage box for identification of the sample and position this box in the

plunge-freezing device transfer container. Fill the liquid nitrogen dewar with (1.5 L) liquid nitrogen. When the ethane container reaches liquid nitrogen temperature, fill the (black) secondary cryogen container with ethane or ethane/propane using the cryogen liquefier.

2. Set the vitrification parameters: using the control panel, set the environmental chamber temperature to 22 °C and the humidity (Hr) to 95 %. Set the blotting time to zero seconds (blotting is used to remove excess liquid from an EM grid before plunging a liquid sample and is not necessary for solid plant samples). Set the liquid ethane temperature to −181 °C when liquid ethane is used, or to −191 °C when a liquid ethane/propane mixture is used.
3. Sample vitrification: Mount a thin sample onto the plunge device forceps (*see* **Note 23**). Using the plunge freezer, rapidly plunge the sample into the liquid ethane, or liquid propane/ethane mixture.
4. After the freezing step, transfer the sample into the cryo-storage box prepared. Plunge more samples, if desired. When finished, store the cryo-storage box in liquid nitrogen before further manipulation (either freeze substitution followed by sectioning or CEMOVIS).

3.3.5 Immunolabeling TEM Protocol

The procedures described are modified from [56] and can be adjusted for basically any new MT-associated protein or for other cytoskeletal components (Fig. 9).

1. Fix samples. Follow the Subheading 3.3.1, **steps 1–5**, except for osmium fixation (**steps 3** and **4**). Osmium fixation would mask the epitopes. After dehydration (**step 5** in Subheading 3.3.1), make LR white resin:acetone mix to start resin infiltration. The resin concentration should gradually increase from 1 drop of resin in 20 mL acetone to 10, 30, 50, 70, 90, 95, and 100 % resin. Exchange 100 % resin three times. Polymerize resin under UV light.
2. Thin sections should be collected on formvar-coated nickel grids.
3. Pre-incubate sections in TBS-blocking solution for 1 h (*see* **Note 24**). This can be done by floating the grids on drops of the solution.
4. Incubate sections in primary antibodies (e.g., monoclonal antibody directed against bovine brain kinesin; Sigma K-1005) diluted 1:50, 1:100 in TBS, 1 % BSA, 0.1 % BSAac, 0.01 % Tween-20 (TBST) for 20–24 h at 4 °C. Antibody concentrations for immunoTEM generally need to be higher than those used for immunofluorescence.

5. Wash in droplets of 50 mM TBS, 1 % BSA, four times, 15 min.
6. Transfer into gold-conjugated secondary antibody in TBS-incubation solution in a dilution of 1:200 or 1:400 and incubate for 1.5 or 2 h at room temperature.
7. For control purposes omit the first antibody on some sections or incubate in antigen-saturated primary antibody.
8. After the incubation with secondary antibody, rinse the sections extensively with TBST to remove any unbound antibody (*see* **Note 25**).
9. Continue as in Subheading 3.3.1, **step 15** (*see* **Note 26**), for uranyl acetate and lead citrate staining.

4 Notes

1. *Fusion protein reporter construction*: Not surprisingly, fusing a fluorescent reporter to a protein can alter its function. The GFP-TUA6 construct (GFP is fused to the N-terminus of alpha tubulin 6) was the first tubulin-based fluorescent reporter to work in plant cells [65]. It generates bright labeling of MTs in aerial tissue (it is expressed but apparently does not incorporate into MTs in root tissues) but causes right-handed organ twisting, a phenomenon attributed to inhibition of the tagging of alpha tubulin at its N-terminus, which may interfere with its GTP hydrolysis-promoting activity [66]. After testing several different fluorescent reporters for imaging MTs, we recommend the constructs listed in the text. These constructs are available upon request from the laboratory of origin or from the *Arabidopsis* stock centres. Most constructs are also available with other fluorescent tags such as yellow fluorescent protein (YFP) and red fluorescent protein (RFP).
2. The expression of *35Spro:GFP-TUB6* in stably transformed lines of the Columbia ecotype is regulated by the cauliflower mosaic virus 35S promoter [58]. In comparison to some other MT reporters, this one generates relatively low fluorescence but there are no detectable developmental or morphological phenotypes [66, 67]. This line cannot be used reliably to examine MTs in root cells though some labeling of MTs can be observed in dividing cells.
3. The *35Spro:GFP-MBD* construct is good for examining MTs in root tissues, where it has been used to document MTs in division [41, 40] and elongation zones [68] as well as during root hair development [15, 69]. Nevertheless, caution needs to be exercised with this reporter. Plants stably expressing GFP-MBD under the CMV35S promoter frequently have severe developmental defects and low seed yield, and MT

bundle formation has been observed under certain conditions (for more details see *Notes on choice of promoter*, **Note 11**). We suggest confining the use of GFP-MBD to roots, and then to select specimens that have low level of expression. This can be achieved by selecting heterozygotes from an F2 segregating population [15].

4. There are many different fluorescent EB1 reporters available but the *EB1bpro:EB1-GFP* construct from the Cyr lab (Pennsylvania State University, USA) uses the promoter element from the *Arabidopsis thaliana* EB1b isoform to drive moderate expression levels [70], which confines comets to the growing plus ends. Even so, we have found that MT growth rates are slightly higher when measured with this reporter compared to GFP-TUB [39], a finding that underscores the importance of carefully regulating the expression of some reporters (for further details see *Notes on choice of promoter*, **Note 11**). EB1 reporters seem to work equally well with the fluorescent tag at either the C- or N-terminus.
5. Using the UBQ1 promoter to drive mRFP-TUB expression prevents gene silencing, which is common with 35S promoters. This reporter also works well in root tissues.
6. Perfluoroperhydrophenanthrene (PP11) is a nontoxic solution, which infiltrates the air space between mesophyll cells of the leaf, enabling a homogenous refractive index through the tissue. The image quality and resolution of the leaf inner tissue is greatly improved with little effect on physiology [71–73].
7. For ultrastructural studies, Spurr's resin (TedPella Inc, USA) [74] may be the best choice among other epoxide resins when different degrees of hardness have to be achieved according to the needs for different plant tissues. Its low viscosity property facilitates its quick infiltration into the plant tissue. The "Agar Low Viscosity Resin (LV)" kit from Agar Scientific is another alternative. For immunogold labeling, use LR white resin (London Resin Company, medium grade). Acrylic resins including LR white resin polymerize under UV light, which prevents epitopes from heat damage that are usually associated with the polymerization of epoxide resin. Their hydrophilic properties and rough surface facilitate epitope recognition by antibodies.
8. Some delicate samples can be effectively permeabilized by repeated freeze-thaw cycles and do not require compression in the frozen state.
9. Excising organs from plants will generate a wound response that may alter MT behavior at a site distal to the wound. Other physically-induced responses include touch responses, gravitropisms, and photoinduction. Forcing cells close to the cover

slip for optimal viewing may generate a touch response that manifests itself in aberrant MT behavior. Some plant organs, such as hypocotyls and roots, are strongly gravity or light responsive and when placed on a conventional horizontal microscope stage will undergo differential flank growth in an attempt to resume vertical growth. Horizontally placed roots, for example, will be stimulated to grow more rapidly on the upper flank and growth will be inhibited on the lower flank. With an inverted stage microscope, this will mean that the growth of cells closest the lens will be reduced or prematurely arrested. Upon placement in the horizontal position, cell division in *Arabidopsis* roots has been shown to be temporarily suspended [40]. Hypocotyls are exquisitely light responsive and rapid switches in MT orientation, likely involving altered dynamics, have been noted when etiolated hypocotyls are exposed to blue light or treated with hormones [75, 76].

10. Phototoxicity: The risk of phototoxicity from the high-intensity light used to excite fluorophores is the first consideration when live cell imaging [77]. Scan times need to be limited to avoid bleaching the fluorescence or impairing the function of the tagged protein. This makes it far more challenging to collect the same level of 3-D information that is possible on fixed, immunolabeled material. Choosing fluorescent proteins that can be excited by low energy, longer wavelength light (such as RFP) is one strategy but special attention to autofluorescence is then required. Probably the most important way to limit phototoxicity is to choose the most light-sensitive camera possible, and to minimize the intensity of the excitation light source. Spinning disc scan heads can also reduce dwell times, reducing both phototoxicity and photobleaching.

11. Choice of promoter: To overcome the need for high-intensity excitation, transgenic lines are typically chosen that have high expression levels, usually achieved by using the 35S cauliflower mosaic virus (CMV) or ubiquitin promoter elements, which drive constitutive and generally high expression in most tissues of higher plants. Depending on where the transgene inserts, and whether multiple insertions take place, the level of expression of MT reporter proteins can vary considerably. The 35S CMV promoter has been used effectively for most MT reporters, but in some cases the high level of expression generates artifacts by changing the dynamics of MTs or promoting bundle formation. For example, 35S-driven EB1-GFP can decorate both the plus and minus ends of growing MTs and is even found along the MT lattice and endomembranes [78, 79], whereas it is distributed to the growing plus end when an endogenous promoter is used to drive its expression [70]. The GFP-MBD reporter fusion was constructed from the

MT-binding domain (amino acid residues 935–1084) of the mammalian MAP4 [80]. Under control of the 35S promoter, GFP-MBD expression decorates MTs in plant cells very effectively (Fig. 2b) but can cause significant developmental defects, suggesting that this heterologous protein (there is no MAP4 homologue known in plant cells) interferes with normal function of MTs. In different genetic backgrounds, the *35Spro:GFP-MBD* has been shown to cause unusual bundle formation that is not observed in the same genotypes with other reporters or by immunofluorescence [81, 82, 40]. Interestingly, in a recent study in which the IRX7 promoter element was used to control GFP-MBD expression, expression was restricted to vascular and epidermal cells, and no developmental defects were reported [83].

12. Culture temperature: It has been reported that temperature shifts significantly alter MT dynamics in *Arabidopsis* epidermal cells [39]. If the intention is to measure MT dynamics (growth and shrinkage velocity, time spent in growth, shrinkage, and pause) then a temperature-controlled stage is necessary. The large proportion of metal components in microscopes makes it difficult to manipulate temperatures. Lenses are good temperature sinks and special lens heaters may be required if the desired culture temperature is above ambient.
13. The buffer concentration has to be selected carefully according to the physiological demand of the cells (10–50 mM is suitable for most plant cell types). When the buffer concentration is too high, osmotic phenomena may occur that lead to a detachment of the plasma membrane from the cell wall.
14. Cells can be alternatively fixed in 50 mM cacodylate buffer and 1 % glutaraldehyde, washed, postfixed for 2 h in a mixture of buffered 1 % osmium tetroxide and 0.8 % potassium hexacyanoferrate, and then post-stained in aqueous 2 % uranyl acetate according to [84].
15. Copper grids rapidly conduct heat away from the support film and help prevent thermal expansion and hence movement of the specimen under thermal radiation. During immunolabeling, however, Cu^{2+} ions may have an inhibitory effect on antibody functioning, so nickel grids are used in immuno-applications. Ni grids, however, are ferromagnetic, and distort the magnetic field of the objective lens, which is especially of concern during tilt series acquisition for tomography, when image alignment is important.
16. Grids can be either coated with formvar in the lab, or purchased pre-coated. Briefly, to prepare formvar coated grids in the lab, dip a clean cover slip in 0.3 % formvar dissolved in chloroform, remove gently, and allow to dry. Next, cut the

edges on top of the cover slip with a sharp pin or razor blade and allow the thin film to float on a water surface (MilliQ water in a glass dish) by gently dipping the cover slip in the water at an angle. Place grids on this film and finally collect from the water surface with ParafilmTM.

17. Prior to counterstaining, spin down (>5 min) the uranyl acetate and lead citrate solutions to remove any precipitates. During incubation in uranyl acetate solution it is necessary to keep the sections in darkness by covering the dish with aluminum foil. Light causes precipitation of the compound.
18. During incubation in lead citrate it is necessary to place some NaOH pellets in the staining dish (near the sections) to prevent precipitation of the lead citrate on the sections due to exposure to CO_2. The following procedure has been found useful for the preparation of Reynolds lead citrate: allow distilled water to boil for 10 min to remove air and carbon dioxide, let cool and use only this water for the subsequent steps. Dissolve 1.33 g lead nitrate and 1.76 g sodium citrate in 30 mL of the boiled distilled water, shake vigorously for 1 min (the salts will not dissolve right away, but will remain in a white solution), and then let stand for 30 min. Next, add 8 mL of 1 M sodium hydroxide until the solution becomes clear. Wait. If necessary, continue adding drops of sodium hydroxide until the solution clears. Finally, fill to 50 mL with boiled distilled water.
19. Samples may be adjusted to increasing sucrose concentrations by cultivating them in medium containing gradually increasing sucrose [54].
20. Alternatively, samples can be frozen in a Leica EMPACT high pressure freezer (Leica Microsystems GmbH, Vienna, Austria).
21. CEMOVIS, or cryo-electron microscopy of vitreous sections, enables direct visualization of high pressure or plunge frozen samples in their native state, avoiding the possible artifacts induced by dehydration and staining during freeze substitution [85]. It requires a cryo-microtome device with which vitreous sections can be cut from the frozen block of tissue. These can then be directly imaged at cryogenic conditions with the TEM.
22. Some researchers recommend adding a certain proportion of water to the samples to improve the quality of the final preservation (e.g., [86, 87]). Added water (5 % (v/v)) saturates the polar solvent during warming, which is beneficial for sample quality because it maintains structural water layers around cellular macromolecules (and stabilizes the hydrophobic effect). It is always necessary to add water (5 % (v/v)) to high-pressure freezing and freeze substitution protocols for STED and PALM/STORM in order to maintain fluorescence of

fluorescent tagged proteins [88]. Moreover, it may be useful to prolong the substitution process for several days or up to a week according to the size of the plant samples (A. Staehelin, personal communication).

23. This method is only suitable when the samples are extremely thin, and it is expected that only the first ~20 μm are preserved well, whereas with high-pressure freezing up to ~200 μm thick samples are preserved well.
24. Depending on the desired blocking, alternatively the following "blocking buffers" can be used: (a) TBS, 1 % BSA, 0.1 % Tween-20 for 20 min followed by incubation in TBS, 1 % acetylated BSA; or (b) TBS containing 1 % BSA, 20 % bovine fetal serum followed by incubation in TBS, 1 % BSA, 1 % BSAac, 0.5 % Tween-20; or (c) TBS, 5 % nonfat milk, 0.1 % Tween-20 followed by TBS, 1 % BSA, 0.2 % gelatine; or (d) 10 mm phosphate-buffered saline (PBS), pH = 7.4, 50 mm glycine (Merck), followed by PBS containing 1 % BSA, 2 % gelatine (Merck).
25. To wash the grids, follow this procedure: rinse the grids containing the sections with a mild spray of TBS from a Pasteur pipette, and then transfer onto small droplets of TBS for 2 min. Next, rinse a second time with TBS and complete the washing step with a mild spray of double-distilled water from a plastic spray bottle.
26. Counterstaining may either be omitted or the duration reduced according to the desired density of staining.

Acknowledgements

This chapter has been supported by Austrian Science Fund (FWF) project P24242-B16 to AH and funding from the Natural Sciences and Engineering Research Council and the Canadian Institutes of Health Research to GOW. The Thieme Verlag KG is acknowledged for their kind permission to reproduce Fig. 1b, c. Figure 6a–c is reproduced with permission of the National Academy of Sciences of the United States of America, and Dr. Jennifer Lippincott-Schwartz, National Institute of Child Health and Human Development, Bethesda, MD. Elsevier Inc. for reproduction of Fig. 10.

References

1. Rasmussen CG, Wright AJ, Muller S (2013) The role of the cytoskeleton and associated proteins in determination of the plant cell division plane. Plant J 75(2):258–269. doi:10.1111/tpj.12177
2. Louveaux M, Hamant O (2013) The mechanics behind cell division. Curr Opin Plant Biol 16(6):774–779. doi:10.1016/j.pbi.2013.10.011

3. Masoud K, Herzog E, Chaboute ME, Schmit AC (2013) Microtubule nucleation and establishment of the mitotic spindle in vascular plant cells. Plant J 75(2):245–257. doi:10.1111/tpj.12179
4. Ruan Y, Wasteneys GO (2014) CLASP: a microtubule-based integrator of the hormone-mediated transitions from cell division to elongation. Curr Opin Plant Biol 22:149–158. doi:10.1016/j.pbi.2014.11.003
5. Ambrose C, Wasteneys GO (2011) Cell edges accumulate gamma tubulin complex components and nucleate microtubules following cytokinesis in Arabidopsis thaliana. PLoS One 6(11):e27423. doi:10.1371/journal.pone.0027423
6. Paredez AR, Somerville CR, Ehrhardt DW (2006) Visualization of cellulose synthase demonstrates functional association with microtubules. Science 312(5779):1491–1495. doi:10.1126/science.1126551
7. Fujita M, Himmelspach R, Hocart CH, Williamson RE, Mansfield SD, Wasteneys GO (2011) Cortical microtubules optimize cell-wall crystallinity to drive unidirectional growth in Arabidopsis. Plant J 66(6):915–928. doi:10.1111/j.1365-313X.2011.04552.x
8. Lei L, Li SD, Bashline L, Gu Y (2014) Dissecting the molecular mechanism underlying the intimate relationship between cellulose microfibrils and cortical microtubules. Front Plant Sci 5:8. doi:10.3389/fpls.2014.00090
9. Fujita M, Lechner B, Barton DA, Overall RL, Wasteneys GO (2012) The missing link: do cortical microtubules define plasma membrane nanodomains that modulate cellulose biosynthesis? Protoplasma 249:S59–S67. doi:10.1007/s00709-011-0332-z
10. Oda Y, Fukuda H (2013) The dynamic interplay of plasma membrane domains and cortical microtubules in secondary cell wall patterning. Front Plant Sci 4:6. doi:10.3389/fpls.2013.00511
11. Sampathkumar A, Yan A, Krupinski P, Meyerowitz EM (2014) Physical forces regulate plant development and morphogenesis. Curr Biol 24(10):R475–R483. doi:10.1016/j.cub.2014.03.014
12. Ivakov A, Persson S (2013) Plant cell shape: modulators and measurements. Front Plant Sci 4:13. doi:10.3389/fpls.2013.00439
13. Zhang CH, Halsey LE, Szymanski DB (2011) The development and geometry of shape change in Arabidopsis thaliana cotyledon pavement cells. BMC Plant Biol 11:13. doi:10.1186/1471-2229-11-27
14. Bibikova TN, Blancaflor EB, Gilroy S (1999) Microtubules regulate tip growth and orientation in root hairs of Arabidopsis thaliana. Plant J 17(6):657–665. doi:10.1046/j.1365-313X.1999.00415.x
15. Sakai T, van der Honing H, Nishioka M, Uehara Y, Takahashi M, Fujisawa N, Saji K, Seki M, Shinozaki K, Jones MA, Smirnoff N, Okada K, Wasteneys GO (2008) Armadillo repeat-containing kinesins and a NIMA-related kinase are required for epidermal-cell morphogenesis in Arabidopsis. Plant J 53(1):157–171. doi:10.1111/j.1365-313X.2007.03327.x
16. Sieberer B, Timmers A (2009) Microtubules in plant root hairs and their role in cell polarity and tip growth in root hairs. In: Emons A, Ketelaar T (eds) Plant cell monograph, vol 12. Springer, Berlin, pp 233–248
17. Ambrose C, Wasteneys GO (2014) Microtubule initiation from the nuclear surface controls cortical microtubule growth polarity and orientation in Arabidopsis thaliana. Plant Cell Physiol 55(9):1636–1645. doi:10.1093/pcp/pcu094
18. Rounds CM, Bezanilla M (2013) Growth mechanisms in tip-growing plant cells. Annu Rev Plant Biol 64:243–265. doi:10.1146/annurev-arplant-050312-120150
19. Chebli Y, Kroeger J, Geitmann A (2013) Transport logistics in pollen tubes. Mol Plant 6(4):1037–1052. doi:10.1093/mp/sst073
20. Yu R, Huang RF, Wang XC, Yuan M (2001) Microtubule dynamics are involved in stomatal movement of Vicia faba L. Protoplasma 216(1-2):113–118. doi:10.1007/Bf02680138
21. Brandizzi F, Wasteneys GO (2013) Cytoskeleton-dependent endomembrane organization in plant cells: an emerging role for microtubules. Plant J 75(2):339–349. doi:10.1111/tpj.12227
22. Foissner I, Menzel D, Wasteneys GO (2009) Microtubule-dependent motility and orientation of the cortical endoplasmic reticulum in elongating characean internodal cells. Cell Motil Cytoskeleton 66(3):142–155. doi:10.1002/Cm.20337
23. Hamada T, Ueda H, Kawase T, Hara-Nishimura I (2014) Microtubules contribute to tubule elongation and anchoring of endoplasmic reticulum, resulting in high network complexity in Arabidopsis. Plant Physiol 166(4):1869–1876. doi:10.1104/pp. 114.252320
24. Holzinger A, Lütz-Meindl U (2002) Kinesin-like proteins are involved in postmitotic nuclear migration of the unicellular green alga *Micrasterias denticulata*. Cell Biol Int 26(8):689–697. doi:10.1006/cbir.2002.0920
25. Holzinger A, Lütz-Meindl U (2003) Evidence for kinesin- and dynein-like protein function in circular nuclear migration in the green alga Pleurenterium tumidum: digital time lapse analysis of inhibitor effects. J Phycol 39(1):106–114. doi:10.1046/j.1529-8817.2003.02074.x
26. Miki T, Nishina M, Goshima G (2015) RNAi screening identifies the armadillo repeat-containing kinesins responsible for microtubule-dependent nuclear positioning in Physcomitrella

patens. Plant Cell Physiol 56:737. doi:10.1093/pcp/pcv002

27. Tatout C, Evans DE, Vanrobays E, Probst AV, Graumann K (2014) The plant LINC complex at the nuclear envelope. Chromosome Res22(2):241–252.doi:10.1007/s10577-014-9419-7
28. Kandasamy MK, Meagher RB (1999) Actin-organelle interaction: association with chloroplast in Arabidopsis leaf mesophyll cells. Cell Motil Cytoskeleton 44(2):110–118. doi:10.1002/(Sici)1097-0169(199910)44:2<110::Aid--Cm3>3.0.Co;2-O
29. Holzinger A, Wasteneys GO, Lütz C (2007) Investigating cytoskeletal function in chloroplast protrusion formation in the arctic-alpine plant *Oxyria digyna*. Plant Biol 9(3):400–410. doi:10.1055/s-2006-924727
30. Holzinger A, Kwok EY, Hanson MR (2008) Effects of arc3, arc5 and arc6 mutations on plastid morphology and stromule formation in green and nongreen tissues of *Arabidopsis thaliana*. Photochem Photobiol 84(6):1324–1335. doi:10.1111/j.1751-1097.2008.00437.x
31. Zaffryar S, Zimerman B, Abu-Abied M, Belausov E, Lurya G, Vainstein A, Kamenetsky R, Sadot E (2007) Development-specific association of amyloplasts with microtubules in scale cells of *Narcissus tazetta*. Protoplasma 230(3-4):153–163. doi:10.1007/s00709-006-0238-3
32. Gilroy S (1997) Fluorescence microscopy of living plant cells. Annu Rev Plant Physiol Plant Mol Biol 48:165–190. doi:10.1146/annurev.arplant.48.1.165
33. Mathur J (2007) The illuminated plant cell. Trends Plant Sci 12(11):506–513. doi:10.1016/j.tplants.2007.08.017
34. Dixit R, Cyr R, Gilroy S (2006) Using intrinsically fluorescent proteins for plant cell imaging. Plant J 45(4):599–615. doi:10.1111/j.1365-313X.2006.02658.x
35. Wasteneys GO, WillingaleTheune J, Menzel D (1997) Freeze shattering: a simple and effective method for permeabilizing higher plant cell walls. J Microsc 188:51–61. doi:10.1046/j.1365-2818.1977.2390796.x
36. Kwok EY, Hanson MR (2003) Microfilaments and microtubules control the morphology and movement of non-green plastids and stromules in *Nicotiana tabacum*. Plant J 35(1):16–26. doi:10.1046/j.1365-313X.2003.01777.x
37. Shaw SL, Kamyar R, Ehrhardt DW (2003) Sustained microtubule treadmilling in *Arabidopsis* cortical arrays. Science 300(5626):1715–1718. doi:10.1126/science.1083529
38. Chan J, Calder G, Fox S, Lloyd C (2007) Cortical microtubule arrays undergo rotary movements in *Arabidopsis* hypocotyl epidermal cells. Nat Cell Biol 9(2):171–U157. doi:10.1038/Ncb1533
39. Kawamura E, Wasteneys GO (2008) MOR1, the *Arabidopsis thaliana* homologue of Xenopus MAP215, promotes rapid growth and shrinkage, and suppresses the pausing of microtubules in vivo. J Cell Sci 121(24):4114–4123. doi:10.1242/Jcs.039065
40. Kawamura E, Himmelspach R, Rashbrooke MC, Whittington AT, Gale KR, Collings DA, Wasteneys GO (2006) MICROTUBULE ORGANIZATION 1 regulates structure and function of microtubule arrays during mitosis and cytokinesis in the *Arabidopsis* root. Plant Physiol 140(1):102–114. doi:10.1104/pp.105.069989
41. Pastuglia M, Azimzadeh J, Goussot M, Camilleri C, Belcram K, Evrard JL, Schmit AC, Guerche P, Bouchez D (2006) gamma-tubulin is essential for microtubule organization and development in *Arabidopsis*. Plant Cell 18(6):1412–1425. doi:10.1105/039644
42. Chang HY, Smertenko AP, Igarashi H, Dixon DP, Hussey PJ (2005) Dynamic interaction of NtMAP65-1a with microtubules in vivo. J Cell Sci 118(14):3195–3201. doi:10.1242/Jcs.02433
43. Shaw SL (2006) Imaging the live plant cell. Plant J 45(4):573–598. doi:10.1111/j.1365-313X.2006.02653.x
44. Hell SW (2003) Toward fluorescence nanoscopy. Nat Biotechnol 21(11):1347–1355. doi:10.1038/nbt895
45. Dyba M, Jakobs S, Hell SW (2003) Immunofluorescence stimulated emission depletion microscopy. Nat Biotechnol 21(11):1303–1304. doi:10.1038/nbt897
46. Hell SW (2007) Far-field optical nanoscopy. Science 316(5828):1153–1158. doi:10.1126/science.1137395
47. Folling J, Bossi M, Bock H, Medda R, Wurm CA, Hein B, Jakobs S, Eggeling C, Hell SW (2008) Fluorescence nanoscopy by ground-state depletion and single-molecule return. Nat Methods 5(11):943–945. doi:10.1038/nmeth.1257
48. Balint S, Verdeny Vilanova I, Sandoval Alvarez A, Lakadamyali M (2013) Correlative live-cell and superresolution microscopy reveals cargo transport dynamics at microtubule intersections. Proc Natl Acad Sci U S A 110(9):3375–3380. doi:10.1073/pnas.1219206110
49. Shtengel G, Galbraith JA, Galbraith CG, Lippincott-Schwartz J, Gillette JM, Manley S, Sougrat R, Waterman CM, Kanchanawong P, Davidson MW, Fetter RD, Hess HF (2009) Interferometric fluorescent super-resolution microscopy resolves 3D cellular ultrastructure. Proc Natl Acad Sci U S A 106(9):3125–3130. doi:10.1073/pnas.0813131106
50. Komis G, Mistrik M, Samajova O, Doskocilova A, Ovecka M, Illes P, Bartek J, Samaj J (2014)

Dynamics and organization of cortical microtubules as revealed by superresolution structured illumination microscopy. Plant Physiol 165(1):129–148. doi:10.1104/pp. 114.238477

51. Ledbetter MC, Porter KR (1963) A microtubule in plant cell fine structure. J Cell Biol 19(1):239. doi:10.1083/Jcb.19.1.239
52. Meindl U, Lancelle S, Hepler PK (1992) Vesicle production and fusion during lobe formation in *Micrasterias* Visualized by high-pressure freeze fixation. Protoplasma 170(3-4):104–114. doi:10.1007/Bf01378786
53. Holzinger A (2000) Aspects of cell development in *Micrasterias muricata* (Desmidiaceae) revealed by cryofixation and freeze substitution. Nova Hedwigia 70(3-4):275–287
54. Segui-Simarro JM, Austin JR, White EA, Staehelin LA (2004) Electron tomographic analysis of somatic cell plate formation in meristematic cells of *Arabidopsis* preserved by high-pressure freezing. Plant Cell 16(4):836–856. doi:10.1105/Tpc.017749
55. Eder M, Lutz-Meindl U (2008) Pectin-like carbohydrates in the green alga Micrasterias characterized by cytochemical analysis and energy filtering TEM. J Microsc 231(2):201–214. doi:10.1111/j.1365-2818.2008.02036.x
56. Holzinger A, Valenta R, Lutz-Meindl U (2000) Profilin is localized in the nucleus-associated microtubule and actin system and is evenly distributed in the cytoplasm of the green alga *Micrasterias denticulata*. Protoplasma 212(3-4):197–205. doi:10.1007/Bf01282920
57. Gaillard J, Neumann E, Van Damme D, Stoppin-Mellet V, Ebel C, Barbier E, Geelen D, Vantard M (2008) Two microtubule-associated proteins of *Arabidopsis* MAP65s promote antiparallel microtubule bundling. Mol Biol Cell 19(10): 4534–4544. doi:10.1091/mbc.E08-04-0341
58. Nakamura M, Naoi K, Shoji T, Hashimoto T (2004) Low concentrations of propyzamide and oryzalin alter microtubule dynamics in *Arabidopsis* epidermal cells. Plant Cell Physiol 45(9):1330–1334. doi:10.1093/Pcp/Pch300
59. Holzinger A, Karsten U, Lutz C, Wiencke C (2006) Ultrastructure and photosynthesis in the supralittoral green macroalga *Prasiola crispa* from Spitsbergen (Norway) under UV exposure. Phycologia 45(2):168–177. doi:10.2216/05-20.1
60. Ambrose C, Allard JF, Cytrynbaum EN, Wasteneys GO (2011) A CLASP-modulated cell edge barrier mechanism drives cell-wide cortical microtubule organization in *Arabidopsis*. Nat Commun 2:430. doi:10.1038/ncomms1444
61. Sugimoto K, Williamson RE, Wasteneys GO (2000) New techniques enable comparative analysis of microtubule orientation, wall texture, and growth rate in intact roots of Arabidopsis. Plant Physiol 124(4):1493–1506
62. Frank J (2006) Electron tomography methods for three-dimensional visualization of structures in the cell. Springer, New York, NY
63. Nitta K, Kaneko Y (2004) Simple plunge freezing applied to plant tissues for capturing the ultrastructure close to the living state. J Electron Microsc (Tokyo) 53(6):677–680. doi:10.1093/jmicro/dfh092
64. Koning RI, Celler K, Willemse J, Bos E, van Wezel GP, Koster AJ (2014) Correlative cryo-fluorescence light microscopy and cryo-electron tomography of *Streptomyces*. Methods Cell Biol 124:217–239. doi:10.1016/B978-0-12-801075-4.00010-0
65. Ueda K, Matsuyama T, Hashimoto T (1999) Visualization of microtubules in living cells of transgenic *Arabidopsis thaliana*. Protoplasma 206(1-3):201–206. doi:10.1007/Bf01279267
66. Abe T, Hashimoto T (2005) Altered microtubule dynamics by expression of modified alpha-tubulin protein causes right-handed helical growth in transgenic *Arabidopsis* plants. Plant J 43(2):191–204. doi:10.1111/j.1365-313X.2005.02442.x
67. Ambrose JC, Wasteneys GO (2008) CLASP modulates microtubule-cortex interaction during self-organization of acentrosomal microtubules. Mol Biol Cell 19(11):4730–4737. doi:10.1091/mbc.E08-06-0665
68. Granger CL, Cyr RJ (2001) Spatiotemporal relationships between growth and microtubule orientation as revealed in living root cells of *Arabidopsis thaliana* transformed with green-fluorescent-protein gene construct GFP-MBD. Protoplasma 216(3-4):201–214
69. Van Bruaene N, Joss G, Van Oostveldt P (2004) Reorganization and in vivo dynamics of microtubules during *Arabidopsis* root hair development. Plant Physiol 136(4):3905–3919. doi:10.1104/pp. 103.031591
70. Dixit R, Chang E, Cyr R (2006) Establishment of polarity during organization of the acentrosomal plant cortical microtubule array. Mol Biol Cell 17(3):1298–1305. doi:10.1091/mbc.E05-09-0864
71. Littlejohn GR, Gouveia JD, Edner C, Smirnoff N, Love J (2010) Perfluorodecalin enhances in vivo confocal microscopy resolution of *Arabidopsis thaliana* mesophyll. New Phytol 186(4):1018–1025. doi:10.1111/j.1469-8137.2010.03244.x
72. Littlejohn GR, Love J (2012) A simple method for imaging *Arabidopsis* leaves using perfluorodecalin as an infiltrative imaging medium. J Vis Exp (59). doi:10.3791/3394
73. Littlejohn GR, Mansfield JC, Christmas JT, Witterick E, Fricker MD, Grant MR, Smirnoff N, Everson RM, Moger J, Love J (2014) An update: improvements in imaging perfluorocarbon-mounted plant leaves with implications for stud-

ies of plant pathology, physiology, development and cell biology. Front Plant Sci 5:140. doi:10.3389/fpls.2014.00140

74. Spurr AR (1969) A low-viscosity epoxy resin embedding medium for electron microscopy. J Ultrastruct Res 26(1):31–43
75. Sambade A, Pratap A, Buschmann H, Morris RJ, Lloyd C (2012) The influence of light on microtubule dynamics and alignment in the *Arabidopsis* hypocotyl. Plant Cell 24(1):192–201. doi:10.1105/tpc.111.093849
76. Vineyard L, Elliott A, Dhingra S, Lucas JR, Shaw SL (2013) Progressive transverse microtubule array organization in hormone-induced Arabidopsis hypocotyl cells. Plant Cell 25(2):662–676. doi:10.1105/tpc.112.107326
77. Dixit R, Cyr R (2003) Cell damage and reactive oxygen species production induced by fluorescence microscopy: effect on mitosis and guidelines for non-invasive fluorescence microscopy. Plant J 36(2):280–290. doi:10.1046/j.1365-313X.2003.01868.x
78. Chan J, Calder GM, Doonan JH, Lloyd CW (2003) EB1 reveals mobile microtubule nucleation sites in Arabidopsis. Nat Cell Biol 5(11):967–971. doi:10.1038/Ncb1057
79. Mathur J, Mathur N, Kernebeck B, Srinivas BP, Hulskamp M (2003) A novel localization pattern for an EB1-like protein links microtubule dynamics to endomembrane organization. Curr Biol 13(22):1991–1997. doi:10.1016/j.cub.2003.10.033
80. Marc J, Granger CL, Brincat J, Fisher DD, Kao TH, McCubbin AG, Cyr RJ (1998) A GFP-MAP4 reporter gene for visualizing cortical microtubule rearrangements in living epidermal cells. Plant Cell 10(11):1927–1939
81. DeBolt S, Gutierrez R, Ehrhardt DW, Melo CV, Ross L, Cutler SR, Somerville C, Bonetta D (2007) Morlin, an inhibitor of cortical microtubule dynamics and cellulose synthase movement. Proc Natl Acad Sci U S A 104(14):5854–5859. doi:10.1073/pnas.0700789104
82. Stoppin-Mellet V, Gaillard J, Vantard M (2006) Katanin's severing activity favors bundling of cortical microtubules in plants. Plant J 46(6):1009–1017. doi:10.1111/j.1365-313X.2006.02761.x
83. Wightman R, Turner SR (2007) Severing at sites of microtubule crossover contributes to microtubule alignment in cortical arrays. Plant J 52(4):742–751. doi:10.1111/j.1365-313X.2007.03271.x
84. Holzinger A, Meindl U (1997) Jasplakinolide, a novel actin targeting peptide, inhibits cell growth and induces actin filament polymerization in the green alga *Micrasterias*. Cell Motil Cytoskeleton 38(4):365–372. doi:10.1002/(Sici)1097-0169(1997)38:4<365::Aid--Cm6>3.0.Co;2-2
85. Al-Amoudi A, Chang JJ, Leforestier A, McDowall A, Salamin LM, Norlen LP, Richter K, Blanc NS, Studer D, Dubochet J (2004) Cryo-electron microscopy of vitreous sections. EMBO J 23(18):3583–3588. doi:10.1038/sj.emboj.7600366
86. Buser C, Walther P (2008) Freeze-substitution: the addition of water to polar solvents enhances the retention of structure and acts at temperatures around -60 degrees C. J Microsc 230(Pt 2):268–277. doi:10.1111/j.1365-2818.2008.01984.x
87. Walther P, Ziegler A (2002) Freeze substitution of high-pressure frozen samples: the visibility of biological membranes is improved when the substitution medium contains water. J Microsc 208:3–10. doi:10.1046/j.1365-2818.2002.01064.x
88. Watanabe S, Punge A, Hollopeter G, Willig KI, Hobson RJ, Davis MW, Hell SW, Jorgensen EM (2011) Protein localization in electron micrographs using fluorescence nanoscopy. Nat Methods 8(1):80–84. doi:10.1038/nmeth.1537
89. Gutierrez R, Lindeboom JJ, Paredez AR, Emons AM, Ehrhardt DW (2009) Arabidopsis cortical microtubules position cellulose synthase delivery to the plasma membrane and interact with cellulose synthase trafficking compartments. Nat Cell Biol 11(7):797–806. doi:10.1038/ncb1886
90. Crowell EF, Bischoff V, Desprez T, Rolland A, Stierhof YD, Schumacher K, Gonneau M, Hofte H, Vernhettes S (2009) Pausing of Golgi bodies on microtubules regulates secretion of cellulose synthase complexes in Arabidopsis. Plant Cell 21(4):1141–1154. doi:10.1105/tpc.108.065334
91. Eng RC, Wasteneys GO (2014) The microtubule plus-end tracking protein ARMADILLO-REPEAT KINESIN1 promotes microtubule catastrophe in *Arabidopsis*. Plant Cell 26(8):3372–3386. doi:10.1105/tpc.114.126789

Part II

Cytoskeleton Dynamics

Chapter 9

Basic Methods to Visualize Actin Filaments In Vitro Using Fluorescence Microscopy for Observation of Filament Severing and Bundling

Shoichiro Ono

Abstract

Dynamics of actin filaments are regulated by a number of actin-binding proteins. To understand the function of an actin-binding protein, it is necessary to characterize effects of the protein on actin filament dynamics in vitro. This chapter describes basic microscopic methods to visualize fluorescently labeled actin filaments using commonly available fluorescence microscope settings. Direct microscopic observation of actin filaments provides strong evidence for severing or bundling of actin filaments.

Key words Actin filament dynamics, Actin-binding proteins, Direct observation, Fluorescence microscopy, Fluorophore labeling

1 Introduction

In vitro characterization of functions of actin-binding proteins involves a variety of biochemical techniques including spectroscopy, centrifugation, chromatography, and electrophoresis. In addition to these methods, recent advancements in digital imaging devices and computer software have made it easier to utilize commonly available fluorescence microscopes to visualize actin filaments at a single-filament level without using advanced equipment such as super-resolution microscope and total internal reflection fluorescence microscope. Under appropriate conditions, single actin filaments can be resolved, and their lengths are measurable. Therefore, by comparing actin filaments in the absence or presence of an actin-binding protein, one can determine changes in the morphology of actin filaments such as shortening, bundling, and branching to determine activity of a protein of interest.

Ray H. Gavin (ed.), *Cytoskeleton: Methods and Protocols*, Methods in Molecular Biology, vol. 1365,
DOI 10.1007/978-1-4939-3124-8_9, © Springer Science+Business Media New York 2016

2 Materials

Use ultrapure water to prepare all solutions. Keep all protein solutions on ice unless indicated otherwise. Avoid repeated freezing and thawing of protein solutions. If a protein solution needs to be stored frozen, make small aliquots before freezing. Protect all fluorophore-containing reagents from lights by covering containers by aluminum foil.

2.1 Buffers

1. G-buffer: 2 mM Tris–HCl (pH 8.0), 0.2 mM $CaCl_2$, 0.2 mM ATP (*see* **Note 1**), 0.2 mM dithiothreitol (*see* **Note 2**).
2. 10× KME buffer: 1 M KCl, 20 mM $MgCl_2$, 10 mM EGTA, 0.2 M HEPES–KOH (pH 7.5).

2.2 Actin Proteins

1. Rabbit muscle G-actin: Reconstitute lyophilized rabbit muscle actin (1 mg) (Life Technologies/Molecular Probes, Catalog # A12375) by adding 1 mL ice-cold water (*see* **Note 3**). Molecular weight of actin is 42,000. Therefore, 1 mg/mL actin equals 23.8 μM.
2. Fluorophore-labeled G-actin: Alexa 488-labeled rabbit muscle G-actin (Life Technologies/Molecular Probes, Catalog # A12373) (*see* **Note 4**). Actin conjugates with other fluorophores are available from Life Technologies/Molecular Probes and Cytoskeleton Inc (*see* **Note 5**). Choose an appropriate conjugate based on availability of fluorescence filters.

2.3 Coverslips and Coating Components

1. Coverslips: 35×50 mm No. 1 rectangular coverslips; 18×18 mm No. 1 square coverslips.
2. 0.1 % nitrocellulose: Mix 25 μL 2 % Collodion (nitrocellulose) (Electron Microscopy Sciences, Catalog # 12620-00) and 475 μL amyl acetate (Electron Microscopy Sciences, Catalog # 10800).

2.4 Microscope and Imaging System

1. A fluorescence microscope (compound or inverted) needs to be equipped with an epifluorescence illumination unit, appropriate fluorescence filter sets, an oil-immersion objective lens of 60–100×, digital camera and computer software for capturing images (*see* **Note 6**).

3 Methods

3.1 Preparation of Nitrocellulose-Coated Coverslips

1. Place 15 μL of 0.1 % nitrocellulose on a 35×50 mm coverslip and spread using a pipet tip. The coating should cover slightly larger area than 18×18 mm (Fig. 1) (*see* **Note 7**).

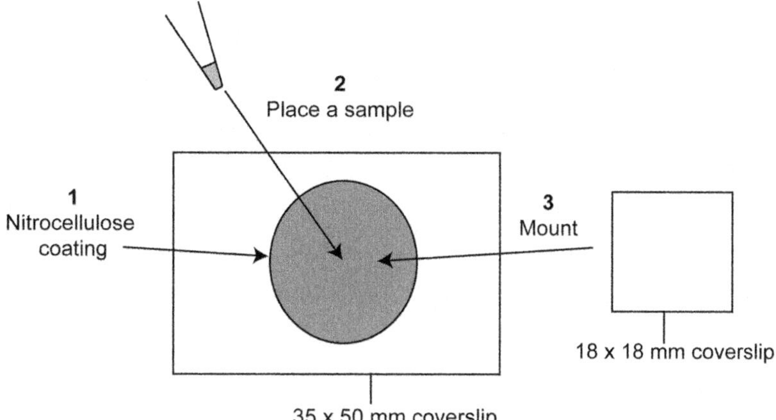

Fig. 1 Sample preparation for observation using an inverted microscope. The 35 × 50 mm coverslip is coated by nitrocellulose (1). For observation using a compound microscope, the 18 × 18 mm coverslip needs to be coated. A sample (1–2 μL) is placed at the center of the coated area (2), and mounted with the 18 × 18 mm coverslip

2. Dry the coverslips on a horizontal surface in a chemical fume hood for at least 10 min. Care should be taken to avoid introduction of dusts on the surface.
3. Store the coated coverslips in a container with a sealable lid. These can be stored for several months.

3.2 Polymerization of Fluorescently Labeled Actin

1. Mix unlabeled rabbit muscle G-actin (final 8 μM), Alexa 488-labeled rabbit muscle G-actin (final 2 μM), and a 1/10 volume of 10× KME buffer (*see* **Note 8**). The ratio of unlabeled actin to labeled actin can be changed to increase or decrease fluorescence intensity (*see* **Note 9**).
2. Cover the tube containing the mixture by aluminum foil, and incubate for 2 h at room temperature (20–25 °C).
3. Keep polymerized actin on ice.

3.3 Incubation of Fluorescently Labeled Actin Filaments with a Protein of Interest

1. Dilute polymerized labeled actin to 2 μM in 1× KME buffer in the absence of any additional proteins (control) or in the presence of a protein of interest (experimental) (*see* **Note 10**). Mix gently, as vigorous pipetting can cause fragmentation of actin filaments. Incubation periods can vary from a few seconds to several hours. Concentrations of a protein of interest can also vary depending on its affinity with actin filaments and will need to be tested experimentally to determine an optimal range.

3.4 Observation by Fluorescence Microscopy

1. Place 1–2 μL of the incubated actin filaments on a nitrocellulose-coated coverslip and mount with an 18 × 18 mm square coverslip (Fig. 1) (*see* **Note 11**). Gently apply a pressure to spread the solution. No sealing is necessary.

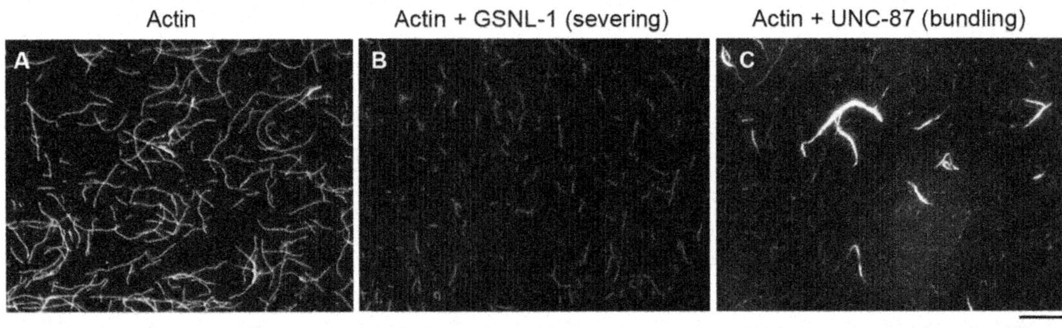

Fig. 2 Micrographs of fluorescently labeled actin filaments. Images of control actin filaments (**a**), actin filaments with *Caenorhabditis elegans* GSNL-1 (a gelsolin-like protein) (**b**), and actin filaments with *C. elegans* UNC-87 (a calponin-like protein) (**c**). The results show that actin filaments are shorter in the presence of GSNL-1 in (**b**) than control actin filaments in (**a**), and that thick bundles are formed in the presence of UNC-87 in (**c**). Bar, 10 μm

2. Observe actin filaments immediately by a fluorescent microscope. Adjust focus only on actin filaments that have been attached to the coated side of the coverslip (*see* **Note 12**).
3. Capture micrographs of representative actin filaments and save them on a computer. Exposure settings can vary significantly depending on the sensitivity of a camera. If less than one second of exposure to epifluorescence illumination is sufficient to capture a micrograph, photobleaching would not be a major issue. If photobleaching prevents capturing of micrographs of good quality, anti-bleaching components should be included in the samples (*see* **Note 13**).

3.5 Data Analysis and Interpretation

1. Compare micrographs of control and experimental actin filaments and determine whether morphological changes are significant (Fig. 2).
2. Quantitative measurements of actin filaments can be performed by a variety of image analysis software. Perhaps, ImageJ (http://rsb.info.nih.gov/ij/) [1] is one of the most useful open-source software for this purpose. Length of an actin filament can be measured by tracing the filament by the Segmented Line tool and measure a pixel count of the line (Analyze > Measure). Then, the pixel count needs to be converted to actual length depending on the magnification factor for each microscope system.
3. Shortening of actin filaments strongly suggests that the filaments are severed (Fig. 2b) [2]. However, one needs to consider that other effects such as sequestration of monomers and depolymerization of filaments from ends can also shorten actin filaments. Capping of pre-formed actin filaments is not likely

to cause shortening in a short period. Alternatively, imaging of immobilized actin filaments in a perfusion chamber allows one to observe the same filaments before and after exposure to a protein of interest [3, 4], which provides stronger evidence for filament severing.

4. Bundles of actin filaments can be clearly distinguished from single filaments by their thicker and brighter appearances (Fig. 2c) [5]. Actin filaments may spontaneously form bundles at acidic pH [6]. Therefore, observation of appropriate control actin filaments is important to interpret results. Glutathione S-transferase (GST) is commonly used as a fusion tag for recombinant proteins. However, GST forms a homodimer [7], and a GST-fusion protein can artificially induce actin bundling through dimerization. Therefore, GST needs to be cleaved off for examination of actin filament bundling. Effects of other fusion tags should also be evaluated critically.
5. This method can be used potentially to characterize filament branching if the branched filaments are stable enough for observation.

4 Notes

1. A stock solution of ATP (0.1 M) should be neutralized to ~pH 7 using NaOH and stored frozen at −20 °C.
2. A freshly prepared dithiothreitol solution works best to protect oxidation of thiols of proteins. Alternatively, a concentrated stock solution of dithiothreitol in water at 0.2–1.0 M can be stored frozen at −20 °C.
3. Rabbit muscle G-actin is also commercially available from Cytoskeleton Inc. However, we have previously encountered a batch of G-actin that is partially aggregated or polymerized even after reconstitution following manufacturer's instruction. To recover G-actin from such actin preparation, add a 1/10 volume of 10× KME buffer, incubate for 2 h at room temperature, ultracentrifuge at 100,000 ×*g* for 1 h, resuspend the pelleted actin in a small volume of G-buffer, dialyze against G-buffer overnight at 4 °C, and then, ultracentrifuge at 100,000 ×*g* for 1 h. The supernatant can be used as G-actin. Alternatively, rabbit muscle G-actin can be relatively easily prepared from rabbit muscle acetone powder. We routinely use rabbit muscle acetone powder from Pel-Freeze Biologicals (Catalog number 41995) for preparation of G-actin as described by Pardee and Spudich [8].
4. Alexa 488 is conjugated to random amines on actin. Protein concentration and labeling efficiency can vary by different lots.

5. Other commercially available labeled actin products are conjugates with Alexa 568, Alexa 594, and Alexa 647 from Life Technologies/Molecular Probes and rhodamine from Cytoskeleton Inc. Alternatively, actin conjugates can be prepared by using reactive dyes. We have used amine-reactive DyLight 549 (Thermo Scientific/Pierce) to label rabbit muscle actin [9]. The labeled G-actin can be snap-frozen in small aliquots in liquid nitrogen and stored at −80 °C.
6. Although most fluorescent microscopes are expected to be suitable for imaging actin filaments, suitability should be tested by pilot experiments to observe control actin filaments. Our lab uses a Nikon TE2000 inverted microscope with an epifluorescence illumination with an HBO mercury lamp, an 60× Plan Apo oil-immersion objective lens (numerical aperture 1.4), SPOT RT Monochrome CCD camera (1 megapixel resolution), and IPLab imaging software (Becton Dickinson, discontinued).
7. Nitrocellulose coating is important to stabilize actin filaments for image capturing. Since the coated surface is used for observation, the bottom coverslip needs to be coated for an inverted microscope, whereas the top coverslip needs to be coated for a compound microscope. If the actin filaments are not efficiently attached, coverslips need to be washed prior to coating by 1 M HCl at 50–60 °C for 2–3 h in a chemical fume hood followed by extensive washing by deionized water.
8. Alternatively, fluorescent labeling of actin filaments can be done by incubating polymerized unlabeled rabbit muscle actin with tetramethylrhodamine-phalloidin (Sigma-Aldrich, catalog number P1951), which has been used in in vitro motility assays for myosin motors. However, phalloidin is known to interfere with actin-filament severing by cofilin and potentially affect functions of other actin-binding proteins. If this method is preferred, effect of phalloidin should be taken into consideration and evaluated by other experimental methods, if necessary.
9. Total concentration of actin can also vary depending on availability of actin proteins. However, note that critical concentration of actin is 0.1–0.2 μM, and this portion of actin remains as G-actin independent of total actin concentrations. Therefore, when total actin concentration is lower, a proportion of G-actin is higher.
10. Actin filaments under control conditions need to be observed for each experiment. Appearance of actin filaments can vary and may affect quantitative values (e.g., lengths). Therefore, appropriate control experiments are important to determine properties of an actin-binding protein.

11. After mounting, actin filaments should be spread out so that individual filaments can be characterized. If there are too many filaments on the coated coverslip, the sample can be diluted immediately prior to mounting. Microscopic observation and data acquisition should be done quickly on diluted samples because dilution can cause actin depolymerization or disassembly of actin bundles.
12. If actin filaments do not attach well to a nitrocellulose-coated coverslip, try acid wash of coverslips before coating (*see* **Note 7**).
13. Anti-bleaching components are 0.036 mg/mL catalase, 0.2 mg/mL glucose oxidase, 6 mg/mL glucose, and 100 mM dithiothreitol as described in [4].

References

1. Schneider CA, Rasband WS, Eliceiri KW (2012) NIH image to ImageJ: 25 years of image analysis. Nat Methods 9:671–675
2. Liu Z, Kanzawa N, Ono S (2011) Calcium-sensitive activity and conformation of *Caenorhabditis elegans* gelsolin-like protein 1 are altered by mutations in the first gelsolin-like domain. J Biol Chem 286:34051–34059
3. Ono S, Mohri K, Ono K (2004) Microscopic evidence that actin-interacting protein 1 actively disassembles actin-depolymerizing factor/cofilin-bound actin filaments. J Biol Chem 279(14):14207–14212
4. Ichetovkin I, Han J, Pang KM, Knecht DA, Condeelis JS (2000) Actin filaments are severed by both native and recombinant *Dictyostelium* cofilin but to different extents. Cell Motil Cytoskeleton 45(4):293–306
5. Yamashiro S, Gimona M, Ono S (2007) UNC-87, a calponin-related protein in *C. elegans*, antagonizes ADF/cofilin-mediated actin filament dynamics. J Cell Sci 120:3022–3033
6. Kawamura M, Maruyama K (1970) Polymorphism of F-actin. I. Three forms of paracrystals. J Biochem 68:885–899
7. Tudyka T, Skerra A (1997) Glutathione S-transferase can be used as a C-terminal, enzymatically active dimerization module for a recombinant protease inhibitor, and functionally secreted into the periplasm of *Escherichia coli*. Protein Sci 6:2180–2187
8. Pardee JD, Spudich JA (1982) Purification of muscle actin. Methods Enzymol 85: 164–181
9. Liu Z, Klaavuniemi T, Ono S (2010) Distinct roles of four gelsolin-like domains of *Caenorhabditis elegans* gelsolin-like protein-1 in actin filament severing, barbed end capping, and phosphoinositide binding. Biochemistry 49:4349–4360

Chapter 10

An In Vitro Model System to Test Mechano-microbiological Interactions Between Bacteria and Host Cells

Luís Carlos Santos, Emilia Laura Munteanu, and Nicolas Biais

Abstract

The aim of this chapter is to present an innovative technique to visualize changes of the f-actin cytoskeleton in response to locally applied force. We developed an in vitro system that combines micromanipulation of force by magnetic tweezers with simultaneous live cell fluorescence microscopy. We applied pulling forces to magnetic beads coated with the *Neisseria gonorrhoeae* Type IV pili in the same order of magnitude than the forces generated by live bacteria. We saw quick and robust f-actin accumulation at the sites where pulling forces were applied. Using the magnetic tweezers we were able to mimic the local response of the f-actin cytoskeleton to bacteria-generated forces. In this chapter we describe our magnetic tweezers system and show how to control it in order to study cellular responses to force.

Key words Magnetic tweezers, Neisseria, f-Actin, Type IV pili, Magnetic beads, Cytoskeleton

1 Introduction

The ability of eukaryotic cells to adapt their shape and sense their environment through their cytoskeleton has been a subject of scientific inquiry for a very long time. Composed of actin microfilaments, microtubules, and intermediate filaments, the cytoskeleton plays a crucial role both in the interior and exterior of cells. On the one hand, the cytoskeleton in conjunction with cell membranes organizes the interior of cells and maintains cell shape by controlling intracellular trafficking and cell division [1–3]. On the other hand, it connects cells to their chemical and physical surroundings, namely to the extracellular matrix and to other cells [4, 5]. Different extracellular matrix proteins transmitting forces to the interior of the cell will lead to different responses from the cytoskeleton [6–8].

The impact of physical force on biological outcomes, usually referred to as mechanobiology, is a key component of many landmark cellular mechanisms such as motility, development, differentiation, and cancer [9–12]. Recent evidence shows that certain bacteria also use mechanical forces as a way to induce

Ray H. Gavin (ed.), *Cytoskeleton: Methods and Protocols*, Methods in Molecular Biology, vol. 1365, DOI 10.1007/978-1-4939-3124-8_10, © Springer Science+Business Media New York 2016

cellular changes in host cells [13, 14]. Bacteria have not ceased to co-evolve with their host and the study of the interactions between bacteria and host cells has often illuminated certain chemical aspects of the cytoskeleton. (e.g., Act A from *Listeria monocytogenes* [15], Toxin from Entero-Pathogenic *Escherichia coli* [16, 17])

The bacterium *Neisseria gonorrhoeae*, the causative agent of gonorrhea, has emerged in recent years as a paradigm for the study of mechano-microbiological interaction [18–20]. Forces generated by cycles of elongation and retraction of Type IV pili—long thin polymers emanating from the surface of *Neisseria gonorrhoeae*—have been shown to have a dramatic impact on the outcome of the interactions between bacteria and human cells [21]. In contrast, in vitro model studies show that bacteria that do not have the ability to exert pulling forces (including mutants that are missing the ATPase PilT, the protein responsible for force generation) exhibit reduced level of invasion compared to their force bearing counterparts [13]. Most importantly, bacteria-generated force is required to remodel the host cell cytoskeleton [13, 22]—a hallmark of infection in vitro as well as in vivo.

How the eukaryotic cytoskeleton responds to force has been the subject of many experimental studies. Central to this question is a technology that allows exerting a controlled force on a single cell. Techniques enabling the application of local force at a subcellular scale usually involve magnetic or optical tweezers [23–25]. Also recently, techniques involving AFM or microplates have been employed to exert global forces on an entire cell [26–28]. Previous studies have used optical tweezers to investigate the impact of forces in the range of hundreds of picoNewton (pN) on remodeling of the cytoskeleton [29–31]. These techniques have provided valuable insight on the response of the eukaryotic cytoskeleton to the magnitude and speed of the local force applied, and on the nature of the physical link to the cytoskeleton. Thus, unraveling the physical cross-talk between eukaryotic cells and their environment is a complex subject of tremendous importance with potential applications ranging from cancer to bioengineering.

To better understand the role of local force in eukaryotic cells we explored the f-actin response to Type IV Pili-mediated force interactions that occur between bacteria from the *Neisseria* species and human skin epithelial cells. Forces generated by *Neisseria gonorrhoeae* bacteria are in the nanoNewton (nN) range and have been shown to affect remodeling of the eukaryotic cytoskeleton [19, 32]. Previous studies have used optical tweezers to provide an in vitro system to look at the interaction between *Neisseria gonorrhoeae* and human cells [33], but optical tweezers are limited in maximum force and cannot reach the nanoNewton (nN) range seen in vivo. We developed an innovative and robust technique by combining magnetic tweezers with live cell fluorescence microscopy. We applied local forces to human skin epithelial cells through magnetic beads coated with Type IV pili from *Neisseria gonorrhoeae* and simultaneously visualized f-actin dynamics. The tweezers create an electromagnetic field gradient that pulls on the pili-coated magnetic beads, which in turn pull on the cell that the beads are attached to. Here, we describe the magnetic tweezers built in our laboratory to study the response of the f-actin cytoskeleton to local pulling forces exerted on the cell body. Our results mimic the early steps of infection of epithelial cells with bacteria from *Neisseria* species. The techniques presented here should be easily implemented for other pili-bearing bacterial species interacting with eukaryotic host cells.

2 Materials

2.1 Live Cell Imaging of f-Actin in Human Epithelial Cells

2.1.1 Cell Culture and DNA Transfection

1. Human skin epithelial cell line A-431 (ATCC CRL-1555).
2. Mammalian expression plasmid DNA vector (pEGFP-N1) encoding GFP-tagged Tractin—an f-actin-associated protein [34].
3. Transfection reagent: Fugene 6 (Promega, WI).
4. DMEM (Gibco, CA) supplemented with 10 % heat inactivated fetal bovine serum, FBS (Gibco).
5. CO_2-independent medium (serum-free) (Gibco).
6. Solution of trypsin 0.05 %, TrypLE Express (Gibco).
7. Phosphate buffer saline (PBS).
8. Glass-bottom cell imaging dishes, round bottom, 50 mm diameter, 7 mm low-height, glass coverslip #0 (MatTek, MA).
9. 60 mm tissue culture plates.
10. Incubator with controlled atmosphere at 37 °C with 5 % CO_2.

2.1.2 Time-Lapse Fluorescence Microscopy

1. Inverted microscope (Olympus IX 70), equipped with a 63×/1.40 NA Oil, Plan Apo Chroma objective (Zeiss).
2. Electron-multiplying charge-coupled device (EMCCD) digital camera PRO-EM 512 (Princeton Instruments, NJ).
3. 100 W high pressure mercury bulb with power supply (BH2-RFL-T3, Olympus).
4. Environmental control chamber to maintain temperature of the sample stable at 37 °C (assembled in-house).
5. Digital image acquisition software: Micromanager (NIH, MD).
6. Digital image processing software: ImageJ (NIH, MD).
7. Data processing software: MatLab (MathWorks, MA).

2.2 Preparing Magnetic Beads Coated with Bacteria Pili

2.2.1 Purifying Pili from Neisseria gonorrhoeae

1. *Neisseria gonorrhoeae* clone MS11 (wild type).
2. 10-cm petri dishes.
3. GonoCoccal Broth (GCB) agar.
4. CHES buffer, pH 9.5 [2-(cyclohexylamino)ethanesulfonic acid].
5. MES buffer, pH 5.0 [2-(*N*-morpholino)ethanesulfonic acid].
6. Micro Ultracentrifuge (RC-M120EX, Sorvall, CT), and fixed angle rotor (RP100AT-289).

2.2.2 Coating Beads with Purified Pili

1. Superparamagnetic beads (Life Technologies, CA): uniform, monosized 1 μm (Dynabeads MyOne) beads, and 3 μm (Dynabeads M-270) in diameter beads, composed of highly cross-linked polystyrene with evenly distributed magnetic material, with a carboxylic acid surface chemistry.
2. EDC [1-(3-Dimethylaminopropyl)-3-ethylcarbodiimide hydrochloride] (Alfa Aesar, MA).
3. MES buffer, pH 5.0 [2-(*N*-morpholino)ethanesulfonic acid].
4. NaOH solution (0.01 M).
5. Phosphate buffer saline (PBS).
6. Tris–HCl, pH 7.4 (50 mM).
7. BSA (0.1 % in PBS).
8. Sonicator Branson #1800 (Branson Ultrasonics, CT).

2.3 Magnetic Tweezers Setup for Applying Local Force to Eukaryotic Cells

2.3.1 Components of the Magnetic Tweezers

1. XYZ 3D micromanipulator MP-285 (Sutter Instrument, CA).
2. Programmable DC power supply and operating software (BK Precision, CA).

2.3.2 Force Calibration

1. Polystyrene beads (2 μm diameter) (Polysciences Inc, PA).
2. Dimethylpolysiloxane (Silicone oil) (Viscosity $\mu = 12{,}500$ centipoise) (Sigma, CO).
3. Poly-L-lysine hydrobromide MW 30,000–70,000 (Sigma, CO).

3 Methods

3.1 Live Cell Imaging of f-Actin in Human Epithelial Cells

Here we describe a detailed protocol to use live fluorescence microscopy for high resolution visualization of fluorescently-tagged tractin—a protein that specifically binds to f-actin—in human epithelial cells [34] (*see* **Note 1**).

3.1.1 Cell Culture and DNA Transfection

We visualized actin dynamics in live cells by time-lapse image acquisition of cells expressing GFP-tagged tractin—an f-actin-associated protein [34]. With this purpose, we introduced a DNA vector carrying the GFP-tagged tractin sequence into A-431 cells—a human epithelial cell line. The CMV promoter and enhancer sequence allow strong and clear visual detection of the tractin-GFP reporter as early as 12 h after transfection, and during the whole course of the experiment. To introduce the tractin-GFP vector into A-431 cells we used the transfection reagent Fugene 6. This method of transfection involves a non-liposomal reagent that achieves a good balance between transfection efficiency—high percentage of cells expressing the fluorescent protein—and low cytotoxicity—low percentage of dead cells due to toxicity of the transfection reagent (*see* **Note 2**).

The steps below describe the optimal transfection protocol that we followed to perform the set of experiments presented in this chapter. All cell culture handling should be performed in sterile conditions inside a flow chamber. For optimal cell performance, A-431 cells should have been kept in culture for at least 1 week, or equivalent to two passages, in 60 mm tissue culture plates, in full medium (DMEM/10 % FBS), and in a 37 °C/5 % CO_2 atmosphere. Follow standard cell passaging protocol using Trypsin solution to detach cells from the culture plate.

1. 48 h before imaging: Prepare one glass-bottom low wall imaging dish for each experimental condition you want to test (e.g., different types of beads, or pili, or force applied, etc.). Seed 5×10^5 cells per imaging dish with 3 mL DMEM/10 % FBS culture medium, and incubate the cells for 24 h (*see* **Note 3**).
2. 24 h before imaging: Mix 20 μL of room temperature Fugene 6 with 175 μL of fresh culture medium and incubate at room temperature for 5 min.
3. Pipette 5–7 μg of DNA into a new Eppendorf tube. The concentration of the DNA stock solution used should be 1–2 μg/μL to avoid dilution of the DNA/Fugene complexes formed in the next step (*see* **Note 4**).
4. Add the Fugene/Medium mix to the DNA in the new tube and incubate at room temperature for 15–30 min to allow formation of the DNA/Fugene complexes.
5. Slowly and dropwise, add the 200 μL DNA/Fugene mix to the cell dish (containing 3 mL of full medium), and mix by gently swirling the dish.
6. Return the dish to the incubator. The tractin-GFP protein is visible as early as 12 h post-transfection, and the expression peak occurs at 24–48 h post-transfection.
7. After at least 12 h of incubation, assess cell viability and transfection efficiency by briefly examining the level of fluorescence

intensity and appropriate subcellular localization of the fluorescent protein (tractin-GFP localizes to f-actin, *see* **Note 1**). If the transfection was successful, proceed to cell imaging.

3.1.2 Time-Lapse Fluorescence Microscopy

Epifluorescence microscopy is a powerful tool to visualize the dynamics of tractin-GFP over time in live cells. We used an inverted microscope, with a 63× oil objective, bright-field illumination, and halogen light source for epifluorescence illumination. Here we provide guidelines to setting up live cell imaging by time-lapse fluorescence microscopy (*see* **Note 5**).

1. Turn on all microscope components. Stabilization of the halogen light source may take a few minutes.
2. Microscope temperature stabilization: Turn on the local heater attached to the microscope's environmental chamber approximately 30 min before image acquisition to make sure all components of the microscope are at a stable temperature at the start of the experiment (*see* **Note 6**).
3. Carefully place the imaging dish on the microscope stage, turn on bright-field illumination, and focus on the cell monolayer.
4. Adjust the bright-field image quality in order to have the sharpest contrast of the cell edges.
5. Adjust the image brightness by controlling the amount of incident light allowed by the diaphragm, and by adding/removing neutral density filters in the light path (*see* **Note 7**).
6. Switch to epifluorescent light, and quickly browse the dish to choose a region of interest (ROI) that contains a transfected cell, i.e. green cell with one or more beads on top. Be swift with this ROI search in order to minimize photobleaching (exposure time-dependent fading) (*see* **Note 8**).
7. Once you have selected the desired ROI, switch to "camera view," and adjust the exposure time on both channels (bright-field and epifluorescence). The aim here is to maximize signal to noise ratio, while simultaneously minimizing photobleaching and phototoxicity. Choose the shortest exposure time possible that provides an optimal balance between these factors.

3.2 Preparing Magnetic Beads Coated with Bacteria Pili

The following protocols involving bacteria handling, streaking and lawning must be performed under sterile conditions and appropriate safety precautions.

3.2.1 Purifying Pili from Neisseria gonorrhoeae

1. Prepare two lawns of *N. gonorrhoeae* MS11 wild type (WT) using two 10 cm plates of GonoCoccal Broth (GCB) agar.
2. Incubate bacteria lawns at 37 °C in 5 % CO_2 atmosphere for 16–20 h. This incubating time window is critical to obtaining

good quality pili with high yields. In general, each 10 cm Petri dish of lawn bacteria yields 0.1 mg of pili.

3. Harvest the bacteria lawns using a polyester swab, and resuspend in 1 mL of 50 mM CHES buffer pH 9.5.
4. Vortex the bacteria suspension for 2 min, and then centrifuge bacteria bodies at 18,000 × *g* for 5 min (~14,000 RPM using a fixed rotor tabletop centrifuge).
5. Collect the supernatant and ultracentrifuge it at 100,000 × *g* for 90 min. The ultracentrifugation step must be performed in suitable ultracentrifuge microtubes. The tubes should be well balanced (<10 μg weighting error), and carefully loaded onto the ultracentrifuge rotor.
6. After centrifugation, carefully remove the tubes from the rotor without disturbing the pili pellet, and discard the supernatant. The pili pellet is almost invisible to the eye.
7. Gently resuspend the pili pellet in 1 mL of 50 mM MES buffer (pH 5.0).

3.2.2 Coating Beads with Purified Pili

In the subheadings below we describe one of the protocols we optimized for coating beads with purified pili from *Neisseria gonorrhoeae* (*see* Subheading 3.2.1). In order to replace the buffer solution in which the magnetic beads are suspended, simply place the tubes containing the beads in contact with a magnet for 2 min, and then carefully remove the liquid supernatant.

Surface Activation of Beads with EDC

1. Wash 100 μL of beads twice with 1 mL of 0.01 M NaOH for 10 min with gentle rotation.
2. Repeat wash three times this time with 100 μL deionized water.
3. Dissolve EDC in cold, deionized water to 50 mg/mL.
4. Add 100 μL of EDC-solution to the beads. Vortex to ensure good mixing.
5. Incubate for 30 min at room temperature with gentle rotation.
6. Wash once with cold, deionized water and once with 50 mM MES pH 5.0 as quickly as possible to avoid hydrolysis of the activated carboxylic acid groups.
7. The beads are now activated and ready for coating with a ligand containing primary amine groups, in this case, the purified pili. Activated beads cannot be stored and you should proceed directly to **step 8**.

Immobilization of Ligand After Activation

8. Remove the wash solution used in **step 6** above. Add 60 μL of pili in 50 mM MES pH 5.0 to the activated beads.
9. Add 40 μL of 50 mM MES pH 5.0 to final volume of 100 μL. Vortex to ensure good mixing.

10. Incubate for 30 min at room temperature with gentle rotation.
11. Wash the pili-coated beads as described below.

Quenching and Washing of Pili-Coated Beads

All immobilization procedures require washing of the pili-coated beads to remove excess pili and to block un-reacted surface.

12. Quench the non-reacted activated carboxylic acid groups by incubating the pili-coated beads with 1 mL 50 mM Tris–HCl pH 7.4 for 15 min at room temperature with gentle rotation.
13. Wash the pili-coated beads four times with 1 mL PBS.
14. Block by adding 0.1 % BSA when it does not interfere with downstream applications of the beads.
15. Resuspend the pili-coated beads in PBS with 0.1 % BSA to the desired concentration. The pili-coated beads are now ready for use (*see* **Note 9**).

3.3 Magnetic Tweezers Setup for Applying Local Force to Eukaryotic Cells

3.3.1 Components of the Magnetic Tweezers

Here we present the magnetic tweezers setup and protocol developed in our lab for studying local f-actin cytoskeleton response to force (Fig. 1). The magnetic tweezers setup we implemented features a solenoid electromagnet as a magnetic field source that is mounted on a microscope stage in order to be brought in close proximity to the magnetic beads attached to cells (Fig. 1). Using a cylindrical holder made of a non-magnetic material, a solenoid electromagnet was handcrafted by coiling a copper wire on the outside of that holder. Typically two to three layers of wire are coiled. Several inches of wire were left at the beginning and end in order to connect the solenoid to the electric current source. A rod-shaped ferromagnetic core was fixed with plastic screws inside of the holder. The core should not touch the inside walls of the holder. This space creates an insulating layer of air that minimizes the heat transfer from the copper wire when an electric current is applied to the solenoid. One end of the rod was machined to be a sharp tip. The role of the ferromagnetic core is to amplify the magnetic field produced by the solenoid electromagnet. The rod's sharp tip creates a high gradient in the magnetic field close to its tip, necessary to apply high forces to the magnetic beads in the sample field. With the use of two electronic connectors (we use alligator-banana end), the beginning and the end of the copper coil are connected to an electric current source. Typically 1 to 3 amperes (A) of continuous current were applied to the solenoid to mimic the range in magnitude of the pulling forces exerted by bacteria in the *Neisseria* species.

3.3.2 Force Calibration

Force magnitude depends on the magnitude of the magnetic field, the size of the bead, and the material the bead is made of. In turn, the magnitude of the magnetic field depends on the geometry of the magnetic tip, the amplitude of the electric current applied to the

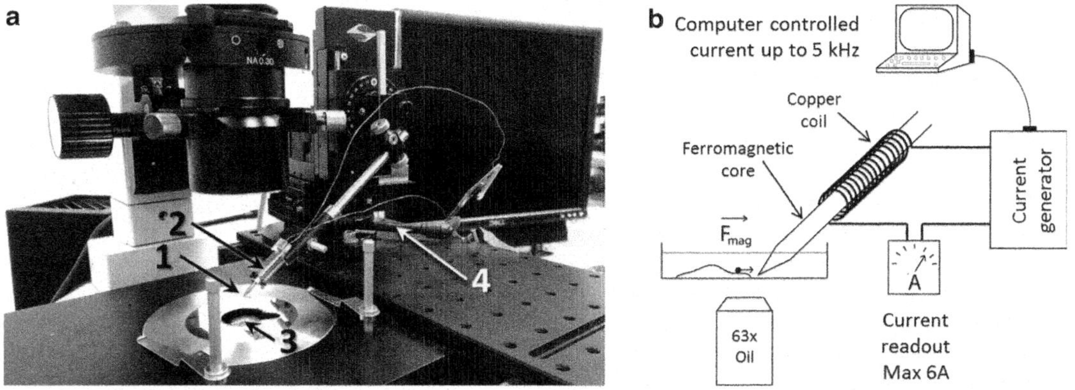

Fig. 1 (**a**) Magnetic tweezers and fluorescence microscopy system used in our lab. (1) Ferromagnetic core tip. (2) Two coiled copper wires. (3) Oil objective (63×). (4) Micromanipulator. (**b**) Schematic representation of the magnetic tweezers and imaging setup. The computer-controlled current generator is connected to a power amplifier (not depicted)

solenoid, and the distance between the tip of the magnetic core and the bead. In our setup, given the high variability in the shape of the magnetic tip and of the complex spatial geometry between the bead and the magnetic tip positions, it is difficult to estimate the magnetic forces mathematically. Therefore, we performed a calibration procedure to obtain "Force vs. Distance" curves for each type of magnetic beads used in our experiment (Fig. 2).

Preparing a Mixture of Magnetic and Non-magnetic Beads for Force Calibration

1. Add a small volume of 100–200 μL of silicone oil to the center of a new glass-bottom imaging dish (*see* **Note 10**).
2. In two separate 1.5 mL tubes prepare silicone oil suspensions of magnetic and non-magnetic polystyrene beads (*see* **Note 11**).
3. *Magnetic beads*: Wash 5 μL of magnetic beads (M-270 or MyOne) in 500 μL deionized water, sonicate 1 min, gather beads on the bottom of the tube with a magnet, and carefully discard the supernatant. Resuspend the washed beads in 500 μL silicone oil by adding the oil to the tube and mix vigorously with a pipette tip. To further homogenize the solution sonicate the bead-silicone oil mixture for 1 h in a heated water bath. The beads do not disperse easily in the viscous oil and repeating the sonication two to three times is often necessary.
4. *Polystyrene (non-magnetic) beads*: Wash 10 μL of polystyrene beads in 1 mL deionized water, vortex briefly, centrifuge for 2 min at 5000 rpm in a tabletop centrifuge (~2500 × *g*), and discard the supernatant. Resuspend the beads in 500 μL silicone oil as described above for the magnetic beads.
5. Mix 50 μL of the magnetic beads with the polystyrene beads to a 3:1 ratio (magnetic beads–polystyrene).

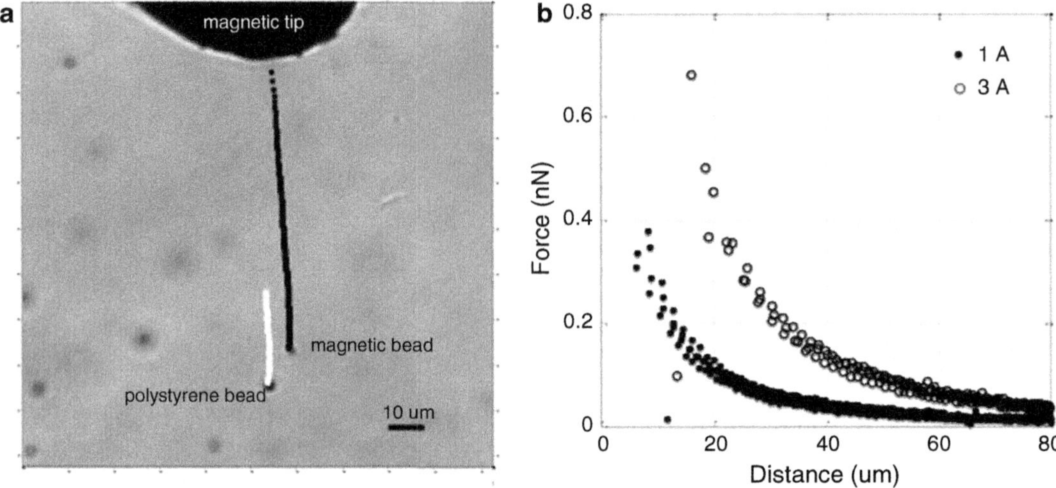

Fig. 2 (**a**) Bright-field micrograph depicting the magnetic tip at the top of the display image and several beads in silicone oil. Trajectories of a magnetic bead (1 μm diameter) and a polystyrene bead (2 μm diameter) are shown in *black dots* and in *white dots*, respectively. The magnetic bead is subject to magnetic force, while the polystyrene bead is only subject to local flow in the oil. Acquisition rate = 10 Hz. Scale bar = 10 μm. (**b**) Calibration curves of "force vs. distance." Distance represents the position of the bead relative to the tip. Values represent three beads for each of two amplitudes of applied current: 1 A (*solid dots*) and 3 A (*open circles*). The force was calculated using Stoke's law $F = 6\pi\mu rv$ ($\mu = 12{,}500$ centipoise, $r = 0.5$ μm, v was calculated from bead displacement between subsequent frames corrected for non-magnetic induced flow)

6. Add the mixture of magnetic and polystyrene beads to the clear silicone oil previously set on the imaging dish (*see* **Note 12**).

Manipulating Tweezers for Force Calibration

7. Mount the imaging dish on the preheated microscope stage (set environmental chamber temperature in advance to 37 °C).
8. Manually position the magnetic tip over the bead-silicone mixture in the dish.
9. Using the XYZ micromanipulator, slowly lower the tweezers tip until it touches the silicone in the dish.
10. Switch to camera view and, using the micromanipulator, slowly lower the tweezers until it is 1–5 μm away from the surface of the dish. Steer it carefully so that the end of its tip shows on the top edge of the displayed image field of view (Fig. 2a) (*see* **Note 13**).
11. Allow the silicone oil to settle for 10 min as the tip manipulation typically causes local flow.
12. Position the magnetic tweezers tip at the edge of a field of view with 5–10 magnetic and polystyrene beads (Fig. 2a).
13. Using BK Precision software, setup the current you wish to apply for your calibration experiment.

14. Using Micromanager, setup the "Multi-Dimensional" acquisition parameters according to your experimental aims.
15. Start image acquisition by clicking "Acquire" on the Micromanager acquisition panel, and subsequently turn on the tweezers by clicking "Run" in the BK software control panel.

Qualitative In Situ Monitoring of the Magnetic Force

16. Tracking of beads was performed using ImageJ plugin 'Particle Tracker', and force analysis was performed using Matla (Fig. 2).
17. During image acquisition, keep visual track of the beads of interest.
18. Correct for focus drift of the sample by choosing fiduciary structures on the glass surface as position references.
19. Maintain the location of the tip of the tweezers the same in all experiments involving cells (*see* **Note 14**).

Plotting "Force vs. Distance" Curves

The motion of a magnetic bead in silicone oil obeys Stokes' law of particle movement in viscous fluids:

$$F = 6\pi\mu rv,$$

where μ is the dynamic viscosity of the silicone oil, r is the bead radius and, v is the speed of the bead. Local flow in the oil can be induced by the magnetic beads that are moving toward the magnetic tip, or by changes in temperature due to the heating of the magnetic tip, and/or by mechanical vibrations. Any motion detected on a polystyrene non-magnetic bead is attributed to local flow in the oil. Therefore, to correct for any motion induced by local flow in the oil, track the motion of nearby polystyrene beads and subtract it from the motion of the magnetic bead. Local speed of the magnetic bead is calculated for each time point as $v = \Delta d/\Delta t$, where Δd is the corrected displacement of the magnetic bead between the current and the subsequent frame and Δt is the time interval between frames. In the calibration shown in Fig. 2, $\Delta t = 0.1$ s (corresponding to a 10 Hz acquisition rate). The force on the magnetic bead at each time point is then calculated using Stoke's law above. Figure 2b shows two calibration curves of "Force vs. Distance" from the tip for magnetic beads of 1 μm in diameter (MyOne) and for two amperages (1 and 3 A) applied on the electromagnet. We can apply forces up to several nN at a distance of 5–10 μm away from the magnetic tip.

Controlling the Amplitude and Speed of Force Pulling on Beads Attached to a Substrate

Magnetic beads subject to a magnetic field are pulled with a force, F, in the direction of the magnetic field. The electric current applied to the electromagnetic coil can be programmed to generate forces up to a few nN, applied in pulses with variable duration and frequency controlled by the BK Precision software (Figs. 3 and

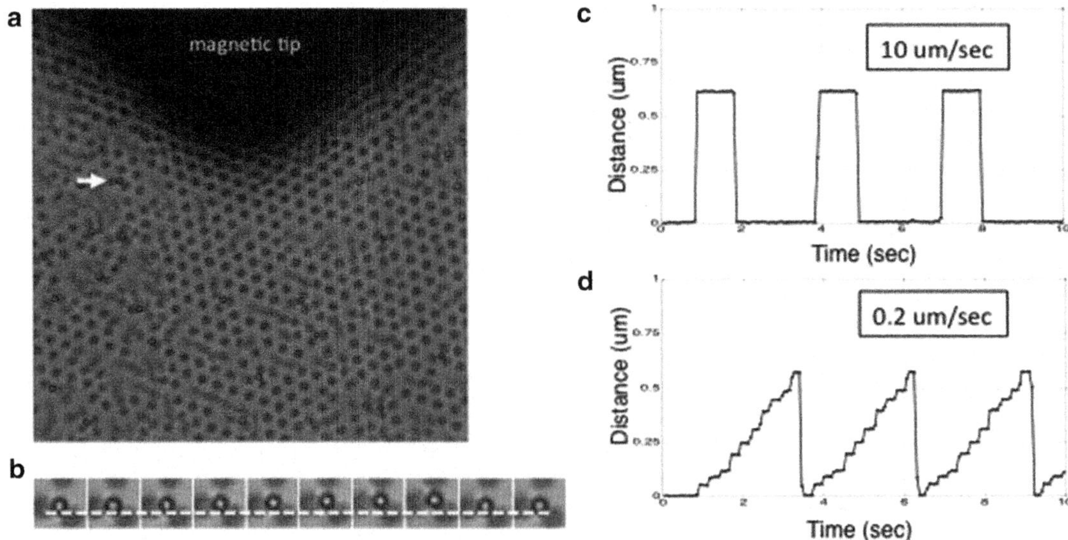

Fig. 3 Magnetic beads attached to an array of flexible polyacrylamide micropillars are subject to cycles of increasing magnetic force. (**a**) The magnetic tip is positioned 5–10 μm above the micropillar tops (Pillars are spaced 3 μm center to center). One bead (*white arrow*) was tracked during two different cycles shown in (**c**) and (**d**). (**b**) Time series montage of images of bead tracked in (**a**) (*white arrow*). *Dashed line* represents initial position before force was applied. (**c**) Bead displacement during cycles of "fast pulling." One cycle comprised the following steps: 1 s at 0 A, 1 s at 3 A, and 1 s at 0 A. The electric current applied to the electromagnet at the second phase produced a sudden magnetic force on the bead that pulled the bead at a high speed of 10 μm/s. (**d**) Bead displacement during cycles of "slow pulling" by gradually increasing the magnetic force. The cycle applied to the electromagnet to produce this force was composed of ten short current pulses of 0.2 s in duration and increasing amplitude from 0 to 3 A. The average speed at which bead moved during one cycle was 0.2 μm/s

4b). By changing the magnitude and duration of the current pulses we control the speed at which the bead moves toward the magnetic tip and, hence, the speed of force pulling on a bead attached to a cell or other substrate. The maximum value of the current during one cycle of pulses is chosen based on the desired maximum force exerted by the bead on the substrate, or cell body. Here, we demonstrate the robustness and versatility of force control of our magnetic tweezers system. To visualize local forces generated by the magnetic tweezers pulling on the magnetic beads we attached magnetic beads coated with bacteria pili to an array of flexible micropillars (Fig. 3a). The bending angle of a micropillar in response to a pulling force is proportional to the force applied. The micropillar array was made according to a protocol described elsewhere [35]. Briefly, the micropillar array is made using a lithography-fabricated silicone master, which was then applied to a mixture of polyacrylamide and crosslinking agent followed by heat curing. The micropillar array was mounted on a glass-bottom imaging dish, the micropillars were coated with poly-l-lysine, and the dish was filled with PBS buffer. Then, pili-coated magnetic beads were added to the micropillar array. As shown in Fig. 3b, the beads attached strongly to the poly-l-lysine coated surface. In Fig. 3c, d we show two examples of a bead pulling at different speeds.

20. In Fig. 3c, a series of 1 s current pulses with a current amplitude of 3 A was applied to the electromagnet. This cycle induced a sudden increase in the magnetic force on the attached bead as indicated by an increasing bending angle of the micropillar, and consequently a high speed of pulling on the substrate.
21. In Fig. 3d, a second series of repeating cycles is shown. Each cycle was composed of ten short current pulses of 0.2 s and with increasing amplitude from 0 to 3 A. As a result, the force exerted on the bead gradually increased, as indicated by an increasing bending angle of the micropillar, and the magnetic bead pulls on the substrate with a slower speed compared with example 1 in Fig. 3c.

3.3.3 Tweezers Manipulation: Applying Force to Pili-Coated Beads on Epithelial Cells

We used magnetic tweezers to study the response of the actin cytoskeleton of A-431 cells to force applied on *N. gonorrhoeae* pili-coated 3 μm diameter beads (Fig. 3). We saw that force exerted by the magnetic tweezers on pili-coated beads mimics the force that bacteria exert on epithelial cells during infection.

Before starting the following protocol, prepare an imaging dish containing A-431 cells expressing tractin-GFP as described in Subheading 3.1.1; turn on all the microscope components in advance as described in Subheading 3.1.2; and prepare pili-coated beads as described in Subheading 3.2.1.

Set Beads onto Epithelial Cells

1. Change the cell medium from full medium to pre-warmed serum-free CO_2-independent medium. The total volume of the cell imaging dish should be 3–4 mL.
2. Remove the bead suspension from the 4 °C rotation apparatus, and warm them up to room temperature.
3. Sonicate the beads for 1 min to break up eventual aggregates and ensure a uniform suspension of single beads.
4. Remove the cell imaging dish from the 37 °C incubator and pipette 1 μL bead suspension onto the cells. Perform this in sterile conditions (*see* **Note 15**).
5. Return the cells to the 37 °C incubator and let the beads settle for 30 min.

Position the Tweezers Over the Cells

6. Remove the cell-imaging dish with beads from the incubator, and mount it on the microscope stage and onto the 63× oil objective. Use the metal clamps attached to the stage to hold the dish in place.

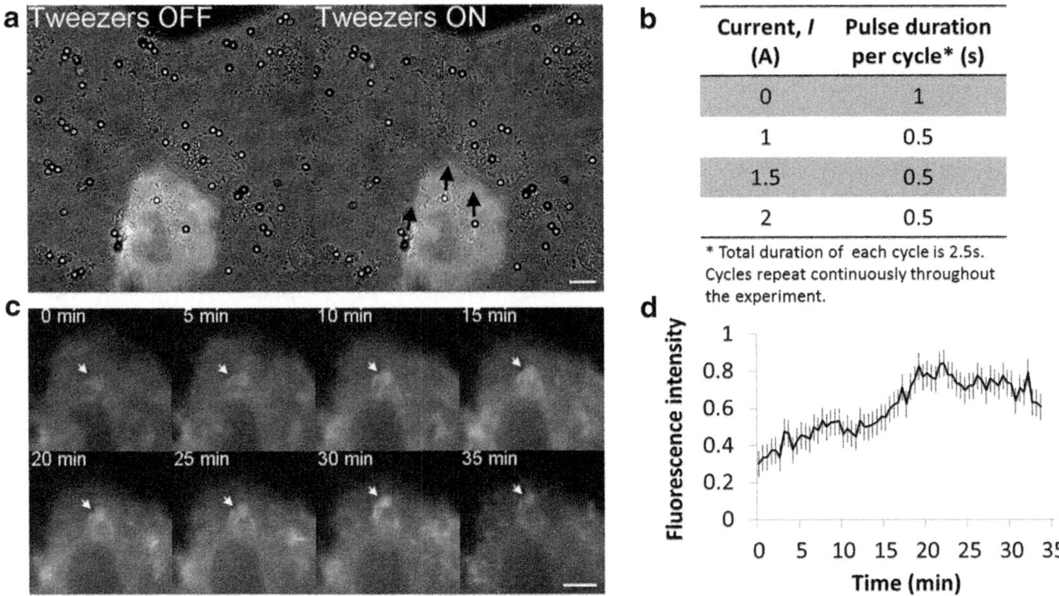

Fig. 4 Pili-coated beads pulled by magnetic tweezers trigger local f-actin accumulation. (**a**) Simultaneous bright-field and fluorescence image acquisition of live epithelial cells. Magnetic beads (high contrast spheres) and tip of the magnetic tweezers (*dark* field on center top of the image) are visible through the bright-field channel. Floating beads abandon the field of view after current is turned on. Beads attached to the epithelium remain in the field of view. *Black arrows* (*right* image, Tweezers ON) point to three beads attached to one cell expressing actin-GFP (center bottom of the image, visible through the fluorescent channel). (**b**) Magnetic force gradient is generated by applying repeated cycles of current pulses of increasing intensity. The table depicts one cycle ranging from no current (0 A) up to 2 A. (**c**) Visualization of actin-GFP in response to local magnetic forces generated by bead pulling. *White arrows* point to bead location at different time-points. Scale bar = 10 μm. (**d**) Mean fluorescence intensity of actin-GFP accumulated at the sites of pulling beads such as shown in (**c**). Error bars represent standard error of the mean of three beads

7. Remove the lid of the cell-imaging dish, and manually bring the magnetic tweezers into place by setting its needle tip above the center of the dish.
8. Using the XYZ micromanipulator, lower the tweezers tip until it touches the medium in the dish. Follow the tweezers manipulation guidelines from the force calibration Subheading 3.3.2.
9. Following the guidelines from Subheading 3.1.2, focus on a region of interest (ROI) that includes a tractin-GFP-expressing cell with several beads on top (*see* **Note 16**).
10. Under bright-field illumination, switch to camera view, and using the micromanipulator, slowly lower the tweezers until it is 1–5 μm away from the cell monolayer. Steer it carefully so that the end of its tip shows on the top edge of the displayed image field of view without touching the cells (Fig. 4a).

Time-Lapse Acquisition and "Current vs. Time" Parameters

11. Using BK Precision software, setup the "Current vs. Time" program you wish to run for your experiment (Fig. 4b) (*see* **Note 17**).
12. Using Micromanager, setup the Multi-Dimensional time-lapse acquisition parameters according to your experimental aims (*see* **Note 18**).
13. Start image acquisition by clicking "Acquire" on the Micromanager acquisition panel, and immediately after the first frame is captured, turn on the tweezers by clicking "Run" in the BK software control panel (*see* **Note 19**).

Qualitative In Situ Monitoring of Bead Pulling Forces

14. After turning on the tweezers, some beads abandon the field of view, either because they were not adhering to the cells or because the force generated by the tweezers overwhelmed the bead-cell adhesion strength and pulled the beads away.
15. Correct for focus drift of the sample by choosing fiduciary structures on the glass surface as position references.

Quantitative Analysis of Force-Dependent f-Actin Accumulation

16. Using ImageJ, measure the fluorescence intensity of the area occupied by one bead (signal) over time by drawing a circle around that area in each frame, and then using the function "measure" from the "analyze" menu.
17. Similarly, measure the fluorescence background (noise) of another area of the cell that is not occupied by a bead.
18. For each frame, subtract the noise from the measured signal to obtain the corrected fluorescence signal, and normalize it to the background for each frame (*see* **Note 20**).

4 Notes

1. Why tractin? Traditional vectors expressing GFP-tagged actin monomers (g-actin) have a drawback: the tendency to suppress assembly of actin filaments (f-actin). An alternative approach is to use GFP-tagged proteins that bind to f-actin, such as tractin [34]. This approach does not interfere with actin polymerization allowing f-actin assembly rates and actin-driven forces to remain unaffected. Furthermore, the signal from actin-GFP represents both g-actin and f-actin, thus yielding a higher fluorescence background. In contrast, tractin-GFP has a specific affinity for f-actin, so it has a higher signal-to-noise ratio than actin-GFP.
2. Higher transfection efficiencies can be achieved using other transfection methods such as Lipofectamine, but they usually result in increased cell death, and higher fluorescence background. Moreover, the Fugene transfection protocol is more

practical, since it does not require removal of serum and does not require washing or changing of culture medium after introducing the reagent/DNA complex.

3. 50 mm imaging dishes with low walls (7 mm height) are suitable for micromanipulation because they provide easy access to the tweezers setup on the microscope stage, without interfering with image acquisition.
4. The amount of DNA used per dish corresponds to a final DNA–Fugene (μg:μL) ratio of 1:4. This ratio should be optimized for each plasmid and cell line. Suboptimal deviations from this ratio will result in decreased transfection efficiency and increased cell death.
5. High quality image acquisition is critical in quantitative fluorescence microscopy. Protein subcellular localization can only be accurately tracked and measured if all steps in the cell preparation protocol, DNA plasmid purification and transfection are optimized and reproducible. Slight variations in the image acquisition settings of the microscope and of the imaging software dramatically impact the quality of results. Thus, optimal image acquisition settings should be saved and reused each time for different experimental runs.
6. This step is critical to avoid focus drift during image acquisition. Due to temperature sensitivity of the materials comprising the microscope stage, objective, and oil, small temperature variations result in relatively large distortion of those materials.
7. Density filters are placed in the light path between the light source and the sample to decrease the amount of light illuminating the sample by partially blocking/reflecting incident light and, thus, help to prevent phototoxicity. This is particularly important for long-term acquisition experiments.
8. It is important to always minimize photobleaching for purposes of fluorescence intensity quantification analyses.
9. Store coated beads at 4 °C with gentle rotation to avoid sedimentation and aggregate formation.
10. To pipette the very viscous silicone oil, cut off the end of a 1000 μL pipette tip and set the pipette to aspirate 500 μL. Allow the silicone oil to settle for 10 min at 37 °C.
11. The two types of beads should have different diameter to be easily identified in the microscope.
12. The final bead-silicone oil mixture should yield about 5–10 beads in the microscope field of view of the 63× oil objective, as shown in Fig. 2a.
13. Practice makes perfect! Get comfortable with the XYZ micromanipulator before performing this experiment. Adjust the sensitivity of the micromanipulator from "course" to "fine" as

the tip moves closer to the surface of the dish. Move the magnetic tip slowly to minimize local flow in the oil.

14. The magnetic force exerted on the beads depends exponentially on the distance of the beads of interest to the tip (Fig. 2b).
15. You may choose to add more beads according to the concentration of your bead suspension. Do this on different regions of the glass-bottom center of the imaging dish in order to have beads uniformly spread out and thus increase chance of having single beads interacting with a tractin-GFP cell.
16. Beads that do not adhere to the cell surface usually display Brownian-like motion and tend to promptly abandon the image field of view as soon as an electromagnetic current is applied.
17. Take into account your tweezers force calibration curve so that the applied force falls into the biologically meaningful range of your experiment. As shown in Fig. 4b we ran continuous cycles of a 4-step force gradient, as defined by the increasingly stronger current values applied.
18. In Fig. 4c we acquired a 35 min image sequence, and tracked tractin-GFP in the fluorescent channel with 30 s interval between frames. We only captured images in the bright-field channel every 5 min to reduce phototoxicity.
19. As shown in Fig. 4a (left image "Tweezers OFF"), we captured the first frame with the tweezers off, then turned on the tweezers, and 30 s later captured the second frame with the "Tweezers ON."
20. The graph in Fig. 4d represents the variation over time of the corrected and normalized mean fluorescence signal of three beads.

References

1. McMahon HT, Boucrot E (2015) Membrane curvature at a glance. J Cell Sci 128:1065–1070
2. Roux A, Cappello G, Cartaud J, Prost J, Goud B, Bassereau P (2002) A minimal system allowing tubulation with molecular motors pulling on giant liposomes. Proc Natl Acad Sci U S A 99:5394–5399
3. Pollard TD (2010) Mechanics of cytokinesis in eukaryotes. Curr Opin Cell Biol 22:50–56
4. Chicurel ME, Chen CS, Ingber DE (1998) Cellular control lies in the balance of forces. Curr Opin Cell Biol 10:232–239
5. Mammoto T, Mammoto A, Ingber DE (2013) Mechanobiology and developmental control. Annu Rev Cell Dev Biol 29:27–61
6. Roca-Cusachs P, del Rio A, Puklin-Faucher E, Gauthier NC, Biais N, Sheetz MP (2013) Integrin-dependent force transmission to the extracellular matrix by α-actinin triggers adhesion maturation. Proc Natl Acad Sci U S A 110: E1361–E1370
7. Kundu AK, Putnam AJ (2006) Vitronectin and collagen I differentially regulate osteogenesis in mesenchymal stem cells. Biochem Biophys Res Commun 347:347–357
8. Vogel V, Sheetz M (2006) Local force and geometry sensing regulate cell functions. Nat Rev Mol Cell Biol 7:265–275
9. Vogel V, Sheetz MP (2009) Cell fate regulation by coupling mechanical cycles to biochemical signaling pathways. Curr Opin Cell Biol 21:38–46
10. Farge E (2003) Mechanical induction of Twist in the Drosophila foregut/stomodeal primordium. Curr Biol 13:1365–1377

11. Engler AJ, Sen S, Sweeney HL, Discher DE (2006) Matrix elasticity directs stem cell lineage specification. Cell 126:677–689
12. Nisticò P, Di Modugno F, Spada S, Bissell MJ (2014) β1 and β4 integrins : from breast development to clinical practice. 1–9, Breast Cancer Research 2014, 16:459
13. Lee SW, Higashi DL, Snyder A, Merz AJ, Potter L, So M (2005) PilT is required for PI(3,4,5)P3-mediated crosstalk between Neisseria gonorrhoeae and epithelial cells. Cell Microbiol 7:1271–1284
14. Bieber D, Ramer SW, Wu CY, Murray WJ, Tobe T, Fernandez R, Schoolnik GK (1998) Type IV pili, transient bacterial aggregates, and virulence of enteropathogenic Escherichia coli. Science 280(5372):2114–2118
15. Lambrechts A, Gevaert K, Cossart P, Vandekerckhove J, Van Troys M (2008) Listeria comet tails: the actin-based motility machinery at work. Trends Cell Biol 18(April):220–227
16. Cleary J, Lai L-C, Shaw RK, Straatman-Iwanowska A, Donnenberg MS, Frankel G, Knutton S (2004) Enteropathogenic Escherichia coli (EPEC) adhesion to intestinal epithelial cells: role of bundle-forming pili (BFP), EspA filaments and intimin. Microbiology 150(Pt 3):527–538
17. Caron E, Crepin VF, Simpson N, Knutton S, Garmendia J, Frankel G (2006) Subversion of actin dynamics by EPEC and EHEC. Curr Opin Microbiol 9:40–45
18. Merz AJ, So M (1997) Attachment of piliated, Opa- and Opc- gonococci and meningococci to epithelial cells elicits cortical actin rearrangements and clustering of tyrosine-phosphorylated proteins. Infect Immun 65:4341–4349
19. Merz AJ, Enns CA, So M (1999) Type IV pili of pathogenic Neisseriae elicit cortical plaque formation in epithelial cells. Mol Microbiol 32:1316–1332
20. Merz AJ, So M (2000) Interactions of pathogenic neisseriae with epithelial cell membranes. Annu Rev Cell Dev Biol 16:423–457
21. Howie HL, Glogauer M, So M (2005) The N. gonorrhoeae type IV pilus stimulates mechanosensitive pathways and cytoprotection through a pilT-dependent mechanism. PLoS Biol 3:e100
22. Higashi DL, Zhang GH, Biais N, Myers LR, Weyand NJ, Elliott DA, So M (2009) Influence of type IV pilus retraction on the architecture of the Neisseria gonorrhoeae-infected cell cortex. Microbiology 155(Pt 12):4084–4092
23. Tanase M, Biais N, Sheetz M (2007) Magnetic tweezers in cell biology. Trends Cell Biol 2:116–118
24. Kuo SC, Sheetz MP (1992) Optical tweezers in cell biology. Trends Cell Biol 2:116–118
25. Kollmannsberger P, Fabry B (2007) High-force magnetic tweezers with force feedback for biological applications. Rev Sci Instrum 78:114301
26. Webster KD, Crow A, Fletcher DA (2011) An AFM-based stiffness clamp for dynamic control of rigidity. PLoS One 6:1–7
27. Desprat N, Richert A, Simeon J, Asnacios A (2005) Creep function of a single living cell. Biophys J 88:2224–2233
28. Fouchard J, Bimbard C, Bufi N, Durand-Smet P, Proag A, Richert A, Cardoso O, Asnacios A (2014) Three-dimensional cell body shape dictates the onset of traction force generation and growth of focal adhesions. Proc Natl Acad Sci U S A 111(36):13075–13080
29. Jiang G, Giannone G, Critchley DR, Kukomoto E, Sheetz MP (2003) Two-piconewton slip bond between fibronectin and the cytoskeleton depends on talin. Nature 424:334-337
30. Dai J, Sheetz MP (1999) Membrane tether formation from blebbing cells. Biophys J 77(6):3363–3370
31. Jiang G, Huang AH, Cai Y, Tanase M, Sheetz MP (2006) Rigidity sensing at the leading edge through alphavbeta3 integrins and RPTPalpha. Biophys J 90:1804–1809
32. Biais N, Ladoux B, Higashi D, So M, Sheetz M. (2008) Cooperative retraction of bundled type IV pili enables nanonewton force generation. PLoS Biology 15;6(4):e87
33. Opitz D, Maier B (2011) Rapid cytoskeletal response of epithelial cells to force generation by type IV pili. PLoS One 6:8
34. Johnson HW, Schell MJ (2010) Neuronal IP3 3-kinase is an F-actin-bundling protein: role in dendritic targeting and regulation of spine morphology. Mol Biol Cell 20:5166–5180
35. Biais N, Higashi D, So M, Ladoux B (2012) Techniques to measure pilus retraction forces. Methods Mol Biol 799:197–216

Chapter 11

Reconstitution of a Minimal Actin Cortex by Coupling Actin Filaments to Reconstituted Membranes

Sven K. Vogel

Abstract

A thin layer of actin filaments in many eukaryotic cell types drives pivotal aspects of cell morphogenesis and is generally cited as the actin cortex. Myosin driven contractility and actin cytoskeleton membrane interactions form the basis of fundamental cellular processes such as cytokinesis, cell migration, and cortical flows. How the interplay between the actin cytoskeleton, the membrane, and actin binding proteins drives these processes is far from being understood. The complexity of the actin cortex in living cells and the hardly feasible manipulation of the omnipotent cellular key players, namely actin, myosin, and the membrane, are challenging in order to gain detailed insights about the underlying mechanisms. Recent progress in developing bottom-up in vitro systems where the actin cytoskeleton is combined with reconstituted membranes may provide a complementary route to reveal general principles underlying actin cortex properties. In this chapter the reconstitution of a minimal actin cortex by coupling actin filaments to a supported membrane is described. This minimal system may be very well suited to study for example protein interactions on membrane bound actin filaments in a very controlled and quantitative manner as it may be difficult to perform in living systems.

Key words Actin, Actin filaments, Actin cortex, Membrane, Supported lipid bilayer, Myosin, Actomyosin, Cytoskeleton, Reconstitution, Synthetic biology

1 Introduction

The actin cytoskeleton is involved in dozens of cellular processes including intracellular cargo transport, exocytosis and endocytosis, cell locomotion, organelle motion, cytoplasmic streaming, polarity formation, and cell division. In vivo studies on the actin cytoskeleton and its interaction partners gave us deep insights about functions of the actin cytoskeleton inside cells and how these functions are aided by other proteins [1]. In parallel, in vitro reconstitution assays of actin and its interaction partners during the last few decades were invented with great success and formed the basis for the detailed molecular understanding about actin filament dynamics and its regulation by interacting proteins as well as the determination of detailed motor proteins properties [2–4]. In spite

Ray H. Gavin (ed.), *Cytoskeleton: Methods and Protocols*, Methods in Molecular Biology, vol. 1365,
DOI 10.1007/978-1-4939-3124-8_11, © Springer Science+Business Media New York 2016

of our vast knowledge about the actin cytoskeleton and its interactions with other proteins, many cellular processes are still ill defined from a mechanical point of view. A key feature of living cells is their ability to control their shape and the intercellular and intracellular communication where the interaction between the cytoskeleton and the cell membrane plays fundamental roles [1, 5]. Here our mechanistic and quantitative understanding is rather limited and begs for further scientific endeavors.

In eukaryotic cells a thin layer of actin filaments interacts with the cell membrane where a part of the actin layer is directly coupled to the cell membrane via anchor proteins and thereby also ensures structural integrity [6]. This system is called the actin cortex and governs cell shape control during cell locomotion and cell division and influences the diffusion behavior of lipids and proteins in the cell membrane [7–9]. During cell morphogenesis the actin meshwork is constantly dynamically reorganized by interacting with the motor protein myosin II and the cell membrane resulting in force transmission to the cell membrane and eventually in cell shape changes [5, 10]. In order to understand these actin motor and membrane interactions in more detail the development of in vitro assays that circumvent the complexity present at the cytoskeleton membrane interface in living cells would be desirable and very well suited to reveal basic principles underlying actin cell cortex mechanics [11, 12]. Nevertheless, looking at the lipid membrane and the actin cytoskeleton independently as isolated systems has dominated the in vitro field the last decades and has been extremely successful leading to the identification of single step sizes of motor proteins on linear actin tracks [13] and the discovery that distinct lipid environments help in protein sorting and influence their functionality [14]. However, to shed light into the underlying principles of how the dynamic interplay between these entities controls lipid protein movements and shape changes of the cell membrane, in vitro assays have been recently developed that combine these two units [15–21]. The combination of the membrane and actin cytoskeleton modules can be subdivided into systems where actin filaments were either reconstituted on planar supported (Figs. 1, 2, and 3) or on free-standing lipid bilayers to mimic a cellular actin cortex [15–21]. Contractile behavior in these systems was induced with the addition of myosin motor assemblies in the presence of ATP. Mica or glass supported lipid bilayers (SLBs) have been successfully used as systems mimicking biological membranes for many years [22, 23]. SLBs are easily accessible to addition and exchange of protein or buffer components in a stepwise manner and are perfectly suited for applying surface-based manipulation and imaging techniques such as Atomic Force Microscopy (AFM), Surface Plasmon Resonance (SPR), and Total Internal Reflection Fluorescence (TIRF) microscopy. Several groups managed to couple filamentous actin to SLBs using pro-

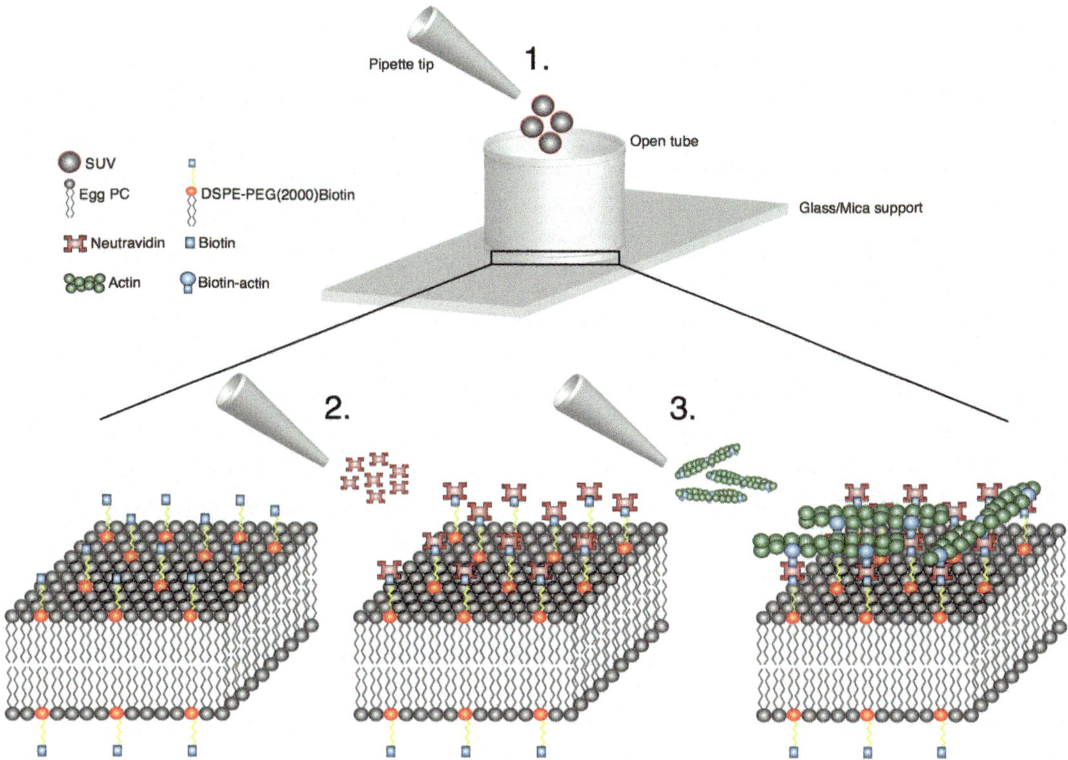

Fig. 1 Steps of minimal actin cortex (MAC) formation. A schematic representation and steps of the lipid bilayer formation and actin coupling are shown. SUVs are added to the glass or mica support (step 1) resulting in the formation of a supported lipid bilayer (EggPC) containing biotinylated lipids (DSPE-PEG(2000)-Biotin). Addition of neutravidin (step 2) and biotinylated actin filaments (step 3) result in their coupling to the lipid bilayer. Adapted from ref. [21] with permission from Copyright © 1999–2015 John Wiley & Sons, Inc.

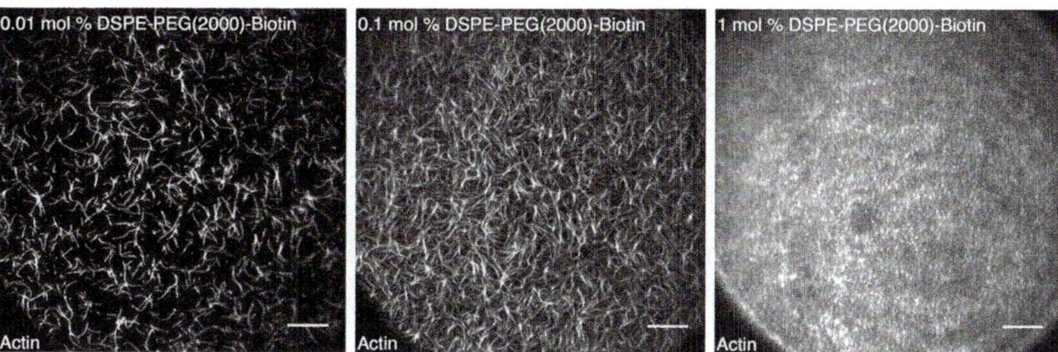

Fig. 2 Different densities of membrane coupled actin filaments. TIRF microscopy images of Alexa 488-phalloidin labeled actin filaments coupled to supported bilayers are shown. The actin filament density increases from the *left* to the *right* images and corresponds to an increase in the amount of DSPE-PEG(2000)-Biotin (0.01 mol %, 0.1 mol %, 1 mol %) in the SLB. Scale bars, 10 μm. Adapted from ref. [21] with permission from Copyright © 1999–2015 John Wiley & Sons, Inc.

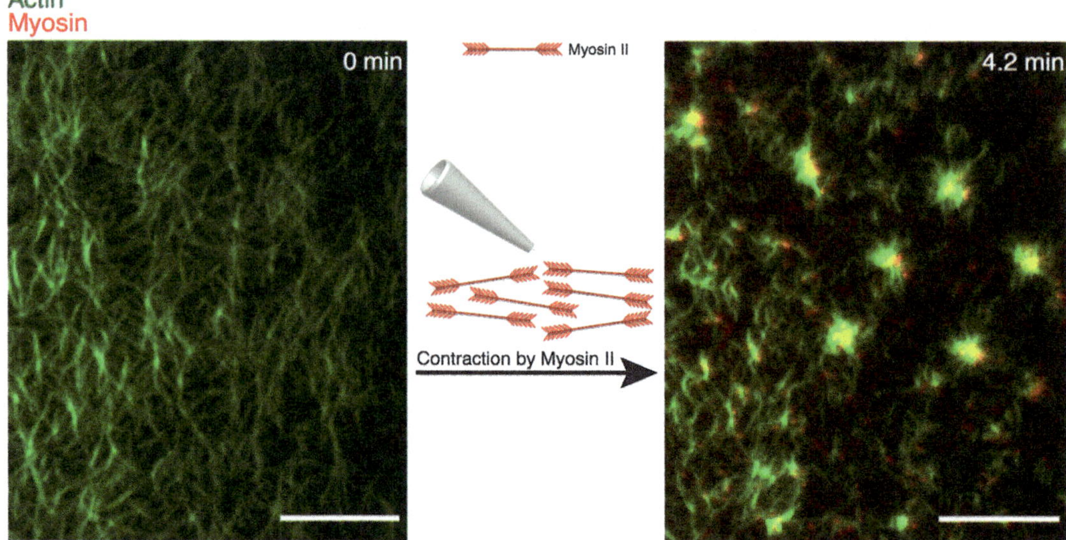

Fig. 3 MAC contraction by myosin motors. TIRF microscopy images of a MAC before (*left*) and after the addition of myosin filaments (*right*) are shown. Myosin filaments (*red*) induce the contraction of the membrane coupled actin layer and the formation of actomyosin clusters (*right* image). Scale bars, 10 μm. Adapted from ref. [15] with permission from Copyright © 2015 eLife Sciences Publications Ltd.

teins that anchor actin filaments to the membrane [24–26]. Recently the development of SLBs with artificial actin anchoring systems turned out to be very useful to study myosin motor activity on membrane coupled actin filaments and revealed a possible mechanism for how compressive stress on actin filaments induced by myosin motors could contribute to actin turnover ([15], Fig. 3) and how different levels of actin meshwork adhesion to the lipid bilayer may control the ratio between meshwork contraction and actin filament severing [18]. Other systems based on free-standing lipid bilayers gave insights on how the presence of a membrane coupled actin meshwork influences the lateral diffusion behavior of membrane lipids and proteins [16] and how the actin meshwork connectivity leads to different behaviors during myosin induced contraction on the membrane [20]. This chapter describes procedures for forming glass or mica supported lipid bilayers, to which actin filaments are subsequently anchored using a lipid-based artificial anchor system.

2 Materials

2.1 Chemicals, Buffers, and Solutions

1. Stock Solutions: 1 M KCl; 1 M $MgCl_2$; 1 M DTT; 0.1 M ATP; 1 M Tris–HCl, pH 7.5; 1 M $CaCl_2$; 0.5 M MOPS, pH 7.0; 0.1 M EGTA, pH 7.4; 3 % NaN_3.

2. Reaction buffer: 50 mM KCl, 2 mM $MgCl_2$, 1 mM DTT, and 10 mM Tris–HCl pH 7.5.
3. Polymerization buffer: 50 mM KCl, 2 mM $MgCl_2$, 1 mM DTT, 1 mM ATP, and 10 mM Tris–HCl, pH 7.5.
4. Labeling buffer: 10 mM MOPS pH 7.0, 0.1 mM EGTA, and 3 mM NaN_3.

2.2 Materials for Membrane Formation

1. Chloroform, ethanol, acetone, Milli-Q water.
2. Lipids (purchased from Avanti Polar Lipids, Alabaster, AL): L-A-PHOSPHATIDYLCHOLINE (EggPC), 1,2-distearoyl-*sn*-glycero-3-phosphoethanolamine-*N*-[biotinyl(polyethylene glycol)-2000] (DSPE-PEG(200)-Biotin).
3. Lipophilic cationic dye DiIC-C18.
4. 22 mm × 22 mm, #1.5, Menzel cover slips (Thermo Fisher, Braunschweig, Germany).
5. 1.5 mL Eppendorf tubes.
6. Kimtech tissue.
7. UV glue: Norland Optical Adhesive 63 (Norland Products Inc.).

2.3 Materials for Actin Preparation

1. Purified or purchased rabbit skeletal muscle actin monomers (Molecular Probes).
2. Biotinylated rabbit actin monomers (tebu-bio (Cytoskeleton Inc.)).
3. Alexa Fluor 488 phalloidin (Molecular Probes).
4. Neutravidin.

2.4 Special Equipment

1. Plasma cleaner.
2. Vacuum centrifuge.
3. Nitrogen or compressed air supply.

3 Methods

3.1 Reconstitution of Supported Lipid Bilayers (SLBs)

3.1.1 Preparation of Small Unilamellar Vesicles (SUV)

1. Herein we describe the preparation of SUVs that are subsequently used for the formation of lipid bilayers on a glass or mica support. Dissolve the lipid L-α-phosphatidylcholine (EggPC) and the biotinylated lipid 1,2-distearoyl-*sn*-glycero-3-phosphoethanolamine-*N*-[biotinyl(polyethylene glycol)-2000] (DSPE-PEG(200)-Biotin) in chloroform in a small glass vial (*see* **Note 1**). Depending on the required actin filament density (Fig. 2), use molar ratios of 99.99, 99.9, and 99 mol% for EggPC and 0.01, 0.1, and 1.0 mol % for DSPE-PEG(2000)-Biotin in order to obtain a total lipid amount of 10 mg/mL. For testing the membrane integrity, add 0.1 mol% of the lipophilic

cationic dye DiIC-C18 or any other reasonable dye to the lipid mixture (*see* **Note 1**). Dry the lipid–chloroform mixture under nitrogen flux for 30 min (*see* **Note 2**), and subsequently put it into a vacuum for 30 min in order to ensure complete solvent evaporation.

2. Rehydrate the Lipids in reaction buffer, and resuspend them by vigorous vortexing (10 min) until the suspension has a milky appearance and no longer contains any visible (lipid) particles. In this step multilamellar vesicles (MLVs) are formed.
3. Transform the MLVs to SUVs by exposing the suspension in the glass vial to ultrasound sonication in a water bath at room temperature until the milky suspension transforms into a translucent solution indicating the transformation of SUVs from MLVs. This process may last 30–45 min (or less) depending on the ultrasound sonication water bath system in use (*see* **Note 3**).

3.1.2 Glass or Mica Preparation

1. In this part we will describe the preparation and treatment of the glass or mica support prior to lipid bilayer formation. In both cases it is recommended to do it freshly as it is advantageous for the formation of a lipid bilayer. Therefore, for example, the time of SUV formation in the water bath may be a good opportunity to prepare the support. We will describe both the treatment of a glass support and the preparation of a mica support. Advantages and disadvantages of both supports are briefly described in Subheading 4 (*see* **Note 4**). For a glass support, rinse a glass coverslip (22 mm × 22 mm, #1.5) with Milli-Q water, and carefully mechanically clean it with Kimtech tissue. Subsequently, sequentially wash the coverslip with ethanol, acetone, and then again with ethanol and Milli-Q water. After the last wash, use a strong nitrogen (or air) flux to remove the remaining liquid and particles and quickly dry the coverslip. Then expose the coverslip to an air plasma treatment for 10 min in a Plasma Cleaner in order to remove left over contaminants and to make the glass surface hydrophilic (*see* **Note 5**).
2. If mica as an alternative support is desired, the cleaning procedure for the coverslip is principally not necessary, although we would recommend a cleaning of the coverslip, which may result in better data acquisition during imaging procedures especially when performing TIRF microscopy. Prior to cleaving a thin sheet of mica, place one drop of the UV glue onto the glass coverslip. If TIRF imaging of the sample is desired, a drop of oil is placed onto the cover slip instead of the UV glue. For mica preparation cut a square approximately 10 mm × 10 mm with a scissor or stamp out a circle of mica with 5 mm radius using a template and a hammer. Place the mica crystal between the sticky side and non-sticky side of tape on a

roll, and quickly pull the tape from the roll. The thick mica crystal will cleave. Part of the mica may not stick to the tape; reapply that part of the mica between the sticky side and non-sticky side of the tape roll, and again quickly pull the tape from the roll. Repeat this procedure several times as necessary to create a very thin mica sheet. Use clean tweezers to place the thin sheet on the UV glue or oil drop.

3. Prepare a chamber on the plasma-treated glass or on the mica surface by cutting a 1.5 mL Eppendorf tube and gluing it to the glass (or mica) surface with Norland UV glue applied to the rim of the tube (Fig. 1). Cure the UV glue by exposing the chamber to long wave ultraviolet light between 320 and 400 nm for 15 min.

3.1.3 Preparation of the Lipid Bilayer

1. After curing the UV glue, mix 10 μL of the SUV suspension with 90 μL of reaction buffer and immediately pipette the mixture onto the glass (or mica) surface of the chamber. Add $CaCl_2$ to a final concentration of 0.1 mM to aid fusion of the SUVs resulting in the formation of a lipid bilayer on the glass (or mica) support (*see* **Note 6**). After 45–60 min, wash the sample with a total volume of 2 mL reaction buffer by gently adding to and removing from the chamber approximately 200 μL aliquots of reaction buffer in order to remove $CaCl_2$ and vesicles that did not fuse with the surface to form the lipid bilayer. Be careful during this step not to touch the lipid bilayer with the pipette tip or to harm the lipid bilayer by too strong pipetting (*see* **Note 7**). Avoid any air contact with the lipid bilayer as it would cause immediate disruption of the membrane. Make sure that the lipid bilayer is always covered with buffer solution (*see* **Note 8**).

3.2 Actin Filament Preparation and Coupling to the SLB

In this section we describe the formation and preparation of biotinylated and phalloidin stabilized (non-dynamic) actin filaments that are eventually coupled to the supported bilayer system. We will not explain the actin purification procedures and will herewith refer to a multitude of existing literature ([27, 28], *see* **Note 9**).

3.2.1 Biotinylated Actin Filament Preparation

1. Mix actin monomers (purified or purchased) and biotinylated actin monomers mixed in a 5:1 (actin–biotin-actin) ratio (*see* **Note 9**) for a final concentration of 39.6 μM. Induce polymerization of the monomer mixture by adjusting the concentrations of: KCl to 50 mM, $MgCl_2$ to 2 mM, DTT to 1 mM, and ATP to 1 mM. Incubate the mixture for 1 h at room temperature for actin filament formation.
2. Prepare an actin-stabilizing solution by placing 60 μL of Alexa Fluor 488 phalloidin in a 1.5 mL Eppendorf tube.
3. Dry the solution in a vacuum centrifuge at room temperature.

4. Redissolve the dried powder in 5 μL methanol.
5. Add 85 μL labeling buffer to the redissolved phalloidin solution.
6. Add sufficient labeling buffer to the 39.6 μM actin mixture to reach a concentration of 20 μM actin.
7. Finally, add 10 μL of 20 μM actin mixture to 90 μL of phalloidin solution, and incubate at room temperature overnight. This procedure yields a final concentration of 2 μM Alexa 488-phalloidin-labeled, biotinylated actin filaments (referred to actin monomers) (*see* **Note 10**).

3.2.2 Actin Filament Binding to the SLB

1. In order to bind biotinylated actin to biotinylated lipids in the SLB, dissolve 2 μg of neutravidin in approximately 200 μL reaction buffer. Put the neutravidin solution through a vortex, and subsequently, pipette it up and down several times to prevent the presence of neutravidin agglomerates. Add the neutravidin solution to the washed SLB and leave it at room temperature for 10 min.
2. Gently wash the sample several times with 200 μL aliquots of reaction buffer (2 mL total volume) to remove unbound neutravidin. Take precautions to avoid disrupting the membrane during the washing steps.
3. Remove at least 100 μL (or if possible, more) of the approximately 200 μL reaction buffer that covers the SLB.
4. Add 10–50 μL of the 2 μM Alexa 488-phalloidin labeled, biotinylated actin filaments to the SLB and incubate for 1 h at room temperature to allow actin filaments to settle down and bind to the neutravidin layer of the SLB.
5. Gently wash the sample several times with 200 μL aliquots of reaction buffer (1–2 mL total volume) to remove unbound actin filaments. It is important to do a gentle washing here as bound actin filaments can be relatively easy ripped of by strong liquid streams induced by non-gentle pipetting (*see* **Note 7**).

4 Notes

1. EggPC can be replaced by other lipids such as DOPC and many others not mentioned here.
2. The lipid–chloroform mixture is dried inside a small glass vial by continuously rotating and holding it in an inclined position. This will result in the formation of a dried lipid film that not only covers the bottom but also the lateral interior wall of the glass vial. The increased dried lipid surface facilitates the

rehydration procedure and avoids the formation of bulky lipid agglomerates, which are difficult to resuspend during vortexing.

3. To ensure a proper and reasonable fast transformation from MLVs to SUVs it is important to expose the glass vial to a maximum amount of energy that is exhibited from the water bath ultrasound system. This is assured by adjusting the water level of the water bath to the point when water drops start to "jump off" the water surface. Place the bottom of the glass vial exactly to this spot close to the water surface. It is possible to freeze the obtained SUV solution and to store the SUVs in small aliquots at −20 °C.
4. Using mica as the lipid support has the advantage that no plasma treatment of the solid surface is needed and it seems that membrane formation is slightly more reproducible as the lipid bilayer formation on the glass depends on the glass surface landscape, accurate cleaning, and proper plasma treatment. However, on the other hand, if TIRF microscopy is desired, the quality of the TIRF images strongly depends on the thickness of the mica sheet. In this case the use of a glass support facilitates subsequent TIRF microscopy. We therefore used mainly glass as support.
5. Successful cleaning and plasma treatment of the glass surface can be observed by a strong decrease of the contact angle of a water droplet placed onto the glass surface.
6. The presence of calcium aids the formation of the lipid bilayer [29].
7. Pipette the washing buffer gently against the interior wall of the cut Eppendorf tube in order to avoid a strong liquid flow towards the bottom of the SLB chamber.
8. The integrity and fluidity of the formed lipid bilayer on the glass or mica surface is confirmed by Fluorescence Recovery after Photobleaching (FRAP) or the direct determination of the diffusion coefficient by Fluorescence Correlation Spectroscopy (FCS). In the case of FRAP the bleached area will recover within approximately 1 min.
9. Note that this protocol works fine with commercially available actin and biotinylated actin. Other actin–biotin-actin ratios such as 20:1 will also work. Whenever desired, the presence of biotinylated actin monomers in the actin filament can be further decreased to weaken the strength of the bond between the biotinylated actin filaments and the membrane.
10. The Alexa 488-phalloidin-labeled biotinylated actin filaments can be stored at 4 °C for as long as 1 month.

Acknowledgements

I am grateful for the financial support by the Daimler und Benz foundation (Project Grant PSBioc8216) and the MaxSynBio consortium, which is jointly funded by the Federal Ministry of Education and Research of Germany and the Max Planck Society.

References

1. Wessells NK, Spooner BS, Ash JF, Bradley MO, Luduena MA, Taylor EL, Wrenn JT, Yamaa K (1971) Microfilaments in cellular and developmental processes. Science 171(3967):135–143
2. Loisel TP, Boujemaa R, Pantaloni D, Carlier MF (1999) Reconstitution of actin-based motility of Listeria and Shigella using pure proteins. Nature 401(6753):613–616. doi:10.1038/44183
3. Spudich JA, Kron SJ, Sheetz MP (1985) Movement of myosin-coated beads on oriented filaments reconstituted from purified actin. Nature 315(6020):584–586
4. Spudich JA (2001) The myosin swinging cross-bridge model. Nat Rev Mol Cell Biol 2(5):387–392, doi: 10.1038/35073086 35073086 [pii]
5. Cramer LP, Mitchison TJ (1995) Myosin is involved in postmitotic cell spreading. J Cell Biol 131(1):179–189
6. Morone N, Fujiwara T, Murase K, Kasai RS, Ike H, Yuasa S, Usukura J, Kusumi A (2006) Three-dimensional reconstruction of the membrane skeleton at the plasma membrane interface by electron tomography. J Cell Biol 174(6):851–862, doi: jcb.200606007 [pii] 10.1083/jcb.200606007
7. Diz-Munoz A, Krieg M, Bergert M, Ibarlucea-Benitez I, Muller DJ, Paluch E, Heisenberg CP (2010) Control of directed cell migration in vivo by membrane-to-cortex attachment. PLoS Biol 8(11):e1000544. doi:10.1371/journal.pbio.1000544
8. Sedzinski J, Biro M, Oswald A, Tinevez JY, Salbreux G, Paluch E (2011) Polar actomyosin contractility destabilizes the position of the cytokinetic furrow. Nature 476(7361):462–466, doi: nature10286 [pii] 10.1038/nature10286
9. Kusumi A, Nakada C, Ritchie K, Murase K, Suzuki K, Murakoshi H, Kasai RS, Kondo J, Fujiwara T (2005) Paradigm shift of the plasma membrane concept from the two-dimensional continuum fluid to the partitioned fluid: high-speed single-molecule tracking of membrane molecules. Annu Rev Biophys Biomol Struct 34:351–378
10. De Lozanne A, Spudich JA (1987) Disruption of the Dictyostelium myosin heavy chain gene by homologous recombination. Science 236(4805):1086–1091
11. Vogel SK, Schwille P (2012) Minimal systems to study membrane-cytoskeleton interactions. Curr Opin Biotechnol. doi: S0958-1669(12)00059-6 [pii] 10.1016/j.copbio.2012.03.012
12. Rivas G, Vogel SK, Schwille P (2014) Reconstitution of cytoskeletal protein assemblies for large-scale membrane transformation. Curr Opin Chem Biol 22:18–26. doi:10.1016/j.cbpa.2014.07.018
13. Howard J (2001) Mechanics of motor proteins and the cytoskeleton. Sinauer Associates, Inc., Sunderland, MA
14. Lingwood D, Simons K (2010) Lipid rafts as a membrane-organizing principle. Science 327(5961):46–50, doi: 327/5961/46 [pii] 10.1126/science.1174621
15. Vogel SK, Petrasek Z, Heinemann F, Schwille P (2013) Myosin motors fragment and compact membrane-bound actin filaments. Elife 2:e00116
16. Heinemann F, Vogel SK, Schwille P (2013) Lateral membrane diffusion modulated by a minimal actin cortex. Biophys J 104(7):1465–1475, doi: S0006-3495(13)00260-9 [pii] 10.1016/j.bpj.2013.02.042
17. Pontani LL, van der Gucht J, Salbreux G, Heuvingh J, Joanny JF, Sykes C (2009) Reconstitution of an actin cortex inside a liposome. Biophys J 96(1):192–198, doi: S0006-3495(08)00038-6 [pii] 10.1016/j.bpj.2008.09.029
18. Murrell MP, Gardel ML (2012) F-actin buckling coordinates contractility and severing in a biomimetic actomyosin cortex. Proc Natl Acad Sci U S A 109(51):20820–20825, doi: 1214753109 [pii] 10.1073/pnas.1214753109
19. Liu AP, Richmond DL, Maibaum L, Pronk S, Geissler PL, Fletcher DA (2008) Membrane-induced bundling of actin filaments. Nat Phys 4:789–793. doi:10.1038/nphys1071
20. Carvalho K, Tsai FC, Lees E, Voituriez R, Koenderink GH, Sykes C (2013) Cell-sized

liposomes reveal how actomyosin cortical tension drives shape change. Proc Natl Acad Sci U S A 110(41):16456–16461. doi:10.1073/pnas.1221524110

21. Vogel SK, Heinemann F, Chwastek G, Schwille P (2013) The design of MACs (minimal actin cortices). Cytoskeleton (Hoboken) 70(11):706–717. doi:10.1002/cm.21136
22. Tamm LK, McConnell HM (1985) Supported phospholipid bilayers. Biophys J 47(1):105–113, doi: S0006-3495(85)83882-0 [pii] 10.1016/S0006-3495(85)83882-0
23. Sackmann E (1996) Supported membranes: scientific and practical applications. Science 271(5245):43–48
24. Johnson BR, Bushby RJ, Colyer J, Evans SD (2006) Self-assembly of actin scaffolds at ponticulin-containing supported phospholipid bilayers. Biophys J 90(3):L21–L23, doi: S0006-3495(06)72259-7 [pii] 10.1529/biophysj.105.076521
25. Barfoot RJ, Sheikh KH, Johnson BR, Colyer J, Miles RE, Jeuken LJ, Bushby RJ, Evans SD (2008) Minimal F-actin cytoskeletal system for planar supported phospholipid bilayers. Langmuir 24(13):6827–6836. doi:10.1021/la800085n
26. Janke M, Herrig A, Austermann J, Gerke V, Steinem C, Janshoff A (2008) Actin binding of ezrin is activated by specific recognition of PIP2-functionalized lipid bilayers. Biochemistry 47(12):3762–3769. doi:10.1021/bi702542s
27. Spudich JA, Watt S (1971) The regulation of rabbit skeletal muscle contraction. I. Biochemical studies of the interaction of the tropomyosin-troponin complex with actin and the proteolytic fragments of myosin. J Biol Chem 246(15):4866–4871
28. Pardee JD, Spudich JA (1982) Purification of muscle actin. Methods Enzymol 85(Pt B):164–181
29. Richter RP, Brisson AR (2005) Following the formation of supported lipid bilayers on mica: a study combining AFM, QCM-D, and ellipsometry. Biophys J 88(5):3422–3433. doi:10.1529/biophysj.104.053728

Chapter 12

Use of Nanobodies to Localize Endogenous Cytoskeletal Proteins and to Determine Their Contribution to Cancer Cell Invasion by Using an ECM Degradation Assay

Isabel Van Audenhove and Jan Gettemans

Abstract

There are numerous ways to study actin cytoskeletal structures, and thereby identify the underlying mechanisms of organization and their regulating proteins. Traditional approaches make use of protein overexpression or siRNA. However to study or modulate resident endogenous proteins, complementary methods are required. Since the discovery of nanobodies in 1993, they have proven to represent interesting tools in a variety of applications due to their high affinity, solubility, and stability. Especially their intracellular functionality makes them ideally suited for the study of actin cytoskeletal regulation.

Here we provide a protocol to clone nanobody cDNAs in frame with an EGFP or mCherry fluorescent tag. We explain how to transfect this fusion protein in eukaryotic (cancer) cells and how to perform immunofluorescence. This allows microscopic analysis of endogenous (cytoskeletal) proteins and gives insight into their endogenous localization. Moreover, we outline an extracellular matrix (ECM) degradation assay as an application of the general protocol. By seeding cells onto a fluorescently labeled gelatin matrix, degradation can be quantified by means of a matrix degradation index. This assay demonstrates the contribution of a protein during cancer cell invasiveness in vitro and the potential of a nanobody to inhibit this degradation through modulation of its target.

Key words Nanobodies, Intrabodies, Immunofluorescence, Endogenous cytoskeletal proteins, ECM degradation assay, Matrix degradation index

1 Introduction

The actin cytoskeleton performs crucial roles in cell maintenance, dynamics and motility. Its ability to form specialized protrusions such as filopodia, lamellipodia, and invadopodia, and its dysregulation in diseases such as cancer inspired cell biologists to study underlying processes leading to the formation of these structures [1–3]. Cytoskeletal structure and function is regulated by a variety of proteins involved in actin nucleation, (dis)assembly, stability, bundling, or cross-linking [4]. These regulating actin binders are often aberrantly expressed in tumor cells to promote their invasive

Ray H. Gavin (ed.), *Cytoskeleton: Methods and Protocols*, Methods in Molecular Biology, vol. 1365,
DOI 10.1007/978-1-4939-3124-8_12,

and metastatic potential [5]. Importantly, cancer cells have acquired the ability to degrade an underlying extracellular matrix (ECM) enabling dissemination from the primary tumor and metastasis [6]. This degradation occurs at proteolytically active invadopodia, which control secretion of several metalloproteases [7].

Since the discovery of GFP [8], fluorescence imaging has become indispensable for actin cytoskeleton studies. Different GFP derivatives became available, such as enhanced GFP (EGFP) with increased brightness, expression capacity, and sensitivity [9]. In addition, ECM degradation studies have only become possible thanks to fluorescently labeled extracellular matrices. This setup allows easy observation of matrix degradation as black areas in the thin fluorescent ECM layer, and this can be transformed into a measure for cancer cell invasiveness.

Traditional approaches to determine protein function in actin-based systems rather make use of overexpression or siRNA-mediated expression downregulation. However, to study and monitor cytoskeletal proteins at the endogenous level, other approaches are desirable, particularly since pharmacological inhibitors targeting structural proteins are generally not available. Nanobodies represent the antigen-binding domains of heavy chain-only antibodies present in species of the *Camelidae* family [10]. Due to their single domain nature, they are easily cloned and produced. Furthermore, their extended CDR loops enable high-affinity recognition of a larger range of epitopes in comparison to conventional antibodies. In addition, typical amino acid substitutions render them more hydrophilic, resulting in higher stability and solubility [11]. This resulted, only recently, in a variety of applications. In research, diagnostics and therapy, nanobodies can be, respectively, used as crystallization aids, noninvasive imaging modules, or infection neutralizers [12]. Of particular interest, nanobodies remain functional in the reducing intracellular environment, making them extremely suitable as intrabodies and allowing intracellular detection or modulation of endogenous proteins. This has been described for L-plastin in cancer cell migration, podosome and immune synapse formation [13–15] and for fascin and cortactin in invadopodium formation [16]. A therapeutic effect of nanobodies has also been demonstrated for CapG and mutant gelsolin targeting nanobodies in cancer metastasis and a rare amyloid disease, respectively [17, 18].

Here, we present a protocol to study cytoskeletal protein localization by means of nanobodies. We describe step by step how to perform cloning, transfection, immunofluorescence, and microscopy with nanobodies as a means to visualize endogenous cytoskeletal proteins. The single-domain nature allows easy cloning and fusion to EGFP or mCherry with only one primer set needed. Transfection further allows EGFP- or mCherry-tagged nanobody expression in a variety of eukaryotic cells. Subsequent immunofluorescence microscopy provides insight into the cellular localization of the cytoskeletal target. Some examples of actin cytoskeletal proteins, i.e., cortactin, fascin, and Arp2/3, and a non-actin binder, i.e., NTF2, are shown in Fig. 1.

As our focus rests upon actin regulating proteins and their contribution to cancer cell invasiveness, we demonstrate additionally how to adapt this standard protocol in order to execute a functional ECM degradation assay. Within this setup, the nanobodies serve as immunomodulators of their target protein. Detailed conclusions obviously require initial characterization of the nanobody epitope, activity, and function [19]. However, this assay quickly demonstrates the active contribution of a cytoskeletal protein in cancer cell invasion and allows identification of interesting therapeutic targets or nanobodies with therapeutic potential. For example, both a fascin and a cortactin nanobody greatly reduce matrix degradation capacity in prostate cancer cells (Fig. 2). As the fascin nanobody has previously been shown to inhibit fascin-mediated actin bundling and the cortactin nanobody has been characterized as a modulator of the C-terminal cortactin SH3 domain [16], this assay revealed the contribution of those protein function/domains in cancer cell invasiveness.

2 Materials

2.1 Nanobody Generation

1. 0.5–1 mg purified protein (antigen) for immunization and subsequent panning.

2.2 cDNA Cloning of Nanobody

1. Nucleospin plasmid EasyPure kit (Macherey-Nagel, Düren, Germany).
2. Primer oligonucleotides (*see* **Note 1**): Forward 5′ CGA GCT CAA GCT TCG GCC ACC ATG CAG GTG CAG CTG CAG GAG 3′; reverse 5′ GGC GAC CGG TGG ATC CTT GCT GGA GAC GGT GAC CTG 3′. Dissolve in MilliQ till a concentration of 50 μM.
3. EasyStart PCR Mix-in-a-tube.
4. HerculaseII Fusion DNA polymerase (Agilent Technologies, Santa Clara, CA, USA).
5. PCR/temperature-controlling device.
6. pEGFP-N1 or pmCherry-N1 vector.
7. Restriction enzymes BamHI-HF and EcoRI-HF (20,000 units/mL) with the corresponding buffer.
8. 7.5 M ammonium acetate (NH_4OAc).
9. Glycogen.
10. Ethanol (100 and 70 %).

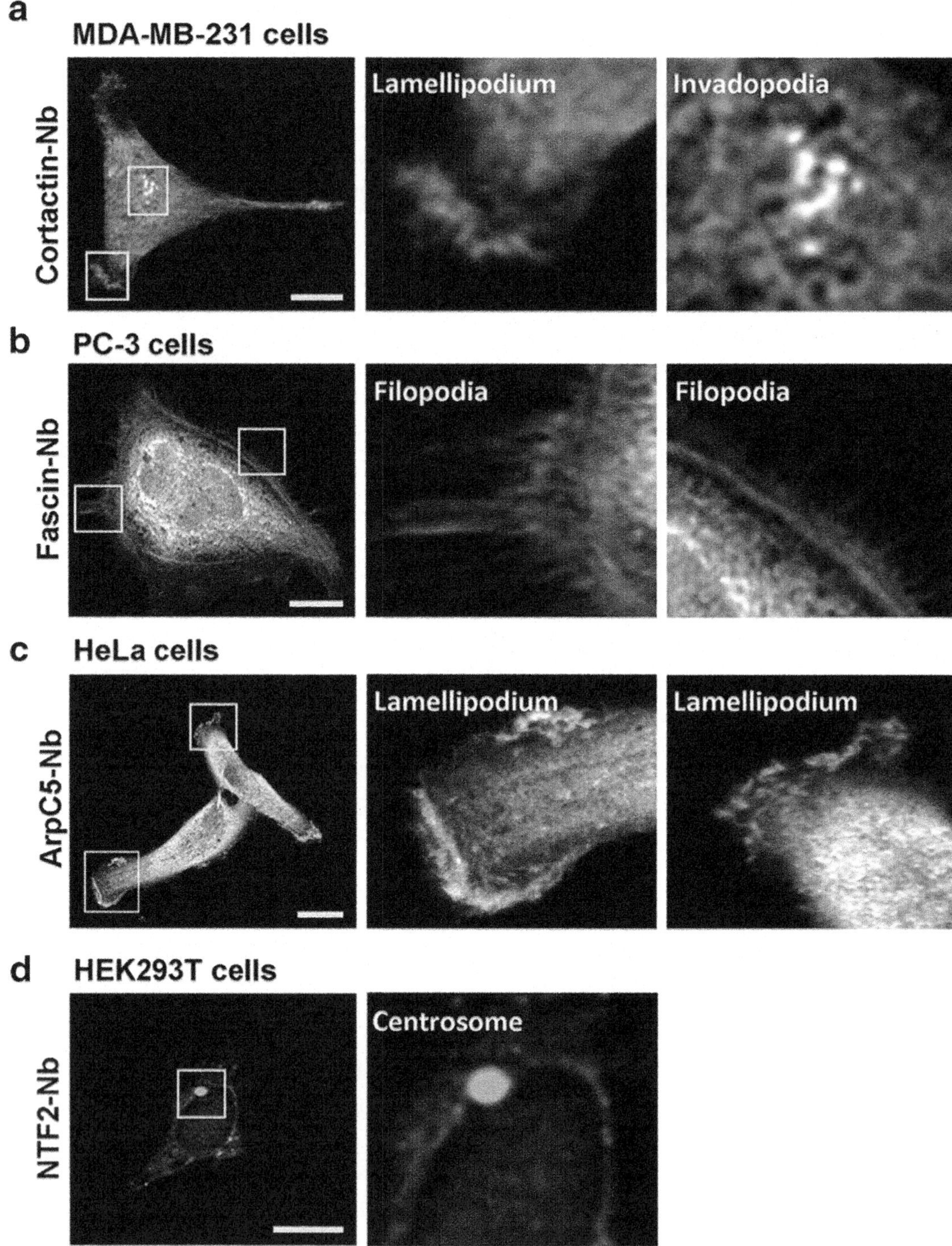

Fig. 1 Epifluorescent images of eukaryotic cells expressing EGFP-tagged nanobodies. (**a**) A nanobody targeting cortactin, a multidomain actin-regulating protein, enables visualization of lamellipodia and invadopodia in MDA-MB-231 breast cancer cells. (**b**) A nanobody raised against the actin bundling protein fascin accumulates in filopodia of PC-3 prostate cancer cells. (**c**) A nanobody targeting the ArpC5 subunit of the Arp2/3 nucleator complex is enriched in lamellipodia in a cervical HeLa cancer cell line. (**d**) An example outside the field of the actin cytoskeleton comes from a nanobody targeting the nuclear transport factor NTF2 [19], which revealed for the first time a centrosomal localization of NTF2, shown here in embryonic kidney cells. Scale bars = 10 μm

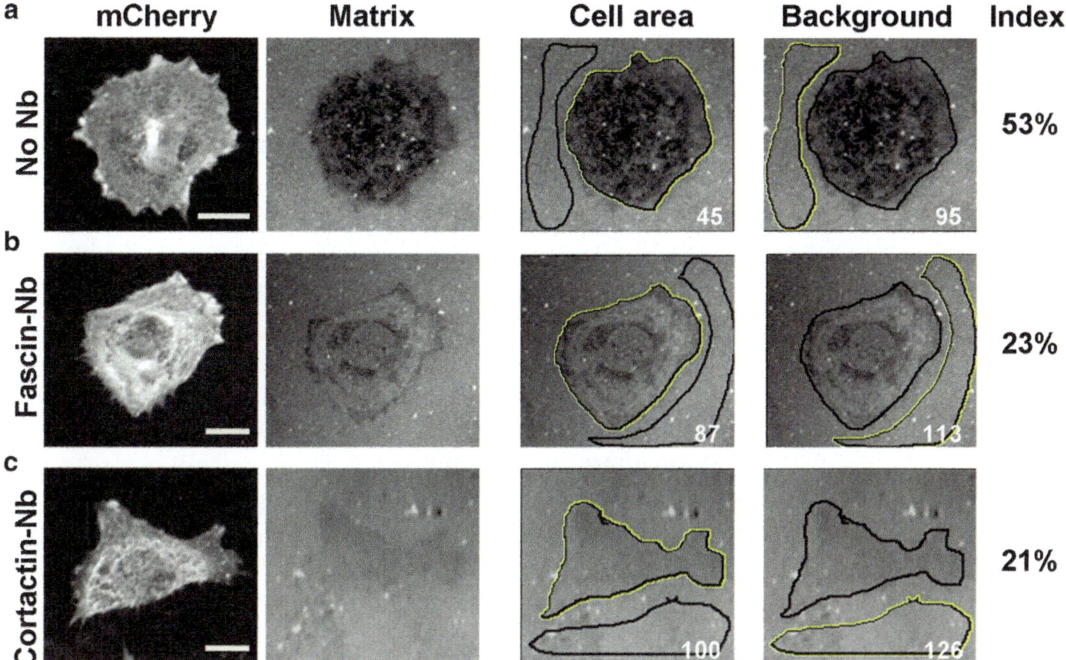

Fig. 2 Epifluorescent images of PC-3 prostate cancer cells expressing (**a**) mCherry only (control), (**b**) an mCherry-tagged fascin nanobody inhibiting fascin actin bundling and (**c**) an mCherry-tagged nanobody targeting the C-terminal cortactin SH3 domain. Cells were seeded onto a green fluorescently labeled gelatin matrix onto which degradation can be observed as black holes. A matrix degradation index is calculated as a measure for degradation area and intensity. The index is defined as the normalized difference between the mean gray value of the background and the mean grey value of the cell area as determined by ImageJ. In the *right panels*, the cell area and a background region are encircled (*yellow*), and the corresponding mean grey value is denoted. This assay reveals that both fascin actin bundling and the cortactin SH3 domain are involved in matrix degradation activity and cancer cell invasion

11. Microcentrifuge.
12. 0.5 M EDTA: Dissolve 18.6 g EDTA in 100 mL dH_2O at a pH of 8.0.
13. TAE (50× stock): 2 M Tris, 1 M acetic acid, 50 mM EDTA. Mix 242 g Tris with 57.1 mL acetic acid and 100 mL 0.5 M EDTA. Make up to 1 L with dH_2O.
14. TAE (diluted from stock): 40 mM Tris, 20 mM acetic acid, 1 mM EDTA. Dilute the stock solution 50 times in dH_2O.
15. 1 % agarose (w/v) in TAE solution: add 3 g agarose to 300 mL diluted TAE (*see* **Note 2**).
16. 20 % SDS (w/v) solution: dilute 20 g SDS in 100 mL dH_2O.
17. 5× DNA Sample buffer: 50 % glycerol (v/v), 50 mM EDTA, 1 % SDS, 0.02 % bromophenol blue (w/v). Add 15 mL glycerol, 3 mL 0.5 M EDTA, 1.5 mL of a 20 % SDS solution, 6 mg bromophenol blue and MilliQ till a total volume of 30 mL.
18. DNA Marker: SmartLadder.

19. GelRed Nucleic Acid Stain.
20. Electrophoresis apparatus.
21. UV illumination apparatus.
22. Nucleospin Gel and PCR cleanup (Macherey-Nagel, Düren, Germany).
23. Cold fusion kit with competent cells.
24. SOC medium: 2 % Bactotrypton (w/v), 0.5 % yeast extract (w/v), 10 mM NaCl, 2.5 mM KCl, 10 mM $MgCl_2$, 20 mM $MgSO_4$, 20 mM glucose. Add 20 g Bactotrypton, 5 g Yeast Extract, 0.58 g NaCl, 0.186 g KCl and MilliQ till a total volume of 900 mL. Autoclave and add a 100 mL filter sterilized solution of 2.03 g $MgCl_2$, 2.46 g $MgSO_4$, 3.96 g glucose-monohydrate, and MilliQ.
25. LB medium: 1 % Bactotrypton (w/v), 0.5 % yeast extract (w/v), 1 % NaCl (w/v). Add 10 g Bactotrypton, 5 g yeast extract, 10 g NaCl and MilliQ till a volume of 1 L. Set pH at 7.
26. Kanamycin.
27. Culture and plate incubator at 37 °C.

2.3 Transfection

1. Laminar flow hood.
2. MDA-MB-231 cells or another cell line of choice (American Type Culture Collection, ATCC, Manassas, VA, USA).
3. Humified 10 % CO_2 incubator at 37 °C.
4. DMEM medium (Dulbecco's modified Eagle medium) with high glucose, glutaMAX supplement, and pyruvate.
5. Penicillin-streptomycin solution (10,000 units/mL).
6. Phosphate-buffered saline (PBS) without Ca^{2+}/Mg^{2+}: 2.67 mM KCl, 1.47 mM KH_2PO_4, 137.93 mM NaCl, 8.06 mM Na_2HPO_4.
7. Trypsin-EDTA.
8. Fetal bovine serum (FBS).
9. Glass cover slips: 15 × 15 mm, thickness nr.1: 0.13–0.16 mm.
10. 12-Well cell culture plates.
11. Rat-Tail collagen type I.
12. Pincet.
13. Jetprime Kit (PolyPlus Transfection, Illkirch, France).
14. Standard bright-field/fluorescence microscope.

2.4 Immunofluorescence

1. 3 % paraformaldehyde (PFA) solution: Add 3 g PFA to 100 mL PBS without Ca^{2+}/Mg^{2+}. Put at 58 °C for 15 min to dissolve. Swirl from time to time. Add NaOH dropwise till the PFA it is completely dissolved (pH around 11). Put again for 15 min at 58 °C. Let cool down for at least 1 h and set pH at 7.4 with HCl. Store at −20 °C.

2. PBS with Ca^{2+}/Mg^{2+}: 0.92 mM $CaCl_2$, 0.49 mM $MgCl_2 \cdot 6H_2O$, 2.67 mM KCl, 1.47 mM KH_2PO_4, 137.93 mM NaCl, 8.06 mM Na_2HPO_4.
3. 0.2 % Triton solution: Dilute 0.2 mL 100 % Triton-X per 100 mL PBS with Ca^{2+}/Mg^{2+} (optional).
4. 0.75 % Glycine solution: Dissolve 0.75 g glycine per 100 mL PBS with Ca^{2+}/Mg^{2+} (optional).
5. 1 % BSA (Albumin from bovine serum) solution: Dissolve 1 g BSA per 100 mL PBS with Ca^{2+}/Mg^{2+} (optional).
6. Alexa Fluor 488 or 594 phalloidin: Dissolve in methanol for a final concentration of 200 units/mL (Molecular Probes, Thermo Fisher Scientific, Waltham, MA, USA) (optional).
7. 4′,6-Diamidino-2-phenylindole (DAPI): Prepare a stock solution of 0.4 mg/mL in MilliQ.
8. VectaShield Mounting medium (Vector Laboratories Inc., Burlingame, CA, USA).
9. Microscope slides 76 × 26 × 1 mm.
10. Nail polish.

2.5 Microscopy

1. Fluorescence microscope equipped with the appropriate filter sets and a camera.
2. Immersion oil.

2.6 Matrix Degradation Assay

1. 6-Well cell culture plates.
2. Gelatin from pig skin. Oregon green 488 conjugate (stock): dissolve in dH_2O for a stock solution of 1 mg/mL (0.1 % gelatin).
3. Gelatin (diluted from stock): 0.01 % solution. Dilute the stock solution ten times in PBS without Ca^{2+}/Mg^{2+}.
4. Glutaraldehyde (stock). 25 % solution.
5. Glutaraldehyde (diluted from stock): 0.5 % solution. Dilute the stock solution 50 times in PBS without Ca^{2+}/Mg^{2+}.
6. Acti-stain 670 phalloidin (optional).
7. ImageJ (National Institutes of Health, Bethesda, MD, USA).

3 Methods

3.1 Nanobody Generation

Nanobodies can be obtained in collaboration with the VIB NSF (Flemish Institute for Biotechnology, Nanobody Service Facility, Brussels University, Belgium). The choice of immunization material is flexible as purified antigen can be either a full-length protein or a structural/functional protein domain or peptide. Briefly, *Camelidae* are subcutaneously injected with antigen on days 0, 7,

14, 21, 28, and 35. On day 39, lymphocytes are prepared from anticoagulated blood. Total RNA is extracted and used as a template for first-strand cDNA synthesis. Antigen-specific nanobodies are isolated and enriched by successive rounds of panning. Nanobodies are finally delivered subcloned in a prokaryotic pHEN4 vector. Additional detailed information about the immunization and panning procedure can be found elsewhere [12, 20].

3.2 cDNA Cloning of Nanobody in pEGFP-N1/pmCherry-N1 Vector

With this protocol, a nanobody can be cloned upstream from an EGFP- or mCherry-tag to obtain a cDNA construct coding for the fusion protein.

1. Prepare plasmid DNA containing the nanobody cDNA with the Nucleospin plasmid EasyPure kit.
2. Add 50 ng nanobody cDNA, 1 μL of forward primer, 1 μL of reverse primer and milliQ up to a total volume of 25 μL to the PCR tubes (*see* **Note 3**). Add 1 μL DNA polymerase to the mixture and perform the PCR reaction for 2 min at 95 °C followed by 30 cycli of 20 s at 95 °C, 20 s at 65 °C and 30 s at 72 °C, and terminate with 3 min at 72 °C.
3. Mix 2 μg pEGFP-N1 or pmCherry-N1 (*see* **Note 4**) vector with 5 μL buffer, 1 μL BamHI-HF (20 units) and MilliQ in a total volume of 50 μL. Allow the digest to proceed for 3 h at 37 °C. After digestion, add 50 μL MilliQ, 50 μL ammonium acetate and 1 μL glycogen together with 300 μL 100 % ethanol. Vortex vigorously and centrifuge for 20 min at full speed at room temperature. Discard the supernatant, add 500 μL 70 % ethanol and centrifuge at full speed for 2 min. Discard the supernatant and let the pellet air-dry (*see* **Note 5**). Resuspend in 10 μL MilliQ, and subsequently add 5 μL buffer, 1 μL EcoRI-HF (20 units) and MilliQ in a total volume of 50 μL. Let digest for 3 h at 37 °C.
4. Prepare a 1 % agarose gel. Add 12.5 μL of 5× DNA sample buffer to both the PCR-product and the vector. Load the samples as depicted in Fig. 3 and let run for approximately 45 min at 90 V. Cut the gel with a clean scalpel as depicted and immerse the middle part (lanes 2–4) in a closed box with GelRed (diluted 10,000 times in dH_2O) for 15 min. Visualize the gel by UV illumination and cut out the corresponding vector (approximately 4700 bp) and PCR product bands (approximately 300–500 bp) with a clean scalpel (light grey parts). Reconstruct the complete gel now by adding lane 1 and 5 again and cut out the corresponding bands of the remaining vector and PCR product samples (dark grey parts) and collect in a 1.5 mL microtube (*see* **Note 6**).
5. Perform gel extraction with the Nucleospin Gel and PCR clean-up kit according to the manufacturer's protocol. Quantify the DNA concentration with a spectrophotometer.

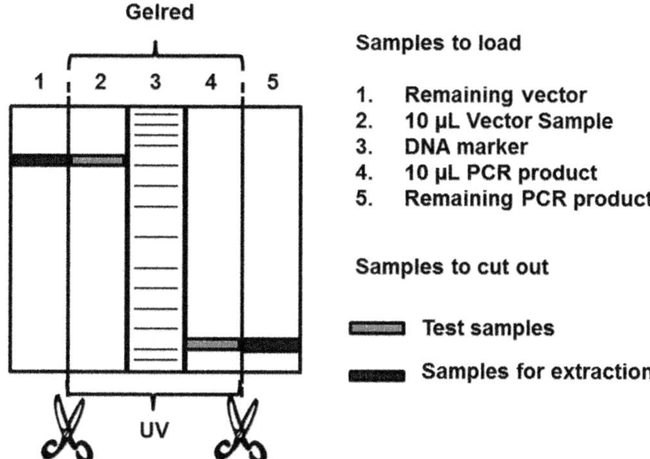

Fig. 3 Set-up of a 1 % DNA agarose gel to enable gel extraction. Use 10 µL of the sample for visualization with GelRed and UV illumination to enable localization of the vector and PCR bands and use the rest of the sample to effectively perform gel extraction and DNA preparation

6. Mix PCR-product (20–200 ng) with vector (10–100 ng) in a 2:1 molar ratio together with 2 µL Cold fusion Mastermix and MilliQ in a total volume of 10 µL in a 1.5 mL microtube. Incubate for 5 min at room temperature followed by 10 min on ice. Add 50 µL cold fusion competent cells and incubate on ice for 20 min. Heat shock at 42 °C for 50 s, transfer on ice for 2 min, and add 250 µL SOC medium. Incubate for 1 h at 37 °C and plate 50 and 200 µL on LB-agar plates containing 25 µg/mL kanamycin. Incubate overnight at 37 °C.
7. Pick 3–6 colonies per construct and grow them overnight in 5 mL LB medium + 25 µg/mL kanamycin in culture tubes.
8. Prepare plasmid DNA with for example the Nucleospin plasmid EasyPure kit and add 0.25 µL EcoRI-HF (5 units) and 0.25 µL BamHI-HF (5 units) to 500–1000 ng of plasmid DNA together with 2 µL buffer and MilliQ in a volume of 20 µL. Allow the digest to proceed for 3 h at 37 °C. Add 5 µL 5× DNA sample buffer and perform agarose gel electrophoresis with GelRed already present within the agarose solution (diluted 10,000-fold). Confirm correct insertion of the nanobody in the vector (*see* **Note 7**): an empty vector should give a single band of approximately 4700 bp while the vector with the nanobody inserted should give a band of approximately 4700 bp and one of 300–500 bp (*see* **Note 8**).

3.3 Transfection in Eukaryotic Cells

This protocol allows expression of the nanobody-EGFP or -mCherry fusion protein in eukaryotic cells with the goal to perform immunofluorescence microscopic analysis. We will describe the protocol for MDA-MB-231 cells, but this is also applicable for many other cell lines, such as HeLa, HEK293T and PC-3 (*see* **Note 9**). All these procedures have to be carried out in a laminar flood hood to guarantee sterility of the cells.

1. Culture MDA-MB-231 cells in DMEM medium with 10 % FBS and 0.1 % PenStrep antibiotics at 37 °C in a humified 10 % CO_2 incubator (*see* **Note 10**). Let them grow till 80–100 % confluency on the day of seeding.
2. Place glass cover slips in a 12-well plate with a sterile pincet. Dilute collagen to 50 μg/mL in PBS without Ca^{2+}/Mg^{2+} and add 1 mL of the solution onto each cover slip. Incubate for at least 1 h and maximum 2 h at 37 °C in the incubator (*see* **Note 11**).
3. In the meantime, wash the cells with PBS without Ca^{2+}/Mg^{2+}, trypsinize them, and determine cell density.
4. Seed 80,000 cells per well in a total volume of 1 mL (*see* **Note 12**). Rock the plate back and forth to distribute the cells evenly.
5. Let the cells adhere overnight at 37 °C in the 10 % CO_2 incubator.
6. Check cell density: cells should cover 60–80 % of the 12 well at the moment of transfection (*see* **Note 13**). Dilute 750 ng DNA into 75 μL Jetprime buffer and mix by vortexing. Add 1.5 μL Jetprime reagent, vortex for 10 s, spin down briefly, and incubate for 10 min at room temperature. Add the transfection mix drop by drop onto the cells. Gently rock the plates back and forth to distribute the mixture evenly. Replace the medium the same day after at least 4 h (*see* **Note 14**).
7. Use a bright-field/fluorescence microscope to check cell vitality (*see* **Note 15**) and transfection efficiency after 24 h. Place the 12-well plate onto the microscope stage and move an appropriate objective into place. We usually check fluorescence with a 20× objective. First use bright-field to focus on the cells and to judge cell viability, then utilize fluorescent illumination to check for EGFP/mCherry fluorescence. Generally, transfection efficiencies of 60–100 % are achieved.

3.4 Immunofluorescence

With this protocol, cells on glass cover slips expressing EGFP- or mCherry-tagged nanobodies will be prepared for immunofluorescence. All procedures can be carried out outside the laminar flow hood. From this point on, cover the 12 well plate at all times with aluminium foil to prevent bleaching of the autofluorescent signal. To treat or wash the cover slips in the 12 well, 1 mL of all solutions is sufficient to completely submerse a cover slip. Let the plate shake back and forth during the washing steps for 3 min each.

1. Fix the cells. First, gently rinse the cover slips with PBS with Ca^{2+}/Mg^{2+}, and then incubate them with thawed 3 % paraformaldehyde solution for 25 min at room temperature. Perform three consecutive washing steps with PBS with Ca^{2+}/Mg^{2+} (*see* **Note 16**).
2. When visualization of the actin cytoskeleton is preferred, one can perform a phalloidin staining (optional). To do so, permeabilize cells by adding 0.2 % Triton solution for 5 min. Wash two times with PBS with Ca^{2+}/Mg^{2+}. Incubate cover slips with 0.75 % glycine solution for 20 min followed by a washing step with PBS with Ca^{2+}/Mg^{2+}. Incubate cover slips for 10 min with 1 % BSA blocking solution. Move the cover slips onto a parafilm present within a box with a humified tissue underneath it. Administer 70 μL of phalloidin staining solution (1/100 dilution of phalloidin in 1 % BSA blocking solution) per cover slip and incubate for 30 min at 37 °C. If a nuclear staining is desired, add DAPI in the staining solution (dilute the stock solution 1000 times) before administering it to the cover slips. After incubation with staining solution, place the cover slips back into the 12-well plate to wash once with 1 % BSA blocking solution followed by two washes with PBS with Ca^{2+}/Mg^{2+}.
3. Mount the cover slips by adding a little drop of Vectashield on a microscope slide, and subsequently gently placing the cover slip, with the cells facing down, onto the Vectashield. Before mounting the cover slip, it is important to remove excess wash buffer and to clean/dry the back of the cover slip. Use a pincet when manipulating a cover slip. Seal the cover slip with nail polish (*see* **Note 17**).

3.5 Microscopy

Since different microscope systems are available, we provide general guidelines to help the reader in the acquisition of fluorescent images.

1. Optimal acquisitions are acquired when using filtersets precisely designed for the fluorochromes applied. EGFP has an excitation maximum at 488 nm and an emission peak at 509 nm. When combining with DAPI (excitation/emission 358/461) or Alexa fluor phalloidin 594 (excitation/emission 581/609), make sure the filter sets are equipped for the acquisition of three distinct spectra to avoid bleed-through from one channel to another (*see* **Note 18**). The same counts for mCherry fluorescence (excitation/emission 587/610) which may be combined with Alexa fluor phalloidin 488 (excitation/emission 495/518) and DAPI.
2. Move the appropriate objective into place. We use 63× oil objectives to visualize various cellular components with the

necessary detail. Place a drop of immersion oil onto your sample and place the slide onto the microscope stage (*see* **Note 19**). Carefully bring your sample into focus.

3. Look for a region of interest and acquire an image by means of the available software and camera device (*see* **Note 20**). Make sure to avoid photobleaching as much as possible. Limit the exposures times for the samples and avoid prolonged unnecessary illumination.

3.6 Extracellular Matrix Degradation Assay

This protocol outlines how to execute a matrix degradation assay as a specific application of the transfection, immunofluorescence, and microscopy protocol described above. It allows for the evaluation of the contribution of the cytoskeletal protein in cancer cell invasion. We normally use PC-3 cells expressing mCherry-labeled nanobodies on a green fluorescent gelatin matrix (*see* **Note 21**).

1. Perform transfection as described in Subheading 3.3 with modifications described in **steps 2–4**.
2. Seed 200,000 cells in a 6 well in a total volume of 2 mL (*see* **Note 12**), no cover slips or coating has to be used.
3. Include an additional 6 well for transfection with the empty vector as a negative control for the assay (*see* **Note 22**).
4. Use 2 μg plasmid, 200 μL buffer, and 4 μL reagent for transfection.
5. Coat cover slips with fluorescently labeled gelatin. All incubation steps (except for the gelatin itself) can be carried out with 1 mL solution per 12 well. Once the gelatin layer is added, protect the cover slips from light at all times to avoid bleaching.
6. Administer a 70 μL drop of the 0.01 % Oregon Green gelatin solution into a 12 well and place a cover slip onto it.
7. Let air-dry in the laminar flow hood for 45 min.
8. Take out the cover slip and invert it into a new 12 well (coating layer is at the upper side now).
9. Add 0.5 % glutaraldehyde to crosslink the gelatin layer. Incubate on ice for 10 min while rocking the plate back and forth.
10. Wash four times with PBS with Ca^{2+}/Mg^{2+} while rocking the plate back and forth (*see* **Note 23**).
11. Quench the cover slips for 1 h by adding cell specific medium (*see* **Note 24**).
12. Trypsinize and count the transfected cells and seed 80,000 cells per gelatin-coated cover slip. This should be at a time point of approximately 24 h after cell transfection.
13. Let incubate for 15–24 h (*see* **Note 25**).

14. Fix, stain and mount cover slips as described in Subheading 3.4. If phalloidin staining is desired use acti-stain 670 phalloidin (dilute 1/50 in 1 % BSA blocking solution) to combine with the green/red labeled matrix/nanobodies. Excitation occurs at 625 nm while the emission peak lies at 675 nm (far red).
15. Perform microscopic analysis as described in Subheading 3.5 (*see* **Note 26**).
16. Determine a matrix degradation index as a measure for cancer cell invasiveness as described in **steps 17–22** (*see* Fig. 2).
17. To calculate the index, an image of the (degraded) matrix is needed as well as an image from which the cell area can clearly be defined (nanobody fluorescence or phalloidin staining).
18. Open the pictures in ImageJ.
19. Accurately draw the contours of the cell onto the matrix picture with the paintbrush tool (brush width 2 pixels).
20. Activate the wand (tracing) tool to click on the contour which will become yellow. Perform Analyze→Measure. Make sure that the "mean gray value" box is ticked in the submenu Analyze→Set Measurements. The output returns the mean gray value (MGV) over the cell area (*see* **Note 27**).
21. Draw a representable contour in the background and determine in the same way the mean gray value (MGV) of the background.
22. The degradation index (*see* **Note 28**) is now defined as

$$\text{Index}(\%) = \frac{\text{MGV Background} - \text{MGV Cell area}}{\text{MGV Background}} \times 100$$

23. When the index is reduced, this may be due to degradation occurring at certain cellular points instead of over the complete cell area (reduced degradation area, *see* Fig. 2b). Another possibility is an overall decreased degradation activity over the total cell body (reduced degradation intensity, *see* Fig. 2c).
24. To draw conclusions, determine the index for at least 25 cells per condition in three independent experiments and perform statistics to define significant differences.

4 Notes

1. Primers are designed for cold fusion cloning according to the manufacturer's guidelines. The forward primer starts with "CGA GCT CAA GCT TCG" corresponding to the vector sequence before the EcoRI restriction site, a Kozak sequence with ATG start site: "GCC ACC ATG," followed by the start

sequence of any nanobody: "CAG GTG CAG CTG CAG GAG." The reverse primer is composed of the vector sequence including the BamHI restriction site "GGC GAC CGG TGG ATC CTT" followed by the complementary last nucleotides of any nanobody "GCT GGA GAC GGT GAC CTG."

2. Dissolve the agarose solution by heating it in a microwave oven. Frequently swirl the solution to avoid boiling. Store at 65 °C to keep it liquid. To make a gel, pour the solution in a gel holder and let solidify.
3. Components can be added on top of the wax layer, which will melt during heating and allow the PCR reaction to take place. After PCR, simply pierce the wax layer with a pipet tip to collect the PCR product.
4. 2 μg vector is enough to perform at least five cloning reactions. If more cloning reactions are required, simply perform additional digest reactions in parallel. Do not exceed the volume needed for the digestion of 2 μg vector.
5. The pellet is ready to be resuspended when its color turns from white to transparent/colorless. Do not over-dry as this will reduce solubility.
6. This protocol ensures that the to-be-extracted DNA is not exposed to UV radiation, which is essential to guarantee DNA quality.
7. Preferably, perform DNA sequencing to verify that the correct nanobody cDNA sequence is inserted in frame with the EGFP or mCherry coding sequence.
8. DNA fragments of intermediate size are often present due to partial digestion. Occasionally, more than two DNA fragments are possible after digestion when the nanobody cDNA sequence itself also contains a BamHI or EcoRI restriction site.
9. Transfection efficiencies with Jetprime in various other cell types are reported by the manufacturer and can be found at http://www.polyplus-transfection.com/2009/08/jetprime%C2%AE/
10. The culture conditions are identical for Hela and HEK293T cells. PC-3 cells are grown in RPMI medium with the same additives.
11. It is also possible to seed cells directly onto a glass cover slip. A collagen coating, however, will stimulate an even spreading of the cells and therefore result in better defined cytoskeletal structures.
12. Respect the total volume exactly. Varying it will lead to other dilutions of the transfection reagent and therefore other transfection parameters.

13. Seeding 80,000 cells normally leads to an optimal confluency of 60–80 % at the moment of transfection. If not, seeding lower or higher numbers of cells is advised as transfection efficiency greatly depends upon cell density at the moment of transfection.
14. It is important to replace the medium on the same day to minimize cell toxicity, but incubating longer than 4 h has been known to enhance transfection efficiency. For that reason, incubation times of 6–8 h are recommended.
15. A fraction of cells dying after transfection is normal. High cell death levels may occur when using nanobodies targeting vital proteins involved in cell proliferation and survival.
16. Alternatively, it is possible to fix cells with 100 % methanol. This may be of interest when you want to combine the nanobody fluorescence with an antibody for immunofluorescence requiring methanol fixation. Perform the rinse and washing steps as described for the PFA fixation but add 1 mL 100 % MeOH onto the cover slip and incubate the plate at −20 °C for 5 min. Keep in mind that phalloidin staining is not possible after methanol fixation.
17. Avoid further movement of the cover slip as this will be disadvantageous for the fluorescence signal. Make sure not to put too much Vectashield as this will promote movement. To keep the cover slip on the same place during polishing, first put nail polish drops on the cover slip edges and then finish the complete contours.
18. By capturing images sequentially instead of simultaneously over the different channels, bleed-through will be avoided. Alternatively, laser-based fluorescence microscopes are known to only excite fluorophores at a particular wavelength and will guarantee a good separation of channel colors.
19. The amount of immersion oil is crucial for a good acquisition. Too little or too much oil will result in a blurry images If this is the case, remove the oil carefully with a clean tissue, clean the objective with lens tissue and start over again.
20. Images with good contrast and a dark image background will be obtained when performing optical sectioning. We therefore use a fluorescence microscope equipped with an Apotome device (Zeiss) enabling structured illumination. Alternatively, confocal microscopy guarantees measurements in one confocal *z*-plane.
21. MDA-MB-231 cells can also be used, but have the tendency to move away after performing degradation. More straightforward analysis can therefore be achieved with PC-3 cells (*see* Fig. 2). As we use green labeled gelatin, we use mCherry-labeled nanobodies to transfect. However, if use

of EGFP-tagged nanobodies is desired, the QCM Gelatin invadopodia assay kit (Millipore, Billerica, MA, USA) containing red-labeled gelation can be used by following the manufacturer's instructions. Slightly longer incubation times are however advised when using this kit as the coating composition differs. Incubate cells for at least 24 h on this coating.

22. A negative control is needed to enable evaluation of effects on matrix degradation. Untransfected cells can also serve as a sort of negative control; however it is better to include cells which were also treated with the transfection reagent. Transfection with empty vector as a control will guarantee that effects are directly linked to the nanobody expression and are independent on the presence of a tag.
23. At this point, it is possible to store the coated cover slips for a limited time at 4 °C before seeding the cells onto it. Keep them with PBS without Ca/Mg for a maximum of 24 h.
24. This is necessary to remove remains of the coating solutions and to prepare the coated cover slip for cell seeding.
25. In our experience, an incubation time of 15–24 h is ideal to generate well defined degradation patterns for analysis in combination with a nice fluorescent-nanobody expression. Shorter incubation will provide less pronounced degradation while longer incubation times will lead to decreased nanobody fluorescent expression, higher chances of cell movement over the matrix and less pronounced differences in degradation patterns.
26. Make sure to analyze matrix degradation patterns at reliable places with a good gelatin coating. If a certain pattern is visible in the gelatin layer independent of the cells resting on it, the cover slip has probably been dehydrated, unevenly coated, or only partially covered by solidified gelatin at that place. Also be aware of stripes caused by handling of the cover slip, which do not at all represent matrix degradation by cells.
27. The grayscale ranges from 0 to 255 with 0 being completely black and 255 representing completely white. The mean gray value reported by ImageJ is determined by averaging the grey values at each pixel over the total area that you depict.
28. When the MGV over the cell is equal to the MGV of the background, the index is 0 % as no degradation occurs. When the MGV over the cell area is 0 (=completely black) the index is 100 %.

References

1. Yamaguchi H, Condeelis J (2007) Regulation of the actin cytoskeleton in cancer cell migration and invasion. Biochim Biophys Acta 1773(5):642–652. doi:10.1016/j.bbamcr.2006.07.001
2. Chhabra ES, Higgs HN (2007) The many faces of actin: matching assembly factors with cellular structures. Nat Cell Biol 9(10):1110–1121. doi:10.1038/ncb1007-1110
3. Ridley AJ (2011) Life at the leading edge. Cell 145(7):1012–1022. doi:10.1016/j.cell.2011.06.010
4. Winder SJ, Ayscough KR (2005) Actin-binding proteins. J Cell Sci 118(Pt 4):651–654. doi:10.1242/jcs.01670
5. Gross SR (2013) Actin binding proteins: their ups and downs in metastatic life. Cell Adh Migr 7(2):199–213. doi:10.4161/cam.23176
6. Yamaguchi H (2012) Pathological roles of invadopodia in cancer invasion and metastasis. Eur J Cell Biol 91(11-12):902–907. doi:10.1016/j.ejcb.2012.04.005
7. Linder S (2007) The matrix corroded: podosomes and invadopodia in extracellular matrix degradation. Trends Cell Biol 17(3):107–117. doi:10.1016/j.tcb.2007.01.002
8. Shimomura O, Johnson FH, Saiga Y (1962) Extraction, purification and properties of aequorin, a bioluminescent protein from the luminous hydromedusan, Aequorea. J Cell Comp Physiol 59:223–239
9. Zhang G, Gurtu V, Kain SR (1996) An enhanced green fluorescent protein allows sensitive detection of gene transfer in mammalian cells. Biochem Biophys Res Commun 227(3):707–711. doi:10.1006/bbrc.1996.1573
10. Hamers-Casterman C, Atarhouch T, Muyldermans S, Robinson G, Hamers C, Songa EB, Bendahman N, Hamers R (1993) Naturally occurring antibodies devoid of light chains. Nature 363(6428):446–448. doi:10.1038/363446a0
11. Muyldermans S (2013) Nanobodies: natural single-domain antibodies. Annu Rev Biochem 82:775–797. doi:10.1146/annurev-biochem-063011-092449
12. Hassanzadeh-Ghassabeh G, Devoogdt N, De Pauw P, Vincke C, Muyldermans S (2013) Nanobodies and their potential applications. Nanomedicine 8(6):1013–1026. doi:10.2217/nnm.13.86
13. Delanote V, Vanloo B, Catillon M, Friederich E, Vandekerckhove J, Gettemans J (2010) An alpaca single-domain antibody blocks filopodia formation by obstructing L-plastin-mediated F-actin bundling. FASEB J 24(1):105–118. doi:10.1096/fj.09-134304
14. De Clercq S, Boucherie C, Vandekerckhove J, Gettemans J, Guillabert A (2013) L-plastin nanobodies perturb matrix degradation, podosome formation, stability and lifetime in THP-1 macrophages. PLoS One 8(11):e78108. doi:10.1371/journal.pone.0078108
15. De Clercq S, Zwaenepoel O, Martens E, Vandekerckhove J, Guillabert A, Gettemans J (2013) Nanobody-induced perturbation of LFA-1/L-plastin phosphorylation impairs MTOC docking, immune synapse formation and T cell activation. Cell Mol Life Sci 70(5):909–922. doi:10.1007/s00018-012-1169-0
16. Van Audenhove I, Boucherie C, Pieters L, Zwaenepoel O, Vanloo B, Martens E, Verbrugge C, Hassanzadeh-Ghassabeh G, Vandekerckhove J, Cornelissen M, De Ganck A, Gettemans J (2014) Stratifying fascin and cortactin function in invadopodium formation using inhibitory nanobodies and targeted subcellular delocalization. FASEB J 28(4):1805–1818. doi:10.1096/fj.13-242537
17. Van Impe K, Bethuyne J, Cool S, Impens F, Ruano-Gallego D, De Wever O, Vanloo B, Van Troys M, Lambein K, Boucherie C, Martens E, Zwaenepoel O, Hassanzadeh-Ghassabeh G, Vandekerckhove J, Gevaert K, Fernandez LA, Sanders NN, Gettemans J (2013) A nanobody targeting the F-actin capping protein CapG restrains breast cancer metastasis. Breast Cancer Res 15(6):R116. doi:10.1186/bcr3585
18. Van Overbeke W, Verhelle A, Everaert I, Zwaenepoel O, Vandekerckhove J, Cuvelier C, Derave W, Gettemans J (2014) Chaperone nanobodies protect gelsolin against MT1-MMP degradation and alleviate amyloid burden in the gelsolin amyloidosis mouse model. Mol Ther 22:1768. doi:10.1038/mt.2014.132
19. Van Audenhove I, Van Impe K, Ruano-Gallego D, De Clercq S, De Muynck K, Vanloo B, Verstraete H, Fernandez LA, Gettemans J (2013) Mapping cytoskeletal protein function in cells by means of nanobodies. Cytoskeleton 70:604. doi:10.1002/cm.21122
20. Ghassabeh GH, Saerens D, Muyldermans S (2010) Isolation of antigen-specific nanobodies. In: Kontermann R, Dübel S (eds) Antibody engineering, vol 2. Springer, Berlin. doi:10.1007/978-3-642-01147-4_20

Chapter 13

Actin-Dynamics in Plant Cells: The Function of Actin-Perturbing Substances: Jasplakinolide, Chondramides, Phalloidin, Cytochalasins, and Latrunculins

Andreas Holzinger and Kathrin Blaas

Abstract

This chapter gives an overview of the most common F-actin-perturbing substances that are used to study actin dynamics in living plant cells in studies on morphogenesis, motility, organelle movement, or when apoptosis has to be induced. These substances can be divided into two major subclasses: F-actin-stabilizing and -polymerizing substances like jasplakinolide and chondramides and F-actin-severing compounds like chytochalasins and latrunculins. Jasplakinolide was originally isolated form a marine sponge, and can now be synthesized and has become commercially available, which is responsible for its wide distribution as membrane-permeable F-actin-stabilizing and -polymerizing agent, which may even have anticancer activities. Cytochalasins, derived from fungi, show an F-actin-severing function and many derivatives are commercially available (A, B, C, D, E, H, J), also making it a widely used compound for F-actin disruption. The same can be stated for latrunculins (A, B), derived from red sea sponges; however the mode of action is different by binding to G-actin and inhibiting incorporation into the filament. In the case of swinholide a stable complex with actin dimers is formed resulting also in severing of F-actin.

For influencing F-actin dynamics in plant cells only membrane permeable drugs are useful in a broad range. We however introduce also the phallotoxins and synthetic derivatives, as they are widely used to visualize F-actin in fixed cells. A particular uptake mechanism has been shown for hepatocytes, but has also been described in siphonal giant algae. In the present chapter the focus is set on F-actin dynamics in plant cells where alterations in cytoplasmic streaming can be particularly well studied; however methods by fluorescence applications including phalloidin and antibody staining as well as immunofluorescence-localization of the inhibitor drugs are given.

Key words Actin filaments, Chondramides, Cytochalasins, Depsipeptide, Jasplakinolide, Latrunculin, Phalloidin, Phallotoxin, Swinholide

1 Introduction

Filamentous actin (F-actin), as a major structural cytoskeletal component, has been targeted by inhibitor drugs over decades and the importance of an undisturbed turnover has been established as prerequisite for living functions in plant cells. Yet, a comprehensive overview of the different categories of F-actin-perturbing

Ray H. Gavin (ed.), *Cytoskeleton: Methods and Protocols*, Methods in Molecular Biology, vol. 1365,
DOI 10.1007/978-1-4939-3124-8_13,

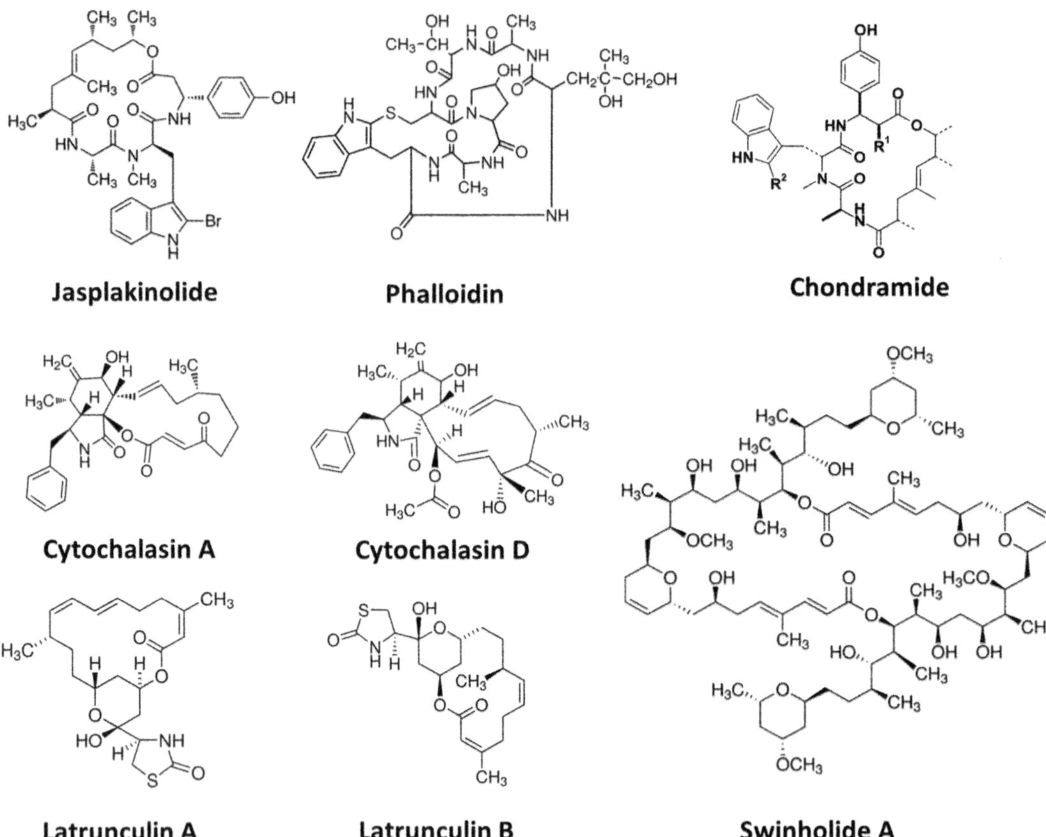

Fig. 1 Molecular structures of a selection of F-actin-stabilizing/polymerizing drugs (jasplakinolide, phalloidin, chondramide) and F-actin disrupting drugs (cytochalasin A, cytochalasin D, latrunculin A, latrunculin B, swinholide A)

substances is still missing. We will here give a short overview on (A) actin-stabilizing and -polymerizing substances like jasplakinolide, chondramide, and phalloidin, and (B) actin-depolymerizing substances like cytochalasin, latrunculin, and swinholide (Fig. 1).

The actin system is involved in different cellular processes like exo- and endocytosis, organelle motions and maintenance of organelle distribution, motility, cell division, and cytoplasmic streaming. Several substances are known to influence these processes by altering intracellular actin organization and have been used extensively for cell biological research [1].

While the mechanism of action is different for the described substances, the effect on the cells might be quite similar—a disruption of the actin-related functions.

This chapter does not cover drugs or inhibitors acting only indirectly on F-actin polymerization (i.e., ATP-synthase blockers like ATPase-blockers *N*-ethylmaleimide (NEM) and 2,3-butanedione monoxime (BDM) [2]) or inhibitors interacting with actin binding proteins like the Arp2/3 complex (i.e., wiskostatin, [3]).

1.1 F-Actin-Stabilizing and -Polymerizing Substances

Actin stabilization is generated mostly by three different compounds: phalloidin [4, 5], jasplakinolide [6, 7], and chondramide [8, 9]. While all three substances have the capacities to stabilize F-actin, phalloidin is not membrane permeable, while jaslplakinolide and chondramides readily enter cells [10, 11]; Therefore phalloidin is used in cell biological research mainly for visualization of F-actin after fluorescence labeling of the compound in fixed tissues [12]. In contrast, the commercially available jasplakinolide can be used for F-actin stabilization in living cells due to its membrane permeability [13].

1.1.1 Jasplakinolide

The common feature of these actin-stabilizing substances is a cyclic depsipeptide (Fig. 1), which is a polymeric compound containing both, amino acids and hydroxy acids, joined by peptide and ester bounds. Chemically jasplakinolide is a cyclo-depsipeptide containing a tripeptide moiety linked to a polyketide chain [14, 15] (Fig. 1). Jasplakinolide is also termed jaspamide [15–18]. For jasplakinolide also F-actin-polymerizing capacities were described [19–23].

Originally jasplakinolide was isolated from the marine sponge *Jaspis* sp. collected at Fiji or the Palau islands [14, 16]. The sponge was authenticated as *Jaspis johnstoni* [15], however taxonomic difficulties were pointed out [24] and the compound can also be isolated form other sponge genera [24, 25]. Detailed procedures for jasplakinolide extraction can be found in [26].

Different approaches were made to synthesize this substance [27,28] or generate nonpeptide mimetics [29]; finally Jasplakinolide became commercially available from Molecular Probes in the late 1990s. An enantioselective total synthesis of (+)-jasplakinolide has been described recently [30].

The biological activities of jasplakinolide are described as anthelminthic, antifungal, insecticidal and selective antimicrobial [14, 16, 31]. In vitro investigations have elucidated the actin polymerizing- and -stabilizing capacities of jasplakinolide [7, 19, 23].

In plant systems the unicellular green alga *Micrasterias* [20], green algae *Acetabularia*, *Pseudobryopsis* and *Nitella* [11], fucoid (brown algal) Zygotes of *Silvetia compressa* and *Pelvetia compressa* [32], *Allium* bulb scale cells and *Sinapis* root hairs [11], tobacco BY-2 cells [33], *Lilium* pollen tubes [34], and *Papaver rhoeas* pollen tubes [35] have been tested for jasplakinolide reaction.

Jaspalkinolide causes tremendous effect already visible at the light microscopic level by malformation of cells after recovery in the development of *Micrasteris* [20] (Fig. 2a). Inhibition or retardation of cell development occurs; during recovery from drug treatment the cells develop a malformed pattern [20]. In pollen tubes the normal bidirectional cytoplasmic streaming is altered, instead a rotary streaming is observed in the swollen apex [34].

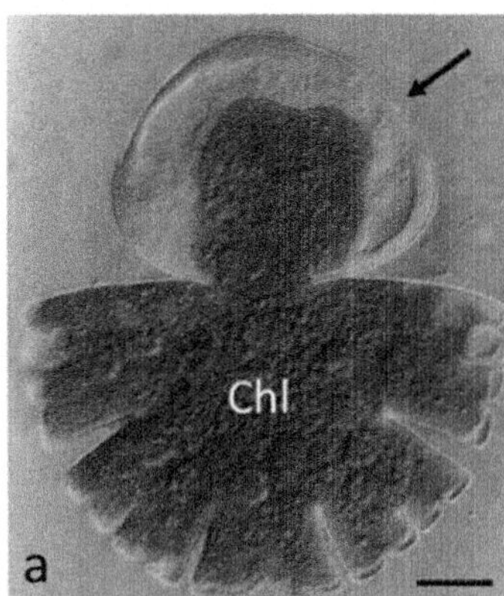

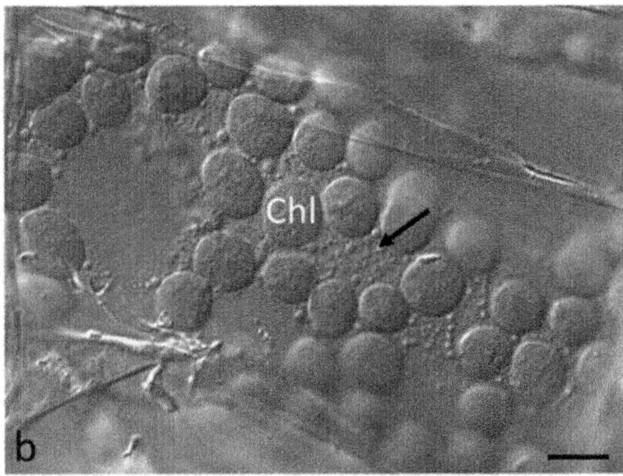

Fig. 2 Light microscopy of F-actin perturbation. (**a**) Effect of jasplakinolide treatment on developing *Micrasterias denticulata* cell young developmental stage treated with 3 μM jasplakinolide for 0.5 h and allowed to recover for 3 h in nutrient solution, young semi-cell exhibits complete loss of normal cell pattern (*arrow*), chloroplast (chl). (**b**) Effect of 5 μm latrunculin B on *Oxyria digyna* mesophyll cell; cytoplasmic streaming is inhibited, between the chloroplasts (chl) granular cytoplasmic portions are visible (*arrow*). Bars: (**a**) 20 μm, (**b**) 5 μm. (**a**) reprinted from ref. 20 with permission. "Copyright Wiley-Liss, Inc., a subsidiary of John Wiley & Sons, Inc." (**b**) Reprinted from ref. 55 with permission of "G. Thieme Verlag KG, Stuttgart"

Moreover altering F-actin levels or dynamics by jasplakinolide plays a functional role in initiating programmed cell death in *Papaver* pollen, triggering a caspase-3-like activity [35]. Polarity establishment is severely changed in fucoid zygotes [32]. When applied during mitosis, binucleated cells are generated as a consequence of jasplakinolide treatment [20].

Jasplakinolide specifically targets F-actin, other cytoskeletal components like desmin or ß-tubulin were not influenced by jasplakinolide [36], microtubule-dependent processes are not affected [37], and the distribution of microtubules and microtubule dependent processes were not altered in green algae [20].

1.1.2 Chondramides

Another group of substances with actin stabilizing and polymerizing capacities are chondramides [9, 38]. These depsipeptides were originally obtained from myxobacteria of *Chondromyces crocatus* [8, 39]. Chemically they are very similar to jasplakinolide, but contain instead of an 19-membered ring an 18-membered macrocyclic ring [40]. Until to date they are underrepresented due to their limited availability. However, synthetic chondramides were produced recently [40], and may become commercially available. In future, chondramides may also be used for cancer therapy [41, 42].

1.2 Actin-Depolymerizing Substances

A prerequisite for an F-actin-perturbing drug is its membrane permeability. Cytochalasins as well as latrunculines represent the most widely used F-actin-sequestering drugs [43].

1.2.1 Cytochalasins

Cytochalasins have been shown to act on cytoplasmic streaming in plant cells [44–47]. A vast number of different cytochalasins (A, B, C, D, E, H, J) have been described, with different inhibitory concentrations and chemical properties, occasionally termed cytochalasanes [48]. It is generally accepted that the cytochalasins slow the rate of filament polymerization by inhibiting the rate of elongation [49]; This action is caused by the high-affinity binding of cytochalasins to the barbed (plus end) of F-actin, thus the monomer addidion is prevented by a "capping" mechanism [43, 49]. Due to the prevention of G-actin incorporation into the filament, a net-depolymerization is provoked [43]. While the main target is the F-actin cytoskeleton, for cytochalasins A and B also inhibit monosaccharide transport across the plasma membrane [5].

Cytochalasins are fungal metabolites, independently discovered and isolated from distinct fungal species (cytochalasin A and B derived from *Helminthosporium dematioideum*; cytochalasin C and D from *Metarhizium anisopliae*) by Aldridge et al. [50]. Previously, Rothweiler and Tamm [51] termed the substance "phomin" as it was detected in *Phomas* sp. (*Fungi imperfecti*, strain S 298). Chemically cytochalasins are a diverse group of polyketide–amino acid hybrid metabolites [48].

Various cytochalasins have a wide range of biological activities, some of which not directly related to actin binding. Some cytochalasins (e.g., cytochalasin B) interfere with the monosaccharide transport systems by binding to high-affinity sites on glucose-transporter proteins and may interfere with hormones.

1.2.2 Latrunculins

Latrunculins (Fig. 1) provide another useful drug for the study of actin polymerization with a complementary mode of action when compared to cytochalasin [52]. The main mechanism of latrunculin action is in an interaction with actin monomers (G-actin) in order to avoid actin polymerization [52]. It has been shown by an in vitro assay that latrunculins bind to purified G-actin resulting in a nonpolymerizable 1:1 complex [6]. Latrunculin A is a toxin purified from the red sea sponge *Latrunculia magnifica* [52, 53]. It has been used in a variety of plant cells inhibiting pollen germination and pollen tube growth [47] and induced plant dwarfism [54]. Both latrunculins arrested cytoplasmic streaming after disrupting the subcortical actin bundles [46]. In a mesophyll cell of the high alpine plant *Oxyria digyna* latrunculin B arrested cytoplasmic streaming (Fig. 2b) [55]. Latrunculin B has a diminished binding capacity when compared to latrunculin A [6].

1.2.3 Swinholide

Swinholide A (Fig. 1) was isolated from the marine sponge *Theonella swinhoei*, collected from the Red Sea [6, 56]. It is membrane permeable and acting as an F-actin-disrupting toxin that stabilizes actin dimers and thus severs F-actin. Chemically it consists of a 44-carbon ring dimeric dilactone macrolide with a twofold axis

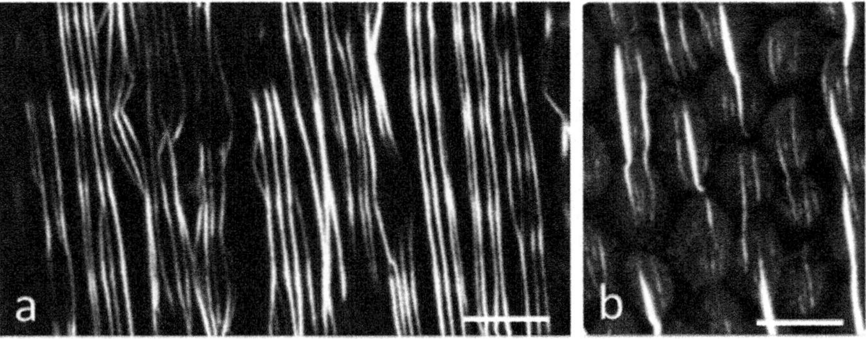

Fig. 3 Subcortical F-actin bundles in internodal cells of *Nitella pseudoflabellata* treated with (**a**) 1 μM cytochalasin A, 1 day, (**b**) cytochalasin H 50 μM, 2 days. Despite these concentrations lead to an arresting of cytoplasmic streaming, the F-actin bundles are clearly visible after phalloidin staining with the perfusion method (*see* Perfusion of *Nitella*). Reprinted from ref. 46 with permission of Oxford University Press

of symmetry [57, 58]. The effects of swinholide A on the actin cytoskeleton and cell morphology are similar to latrunculins rather than of cytochalasins [57].

1.3 Visualization of Inhibitory Effects

The function of these drugs can be either studied by arresting or inhibiting the cytoplasmic streaming (Fig. 2b) [55]. The velocity of cytoplasmic streaming can provide useful insights into the function of the different drugs [46].

Visualization of the changes in F-actin is usually performed by phalloidin staining [46]. It has been shown in the charophyte green alga *Nitella pseudoflabellata*, that thick F-actin bundles may not appear influenced by e.g. cytochalasin A and H (Fig. 3), despite that the cytoplasmic streaming is arrested. After jasplakinolide or chondramide treatment an altered F-actin distribution was found, characterized by a patchy appearance of cortical actin [59], the formation of actin dots [60], or a disruption of the actin cytoskeleton [11, 36].

As competitive binding inhibition of phalloidin to F-actin by jasplakinolide is described [19], and decreased FITC-phalloidin labeling is noted as a consequence of jasplakinolide treatment in MDCK cells [61], the problems visualizing jasplakinolide effects by fluorescently labeled phalloidin are obvious. Moreover, phalloidin and jasplakinolide have the same effect when applied to living *Dictyostelium* cells, namely the formation of actin aggregates [22].

In order to avoid the difficulties mentioned, only electron microscopy appears to be a sufficient technique for detection of jasplakinolide effects on F-actin [20, 62].

At electron micrographs tangential sections of F-actin can easily be recognized by the filament diameter of about 4–5 nm. In organisms like *Micrasterias*, where normally almost no bundling of F-actin occurs, and thus F-actin is hardly found at electron

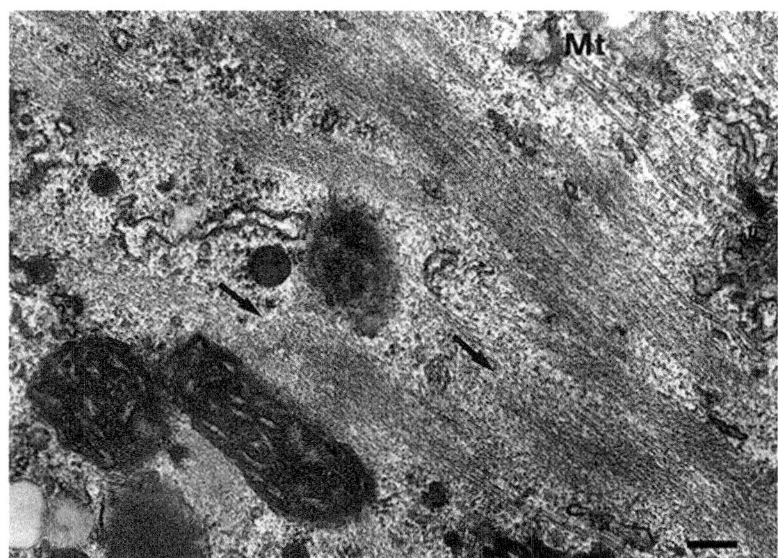

Fig. 4 Detail of the ultrastructure of *Micrasterias denticulata* cell treated with 3 μM jasplakinolide for 2 h and fixed for electron microscopy according tothe Chemical Fixation Protcol, dense accumulations of F-actin are visible in the cytoplasm (*arrows*). Bar = 0.2 μm. Reprinted from ref. 20 with permission "Copyright Wiley-Liss, Inc., a subsidiary of John Wiley & Sons, Inc."

microscopical images, alterations causing extensive bundling of actin are easily detected [20] (Fig. 3). Also immunodetection of the altered F-actin system after jasplakinolide treatment is successful [59] and allows a comprehensive picture to be drawn in combination with other techniques (Fig. 4).

Herein different procedures shall be described for the detection of actin filament aggregates generated by application of jasplakinolide. Due to the advantages of electron microscopy and/or immunomethods, these will be mentioned more in detail; however different protocols for phalloidin stainings will be given as they have been used for some organisms in combination with jasplakinolide treatment [22, 59, 61].

2 Materials

2.1 Cell Cultures

1. *Micrasterias denticulata* Bréb. (unicellular green algae) are used for light microscopy and electron microscopy [20] and immuno-electron microscopy since alterations in the cell shape caused by drug treatments are easily detected. The cells are grown in culture medium under semisterile conditions in a light/dark regime of 14/10 h at 20 °C.

2. Characean green algae (*Chara* sp., *Nitella* sp.), fresh collected or culture grown material.
3. *Acetabularia acetabulum.*
4. Stably transformed *Arabidopsis thaliana* expressing a fusion construct of GFP with the actin binding domain 2 (ABD2) of the plant actin-binding protein fimbrin [63, 64].

2.2 Chemicals, Buffers, Solutions

2.2.1 Actin Perturbing Drugs and Solvents

1. Jasplakinolide, Mol. Wt. 709.68, (Molecular Weight 709.67, J4580-100UG Sigma-Aldrich). 100 μg diluted upon arrival in methanol to achieve 10 mM stock solution and kept refrigerated at −20 °C (*see* **Note 1**).
2. Chondramides (while the original isolate from myxobacteria [8, 39] is not commercially available, synthetic chondramide A analogous have been produced, *see* ref. 40).
3. Cytochalasins (all available from Sigma-Aldrich): cytochalasin A: M. Wt. 477.59, C6637-1MG; cytochalasin B: M. Wt. 479.61, C6762-1MG; cytochalasin C: M. Wt. 507.62, C30382-1MG; cytochalasin D: M. Wt. 507.62, C8273-1MG; cytochalasin E: M. Wt. 495.56, C2149-1MG; cytochalasin H: M. Wt. 493.63, C0889-5MG; cytochalasin J: M. Wt. 451.60 (currently not available).
4. Latrunculin (A, B): LAT A: M. Wt. 421.55, L5163-100UG (Sigma-Aldrich); LAT B: M. Wt. 395.51, L5288-1MG (Sigma-Aldrich).
5. Swinholide A, M. Wt. 1,389.87, S9810-10UG (Sigma-Aldrich).
6. Phalloidin M. Wt. 788.87, P2141-1MG (Sigma-Aldrich).
7. Methanol reagent.
8. Dimethyl sulfoxide (DMSO) reagent.

2.2.2 Chemicals, Buffers, and Solutions for Electron Microscopy

1. Cacodylic buffer: 50 mM cacodylic acid sodium salt, pH 7.2.
2. Cacodylate-glutaraldehyde: 50 mM cacodylic buffer, 1 % glutaraldehyde, pH 7.2.
3. Cacodylate-osmium: 50 mM cacodylic acid buffer, 1 % osmium tetroxide, 0.8 %, potassium hexacyanoferrate III.
4. Phosphate buffer fixative: 25 mM sodium hydrogen phosphate, 25 mM potassium dihydrogen phosphate, 1 % glutaraldehyde, 1 % osmium tetroxide, pH 6.2.
5. Calcium chloride.
6. 2 % Aqueous uranyl acetate dihydrate.
7. Ethanol solutions: 15, 30, 40, 50, 60, 70, 80, 90, 95, 100 %.
8. 1,2-Propylene oxide reagent.
9. Propylene oxide-ethanol mixture: 1,2-Propylene oxide and ethanol (1:1 v/v).

10. A suitable embedding resin: Embed 812, Araldite 502, Spurr's resin or others.
11. Propylene oxide-embedding resin mixture: Mixture of 1,2 Propylene oxide and embedding resin (1:1 v/v).
12. Formvar 1595 E.
13. Chloroform reagent.
14. Lead citrate.

2.2.3 Chemicals, Buffers, and Solutions for Immunoelectron Microscopy

1. Lecithin (10 %) dissolved in chloroform.
2. Liquid nitrogen.
3. Acetone-tannic acid: acetone containing 0.1 % tannic acid (v/v).
4. Acetone-osmium: acetone containing 2 % osmium tetroxide and 0.05 % uranyl acetate.
5. LR white resin (London resin company, medium grade).
6. TBS: Tris-buffered saline, 50 mM, pH 7.5.
7. Blocking solution: TBS containing 1 % BSA (fraction V), 1 % acetylated BSA, 01 % Tween-20.
8. 10 % bovine calf fetal serum.
9. Primary antisera/antibodies, e.g., monoclonal anti actin N 350 monoclonal anti actin (Amersham).
10. 10 nm colloidal gold conjugated, secondary antibody appropriate for the primary antibody.

2.2.4 Chemicals, Buffers, and Solutions for Fluorescence Microscopy

1. MT-stabilizing buffer (MTSB): 50 mM PIPES buffer, 50 mM $MgSO_4$, 50 mM EGTA, pH 6.9.
2. Dimethylsulfoxide (DMSO) reagent.
3. Paraformaldehyde fixative: 4 % paraformaldehyde and cytospin (Cytospin 3 preparation systems, Shandon Scientific Ltd. Cheshire, UK).
4. 50 μM ammonium chloride.
5. PBS: Phosphate buffered saline: 137 mM sodium chloride, 2.7 mM potassium chloride, 4.3 mM sodium hydrogen phosphate, 1.4 mM potassium dihydrogen phosphate.
6. Triton solution: 0.1 % Triton X-100 in PBS.
7. 2 % bovine serum albumin (BSA).
8. Anti actin monoclonal antibody N 350 (Amersham) (*see* **Note 2**).
9. Anti actin monoclonal antibody (clone C4).
10. Phalloidin: 0.4 mg/ml FITC (or alexa) conjugated phalloidin Alexa phalloidin (Molecular Probes, Leiden, The Netherlands; prepare from a 6.6 mM stock solution in methanol) at a concentration of 0.16 mM.

11. Phenylenediamine.
12. Texas red conjugated antimouse IgM antibody (Jackson Immuno Research).
13. Steedman's wax: PEG 400 distearate and 1-hexadecanol mixed in proportions 9:1 (w/w) [68].
14. Isotonic perfusion solution: 200 mM sucrose, 70 mM KCl, 4.5 mM $MgCl_2$, 5 mM ethyleneglycoltetraacetic acid (EGTA), 1.48 mM $CaCl_2$, 10 mM piperazine-*N*,*N*′-bis (2-ethanesulfonic acid) (PIPES, pH 7.0).

3 Methods

3.1 Treatment and Recovery Experiments

1. Incubate cells (from the above mentioned plant cell types or tissue *see Cell Cultures*) in 1 ml of culture medium containing respective inhibitory drug concentrations for time periods of 0.5, 1, 2, 3, 4, 24, and 48 h and observe the effect on cytoplasmic streaming (Fig. 2, *see* **Note 3**).
2. Incubate cells in 1 ml of culture medium containing the respective drug (*see* **step 1**), wash several times with culture medium and allow developing in culture medium for up to 24 h. This is regarded as "recovery experiment."
3. Investigate the cells treated as in **steps 2** and **3** under a regular light microscope or fix for electron microscopy.

3.2 Visualization of F-Actin After Drug Treatment

The procedure described herein has first been used for the unicellular green alga *Micrasterias* by Meindl (1990) [65].

3.2.1 Chemical Fixation/Electron Microscopy of Algal Cells After Holzinger and Meindl [20]

1. Fix treated cells, cells allowed to recover in nutrient solution and untreated control cells in cacodylate-glutaraldehyde for 15 min (*see* **Note 4**).
2. Wash cells three times 5 min each in cacodylic buffer.
3. Postfix cells in cacodylate-osmium for 2 h.
4. Wash cells, as in **step 2**, three times 5 min each in distilled water.
5. Incubate cells in 2 % uranyl acetate for 2 h (*see* **Note 5**).
6. Wash cells, as in **step 2**, three times 5 min each in distilled water.
7. Dehydrate the cells in increasing concentrations of cold ethanol (*see* **Note 6**), 15 min per concentration:
   15, 30, 40, 50, 70, 80, 90, 95, and 100 %. Keep in 100 % for 30 min.
8. Transfer the cells to the propylene oxide-ethanol mixture, allow to equilibrate for 10 min, and transfer to propylene oxide.

9. Transfer the cells to the propylene oxide-embedding resin mixture (*see* **Note 7**).
10. Rotate the cells in this mixture for 48 h in order to allow the propylene oxide to evaporate and a sufficient penetration of the cells with the resin to be achieved.
11. Transfer the cells to freshly prepared resin in aluminum dishes and orientate them by the use of eyelashes connected to holders.
12. Incubate in an exsiccator for 4 days.
13. Polymerize the resin for 24 h at 60 °C.
14. Select cells and section at an ultramicrotome.
15. Collect sections on formvar-coated copper grids (*see* **Note 8**).
16. Counterstain the sections with 2 % uranyl acetate and Reynold's lead citrate (*see* **Note 9**).
17. Investigate and photograph at a transmission electron microscope at 60–80 kV.

3.2.2 Immuno-Electron Microscopy

The following fixation technique has been carried out for *Micrasterias* for the first time by Meindl et al. (1992) [66]. The recipe for actin detection by means of immunogold localization is based on Holzinger et al. (1999) [67].

1. Prepare gold or aluminum specimen holders with bed depths of 100–300 µm by dipping in lecithin.
2. Collect treated cells to be transferred to specimen holders. Large cells like *Micrasterias* (200 µm in diameter) can be easily collected under a stereo microscope by wrapping with cotton fibers.
3. Fix treated cells by high-pressure freeze fixation by the use of a hyperbaric freeze device (formally Balzers HPM 010, later taken over by BAL-TEC; then produced by ABRA Fluid AG, Widnau, Switzerland; currently available as "High Pressure Freezing Machine HPM 010" by Boeckeler Instruments Inc., Tucson/AZ, USA), for other methods (*see* **Note 10**).
4. Collect and store samples under liquid nitrogen.
5. Transfer samples to a freeze substitution device (Reichert-Jung, cs auto or LEICA EM AFS, Leica Microsysteme GmbH, Vienna, Austria).
6. Substitute samples at −80 °C for 24 h in acetone-tannic acid solution.
7. Wash several times with acetone.
8. Substitute samples at −80 °C for 24 h in acetone-osmium solution (*see* **Note 11**).
9. Allow samples to reach −30 °C within 5 h.

10. Continue substitution at −30 °C for 10 h.
11. Allow samples to reach room temperature (20 °C) within 5 h.
12. Remove osmium tetroxide/uranyl acetate solution and rinse several times with acetone.
13. Change acetone for ethanol by several rinses.
14. Transfer samples to LR white in aluminum dishes and cover with cellophane foil.
15. Allow samples to infiltrate in an exsiccator for 24 h.
16. Polymerize under UV light at room temperature for 24 h.
17. Prepare sections for electron microscopy and transfer to formvar-coated gold or gilded copper grids.
18. Incubate in blocking solution for 30 min up to 1 h. A blocking step with 10 % bovine calf fetal serum may be applied if blocking is not sufficient.
19. Transfer into primary antibody against actin. Purified antibody may be diluted in blocking solution. Incubation should last for 1.5 h at room temperature or for up to 24 h at 4 °C.
20. Wash four times 15 min in TBS by transferring grids to 50 μl droplets of TBS.
21. Incubate in 10 nm-gold (*see* **Note 12**) labeled secondary antibody diluted in blocking solution (*see* **Note 13**).
22. Wash by rinsing with TBS, followed by incubation in droplets of TBS for 2 min, followed by a brief rinse with distilled water.
23. Counterstain with uranyl acetate and lead citrate if necessary (*see Chemical Fixation*).
24. Investigate at a transmission electron microscope at 80 kV.

3.2.3 Immunofluorescence Microscopy of Latrunculin B-Treated Maize Roots After Baluska et al. [54, 68]

1. Grow maize root tips for 48 h.
2. Cut apical segments (6–8 mm) and transfer into 4.6 ml MTSB buffer mixed with 0.5 ml of DMSO for 15 min.
3. Fix for 1 h in MTSB/DMSO solution containing 4 % paraformaldehyde.
4. Dehydrate in increasing ethanol series in PBS.
5. Embed in Steedman's wax by stepwise infiltration in proportions 2:1, 1:1, and 1:2 (v/v) for 2 h each step at 37 °C.
6. Infiltrate under vacuum three changes in pure wax.
7. Polymerize wax at room temperature.
8. Prepare longitudinal sections with a thickness between 4 and 8 μm.
9. Mount sections on poly-L-lysine-coated slides.
10. Dewax the sections in ethanol.

11. Rehydrate in ethanol/PBS series and incubate in MTSB for 45 min.
12. Digest cell walls in 1 % hemicellulose dissolved in 0.5 M EGTA, 0.4 M manniol, 1 % Triton X-100, 0.3 mM phenlymethlysulfonyl fluoride (dissolved in MTSB).
13. Incubate in blocking solution (2 % BSA) for 30 min.
14. Incubate in anti actin monoclonal anti-actin antibody (clone C4, ICN, Costa Mesa, CA) for 2 h at room temperature. Rinse three times 15 min in MTSB.
15. Incubate in FITC-conjugated anti-mouse IgGs (diluted 1:100–1:200) for 1 h at room temperature.
16. Remove secondary antibody and mount in phenylenediamine.
17. Examine at a confocal laser scanning microscope at excitation with an argon laser at 488 nm.

3.2.4 Perfusion of Nitella Internodal Cells for FITC-Phalloidin Staining of F-Actin After Foissner and Wasteneys (2007) [46] (Fig. 3; See **Note 14***)*

1. Place an internodal cell on the cover slip bottom of a perfusion chamber and press into vacuum grease lines; place small reservoirs with grooves over the grease and firmly press down without damaging the cells.
2. Cover the central portion of the cell between the reservoirs with silicon fluid to avoid evaporation.
3. Bath the ends of the cells with isotonic perfusion solution.
4. Reduce the turgor pressure of the cells in hypertonic solution.
5. Cut the ends with small scissors and allow the perfusion solution to enter the cells.
6. After 1 min replace the perfusion solution by an F-actin staining solution containing Alexa-phalloidin.
7. After 20 min examine at the CLSM (*see* Immunofluorescence Microscopy).

4 Notes

1. Jasplakinolide is also dissolvable in DMSO [31].
2. The choice of the anti actin antibody is very important. The epitope of the N 350 (a mouse monoclonal IgM antibody directed against chicken gizzard actin) binding site is obviously not changed by jasplakinolide [11], whereas binding sites for other actin antibodies (e.g., clone C4—a mouse monoclonal IgG antibody directed against chicken gizzard actin, ICN) seem to be altered substantially which prevents visualization of F-actin after jasplakinolide treatment [11]. This antibody may be used without problems in combination with latrunculine B treatment [54].

3. The concentrations of the inhibitory drugs may vary drastically. For jasplakinolide was found to be effective at 1.5–3 μM in producing F-actin aggregates for *Micrasterias denticulate* [20]. Concentrations above 250 μM of jasplakinolide were lethal. The concentrations of jasplakinolide needed for other cells have to be found empirically. Variations in the literature range from as low as 100 nM to 1 μM, 2.5–10 μM [11, 32, 59, 69]. However they should be in the same range as described herein, since the effective concentrations correlate with data obtained from the literature [22, 31, 61]. Latrunculins had a different lowest effective concentrations in *Micrasterias denticulata*: chondramide A 20 μM, chondramide B 15 μM, chondramide C 5 μM, chondramide D 10 μM [38]. In contrast in human tumor cell lines, 3–85 nM chondramide effectively inhibited proliferation [9]. For cytochalsins [46] found streaming-arresting concentration in *Nitella pseudoflabellata* between 1 and 200 μM. The lowest concentrations to arrest streaming were at 1 μM in cytochalasins A and E, 30 μM in cytochalasin H, 60 μM cytochalasin D and 200–220 μM in cytochalasins C, J, and B. These authors distinguish between streaming arresting (i.e. fully arresting streaming within 1 h, Fig. 3) and streaming inhibiting (i.e., reducing streaming velocity by 15–80 % within 1 h) concentrations. In another green alga *Xanthidium armatum*, cytochalasin D arrested cytoplasmic streaming at concentrations below 10 μM.

   In contrast, latrunculins have lower effective concentrations; Latruncunlin B may cause severe effects even at extremely low concentrations in the picomolar to nanomolar range [45]. Several authors use it in higher concentration in the μM range (10 μM [54]; 5, 10 μM [55]). Moreover, a combined treatment with latrunculins and cytochalasins, rapidly arrested cytoplasmic streaming even at concentrations that had only mild effects on the streaming rate when used separately [46]. Swinholide has not been extensively applied to plant cells, only IC 50 values for human nasopharynx cancer cells have been determined at 6 nM [70].

4. The fixation procedure is performed in a “Balzers Fixomat”—a device not commercially available anymore—consisting of sintered glass suction filters connected via a valve to a pump station, allowing to remove solutions by underpressure while cells remain in the filter. Moreover the temperature can be adjusted via a cooling bath. Alternatively the cells can be fixed in glass dishes, ideally in the shape of a hemisphere. In this case it is best to remove only the solutions with a pipette and not to transfer the cells.

   For other organisms more specialized methods might be necessary—for example embedding of the specimens in agarose prior to the fixation procedure.

5. During incubation in uranyl acetate solution it is necessary to keep the cells in darkness, by covering the filter or dish with aluminum foil.
6. During the whole procedure of dehydration the temperature should be kept at 6 °C. At the steps 50 and 70 % ethanol temperature might be lowered to 4 °C. In case of fixation in round glass dishes they should be kept over ice. At the step of 100 % ethanol the temperature should be increased to room temperature.
7. Very good results are obtained with a 1:1 mixture of glycid ether 100 (a substance equivalent to Epon 812) and MNA. Prior to use DMP must be added (5 droplets to 10 ml of resin). However a mixture of Embed-812/Araldite 502/ DDSA (DMP-30 or BDMA must be added prior to use) will also give adequate results. When different hardness has to be achieved, Spurr's resin may be the best choice: As several components of the original mixture are not deliverable any more, we now use the kit "Agar Low Viscosity Resin (LV)" by Agar Scientific Ltd. Essex CM24 8DA, England.
8. Coating of grids with formvar may not be necessary for all objects. We use the following procedure: A light microscopic slide is cleaned with lens tissue, tipped into 0.3 % formvar dissolved in chloroform. The resulting formvar film is cut at the edges of the slide and allowed to float on distilled water. Grids are placed on the film which is then removed with parafilm.
9. The time periods needed for counterstaining depend on the staining already achieved during the fixation/staining procedure and are typically 5–30 min for uranyl acetate and 1–5 min for Reynold's lead citrate.
10. For large cells like *Micrasterias* (about 200 μm in diameter) high pressure freezing has been found the only appropriate technique for freeze fixation. Alternatively to the HPM 010, also the Leica EMPACT high-pressure freezer (Leica Microsysteme GmbH, Vienna, Austria) gives reasonable results. However for small and less vacuolated cells plunge freezing which might be achieved with a relative simple equipment, will also give reasonable results.
11. Osmium tetroxide dissolves well in acetone. Uranylacetat is best to be sonicated to dissolve it. For other organisms it might be appropriate to substitute without osmium tetroxide and uranyl acetate.
12. Gold particles with 10 nm diameter are used commonly. The usage of smaller gold particles might enhance the accuracy of the detected locus.
13. The actual dilution factor has to be found empirically for each system, but it should be in the range of 1:50–1:200.

14. As mentioned in the introduction, the use of fluorescently labeled phalloidin is problematic due to competitive binding with jasplakinolide [19, 23]. It has been reported that simultaneous addition of jasplakinolide and FITC-phalloidin (1:1) did not result in F-actin labeling, whereas addition of the inactive analogue jasplakinolide B and FITC-phalloidin resulted in staining as strong as in controls [11].

Acknowledgements

We would like to thank Dr. Ilse Foissner, University of Salzburg, for providing the images in Fig. 3. This chapter was supported by FWF grant P24242-B16 to A.H.

References

1. Allingham JS, Klenchin VA, Rayment I (2006) Actin-targeting natural products: structures, properties and mechanisms of action. Cell Mol Life Sci 63(18):2119–2134. doi:10.1007/s00018-006-6157-9
2. Oertel A, Holzinger A, Lütz-Meindl U (2003) Involvement of myosin in intracellular motility and cytomorphogenesis in *Micrasterias*. Cell Biol Int 27(12):977–986. doi:10.1016/j.cellbi.2003.07.004
3. Guerriero CJ, Weisz OA (2007) N-WASP inhibitor wiskostatin nonselectively perturbs membrane transport by decreasing cellular ATP levels. Am J Physiol Cell Physiol 292(4):C1562–C1566. doi:10.1152/ajpcell.00426.2006
4. Lynen F, Wieland U (1937) Über die Giftstoffe des Knollenblätterpilzes IV. Justus Liebigs Ann Chem 533(1):93–117. doi:10.1002/jlac.19385330105
5. Cooper JA (1987) Effects of cytochalasin and phalloidin on actin. J Cell Biol 105(4):1473–1478. doi:10.1083/jcb.105.4.1473
6. Spector I, Braet F, Shochet NR et al (1999) New anti-actin drugs in the study of the organization and function of the actin cytoskeleton. Microsc Res Tech 47(1):18–37. doi:10.1002/(SICI)1097-0029(19991001)47:1<18:AID-JEMT3>3.0.CO;2-E
7. Bubb MR, Spector I, Beyer BB et al (2000) Effects of jasplakinolide on the kinetics of actin polymerization. An explanation for certain in vivo observations. J Biol Chem 275(7):5163–5170
8. Kunze B, Jansen R, Sasse F et al (1995) Chondramides A-D, new antifungal and cytostatic depsipeptides from *Chondromyces crocatus* (Myxobacteria). Production, physico-chemical and biological properties. J Antibiot 48(11):1262–1266. doi:10.7164/antibiotics.48.1262
9. Sasse F, Kunze B, Gronewold TMA et al (1998) The chondramides: cytostatic agents from myxobacteria acting on the actin cytoskeleton. J Natl Cancer Inst 90(20):1559–1563. doi:10.1093/jnci/90.20.1559
10. Waldmann H, Hu T, Renner S et al (2008) Totalsynthese von chondramid C und bindung an F-aktin. Angew Chem 120(34):6573–6577. doi:10.1002/ange.200801010
11. Sawitzky H, Liebe S, Willingale-Theune J et al (1999) The anti-proliferative agent jasplakinolide rearranges the actin cytoskeleton of plant cells. Eur J Cell Biol 78(6):424–433. doi:10.1016/S0171-9335(99)80085-5
12. Peterson JR, Mitchison TJ (2002) Small molecules, big impact: a history of chemical inhibitors and the cytoskeleton. Chem Biol 9(12):1275–1285. doi:10.1016/S1074-5521(02)00284-3
13. Matthews JB, Smith JA, Hrnjez BJ (1997) Effects of F-actin stabilization or disassembly on epithelial Cl- secretion and Na-K-2Cl cotransport. Am J Physiol Cell Physiol 272(1):C254–C262
14. Crews P, Manes LV, Boehler M (1986) Jasplakinolide, a cyclodepsipeptide from the marine sponge, *Jaspis* sp. Tetrahedron Lett 27(25):2797–2800. doi:10.1016/S0040-4039(00)84645-6
15. Braekman JC, Daloze D, Moussiaux B et al (1987) Jaspamide from the marine sponge *Jaspis johnstoni*. J Nat Prod 50(5):994–995. doi:10.1021/np50053a048
16. Zabriskie TM, Klocke JA, Ireland CM et al (1986) Jaspamide, a modified peptide from a *Jaspis* sponge, with insecticidal and antifungal

activity. J Am Chem Soc 108(11):3123–3124. doi:10.1021/ja00271a062

17. Fabian I, Halperin D, Lefter S et al (1999) Alteration of actin organization by Jaspamide inhibits ruffling, but not phagocytosis or oxidative burst, in HL-60 cells and human monocytes. Blood 93(11):3994–4005
18. Terracciano S, Bruno I, D'Amico E et al (2008) Synthetic and pharmacological studies on new simplified analogues of the potent actin-targeting Jaspamide. Bioorg Med Chem 16(13):6580–6588. doi:10.1016/j.bmc.2008.05.019
19. Bubb MR, Senderowicz AM, Sausville EA et al (1994) Jasplakinolide, a cytotoxic natural product, induces actin polymerization and competitively inhibits the binding of phalloidin to F-actin. J Biol Chem 269(21):14869–14871
20. Holzinger A, Meindl U (1997) Jasplakinolide, a novel actin targeting peptide, inhibits cell growth and induces actin filament polymerization in the green alga *Micrasterias*. Cell Motil Cytoskeleton 38(4):365–372. doi:10.1002/(SICI)1097-0169(1997)38:4<365:AID--CM6>3.0.CO;2-2
21. Sheikh S, Gratzer WB, Pinder JC et al (1997) Actin polymerisation regulates integrin-mediated adhesion as well as rigidity of neutrophils. Biochem Biophys Res Commun 238(3):910–915. doi:10.1006/bbrc.1997.7407
22. Lee E, Shelden EA, Knecht DA (1998) Formation of F-actin aggregates in cells treated with actin stabilizing drugs. Cell Motil Cytoskeleton 39(2):122–133. doi:10.1002/(SICI)1097-0169(1998)39:2<122:AID--CM3>3.0.CO;2-8
23. Visegrády B, Lorinczy D, Hild G et al (2005) A simple model for the cooperative stabilisation of actin filaments by phalloidin and jasplakinolide. FEBS Lett 579(1):6–10. doi:10.1016/j.febslet.2004.11.023
24. Murray LM, Johnson A, Diaz MC et al (1997) Geographic variation in the tropical marine sponge *Jaspis* cf. *johnstoni*: an unexpected source of new terpene-benzenoids. J Org Chem 62:5638–5641
25. Fabian I, Shur I, Bleiberg I et al (1995) Growth modulation and differentiation of acute myeloid leukemia cells by jaspamide. Exp Hematol 23(7):583–587
26. Ayscough KR, Stryker J, Pokala N et al (1997) High rates of actin filament turnover in budding yeast and roles for actin in establishment and maintenance of cell polarity revealed using the actin inhibitor latrunculin-A. J Cell Biol 137(2):399–416
27. Chu KS, Negrete GR, Konopelski JP (1991) Asymmetric total synthesis of (+)-jasplakinolide. J Org Chem 56(17):5196–5202. doi:10.1021/jo00017a037
28. Inman W, Crews P (1989) Novel marine sponge-derived amino acids. 8. Conformational analysis of jasplakinolide. J Am Chem Soc 111(8):2822–2829. doi:10.1021/ja00190a013
29. Kahn M, Nakanishi H, Su T et al (1991) Design and synthesis of nonpeptide mimetics of jaspamide. Int J Pept Protein Res 38(4):324–334
30. Ghosh AK, Moon DK (2007) Enantioselective total synthesis of +-jasplakinolide. Org Lett 9(12):2425–2427. doi:10.1021/ol070855h
31. Scott VR, Boehme R, Matthews TR (1988) New class of antifungal agents: jasplakinolide, a cyclodepsipeptide from the marine sponge, *Jaspis* species. Antimicrob Agents Chemother 32(8):1154–1157
32. Hable WE, Miller NR, Kropf DL (2003) Polarity establishment requires dynamic actin in fucoid zygotes. Protoplasma 221(3-4):193–204. doi:10.1007/s00709-002-0081-0
33. Ou GS, Chen ZL, Yuan M (2002) Jasplakinolide reversibly disrupts actin filaments in suspension-cultured tobacco BY-2 cells. Protoplasma 219(3-4):168–175. doi:10.1007/s007090200018
34. Cárdenas L, Lovy-Wheeler A, Wilsen KL et al (2005) Actin polymerization promotes the reversal of streaming in the apex of pollen tubes. Cell Motil Cytoskeleton 61(2):112–127. doi:10.1002/cm.20068
35. Thomas SG, Huang S, Li S et al (2006) Actin depolymerization is sufficient to induce programmed cell death in self-incompatible pollen. J Cell Biol 174(2):221–229. doi:10.1083/jcb.200604011
36. Senderowicz AM, Kaur G, Sainz E et al (1995) Jasplakinolide's inhibition of the growth of prostate carcinoma cells in vitro with disruption of the actin cytoskeleton. J Natl Cancer Inst 87(1):46–51
37. da Costa SR, Yarber FA, Zhang L et al (1998) Microtubules facilitate the stimulated secretion of beta-hexosaminidase in lacrimal acinar cells. J Cell Sci 111(Pt 9):1267–1276
38. Holzinger A, Lütz-Meindl U (2001) Chondramides, novel cyclodepsipeptides from myxobacteria, influence cell development and induce actin filament polymerization in the green alga *Micrasterias*. Cell Motil Cytoskeleton 48(2):87–95. doi:10.1002/1097-0169(200102)48:2<87:AID-CM1000>3.0.CO;2-C

39. Jansen R, Kunze B, Reichenbach H et al. (1996) Chondramides A-D, new cytostatic and antifungal cyclodepsipeptides from Chondromyces crocatus (myxobacteria): isolation and structure elucidation. Liebigs Ann 2:285–290
40. Ma CI, Diraviyam K, Maier ME et al (2013) Synthetic chondramide A analogues stabilize filamentous actin and block invasion by *Toxoplasma gondii*. J Nat Prod 76(9):1565–1572. doi:10.1021/np400196w
41. Foerster F, Braig S, Moser C et al (2014) Targeting the actin cytoskeleton: selective antitumor action via trapping PKCε. Cell Death Dis 5:e1398. doi:10.1038/cddis.2014.363
42. Menhofer MH, Kubisch R, Schreiner L et al (2014) The actin targeting compound Chondramide inhibits breast cancer metastasis via reduction of cellular contractility. PLoS One 9(11):e112542. doi:10.1371/journal.pone.0112542
43. Hussey PJ, Ketelaar T, Deeks MJ (2006) Control of the actin cytoskeleton in plant cell growth. Annu Rev Plant Biol 57:109–125. doi:10.1146/annurev.arplant.57.032905.105206
44. Url T, Höftberger M, Meindl U (1993) Cytochalasin B influences dictyosomal vesicle production and mophogenesis in the desmid *Euastrum*. J Phycol 29(5):667–674. doi:10.1111/j.0022-3646.1993.00667.x
45. Höftberger M, Lütz-Meindl U (1999) Septum formation in the desmid *Xanthidium* (Chlorophyta): effects of cytochalasin D and latruculin B suggest the involvement of actin microfilaments. J Phycol 35(4):768–777. doi:10.1046/j.1529-8817.1999.3540768.x
46. Foissner I, Wasteneys GO (2007) Wide-ranging effects of eight cytochalasins and latrunculin A and B on intracellular motility and actin filament reorganization in characean internodal cells. Plant Cell Physiol 48(4):585–597. doi:10.1093/pcp/pcm030
47. Gibbon BC, Kovar DR, Staiger CJ (1999) Latrunculin B has different effects on pollen germination and tube growth. Plant Cell 11(12):2349–2363
48. Scherlach K, Boettger D, Remme N et al (2010) The chemistry and biology of cytochalasans. Nat Prod Rep 27(6):869–886. doi:10.1039/b903913a
49. Bonder EM (1986) Cytochalasin B slows but does not prevent monomer addition at the barbed end of the actin filament. J Cell Biol 102(1):282–288. doi:10.1083/jcb.102.1.282
50. Aldridge DC, Armstrong JJ, Speake RN et al. (1967) The cytochalasins, a new class of biologically active mould metabolites. Chem. Commun. (London) (1): 26. doi: 10.1039/c19670000026
51. Rothweiler W, Tamm C (1966) Isolation and structure of Phomin. Experientia 22(11):750–752. doi:10.1007/BF01901360
52. Coué M, Brenner SL, Spector I et al (1987) Inhibition of actin polymerization by latrunculin A. FEBS Lett 213(2):316–318. doi:10.1016/0014-5793(87)81513-2
53. Spector I, Shochet N, Kashman Y et al (1983) Latrunculins: novel marine toxins that disrupt microfilament organization in cultured cells. Science 219(4584):493–495. doi:10.1126/science.6681676
54. Baluska F, Jasik J, Edelmann HG et al (2001) Latrunculin B-induced plant dwarfism: plant cell elongation is F-actin-dependent. Dev Biol 231(1):113–124. doi:10.1006/dbio.2000.0115
55. Holzinger A, Wasteneys GO, Lütz C (2007) Investigating cytoskeletal function in chloroplast protrusion formation in the arctic-alpine plant *Oxyria digyna*. Plant Biol (Stuttg) 9(3):400–410. doi:10.1055/s-2006-924727
56. Carmely S, Kashman Y (1985) Structure of swinholide-a, a new macrolide from the marine sponge *Theonella swinhoei*. Tetrahedron Lett 26(4):511–514. doi:10.1016/S0040-4039(00)61925-1
57. Bubb MR, Spector I, Bershadsky AD et al (1995) Swinholide A is a microfilament disrupting marine toxin that stabilizes actin dimers and severs actin filaments. J Biol Chem 270(8):3463–3466
58. Klenchin VA, King R, Tanaka J et al (2005) Structural basis of swinholide A binding to actin. Chem Biol 12(3):287–291. doi:10.1016/j.chembiol.2005.02.011
59. Posey SC, Bierer BE (1999) Actin stabilization by jasplakinolide enhances apoptosis induced by cytokine deprivation. J Biol Chem 274(7):4259–4265
60. Braet F, Spector I, de Zanger R et al (1998) A novel structure involved in the formation of liver endothelial cell fenestrae revealed by using the actin inhibitor misakinolide. Proc Natl Acad Sci U S A 95(23):13635–13640
61. Shurety W, Stewart NL, Stow JL (1998) Fluid-phase markers in the basolateral endocytic pathway accumulate in response to the actin assembly-promoting drug Jasplakinolide. Mol Biol Cell 9(4):957–975
62. Shaw MK, Tilney LG (1999) Induction of an acrosomal process in *Toxoplasma gondii*: visualization of actin filaments in a protozoan parasite. Proc Natl Acad Sci U S A 96(16):9095–9099
63. Sheahan MB, Staiger CJ, Rose RJ et al (2004) A green fluorescent protein fusion to actin-binding domain 2 of *Arabidopsis* fimbrin highlights new features of a dynamic actin cytoskeleton in live

plant cells. Plant Physiol 136(4):3968–3978. doi:10.1104/pp.104.049411

64. Voigt B, Timmers AC, Samaj J et al (2005) GFP-FABD2 fusion construct allows in vivo visualization of the dynamic actin cytoskeleton in all cells of *Arabidopsis* seedlings. Eur J Cell Biol 84(6):595–608. doi:10.1016/j.ejcb.2004.11.011
65. Meindl U (1990) Effects of temperature on cytomomorphogenesis and ultrastructure of *Micrasterias denticulata* Bréb. Protoplasma 157:3–18
66. Meindl U, Lancelle S, Hepler PK (1992) Vesicle production and fusion during lobe formation in *Micrasterias* visualized by high-pressure freeze fixation. Protoplasma 170(3-4):104–114. doi:10.1007/BF01378786
67. Holzinger A, de Ruijter N, Emons AM et al (1999) Spectrin-like proteins in green algae (Desmidiaceae). Cell Biol Int 23(5):335–344. doi:10.1006/cbir.1999.0365
68. Baluska F, Parker JS, Barlow PW (1992) Specific patterns of cortical and endoplasmic microtubules associated with cell growth and tissue differentiation in roots of maize (*Zea mays* L.). J Cell Sci 103(1):191–200
69. Chichili GR, Rodgers W (2007) Clustering of membrane raft proteins by the actin cytoskeleton. J Biol Chem 282(50):36682–36691. doi:10.1074/jbc.M702959200
70. de Marino S, Festa C, D'Auria MV et al (2011) Swinholide J, a potent cytotoxin from the marine sponge *Theonella swinhoei*. Mar Drugs 9(6):1133–1141. doi:10.3390/md9061133

Part III

Cell and Organelle Motility

Chapter 14

Quantitative Motion Analysis in Two and Three Dimensions

Deborah J. Wessels, Daniel F. Lusche, Spencer Kuhl, Amanda Scherer, Edward Voss, and David R. Soll

Abstract

This chapter describes 2D quantitative methods for motion analysis as well as 3D motion analysis and reconstruction methods. Emphasis is placed on the analysis of dynamic cell shape changes that occur through extension and retraction of force generating structures such as pseudopodia and lamellipodia. Quantitative analysis of these structures is an underutilized tool in the field of cell migration. Our intent, therefore, is to present methods that we developed in an effort to elucidate mechanisms of basic cell motility, directed cell motion during chemotaxis, and metastasis. We hope to demonstrate how application of these methods can more clearly define alterations in motility that arise due to specific mutations or disease and hence, suggest mechanisms or pathways involved in normal cell crawling and treatment strategies in the case of disease. In addition, we present a 4D tumorigenesis model for high-resolution analysis of cancer cells from cell lines and human cancer tissue in a 3D matrix. Use of this model led to the discovery of the coalescence of cancer cell aggregates and unique cell behaviors not seen in normal cells or normal tissue. Graphic illustrations to visually display and quantify cell shape are presented along with algorithms and formulae for calculating select 2D and 3D motion analysis parameters.

Key words Cell motility, 2D motion analysis, 3D reconstruction, Cell migration, 3D data, 4D tumorigenesis model

1 Introduction

The importance of cell migration in developmental biology, immunity, and disease has oft been repeated [1–4]. Indeed, it is becoming increasingly apparent that basic cell motility, that is, crawling in the absence of chemoattractant or other directionally stimulatory signals, may be an inherent cellular property [5]. Molecular mechanisms that drive cell motility have been studied extensively in the model system *Dictyostelium discoideum* [6–10] as well as in the so-called professional migratory cells such as neutrophils [1, 11–14] and macrophages [15, 16]. Considerable attention has also been focused on cell migration in the context of metastatic cancer [17–20] and understandably so since metastasis is the primary cause of death from this disease [21–23].

Ray H. Gavin (ed.), *Cytoskeleton: Methods and Protocols*, Methods in Molecular Biology, vol. 1365, DOI 10.1007/978-1-4939-3124-8_14, © Springer Science+Business Media New York 2016

Basic cell motility may be modulated by chemotactic and haptotactic signals [24]. Hence, the failure of certain mutants to respond to an exogenous signal may actually be due to a mutation in the basic motility apparatus rather than a mutation in a signal transduction pathway [25] and this important distinction can only be made using appropriate quantitative methods. In addition, different migratory behaviors are manifested on 1D [3], 2D [3, 26] and 3D surfaces [27] due to physical properties of the substrate [27, 28], interactions between the substrate components and cell adhesion receptors [29], and the activity of extracellular proteases such as the metalloproteinases that degrade or remodel the matrix [30]. Again, deciphering the underlying mechanisms responsible for these characteristics requires accurate quantitation. It likewise is important to know the specific motility defects of abnormally behaving immune cells [24, 31, 32] in order to accurately diagnose the disease and provide successful treatment. Finally, it should be noted that although transwell assays are widely used to assay motility and chemotactic defects [33–35], they may not distinguish between chemotaxis and chemokinesis, provide only a limited amount of all-or-none information and no quantitative information on cytoskeletal dynamics or mechanisms of cell migration.

The aforementioned points speak to the need for high resolution 2D and 3D quantitative methods in which cell shape changes mediated by pseudopod or lamellipod dynamics [24, 36, 37], invadopodia [38, 39], podosomes [40, 41] and/or filopodia [37, 42] are analyzed in order to distinguish normal from aberrant motility and normal chemotaxis or other forms of directed motility on one hand from directed crawling that has been dampened or obliterated by signal transduction deficiencies on the other. Here we will describe high-resolution 2D and 3D methods for motion analysis of individual cells. We include protocols for 2D motion analysis of basic cell motility as well as chemotaxis on a 2D substrate, 3D motion analysis of basic cell motility and reconstruction on a 2D substrate and finally, 3D motion analysis of basic cell motility and reconstruction in a 3D matrix (i.e., 4D). As an example of the validity of this approach, we will show how high-resolution quantitation of cell motility led to the discovery in the model system *Dictyostelium discoideum* that upregulation of *lpten*, a homolog of *ptenA* that is, in turn, an ortholog of the human *PTEN* mutated in many human cancers, rescues defects in *ptenA*$^{-}$ mutants, raising the possibility of a new stratagem for cancer therapy [43]. In addition, accurate quantitation of cell shape changes in basic cell motility as well as in chemotaxis has provided insights into human diseases [31, 44] and, as we show here, can be used to evaluate coalescence of cancer cells, cell-cell interactions within the tumor microenvironment as well as effects of potential cancer treatments [41].

2 Materials

2.1 Culturing Dictyostelium discoideum

1. *D. discoideum* wild-type and mutant strains, as well as many plasmids, can be obtained from the Dictybase Stock Center. (http://www.dictybase.org/StockCenter/StockCenter.html).
2. HL-5 nutrient growth media supplemented with the appropriate antibiotics for transformed strain selection (http://dictybase.org).
3. Amoebae maintained at the low-log phase of growth (~2×10^6 cells/mL in HL-5) (*see* **Note 1**).

2.2 Development of D. discoideum for Basic Cell Motility and Chemotaxis Assays

1. BSS: 20 mM KCl, 2.5 mM $MgCl_2$, 20 mM KH_2PO_4, 5 mM Na_2HPO_4 (pH 6.4), 0.34 mM streptomycin sulfate, filter sterilized [22, 23].
2. HAB040700 nitrocellulose filters (Millipore, Billerica, MA, USA).
3. Millipore support filters (catalog AP1004700; www.millipore.com).
4. Humidified incubation chamber at 22 °C.
5. cAMP: 1 mM stock solution of adenosine 3′,5′-cyclic monophosphate (cAMP; Sigma-Aldrich, Inc. http://www.sigmaaldrich.com) diluted in BSS and frozen until use.

2.3 Isolation of Primary Human PMNs for Basic Cell Motility and Chemotaxis Assays

1. 10 mL whole venous blood drawn from healthy donors using IRB-approved informed consent protocols into blood collection tubes coated with anticoagulant, preferably EDTA for use with Polymorphprep. Heparin-coated tubes are not recommended for use with Polymorphprep.
2. Polymorphprep Axis Shield Density Gradient Media (www.cosmobiousa.com) for separation of neutrophils from whole blood.
3. Hepes-buffered Hanks' balanced salt solution (H-HBSS) with 1.26 mM $CaCl_2$ and 0.49 mM $MgCl_2$ (Life Technologies, Carlsbad, CA), pH 7.4 and 10 mM Hepes.
4. Centrifuge capable of 400–900 × *g* forces at room temperature with the option of no-brake deceleration.
5. *N*-formyl peptide (fMLP): 500 μM stock solution of *N-formyl-met*-leu-phe (Sigma-Aldrich, Inc.) in DMSO frozen until use.

2.4 Culturing Cell Lines

1. Laboratory personnel who have contact with cells, tissue, and/or blood products, particularly those of human origin, should be trained in the proper use of personal protective equipment (PPE) and the safe handling of material that may harbor bloodborne pathogens (*see* **Note 2**).
2. Certified biological safety cabinet, biohazardous waste decontamination, and biohazardous waste disposal systems.

3. Cell culture media: Dulbecco's modified Eagle's medium (DMEM), RPMI 1640, (Life Technologies, Carlsbad, CA, http://www.lifetechnologies.com) or supplier's media recommendation.
4. Fetal bovine serum (FBS, Atlanta Biologicals https://www.atlantabio.com/) (*see* **Note 3**).
5. Penicillin/streptomycin mixture (Life Technologies, Carlsbad, CA, http://www.lifetechnologies.com).
6. DPBS (Dulbecco's phosphate-buffered saline; Life Technologies, Carlsbad, CA http://www.lifetechnologies.com) without magnesium and without calcium for cell dissociation: 2.67 mM potassium chloride (KCl), 1.47 mM potassium phosphate monobasic (KH_2PO_4), sodium chloride (NaCl), and sodium phosphate dibasic ($Na_2HPO_4 \cdot 7H_2O$).
7. Trypsin-0.25 % EDTA with phenol red (Life Technologies, Carlsbad, CA, http://www.lifetechnologies.com).
8. CO_2 incubator at 37 °C and 5 % CO_2 (http://www.thermofisher.com/en/home.html).
9. Tissue culture flasks for cell culturing (http://www.usascientific.com/).

2.5 Culturing Cells from Tissue

1. DMEM/F12, M199, and/or RPMI 1640 (Life Technologies, Carlsbad, CA, http://www.lifetechnologies.com).
2. Horse serum and fetal calf serum (Life Technologies, Carlsbad, CA, http://www.lifetechnologies.com).
3. Epidermal growth factor human (EGF) (Sigma-Aldrich, St. Louis, MO, http://www.sigmaaldrich.com).
4. Hydrocortisone (Sigma-Aldrich, St. Louis, MO, http://www.sigmaaldrich.com).
5. Insulin (Sigma-Aldrich, St. Louis, MO).
6. Cholera toxin (Sigma-Aldrich, St. Louis, MO).
7. Penicillin/streptomycin (Gemini Bio-products).
8. 6-Well plates (Fisher Scientific, http://www.fishersci.com).
9. Sterile scissors, scalpels, and forceps.
10. CO_2 incubator at 37 °C and 5 % CO_2.

2.6 Preparation of Matrigel Cultures in Petri Dishes

1. Aliquoted Matrigel matrix (Becton Dickinson Bioscience, Franklin Lakes, NJ) stored at −20 °C (*see* **Note 4**).
2. Chilled pipette tips.
3. Chilled 35 mm and/or 65 mm plastic Petri dishes with glass insert in dish bottom (InVitro Scientific, www.invitrosci.com).

4. For 3D experiments, the Petri dishes may have to be modified. Modifications require sharp scissors sterilized in 100 % ethanol to trim the height of the dish bottom (*see* **Note 5**).
5. Modified Petri dish lid for use with DIC optics in 3D experiments (*see* **Note 6**).

2.7 2D Experiment on 2D substrate

1. Upright or inverted microscope (depending on the chamber used) equipped with phase or bright-field optics and 10–40× objectives.
2. When working with cell lines, tissues, and/or long-term experiments with immune cells isolated from freshly drawn blood, the microscope should be fitted with environmental controls or contained within a CO_2 incubator at 37 °C and 5 % CO_2.
3. Sony XCD-V50 1/3-type Progressive Scan IT CCD camera (or comparable) with a IEEE1394B high-speed digital interface that operates at 60 fps and 640×480 resolution, with C-mount and controlled by FireWire compatible software such as FireI (www.unibrain.com), Adobe Premiere®, QuickTime Pro®, iStopMotion, or iMovie.
4. Sykes-Moore perfusion chamber (Bellco Glass, Inc., http://www.bellcoglass.com) with 2.5 mm gasket, holder, and 25 mm #2 cover slips coupled to programmable pumps to assay basic cell motility, to assay response to global exposure to chemoattractant or to generate a dynamic gradient of temporal waves [46–48] (Fig. 1a).
5. For perfusion with the Sykes-Moore chamber, two pieces of Tygon tubing (1/16 in. ID), each connected to a 21-gauge needle via a luer lock for inlet and outlet ports and a 60 cm^3 syringe connected to the other end of the inlet tubing via a luer fitting (Fig. 1a, b).
6. For perfusion experiments, NE-1000 Multiphase Programmable Pumps (New Era Pump Systems, Farmingdale, NY) [49, 50] or other continuous pump system that does not introduce oscillations (Fig. 1b).
7. For quantifying the response of single cells to chemoattractant: Dunn chamber (http://www.hawksley.co.uk/), Zigmond chamber (Fig. 1c, d). (http://www.neuroprobe.com/products/zigmond.html), or modified Zigmond chamber [44, 51].
8. Single cells can also be imaged in gradients of chemoattractant generated in a microfluidic device [10, 52–54]. Various perfusion and chemotaxis chambers are also available from Ibidi (www.ibidi.com).
9. Coated cover slip within a Petri dish for mammalian cells (Fig. 1e).
10. An environmental chamber if pH and temperature need to be regulated; alternatively, microscope can be housed within a 37 °C, 5 % CO_2 incubator (Fig. 1f).

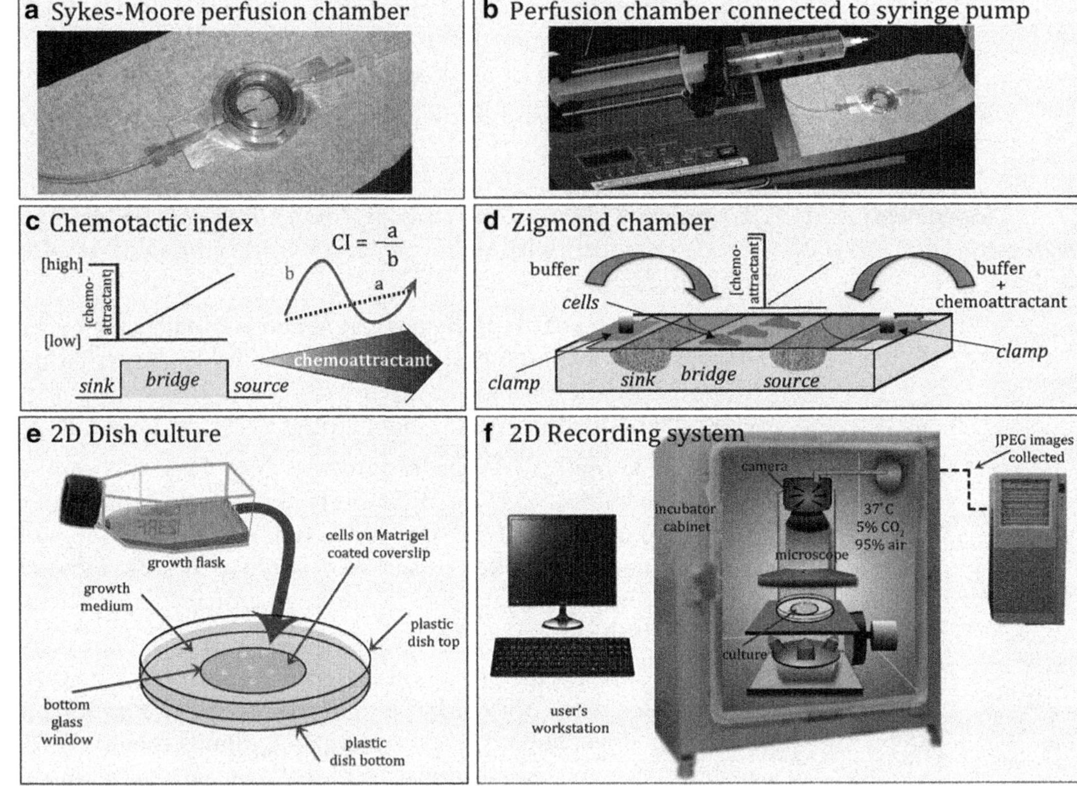

Fig. 1 Methods for 2D motion analysis of cells crawling on a 2D substrate. (*a*) The assembled Sykes-Moore perfusion chamber with inlet and out tubes inserted into their respective ports. (*b*) The Sykes-Moore perfusion chamber connected to a 60 cm^3 syringe containing the perfusion solution and placed in a programmable syringe pump. (*c*) Diagram of the gradient of chemoattractant that forms by diffusion across the bridge of the Zigmond chamber and the method for calculating the chemotactic index (CI). (*d*) Diagram of the Zigmond chemotaxis chamber. Cells are attached to the cover slip that is inverted over the bridge. Buffer is placed in the "sink" trough and buffer plus chemoattractant in the "source" trough. (*e*) Cells can be spread onto a thin coat of Matrigel or other matrix cast onto the glass window (cover slip) inserted into the bottom of a Petri dish for 2D analysis of mammalian cells crawling on a 2D substrate. (*f*) In cases requiring environmental regulation, image acquisition can be performed on a microscope with camera housed in an incubator

11. Dedicated acquisition computer capable of storing hundreds of gigabytes of data.
12. Motion analysis software such as ImageJ (http://imagej.nih.gov/ij/) or J3D-DIAS4.1 (*see* **Note 7**).

2.8 3D Experiment on a 2D substrate

1. Short-working-distance perfusion chamber such as the one available from Harvard Apparatus (http://www.harvardapparatus.com) or from Life Imaging Services (http://www.lis.ch/download/LIS_Ludin_Chamber_v8.pdf).
2. Microscope equipped with differential interference (DIC) optics, high numerical aperture objectives, high numerical aperture condenser, and motorized stage capable of moving

the focus through 10–20 μm in 2–3 s every 4–5 s (*see* **Note 8**) for rapidly moving cells such as *D. discoideum* and PMNs.

3. Environmental chambers are required to maintain temperature and pH for long-term imaging of mammalian cells. Alternatively, a microscope may be housed within a 37 °C, 5 % CO_2 incubator (*see* **Note 9**).
4. Digital camera controlled by video acquisition software such as the Zeiss AxioCam MRc5 and Laser Vision 6 software as described in **item 3**, Subheading 2.7, synchronized with the LED light engine and motorized stage.
5. A computer capable of running Microsoft Windows XP, equipped with serial ports (RS-232) to control the microscope *Z*-axis stepper motor, a firewire port (IEEE 1394) to capture images from the AxioCam MRc5 camera, and a storage disk capable of storing hundreds of gigabytes of data.
6. Software for 3D reconstruction (*see* **Note 7**).

2.9 3D Experiment on a 3D substrate

1. Cell cultures embedded in Matrigel in modified Petri dishes (Fig. 2a) as described in Subheading 2.6, **Notes 5** and **6**.
2. Microscope equipped with differential interference contrast (DIC) optics, programmable motorized stage, 20× or 40× long-working-distance objectives and LED light engine (*see* **Note 10**) contained within a 37 °C, 5 % CO_2 incubator or other means of regulating temperature and pH for 30 days or more of continual imaging (Fig. 2b).
3. Digital camera controlled by video acquisition software such as the Zeiss AxioCam MRc5 and Laser Vision 6 software, synchronized with the LED light engine and motorized stage.
4. A computer capable of running Microsoft Windows XP, equipped with serial ports (RS-232) to control the microscope *Z*-axis stepper motor, a firewire port (IEEE 1394) to capture images from the AxioCam MRc5 camera, and a storage disk capable of storing hundreds of gigabytes of data.
5. Software for 3D reconstruction (*see* **Note 7**).

3 Methods

3.1 Preparation of D. discoideum amoebae for 2D Analysis of Basic Motile Behavior and Chemotaxis

1. Harvest 5×10^7 cells in the low log phase of growth (~2×10^6 cells/mL), wash three times in BSS, and disperse on a nitrocellulose filter supported by two Millipore pre-filters pre-saturated with BSS. Incubate in a humidified chamber at 22 °C until the onset of aggregation [55] when motility and chemotactic responsiveness peak [55, 56]. Wash cells from the filter pads and dilute to 3×10^4 cells/mL (*see* **Note 10**) for inoculation into chamber.

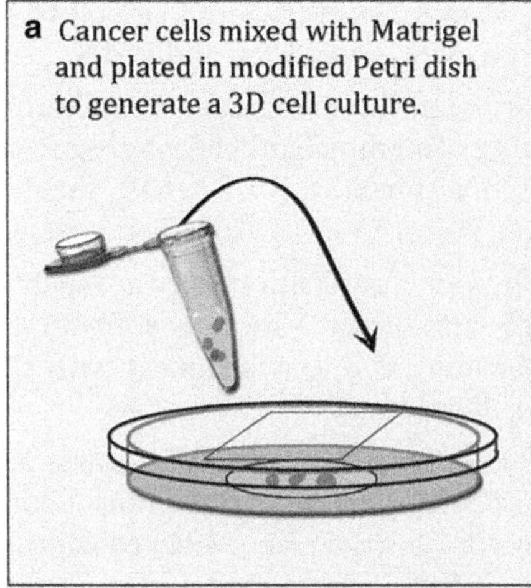

a Cancer cells mixed with Matrigel and plated in modified Petri dish to generate a 3D cell culture.

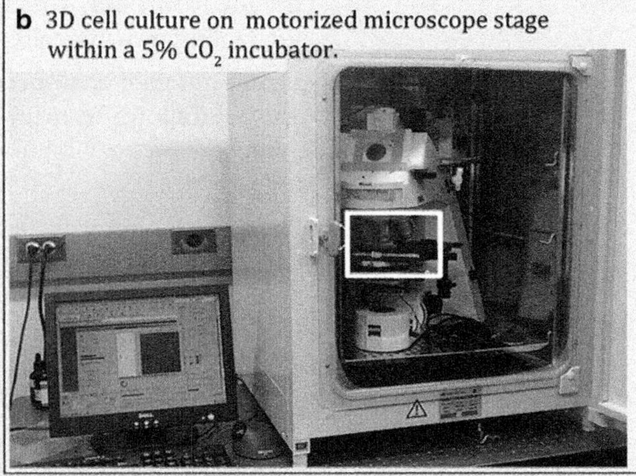

b 3D cell culture on motorized microscope stage within a 5% CO2 incubator.

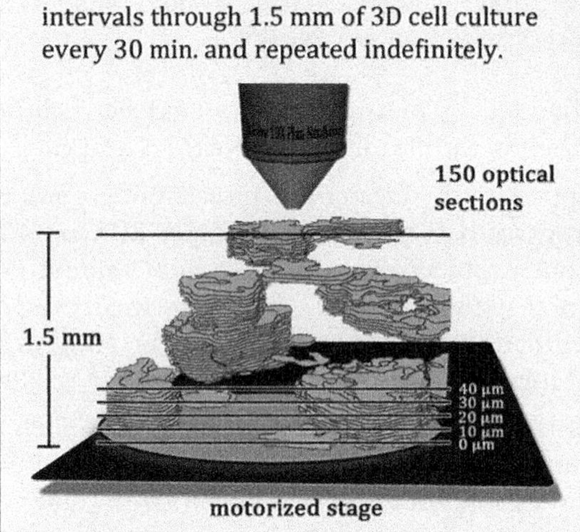

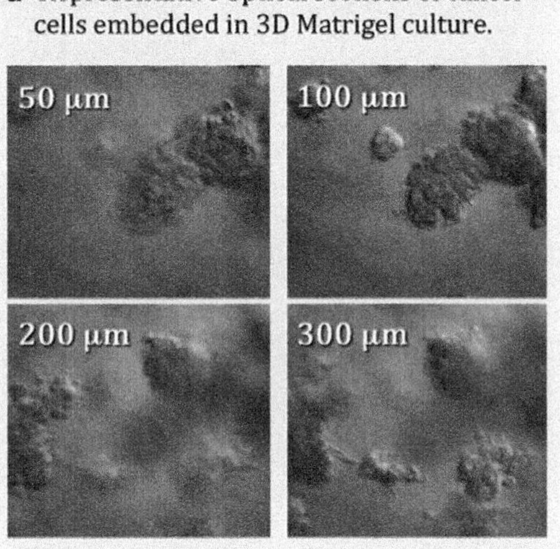

Fig. 2 Methods for 3D motion analysis of cells crawling on a 3D substrate. (*a*) Modified Petri dish for 3D analysis of cells embedded in a 3D matrix. Cells are mixed with Matrigel and plated onto a glass window in the bottom section of the Petri dish that has been trimmed to fit within the allowable working distance of the microscope. A glass window in the lid is necessary to achieve optical sections using DIC microscopy. (*b*) Microscope fitted with DIC optics and a motorized stage housed inside a 5 % CO_2, 37 °C incubator for long-term imaging of living cells in 3D cultures. Camera and motor are synchronized by controlling software. A timer regulates the LED light so that the specimen is only illuminated during the 45 s in which the z-series is actively being acquired. The *white box* indicates the area of the stage that is diagrammed at a larger scale in (*c*). (*c*) Diagrammatical summary of optical sectioning using DIC microscopy. A typical z-series of 150 optical sections is acquired through 1.5 mm of Matrigel matrix at 5–10 μm increments. (*d*) Representative optical sections through a culture of aggregating cancer cells embedded in a 3D Matrigel matrix. The level of each section is given in the upper left hand corner of each panel

2. Mutant *Dictyostelium* strains may be defective in cAMP signaling and cAMP induced gene expression, and therefore require exogenously applied pulses of cAMP in order to become motile and/or chemoresponsive [48]. To exogenously pulse cells,

harvest *D. discoideum* amoebae at the low log phase of growth, wash in BSS and resuspend at a density of 2×10^7 cells/mL in BSS. Maintain the flask containing the cell suspension on a rotary shaker at 180 rpm for 1 h. Cells are then pulsed by programming the NE-1000 Pumps to deliver 80 nM cAMP (final) from a 10 cm^3 syringe every 6 min for 6 subsequent hours [43, 48, 57]. At the completion of pulsing, cells are pelleted, washed in BSS, and diluted to 3×10^4 cells/mL for inoculation into a chamber.

3. For analysis of basic motile behavior, inoculate 1.1 mL of the dilute cell suspension into a Sykes-Moore perfusion chamber. Allow cells to adhere to the cover slip for 5–10 min, seal the chamber, and insert inlet and outlet needles into the respective ports. Connect inlet tube to a 60 cm^3 syringe filled with BSS and insert the syringe into a pump programmed to deliver 4 mL/min (Fig. 1a; *see* **Notes 10** and **11**).
4. For analysis of chemotaxis, streak 20 μL of the dilute cell suspension across the center of a clean 24×30 mm cover slip and allow 5 min for the cells to adhere. Invert the cover slip so that the cells contact the bridge then lightly fasten the clamps. Carefully pipette 55–60 μL of BSS into one trough (sink) and 55–60 μL of 10^{-6} M cAMP (source) into the other trough. Chamber assembly is diagrammed in Fig. 1d.

3.2 Preparation of Human Primary Polymorphonuclear Neutrophils (PMNs) for 2D Analysis of Basic Motile Behavior and Chemotaxis on a 2D Substrate

1. Whole venous blood is drawn into a blood collection tube coated with anticoagulant. The tube is gently inverted several times, brought to room temperature by gently rocking for about 30 min and 5 mL of blood is layered onto the top of 5 mL of Polymorphprep, also at room temperature, in a 15 mL Falcon centrifuge tube according to the manufacturer's instructions [49].
2. Centrifuge at room temperature for 30–35 min at $500 \times g$ with no brake.
3. The PMN band is collected, washed in H-HBSS by centrifugation at $400 \times g$ for 10 min, and resuspended to a concentration of 1.5×10^6 cells/mL in H-HBSS. Brake can be used during washing steps.
4. Inoculate into the Sykes-Moore perfusion chamber for motion analysis of basic motile behavior as described in Subheading 3.1, **step 3**, using H-HBSS as the perfusion buffer (Fig. 1a, b).
5. Inoculate into a chemotaxis chamber for motion analysis of chemotaxis as described in Subheading 3.1, **step 4**, using H-HBSS as the sink buffer and H-HBSS plus 0.1 μM fMLP as the chemoattractant in the source trough as illustrated in Fig. 1d.

3.3 Preparation of Cell Lines for 2D Analysis of Basic Motile Behavior on a 2D Substrate

1. Using a sterile, pre-chilled pipet tip, distribute 50 μL/cm^2 of ice-cold Matrigel onto the glass insert of a pre-chilled Petri dish, taking care not to introduce bubbles.
2. Incubate the Petri dish for 30 min at 37 °C.
3. Dissociate cells from tissue culture flask, resuspend in media, wash, resuspend in fresh media, and count.
4. Withdraw the volume required to yield 5×10^5 cells, adjust to 100 μL, and spread onto the pre-coated, gelled Matrigel coat.
5. Incubate for 30 min to allow cells to attach.
6. Add 1–3 mL of appropriate media to the dish.
7. Place preparation on the stage of the 2D microscope and begin image acquisition.

3.4 Image Capture and Outlining in 2D

1. For *D. discoideum* or for short-term recording of PMNs, place cells in the appropriate chamber on the stage of microscope equipped with phase optics and a camera controlled by software. Set the time interval and duration of experiment.
2. For mammalian cells, place the chamber on a microscope within a 37 °C, CO_2 incubator or in an environmental chamber.
3. Images are saved as a series of sequential JPEG files that can be opened in J3D-DIAS4.1 and saved in native format (Fig. 3a).
4. Imaging processing can facilitate automatic outlining (Fig. 3b) (*see* **Note 12**).
5. The edge of cells can be automatically detected in each frame employing a grayscale threshold algorithm (Fig. 3c).
6. The perimeter is converted to a mathematically precise beta-spline replacement image [58–61] and the centroid, or geometric center [59–61], at each time point (Fig. 3d) is determined (*see* **Note 13**).

3.5 2D Cell Movement and Shape Parameters

1. Cell motility parameters can be computed over time based on the changing position of the centroid [59–62] (Table 1a; Figs. 4d and 5b) and cell shape parameters can be quantified from the beta spline replacement images (Table 1b; *see* **Notes 14–16**).
2. The "chemotactic index" is computed as the net distance moved towards the source of chemoattractant divided by the total distance the cell migrated [63, 64] (Fig. 1c).
3. Cell shape changes can be evaluated by "unwrapping" and stacking. (Fig. 3b) Polar coordinates are converted to rectangular coordinates and stacked with each unwrapped perimeter having a lower offset than the previous one. This display accentuates the position and frequency of protrusions over time and therefore is particularly useful in determining cytoskeletal contributions to lamellipodial behavior [24, 36]

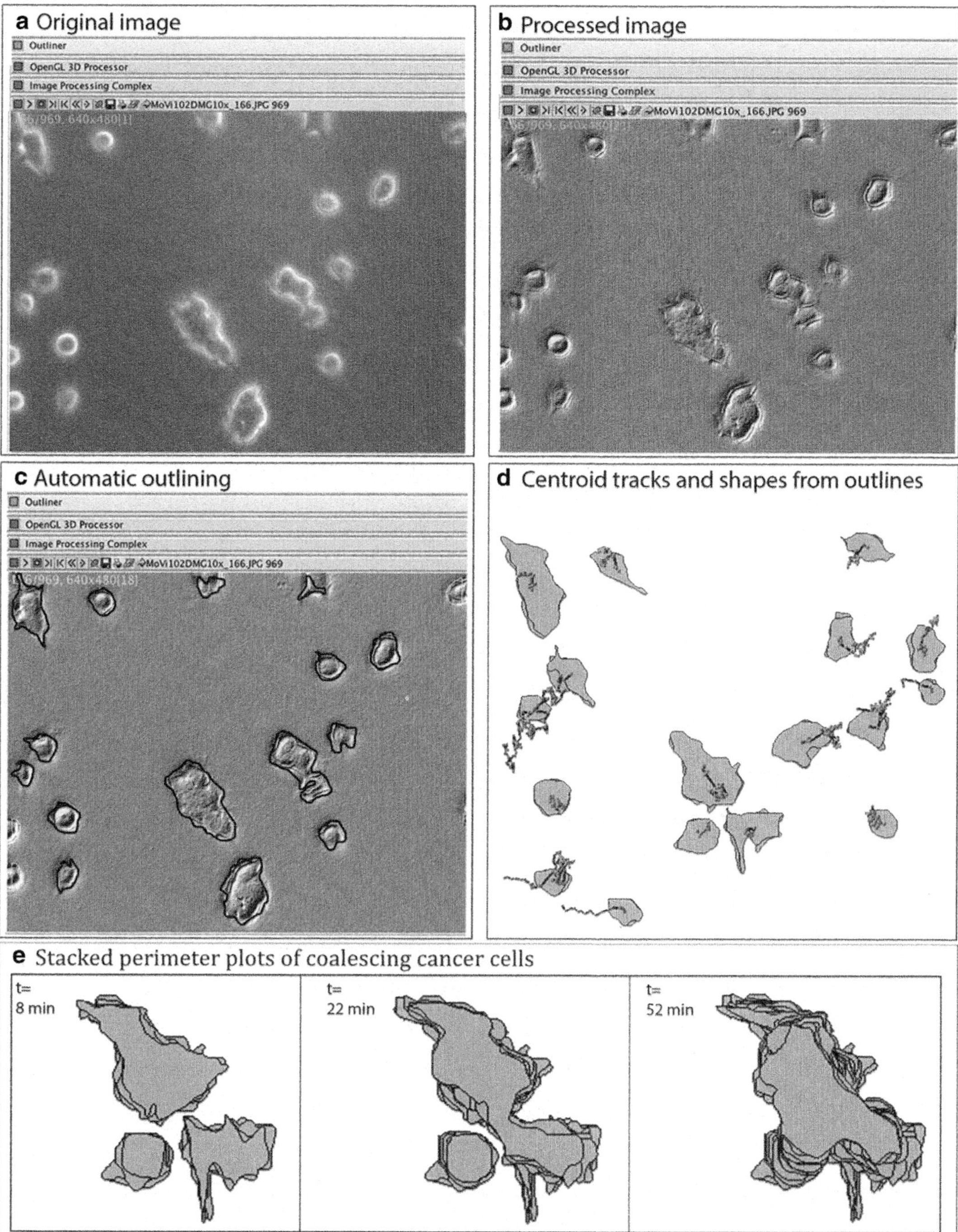

Fig. 3 2D automatic outlining, path and shape generation in cancer cells crawling on top of a 2D, thin coat of matrix. (*a*) Original phase image. (*b*) The image can be processed to facilitate automatic outlining. (*c*) Automatic outlining. (*d*) Centroid tracks and shapes generated from the outlines. (*e*) Stacked perimeter plots of coalescing cancer cells. Time is indicated in the *upper left hand corner*

Table 1
2D motility and shape parameters

A. Centroid-based 2D motility parameters														
Object	Frames	Time (min)	Total elapsed time (min)	Total path length (μm)	Net path length (μm)	Direct. total	Direct. upward	Direct. right	Speed (μm/min)	Direction (deg)	Direction change (deg)	Accel. (μm/sq min)	Persist. (μm/min deg)	Axis tilt (deg)
1	1–33	0.000–64.000	64	68.63	8.48	0.12	0.04	0.12	0.80	228.2	48.0	0.14	0.15	28.0
2	1–52	0.000–102.000	102	110.66	15.56	0.14	−0.08	0.12	0.69	297.7	40.8	0.19	0.13	333.0
3	1–102	0.000–202.000	202	137.61	10.80	0.08	−0.02	0.08	0.42	225.8	48.9	0.13	0.09	329.0
4	1–20	0.000–38.000	38	32.86	18.88	0.58	0.40	−0.41	0.76	135.5	70.5	0.06	0.22	339.6
5	1–16	0.000–30.000	30	21.19	3.78	0.18	−0.04	0.18	0.58	259.1	54.2	0.06	0.11	331.0
6	1–102	0.000–202.000	202	287.43	57.39	0.20	0.17	0.10	0.99	203.8	51.7	0.21	0.17	347.5
7	1–16	0.000–30.000	30	36.69	23.98	0.65	−0.19	0.63	1.15	292.1	74.3	0.07	0.31	337.1
8	1–102	0.000–202.000	202	237.02	36.95	0.16	0.14	0.08	0.83	210.5	50.0	0.18	0.16	342.5
9	1–16	0.000–30.000	30	61.53	13.43	0.22	−0.18	0.12	1.67	282.9	58.1	0.22	0.48	20.9
10	1–102	0.000–202.000	202	199.59	17.87	0.09	−0.05	−0.08	0.74	148.4	49.6	0.13	0.16	332.6

11	1–73	0.000–144.000	144	165.17	14.98	0.09	0.07	−0.06	0.80	49.1	49.4	0.17	0.16	319.3
12	1–102	0.000–202.000	202	165.93	8.74	0.05	0.05	−0.02	0.57	142.0	50.7	0.13	0.12	318.7
13	1–102	0.000–202.000	202	171.37	18.15	0.11	−0.09	0.06	0.67	311.3	59.9	0.09	0.16	318.6
14	1–11	0.000–20.000	20	32.73	6.56	0.20	−0.14	0.15	0.93	276.8	38.4	0.41	0.23	345.2
15	1–26	0.000–50.000	50	49.23	13.43	0.27	−0.21	−0.17	0.65	41.4	44.0	0.16	0.10	326.4
16	1–82	0.000–162.000	162	161.33	36.84	0.23	0.00	0.23	0.84	284.2	60.9	0.08	0.22	342.4
17	1–39	0.000–76.000	76	62.32	47.64	0.76	−0.07	0.76	0.74	274.8	68.1	0.05	0.20	312.5
19	21–102	40.000–202.000	162	90.32	24.27	0.27	−0.23	−0.14	0.41	21.7	55.0	0.07	0.07	7.1
20	25–102	48.000–202.000	154	110.23	12.85	0.12	−0.08	−0.08	0.52	49.1	56.2	0.10	0.13	316.1
21	25–102	48.000–202.000	154	178.85	12.32	0.07	0.07	−0.01	0.95	143.3	61.8	0.11	0.24	355.9
23	52–102	102.000–202.000	100	88.20	5.06	0.06	0.05	−0.02	0.73	9.5	61.1	0.08	0.22	326.8
26	74–102	146.000–202.000	56	44.57	2.44	0.06	0.00	0.06	0.58	301.9	52.5	0.11	0.16	331.0
			MEAN	114.2	18.7	0.2	0.0	0.1	0.8	190.4	54.7	0.1	0.2	289.1
			± S.D.	73.6	14.3	0.2	0.1	0.2	0.3	102.7	9.1	0.1	0.1	110.6

(continued)

Table 1
(continued)

B. 2D shape parameters

Object	Frames	Time (min)	Total elapsed time (min)	Maximum length (μm)	Mean width (μm)	Area (sq μm)	Per. (μm)	Rnd (%)	Mean radial length (μm)	Radial deviation (%)	Mean Conv. (deg)	Mean Conc. (deg)	Pos. flow (%)	Neg. flow (%)
1	1–33	0.000–64.000	64	41.8	19.6	823.4	119.9	71.9	16.4	18.9	725.5	365.5	12.8	13.0
2	1–52	0.000–102.000	102	69.2	21.6	1493.9	189.9	52.3	23.3	31.2	1058.0	698.0	9.4	10.5
3	1–102	0.000–202.000	202	28.1	12.4	344.1	75.1	78.1	10.8	18.8	518.5	1585	10.4	11.8
4	1–20	0.000–38.000	38	31.9	13.9	439.5	87.2	75.1	12.3	19.0	563.4	203.4	12.7	15.2
5	1–16	0.000–30.000	30	35.2	18.3	640.4	101.9	77.7	14.4	17.1	601.8	241.8	9.9	12.0
6	1–102	0.000–202.000	202	32.9	13.9	451.8	90.2	72.2	12.4	19.7	620.2	260.2	18.2	20.5
7	1–16	0.000–30.000	30	22.8	10.7	243.4	61.9	81.9	9.0	17.0	473.7	113.7	20.3	19.4
8	1–102	0.000–202.000	202	23.6	10.7	254.2	63.6	79.8	9.2	18.3	493.2	136.8	20.3	25.1
9	1–16	0.000–30.000	30	35.7	16.0	571.1	100.5	71.4	13.8	20.0	625.3	265.3	20.8	24.2
10	1–102	0.000–202.000	202	54.9	20.4	1117.9	148.8	65.8	19.7	25.1	819.5	466.7	9.9	10.7
11	1–73	0.000–144.000	144	88.2	24.5	2152.7	236.4	49.1	28.8	33.6	1117.7	762.7	9.9	9.7

12	1–102	0.000–202.000	202	27.7	14.1	394.2	78.8	78.8	11.3	15.4	565.4	205.4	14.5	16.8
13	1–102	0.000–202.000	202	24.8	12.2	307.0	67.4	84.2	10.0	15.7	466.2	106.2	15.3	18.0
14	1–11	0.000–20.000	20	51.2	16.9	862.4	165.0	41.5	17.5	30.2	1015.5	655.5	13.6	15.2
15	1–26	0.000–50.000	50	25.7	15.0	387.5	72.6	92.0	11.1	9.2	440.0	80.0	12.6	17.3
16	1–82	0.000–162.000	162	26.8	11.8	314.3	74.3	75.7	10.3	19.5	567.8	207.8	17.7	20.4
17	1–39	0.000–76.000	76	24.2	11.8	286.0	66.7	81.2	9.6	17.1	474.0	114.0	14.3	17.5
19	21–102	40.000–202.000	162	35.6	17.5	621.8	98.5	81.4	14.2	17.5	585.9	225.9	7.3	7.8
20	25–102	48.000–202.000	154	29.5	12.5	373.5	78.3	76.3	11.2	20.1	520.8	160.8	12.3	12.9
21	25–102	48.000–202.000	154	27.5	12.9	357.8	76.1	77.1	10.8	17.9	553.8	193.8	17.8	18.9
23	52–102	102.000–202.000	100	25.5	13.1	337.1	70.8	83.9	10.5	14.1	503.6	143.6	14.9	17.0
26	74–102	146.000–202.000	56	152.7	26.3	4017.7	395.0	32.4	44.0	51.1	1599.7	1239.7	5.2	6.1
			MEAN	41.6	15.7	763.3	114.5	71.8	15.0	21.2	677.7	318.4	13.6	15.5
			±S.D.	29.7	4.4	861.5	77.4	14.9	8.1	8.8	284.3	284.7	4.3	5.0

4. Stacked perimeter plots (Figs. 3e, 4c and 5a) are another means of displaying dynamic cell shapes over time and are useful in evaluating the frequency of lateral pseudopod extension and directional persistence, especially in the comparison of mutant or defective cells with their normal counterparts (Figs. 4d and 5b) [10, 65, 66].
5. Cytoskeletal defects can often be detected in measurements of membrane protrusions [24, 36]. This information can be visualized quickly and effectively on large numbers of cells through the use of "difference pictures" [60, 61, 63] (Fig. 4a). Difference pictures are constructed by overlaying the outline of the cell at frame *n-x* with the outline of the current frame *n* where *x* is the user-specified interval. Expansion zones are regions into which the cell expanded during the specified interval (black areas in Fig. 4a) and retraction zones are regions from which the cell retracted (hatched areas in Fig. 4a). Areas shared in the two images are color-coded gray.
6. The area contained within expansion and retraction zones as determined from differencing can be quantified in J3D-DIAS4.1. This measurement can be used to determine if cytoskeletal mutants are defective in force generation required to extend lamellipodia or pseudopod or to retract the uropod [67–69]. The user windows the localized expansion or retraction zone from a series of difference pictures (Fig. 4a). The percent of the total area within the window is then calculated.

3.6 Sample Preparation for 3D Analysis of 3D Culture

1. Modify the Petri dish for use with DIC optics and to fit within the allowable working distance of the microscope (*see* **Notes 6** and **7**).
2. Using a sterile, pre-chilled pipet tip, apply 100 μL of ice cold Matrigel onto the 30 mm glass insert of a pre-chilled 65 mm Petri dish, taking care not to introduce bubbles.
3. Incubate the Petri dish for 30 min at 37 °C.

Fig. 4 (continued) (*black*) and retraction zones (*hatched*) can be visualized in difference pictures. Relative flow is measured from a difference picture by windowing the area of interest (*boxes*) at 0, 4, and 14 min. The percent of the total area contained within the window is calculated and displayed alongside the window. (*b*) Unwrapping and stacking the cell perimeter over time provides a history of expansions and can be used to detect cyclical shape changes. (*c*) Comparison of stacked perimeter plots of control, *ptenA*$^-$, *lpten*$^-$, and *ptenA*$^-$ cells overexpressing *lpten* (*ptenA*$^-$/*lpten*^{oe}) in the model system *D. discoideum* reveal significant defects in instantaneous velocity in *ptenA*$^-$ cells and increased lateral pseudopod extension in *lpten*$^-$ cells [43]. The final shape in the time series of the stacked perimeter plots is shaded *dark gray*. Interestingly, overexpression of *lpten* in *ptenA*$^-$ background rescues the velocity and shape defects in *ptenA*$^-$, conclusions confirmed by quantitative parameters (*d*). (*e*) 3D reconstructions of control and *lpten*$^-$ cells migrating on a 2D substrate reveal that the majority of lateral pseudopods extended by *lpten*$^-$ cells (*arrows*) are off the surface and therefore do not produce turns [71]

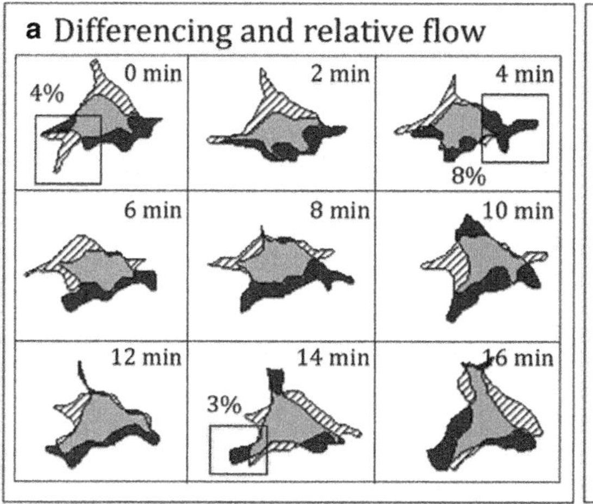

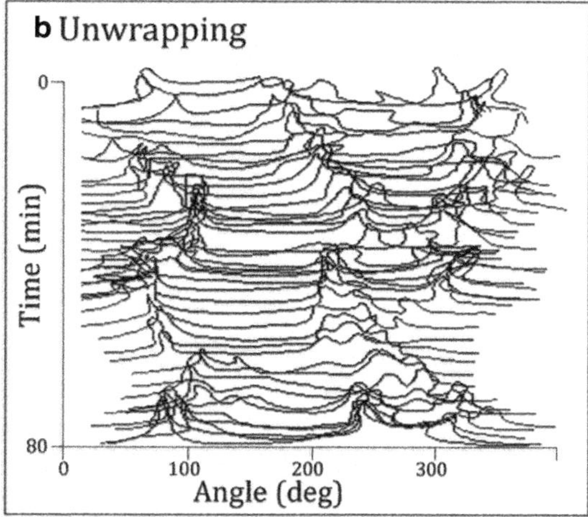

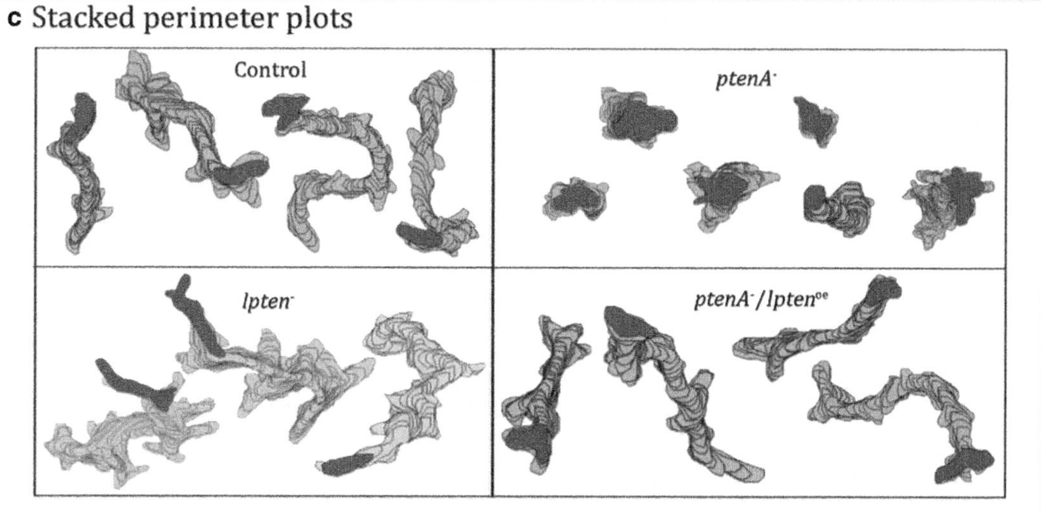

d Sample parameters

Cell type	Inst. Vel. (μm /min)	Percent cells ≥ 9 μm/min	Percent motile cells	Lateral pseudopods per 10 min
Control	8.3 ± 3.6	43%	83%	5.3 ± 1.5
ptenA⁻	3.6 ± 0.7	0%	21%	9.5 ± 2.2
lpten⁻	7.7 ± 2.7	25%	98%	12.4 ± 2.9
ptenA⁻/*lpten*ᵒᵉ	7.5 ± 2.1	21%	90%	5.8 ± 1.7

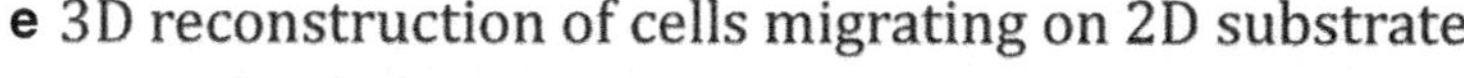

e 3D reconstruction of cells migrating on 2D substrate

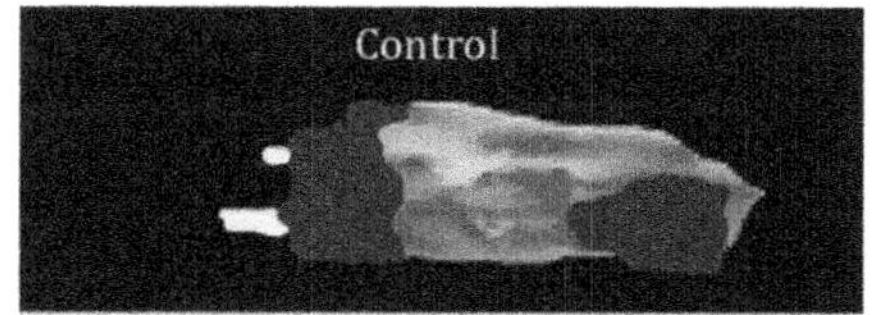

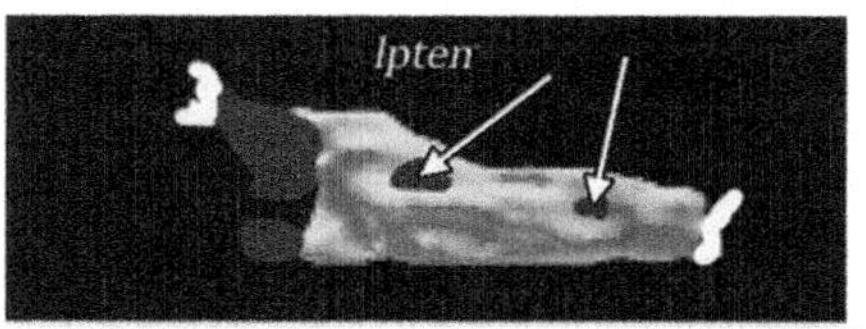

Fig. 4 2D cell shape parameters available in J3D-DIAS4.1 provide dynamical information. (*a*) Difference pictures at 2 min intervals of a cell from a human cancer cell line migrating on a 2D Matrigel substrate. Expansion zones

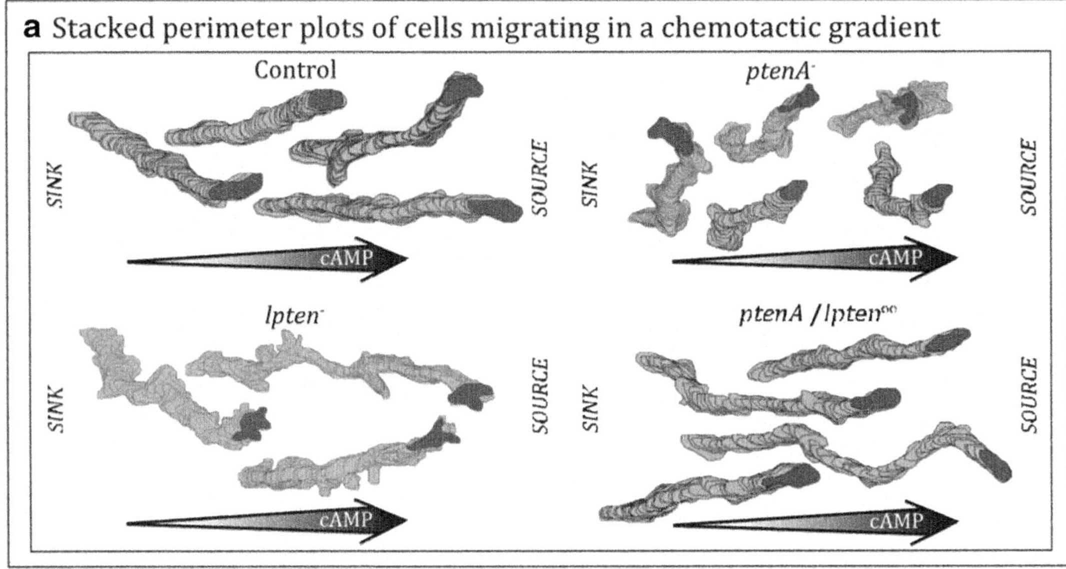

b Motion analysis parameters

Cell type	Inst. Vel. (μm /min)	Percent cells ≥ 9 μm/min	Chem. Index (C.I.)	Percent Positive C.I.	Percent motile cells	Lateral pseudopods per 10 min
Control	9.0 ± 4.0	41%	+0.56 ± 0.22	100%	93%	1.0 ± 0.7
ptenA⁻	5.3 ± 1.5	0%	+0.18 ± 0.26	80%	72%	3.8 ± 1.7
lpten⁻	9.8 ± 3.0	51%	+0.63 ± 0.33	99%	100%	5.2 ± 1.4
ptenA⁻ /*lpten*^oe	7.6 ± 2.4	27%	+0.66 ± 0.16	100%	96%	1.1 ± 0.6

Fig. 5 Comparison of 2D cell shape and motion parameters in control, *ptenA*⁻, *lpten*⁻, and *ptenA*⁻/*lpten*^oe strains in *D. discoideum* demonstrates that cells from all strains, with the exception of *ptenA*⁻, are capable of efficient chemotaxis. (*a*) Stacked perimeter plots of control, *ptenA*⁻, *lpten*⁻, and *ptenA*⁻/*lpten*^oe migrating in a Zigmond chamber in the presence of a spatial gradient of cAMP. The source of the gradient is to the *right*, as indicated by the direction of the *large arrow* at the *bottom* of *each panel*. The final shape in the time series of the stacked perimeter plots is shaded *dark gray*. (*b*) Sample motion analysis parameters computed by J3D-DIAS4.1 demonstrate that *ptenA*⁻ defects in velocity and later pseudopod suppression in a spatial gradient of cAMP are rescued by over-expression of *lpten* in *ptenA*⁻ background

4. Dissociate cells from tissue culture flask, resuspend in media, wash, resuspend in fresh media, and count.
5. Withdraw the volume required to yield 5×10^5 cells, bring the total volume up to 250 μL and chill.
6. Add 500 μL of ice cold Matrigel to the 100 μL of chilled cell suspension and mix by gently pipetting up and down. Distribute the entire 750 μL onto the Matrigel-coated cover slip (*see* Fig. 2a).

7. Incubate for 30 min to allow gelation.
8. Add 2–3 mL of media to the dish.
9. Place preparation on the stage of the 3D microscope and begin image acquisition (Fig. 2b).

3.7 Optical Sectioning

1. For 3D analysis of basic motile behavior on a 2D substrate, samples are inoculated into a perfusion chamber. The chamber is connected to a perfusion pump and positioned on the stage of an inverted microscope equipped with DIC optics and a high numerical aperture objective such as the Zeiss 63× Planapochromat.
2. To obtain optical sections of single cells such as *D. discoideum* amoebae or human PMNs migrating on a 2D substrate, program the motor to move through the *z*-axis height, generally 10–20 μm, in 2–3 s and repeat the process every 4–5 s for 5–10 min per cell (Fig. 4e) (*see* **Note 17**).
3. For 3D analysis of cells embedded in a 3D matrix, place the 3D culture on the stage of a microscope with environmental controls or contained within a 5 % CO_2, 37 °C incubator. Set the top and bottom of the z range, the z increment, the time interval between acquisition of each z-series, and the length of the experiment as shown in Fig. 2c (*see* **Note 18**).
4. The acquisition software typically generates a numbered JPEG image stack that can be played and saved as a movie (Fig. 2c). Objects are then detected from this movie as described below.

3.8 Object Detection and Reconstruction in 4D Experiment

1. Bitmapping is a rapid and accurate method to detect objects (Fig. 6a). This method distinguishes a cell or group of cells based on the fact that they exhibit a higher grayscale variation than the background. More specifically, bitmapping assigns a 0–256 gray scale value to individual pixels within a user-defined kernel; i.e., a 3 × 3, 5 × 5 or 7 × 7 pixel matrix. The gray scale values within the kernel are averaged, the standard deviation (sd) computed and that sd is assigned to a reference pixel within the kernel. If the referenced sd is above the user-determined threshold value, then the pixel is considered part of the object and is retained as such. The kernel moves by one pixel and the process is repeated until every pixel within the field is scanned. Background or out of focus objects in each optical section will have more uniform gray scale values, lower sd values and thus be discarded. One advantage of bitmap tracing over outlining is that holes, gaps or concave indentations present within the real image are maintained.
2. Manual outlining can be used for single cells and fine structures such as filopodia (Fig. 6b).

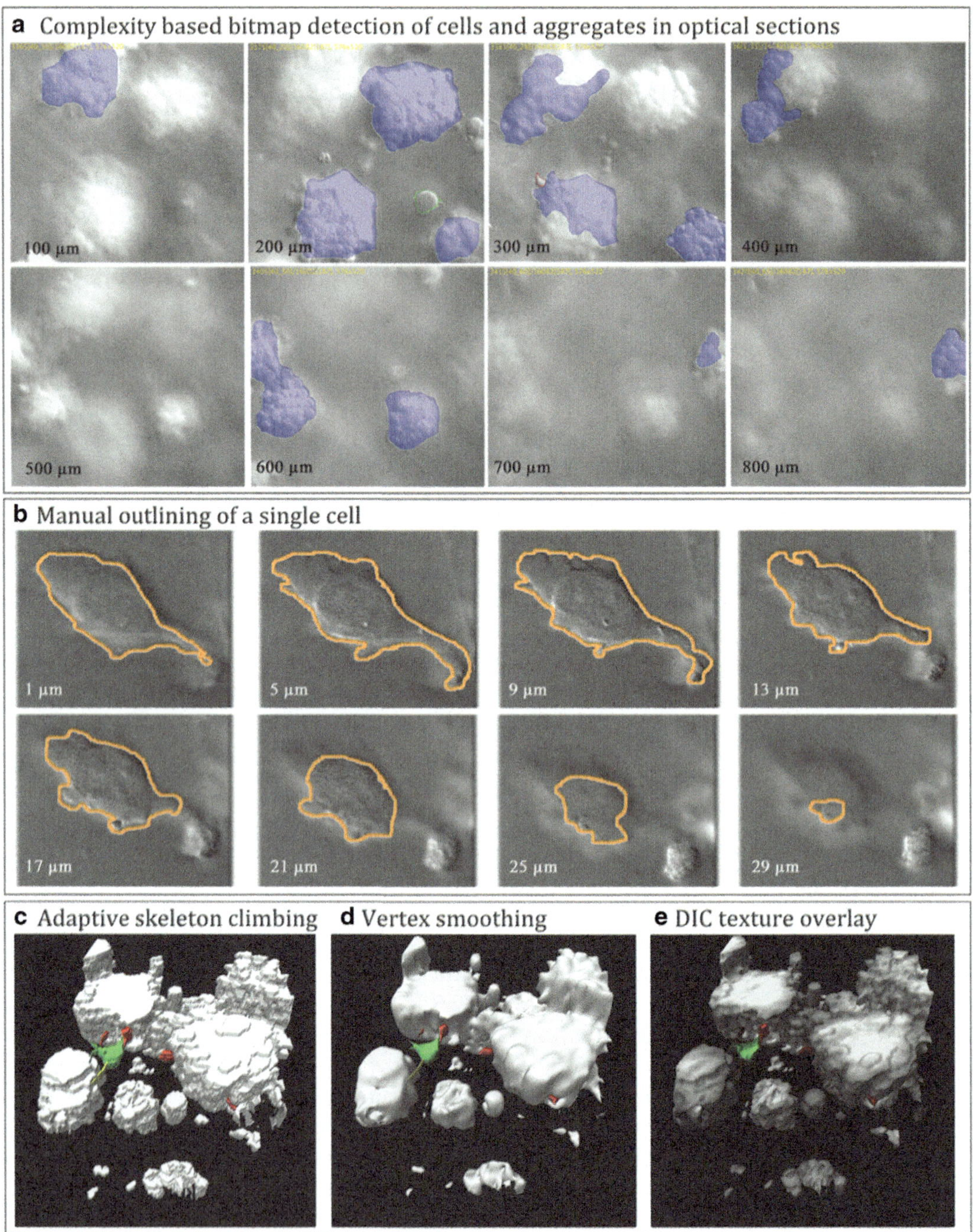

Fig. 6 Examples of automatic bitmap object detection, manual outlining of single cells, and 3D reconstruction methods. (*a*) Fields of cells and aggregates within a 3D matrix optically sectioned at 30-min intervals for 30 days generates movies comprised of over 100,000 frames. Cells and aggregates can be identified in these movies by complexity based automatic bitmap detection. (*b*) Single cells and organelles can be manually traced in optical sections. (*c–e*) Methods for 3D reconstructions combining automatically detected and manually detected objects. Coalescing aggregates in preparations of cancer cells are color-coded gray, facilitator cell and filopodia are color-coded *green* and *yellow*, respectively, and probe cells, *red*

3. Overlapping traces are stacked into a 3D z-series, either as a complexity stack in the case of bitmap tracings or as a series of beta-spline replacements (*see* **Note 19**), in the case of outlines.
4. Outlines are filled with voxels (3D pixels) and bitmap pixels are expanded into voxels. In both cases, the *x*, *y*, and *z* scales determine the voxel dimensions.
5. Java OpenGL (JOGL) implemented in J3D-DIAS4.1 performs 4D reconstructions using the "adaptive skeleton climbing isoform extraction" method [70], a variation of the "marching cubes" algorithm (Fig. 6c; *see* **Note 20**).
6. Vertices are smoothed three times to attenuate the sharp edges in the surface of the object (Fig. 6d) (*see* **Note 21**). DIC texture overlay can be added to restore details present in the original image (Fig. 6e).
7. Manually and automatically traced cells can be combined into a single movie file, reconstructed as wireframe images or as solid, shaded nontransparent objects, such as the cells shown in Fig. 6d, e. Filopodia, lamellipodia, and nuclei or other organelles can be independently color-coded.
8. Images can be rotated and viewed from different angles and saved as a movie.
9. Mechanisms of coalescence and the presence of different cell types can be analyzed in 4D in preparations of cancer cells. In Fig. 6c a facilitator cell (color-coded green) that emerged from an aggregate of a tumorigenic cell line contacted a probe cell (color-coded red) that emerged from another aggregate. The facilitator and probe then pulled the two aggregates together, resulting in coalescence.
10. The effects of potential anticancer therapies including monoclonal antibodies can be evaluated in 4D as shown in Fig. 7.

3.9 4D Motility and Dynamic Morphology Parameters

1. The 3D cell center (3D centroid) can be determined by averaging the *x*-, *y*-, and *z*-coordinates of the interior points of the faceted object and used to generate 4D paths.
2. Instantaneous velocity is calculated from the 3D position of the centroid by the central difference method [62] using the 2D formulae described earlier [60] (*see* **Note 14**). Other 2D parameters calculated from the centroid position can likewise be extrapolated to 3D.
3. The triangularized faceted surface is a polyhedron. *See*http://wwwf.imperial.ac.uk/~rn/centroid.pdf to find the standard calculus formulae to determine the volume of a polyhedron.
4. The "surface complexity" function was introduced into J3D-DIAS4.1 as a means to quantify the activity around the periph-

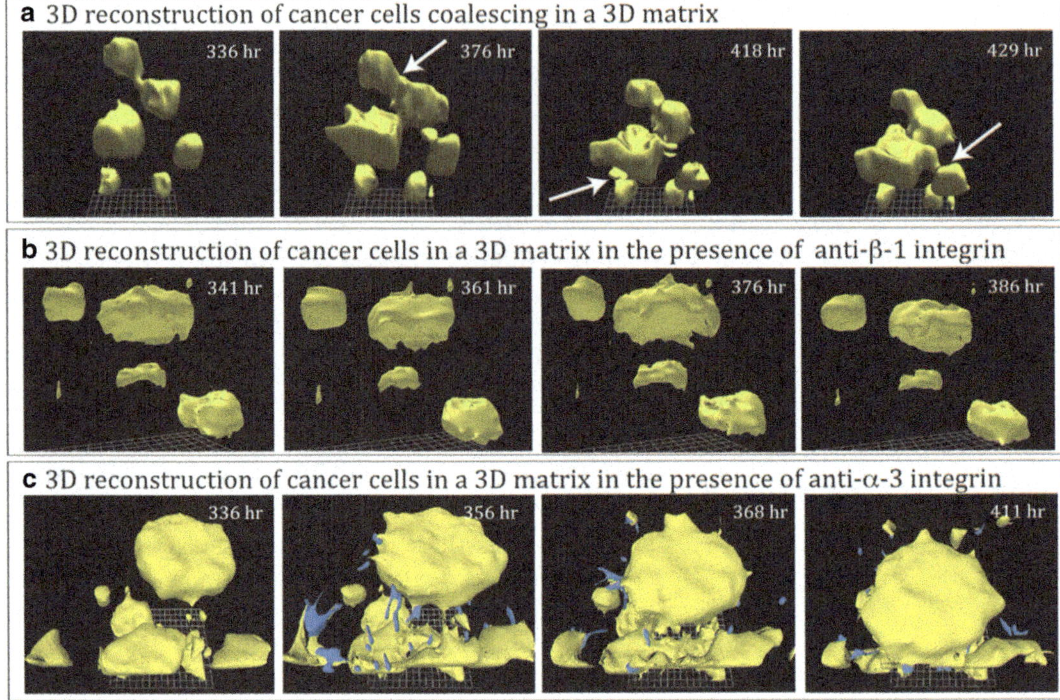

Fig. 7 J3D-DIAS4.1 generated 3D reconstructions of cancer cells and aggregates embedded in a 3D Matrigel matrix reveal effects of potential anticancer treatments such as monoclonal antibodies on cell growth, coalescence and single cell behavior. (*a*) Untreated cancer cells coalesce into aggregates in a 3D matrix as indicated by the *white arrows*. (*b*) Aggregate coalescence is inhibited in identical cultures treated with anti-β-1 integrin monoclonal antibody (AIIB2). (*c*) Aggregate coalescence and rapid exit of single cells from the aggregates (color-coded *blue*) are stimulated in identical cultures treated with anti-α-3 integrin monoclonal antibody (P1B5), demonstrating that differential effects of different treatments are detected in the 4D analysis

ery of aggregates formed in cancer cell cultures (Scherer et al. submitted). The surface complexity parameter is the reciprocal of 3D roundness (*see* **Note 22**).

4 Notes

1. To insure that experiments are performed on exponentially dividing populations of cells and to minimize accumulation of spontaneous mutations, *D. discoideum* cells should be routinely passaged before reaching stationary phase (~1×10^7 cells/mL). In addition, fresh cultures should be reconstituted from frozen stocks every 4–6 weeks.
2. Protocols can be found at the web site for the Center for Disease Control http://www.cdc.gov/biosafety/publications/bmbl5/index.htm and/or through the Research Office at the institution where the work is being performed. All human tissues

and/or blood products must be obtained under informed consent using protocols approved by the institution's IRB.

3. Heat inactivation of serum is no longer recommended for most cell culture applications (http://www.sigmaaldrich.com/technical-documents/protocols/biology/the-cell-environment.html).
4. Matrigel matrix will gel quickly, so care must be taken during preparation to keep it, as well as the surfaces and media that it contacts, chilled.
5. If the height of the dish exceeds the working distance of the optics, then approximately 5 mm can be trimmed from the dish bottom with sterile scissors before media is added.
6. For optical sectioning using DIC, a glass window should replace the plastic in the light path through the lid of the dish. An area with a diameter of 20–30 mm can be removed from the center of the plastic lid with a drill bit or laser cutter. A glass cover slip of larger dimensions can then be placed over the hole and sealed with Vaseline. The lid with the glass window is sterilized with ethanol and/or UV exposure within the biosafety cabinet.
7. J3D-DIAS4.1 can be accessed by collaboration at the W.M. Keck Dynamic Image Analysis Facility, David R. Soll, director, Biology Department, University of Iowa, Iowa City, IA 52245 USA.
8. The z-height is empirically determined and depends on the cells being studied. The acquisition time for an individual z-series and the interval between each z-series can be increased for slower moving cells.
9. LED light significantly reduces phototoxicity during long-term imaging but is generally not necessary for *D. discoideum* and PMNs.
10. Inoculation of a dilute cell suspension along with continual perfusion prevents accumulation of cAMP in the chamber, thereby eliminating its effects on cell behavior.
11. Pre-fill syringe, inlet tube, and needle with appropriate solution so as to avoid air bubbles.
12. Image processing algorithms such as differencing, black filtering and white filtering can facilitate automatic edge detection, particularly in images with a "phase halo." Differencing copies the image, shifts the copy by one pixel in the horizontal direction, one pixel in the vertical direction, and then subtracts the two images. Neutral gray (128) is added to the resulting intensities. To black filter, pixels darker than a user determined threshold are detected and removed from a bitmap outline. Conversely, to white filter, pixels lighter than user-determined threshold are detected and removed from the bitmap outline.

13. The initial outline is a connected circuit of pixels determine by the threshold setting. Beta splines replace the outline with curves approximating the original pixels. Because the original pixel outline is likely to be jagged, "spline resolution" functions can be applied. These functions remove some of the pixels, allowing the calculation of splines and generation of a smooth, rather than stair-stepped, final outline. Bias and tension functions also can be applied to adjust the accuracy of the final outline.

14. The speed of cellular translocation (instantaneous velocity) is calculated by the central difference method [62] as follows [60]:

    Speed[f] = (**scale** x **frate**)x **sqrt**(((**x**[f+**I**]-**x**[f-**I**])/2**I**)^2 + ((**y**[f+**I**]-**y**[f-**I**])/2**I**)^2 + ((**z**f+**I**]-**z**[f-**I**])/2**I**)^2))) when 1 < = f-**I** and f+**I** < =**F**

    Speed[f] = (**scale** x **frate**)x **sqrt**(((**x**[f+**I**]-**x**[f])/**I**)^2 + ((**y**[f+**I**]-**y**[f])/**I**)^2 + ((**z**f+**I**]-**z**[f-I])/2**I**)^2))) when f-**I** < 1 and f+**I** < = **F** (first frame)

    Speed [f] = (**scale** x **frate**)x **sqrt**(((**x**[f]-**x**[f-**I**])/**I**)^2 + ((**y**[f]-**y**[f-**I**])/**I**)^2 + ((**z**f+**I**]-**z**[f-I])/2**I**)^2))) when 1 < =f-**I** and f+**I** > **F** (last frame),

    where **F** is the total number of frames,

    f is the "current" frame,

    (**x**[f],**y**[f]) are the coordinates of the centroid of an object in frame f, 1 < =f< =**F**,

    I is the centroid increment,

    frate is the frame rate in # of frames per unit time,

    scale is the scale factor in distance units per pixel,

    sqrt is the square root function, and

    x denotes multiplication.

    Speed [f] = 0

15. Direction of travel using the central difference method is calculated as follows:

    Dir[f] = **angle**((**x**[f+**I**]-**x**[f-**I**]), (**y**[f+**I**]-**y**[f-**I**]))

    when 1 < =f-**I** and f+**I** < =**F**

    Dir[f] = **angle**((**x**[f+**I**]-**x**[f]), (**y**[f+**I**]-**y**[f])) when f-**I** < 1 and f+**I** < =**F**

    Dir[f] = **angle**((**x**[f]-**x**[f-**I**]), (**y**[f]-**y**[f-**I**])) when 1 < =f-**I** and f+**I** > **F**

    Dir[f] = 0 otherwise

    Multiples of ±360° are added to the direction to make the graph continuous.

16. Direction change using the central difference method is calculated as follows:

DirCh[f] = 0 when f-**I** < 1 **DirCh**[f] = **abs**(**Dir**[f]-**Dir**[f-**I**]) otherwise

If the direction change is greater than 180°, it is subtracted from 360. This always gives values between 0 and 180°

17. 3D movies of single cells migrating on a 2D substrate are typically comprised of 30–60 optical sections per individual z-series. The height is due to the fact that these cells often lift the anterior off the substrate and can extend pseudopodia in the *z* plane [71–73]. Ten minutes of optical sectioning, therefore, will generate a movie that is 3600–7200 individual frames.
18. We typically optically section cells and cell aggregates through 1 mm of Matrigel at 5–10 μm increments every 30 min over a period of 14–30 days or longer if media is refreshed. This process can therefore yield 122,000 optical sections per preparation per period of analysis.
19. Beta-splines replace the pixels in the outline with curves that approximate the original pixels [58]. In essence, beta-splines smooth the outline by applying bias, tension, and resolution [60] and generate a mathematical model from which data can then be calculated [59, 60, 63, 74].
20. Descriptively, in adaptive skeleton climbing a triangular facet is assigned to the raw voxel block in the most stable possible configuration. The user determines the size of the facet; the smaller the size, the greater the complexity of the surface. The remaining facets are then placed in the most stable possible configuration. Finally, irregularly shaped facets are added to any spaces between facets.
21. The *x*, *y*, and *z* coordinates of each vertex are averaged with neighboring *x*, *y*, and *z* coordinates [75].
22. 1/(6x **sqrt** xVol/(surface area$^{3/2}$)) where sqrt is the square root, Vol is the volume and surface area is the sum of the area of all facets. This yields a value for surface complexity that increases, rather than decreases, with increasing complexity.

Acknowledgments

This work was supported by the Developmental Studies Hybridoma Bank (DSHB), a National Resource created by the NICHD of the NIH and maintained at The University of Iowa, Department of Biology, Iowa City, IA 52242. We thank Brett Hanson, Joseph Ambrose, Kanoe Russell, Emma Buchele, Brian Kroll, Michele Livitz, Benjamin Soll, and Nicole Richardson for technical assistance. The monoclonal antibodies AIIB2, developed by C.H. Damsky, and P1B5, developed by E.A. Wayner and W.G. Carter, were obtained from the DSHB.

References

1. Wang MJ, Artemenko Y, Cai WJ, Iglesias PA, Devreotes PN (2014) The directional response of chemotactic cells depends on a balance between cytoskeletal architecture and the external gradient. Cell Rep 9:1110–1121
2. Danuser G, Allard J, Mogilner A (2013) Mathematical modeling of eukaryotic cell migration: insights beyond experiments. Annu Rev Cell Dev Biol 29:501–528
3. Doyle AD, Petrie RJ, Kutys ML, Yamada KM (2013) Dimensions in cell migration. Curr Opin Cell Biol 25:642–649
4. Ridley AJ, Schwartz MA, Burridge K, Firtel RA, Ginsberg MH, Borisy G, Parsons JT, Horwitz AR (2003) Cell migration: integrating signals from front to back. Science 302:1704–1709
5. Biname F, Pawlak G, Roux P, Hibner U (2010) What makes cells move: requirements and obstacles for spontaneous cell motility. Mol Biosyst 6:648–661
6. Annesley SJ, Fisher PR (2009) Dictyostelium discoideum--a model for many reasons. Mol Cell Biochem 329:73–91
7. Alvarez-Gonzalez B, Meili R, Firtel R, Bastounis E, Del Alamo JC, Lasheras JC (2014) Cytoskeletal mechanics regulating amoeboid cell locomotion. Appl Mech Rev 66:pii: 050804
8. Davidson AJ, Insall RH (2013) SCAR/WAVE: a complex issue. Commun Integr Biol 6:e27033
9. Soll DR, Wessels D, Kuhl S, Lusche DF (2009) How a cell crawls and the role of cortical myosin II. Eukaryot Cell 8:1381–1396
10. Wessels D, Lusche DF, Scherer A, Kuhl S, Myre MA, Soll DR (2014) Huntingtin regulates Ca(2+) chemotaxis and K(+)-facilitated cAMP chemotaxis, in conjunction with the monovalent cation/H(+) exchanger Nhe1, in a model developmental system: insights into its possible role in Huntingtons disease. Dev Biol 394:24–38
11. Futosi K, Fodor S, Mocsai A (2013) Neutrophil cell surface receptors and their intracellular signal transduction pathways. Int Immunopharmacol 17:638–650
12. Kumar S, Xu J, Kumar RS, Lakshmikanthan S, Kapur R, Kofron M, Chrzanowska-Wodnicka M, Filippi MD (2014) The small GTPase Rap1b negatively regulates neutrophil chemotaxis and transcellular diapedesis by inhibiting Akt activation. J Exp Med 211:1741–1758
13. Itakura A, Aslan JE, Kusanto BT, Phillips KG, Porter JE, Newton PK, Nan X, Insall RH, Chernoff J, McCarty OJ (2013) p21-Activated kinase (PAK) regulates cytoskeletal reorganization and directional migration in human neutrophils. PLoS One 8:e73063
14. Niggli V (2003) Signaling to migration in neutrophils: importance of localized pathways. Int J Biochem Cell Biol 35:1619–1638
15. Rougerie P, Miskolci V, Cox D (2013) Generation of membrane structures during phagocytosis and chemotaxis of macrophages: role and regulation of the actin cytoskeleton. Immunol Rev 256:222–239
16. Pixley FJ (2012) Macrophage migration and its regulation by CSF-1. Int J Cell Biol 2012: 501962
17. Geiger TR, Peeper DS (2009) Metastasis mechanisms. Biochim Biophys Acta 1796: 293–308
18. Yilmaz M, Christofori G, Lehembre F (2007) Distinct mechanisms of tumor invasion and metastasis. Trends Mol Med 13:535–541
19. Safdari Y, Khalili M, Ebrahimzadeh MA, Yazdani Y, Farajnia S (2015) Natural inhibitors of PI3K/AKT signaling in breast cancer: emphasis on newly-discovered molecular mechanisms of action. Pharmacol Res 93:1
20. Castillo-Pichardo L, Humphries-Bickley T, De La Parra C, Forestier-Roman I, Martinez-Ferrer M, Hernandez E, Vlaar C, Ferrer-Acosta Y, Washington AV, Cubano LA et al (2014) The rac inhibitor EHop-016 inhibits mammary tumor growth and metastasis in a nude mouse model. Transl Oncol 7:546–555
21. Nguyen DX, Bos PD, Massague J (2009) Metastasis: from dissemination to organ-specific colonization. Nat Rev Cancer 9:274–284
22. Fidler IJ (2002) Critical determinants of metastasis. Semin Cancer Biol 12:89–96
23. Eccles SA, Welch DR (2007) Metastasis: recent discoveries and novel treatment strategies. Lancet 369:1742–1757
24. Ishihara D, Dovas A, Park H, Isaac BM, Cox D (2012) The chemotactic defect in wiskott-Aldrich syndrome macrophages is due to the reduced persistence of directional protrusions. PLoS One 7:e30033
25. Soll DR, Wessels D, Heid PJ, Zhang H (2002) A contextual framework for characterizing motility and chemotaxis mutants in Dictyostelium discoideum. J Muscle Res Cell Motil 23:659–672
26. Hind LE, Mackay JL, Cox D, Hammer DA (2014) Two-dimensional motility of a macrophage cell line on microcontact-printed fibronectin. Cytoskeleton 71:542–554

27. Charras G, Sahai E (2014) Physical influences of the extracellular environment on cell migration. Nat Rev Mol Cell Biol 15:813–824
28. Chang Stephanie S, Guo W-h, Kim Y, Wang Y-l (2013) Guidance of cell migration by substrate dimension. Biophys J 104:313–321
29. Corall S, Haraszti T, Bartoschik T, Spatz J, Ludwig T, Cavalcanti-Adam E (2014) \upalpha 5\upbeta 1-integrin and MT1-MMP promote tumor cell migration in 2D but not in 3D fibronectin microenvironments. Comput Mech 53:499–510
30. Page-McCaw A, Ewald AJ, Werb Z (2007) Matrix metalloproteinases and the regulation of tissue remodelling. Nat Rev Mol Cell Biol 8:221–233
31. Stepanovic V, Wessels D, Goldman FD, Geiger J, Soll DR (2004) The chemotaxis defect of Shwachman-Diamond Syndrome leukocytes. Cell Motil Cytoskeleton 57:158–174
32. Williams RS, Boeckeler K, Graf R, Muller-Taubenberger A, Li Z, Isberg RR, Wessels D, Soll DR, Alexander H, Alexander S (2006) Towards a molecular understanding of human diseases using Dictyostelium discoideum. Trends Mol Med 12:415–424
33. Van Goethem E, Poincloux R, Gauffre F, Maridonneau-Parini I, Le Cabec V (2010) Matrix architecture dictates three-dimensional migration modes of human macrophages: differential involvement of proteases and podosome-like structures. J Immunol 184:1049–1061
34. Roh-Johnson M, Bravo-Cordero JJ, Patsialou A, Sharma VP, Guo P, Liu H, Hodgson L, Condeelis J (2014) Macrophage contact induces RhoA GTPase signaling to trigger tumor cell intravasation. Oncogene 33:4203–4212
35. Zhang Y, Choksi S, Chen K, Pobezinskaya Y, Linnoila I, Liu Z-G (2013) ROS play a critical role in the differentiation of alternatively activated macrophages and the occurrence of tumor-associated macrophages. Cell Res 23:898–914
36. Sidani M, Wessels D, Mouneimne G, Ghosh M, Goswami S, Sarmiento C, Wang W, Kuhl S, El-Sibai M, Backer JM (2007) Cofilin determines the migration behavior and turning frequency of metastatic cancer cells. J Cell Biol 179:777–791
37. Starke J, Maaser K, Wehrle-Haller B, Friedl P (2013) Mechanotransduction of mesenchymal melanoma cell invasion into 3D collagen lattices: filopod-mediated extension–relaxation cycles and force anisotropy. Exp Cell Res 319:2424–2433
38. Wang Y, McNiven MA (2012) Invasive matrix degradation at focal adhesions occurs via protease recruitment by a FAK-p130Cas complex. J Cell Biol 196:375–385
39. Gligorijevic B, Bergman A, Condeelis J (2014) Multiparametric classification links tumor microenvironments with tumor cell phenotype. PLoS Biol 12:e1001995
40. Murphy DA, Courtneidge SA (2011) The 'ins' and 'outs' of podosomes and invadopodia: characteristics, formation and function. Nat Rev Mol Cell Biol 12:413–426
41. Linder S, Wiesner C (2015) Tools of the trade: podosomes as multipurpose organelles of monocytic cells. Cell Mol Life Sci 72:121
42. Heid PJ, Geiger J, Wessels D, Voss E, Soll DR (2005) Computer-assisted analysis of filopod formation and the role of myosin II heavy chain phosphorylation in Dictyostelium. J Cell Sci 118:2225–2237
43. Lusche DF, Wessels D, Richardson NA, Russell KB, Hanson BM, Soll BA, Lin BH, Soll DR (2014) *PTEN* redundancy: overexpressing *lpten*, a homolog of *Dictyostelium discoideum ptenA*, the ortholog of human *PTEN*, rescues all behavioral defects of the mutant ptenA-. PLoS One 9:e108495
44. Volk AP, Heise CK, Hougen JL, Artman CM, Volk KA, Wessels D, Soll DR, Nauseef WM, Lamb FS, Moreland JG (2008) ClC-3 and IClswell are required for normal neutrophil chemotaxis and shape change. J Biol Chem 283:34315–34326
45. Scherer A, Kuhl S, Wessels D, Lusche DF, Hanson B, Ambrose J, Voss E, Fletcher E, Goldman C. and Soll DR (2015) A Computer-Assisted 3D Model for Analyzing the Aggregation of Tumorigenic Cells Reveals Specialized Behaviors and Unique Cell Types that Facilitate Aggregate Coalescence. PLoS ONE. 10.1371/journal.pone.0118628
46. Soll DR, Wessels D, Heid PJ, Voss E (2003) Computer-assisted reconstruction and motion analysis of the three-dimensional cell. ScientificWorldJournal 3:827–841
47. Varnum B, Edwards KB, Soll DR (1985) Dictyostelium amebae alter motility differently in response to increasing versus decreasing temporal gradients of cAMP. J Cell Biol 101:1–5
48. Wessels D, Brincks R, Kuhl S, Stepanovic V, Daniels KJ, Weeks G, Lim CJ, Spiegelman G, Fuller D, Iranfar N et al (2004) RasC plays a role in transduction of temporal gradient information in the cyclic-AMP wave of Dictyostelium discoideum. Eukaryot Cell 3:646–662
49. Geiger J, Wessels D, Soll DR (2003) Human polymorphonuclear leukocytes respond to waves of chemoattractant, like Dictyostelium. Cell Motil Cytoskeleton 56:27–44

50. Heid PJ, Wessels D, Daniels KJ, Gibson DP, Zhang H, Voss E, Soll DR (2004) The role of myosin heavy chain phosphorylation in Dictyostelium motility, chemotaxis and F-actin localization. J Cell Sci 117:4819–4835
51. Shutt DC, Jenkins LM, Carolan EJ, Stapleton J, Daniels KJ, Kennedy RC, Soll DR (1998) T cell syncytia induced by HIV release. T cell chemoattractants: demonstration with a newly developed single cell chemotaxis chamber. J Cell Sci 111(Pt 1):99–109
52. Lusche DF, Wessels D, Scherer A, Daniels K, Kuhl S, Soll DR (2012) The IplA Ca2+ channel of Dictyostelium discoideum is necessary for chemotaxis mediated through Ca2+, but not through cAMP, and has a fundamental role in natural aggregation. J Cell Sci 125: 1770–1783
53. Scherer A, Kuhl S, Wessels D, Lusche DF, Raisley B, Soll DR (2010) Ca2+ chemotaxis in Dictyostelium discoideum. J Cell Sci 123:3756–3767
54. Skoge M, Adler M, Groisman A, Levine H, Loomis WF, Rappel WJ (2010) Gradient sensing in defined chemotactic fields. Integr Biol (Camb) 2:659–668
55. Varnum B, Edwards KB, Soll DR (1986) The developmental regulation of single-cell motility in Dictyostelium discoideum. Dev Biol 113:218–227
56. Varnum B, Soll DR (1981) Chemoresponsiveness to cAMP and folic acid during growth, development, and dedifferentiation in Dictyostelium discoideum. Differentiation 18:151–160
57. Hoeller O, Kay RR (2007) Chemotaxis in the absence of PIP3 gradients. Curr Biol 17:813–817
58. Barsky B (1988) Computer graphics and geometric modeling using beta-splines. Springer, New York, NY
59. Soll D, Voss E (1998) Two and three dimensional computer systems for analyzing how cells crawl. In: Soll D, Wessels D (eds) Motion analysis of living cells. John Wiley, Inc., New York, NY, pp 25–52
60. Soll DR (1995) The use of computers in understanding how animal cells crawl. Int Rev Cytol 163:43–104
61. Soll DR, Wessels D, Voss E, Johnson O (2001) Computer-assisted systems for the analysis of amoeboid cell motility. Methods Mol Biol 161:45–58
62. Maron MJ (1982) Numerical analysis. Macmillan, New York, NY
63. Soll DR, Voss E, Johnson O, Wessels D (2000) Three-dimensional reconstruction and motion analysis of living, crawling cells. Scanning 22:249–257
64. Soll DR, Voss E, Varnum-Finney B, Wessels D (1988) "Dynamic Morphology System": a method for quantitating changes in shape, pseudopod formation, and motion in normal and mutant amoebae of Dictyostelium discoideum. J Cell Biochem 37:177–192
65. Chung CY, Lee S, Briscoe C, Ellsworth C, Firtel RA (2000) Role of Rac in controlling the actin cytoskeleton and chemotaxis in motile cells. Proc Natl Acad Sci 97:5225–5230
66. Wessels D, Lusche DF, Steimle PA, Scherer A, Kuhl S, Wood K, Hanson B, Egelhoff TT, Soll DR (2012) Myosin heavy chain kinases play essential roles in Ca2+, but not cAMP, chemotaxis and the natural aggregation of Dictyostelium discoideum. J Cell Sci 125:4934–4944
67. Wessels D, Murray J, Jung G, Hammer JA 3rd, Soll DR (1991) Myosin IB null mutants of Dictyostelium exhibit abnormalities in motility. Cell Motil Cytoskeleton 20:301–315
68. Soll D, Wessels D (1998) Computer-assisted characterization of the behavioral defects of cytoskeletal mutants of Dictyostelium discoideum. In: Soll D, Wessels D (eds) Motion analysis of living cells. John Wiley, Inc., New York, NY, pp 101–140
69. Wessels D, Soll DR, Knecht D, Loomis WF, De Lozanne A, Spudich J (1988) Cell motility and chemotaxis in Dictyostelium amebae lacking myosin heavy chain. Dev Biol 128:164–177
70. Poston T, Wong TT, Heng PA (1998) Multiresolution isosurface extraction with adaptive skeleton climbing. Comput Graph Forum 17:137
71. Wessels D, Reynolds J, Johnson O, Voss E, Burns R, Daniels K, Garrard E, O'Halloran TJ, Soll DR (2000) Clathrin plays a novel role in the regulation of cell polarity, pseudopod formation, uropod stability and motility in Dictyostelium. J Cell Sci 113(Pt 1):21–36
72. Wessels D, Vawter-Hugart H, Murray J, Soll DR (1994) Three-dimensional dynamics of pseudopod formation and the regulation of turning during the motility cycle of Dictyostelium. Cell Motil Cytoskeleton 27:1–12
73. Wessels D, Voss E, Von Bergen N, Burns R, Stites J, Soll DR (1998) A computer-assisted system for reconstructing and interpreting the dynamic three-dimensional relationships of the outer surface, nucleus and pseudopods of crawling cells. Cell Motil Cytoskeleton 41:225–246
74. Soll DR (1999) Computer-assisted three-dimensional reconstruction and motion analysis of living, crawling cells. Comput Med Imaging Graph 23:3–14
75. Hermann L (1976) Laplacian-isoparametric grid generation scheme. J Eng Mech Div 102:749–756

Chapter 15

Measurement of Cell Motility Using Microgrooved Substrates

James H.-C. Wang, Guangyi Zhao, and Bin Li

Abstract

Cells migrate to perform specific functions such as tissue/organ formation during development or to repair injured tissue. Therefore, the study of cell migration is important not only for fundamental cell biology research, but also to understand tissue morphogenesis and wound healing. In this chapter, we describe a method developed in our laboratory to measure cell migration in a uniaxial direction. In this approach, linear microgrooves, fabricated on a transparent poly(dimethylsiloxane) (PDMS) substrate, allow cells to move in a uniaxial direction, in contrast to the random movement of cells in all directions on a smooth substrate surface. This chapter describes in detail the materials and methods needed to measure cell movement using microgrooved substrates, as well as applications of this method.

Key words Cell migration, Micropatterning, Microgrooves, PDMS, Tendon fibroblasts

1 Introduction

Cell migration plays a fundamental role in the normal development and homeostasis of tissues and organs. It is also an essential cellular process for both physiological and pathological processes, including wound healing, tissue morphogenesis, angiogenesis, and metastasis. Several methods have been developed to study cell migration in vitro. In general, these methods involve plating cells on plain glass or plastic surfaces coated with matrix proteins such as fibronectin or poly-L-lysine [1]. On these smooth culture surfaces, cells move around in a random fashion, also called "random walk," which can be recorded and analyzed using time-lapse microscopy. The analysis and interpretation of cell movement using this method is complicated because cells move on a two-dimensional (2D) surface and in a "random walk" fashion rather than moving unidirectionally, which is easier to analyze.

Recently, researchers have started using microgrooved substrates, inducing cells to move in only one dimension (1D), to better analyze cell movement [2]. With this 1D approach, observation

Ray H. Gavin (ed.), *Cytoskeleton: Methods and Protocols*, Methods in Molecular Biology, vol. 1365,
DOI 10.1007/978-1-4939-3124-8_15, © Springer Science+Business Media New York 2016

of cell movement is easier, and analysis and interpretation of cell migration is simpler than with 2D surface-based methods. In this chapter, we present details that will enable implementation of this novel procedure to measure cell movement on "trails" and provide a few examples to demonstrate the effectiveness of the method.

In brief, microgrooved surfaces (Fig. 1) can be constructed and used to measure cell migration by performing the following four steps: (1) Preparation of microgrooved silicon wafers using photolithographic and reactive ion etching techniques; (2) fabrication of silicone membranes made of poly(dimethylsiloxane) (PDMS) using microgrooved silicon wafers as molds; (3) marking membranes with India ink to monitor cell movement; and (4) analyzing cell migration on a sequence of cell images captured by a CCD camera.

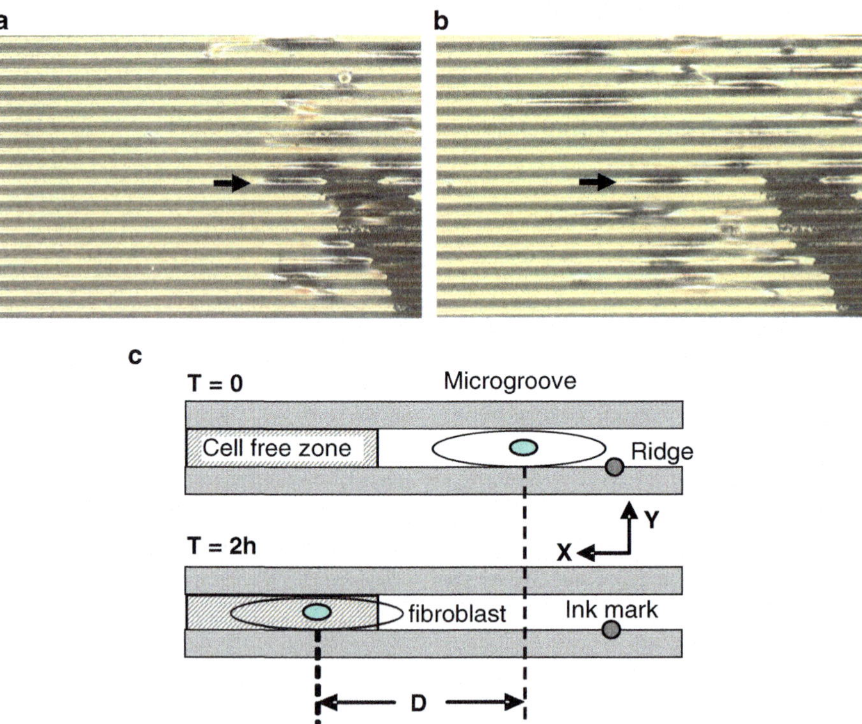

Fig. 1 (**a**, **b**) Cell mobility measurement on a microgrooved silicone membrane. Human tendon fibroblasts are observed as dark stretched spots with a *white* border. Cells on the *left* side of the membrane were removed by scratching. Ink markings were made on the *right* side (*dark area*). *Arrows* indicate the same cell that moved from close to the ink mark in (**a**) to a further point in (**b**) in 12 h. (Adopted with permission from Figs. 1, 2 in Thampatty et al. 2007, *Cell MotilCytoskeleton* 64(1):1–5). (**c**) Illustration of the paired cell images obtained from the CCD camera. The distance traveled by a cell (*blue dot*) from time $T=0$ to $T=2$ h through a ridge on the microgrooved substrate is calculated as the distance *D*. The grey circle represents the India ink mark used as a reference point

2 Materials

1. Shipley 1827 positive photoresist (Shipley Company, Marlborough, MA).
2. Silicon wafer (University Wafer, South Boston, MA).
3. Silicone fluid components, 601A and 601B (Wacker Silicones Corporation, Adrian, MI).
4. ProNectin-F.
5. Phosphate buffered saline (PBS): 3.2 mM Na_2HPO_4, 0.5 mM KH_2PO_4, 1.3 mM KCl, 135 mM NaCl, pH 7.4.
6. Dulbecco's Modified Eagle Medium (DMEM).
7. Fetal bovine serum (FBS).
8. Penicillin.
9. Streptomycin.
10. Transforming growth factor β1 (TGF-β1).
11. Microgroove photomask (MEMS & Nanotechnology Exchange, Reston, VA).
12. Waterproof black India ink.

3 Methods

3.1 Preparation

3.1.1 Fabrication of Silicon Wafer Molds

1. Spin-coat silicon wafers with Shipley 1827 positive photoresist.
2. Expose the wafers to UV light through a stripe photomask on the mask aligner, which creates the microgroove features on the wafers.
3. Use these molds in subsequent steps to create microgrooved silicone membranes.

3.1.2 Fabrication of Microgrooved Membranes

1. Prepare a silicone precursor liquid by mixing the silicone fluid components 601A and 601B at a weight ratio of 10:1.
2. Vigorously mix by stirring. Remove air bubbles by vacuum suction for 0.5–1 h (*see* **Note 1**).
3. Slowly pour the mixture onto the silicon wafer molds and allow them to cure at room temperature for 24 h.
4. Remove the cured PDMS silicone membrane from the mold to obtain a microgrooved silicone membrane (1.5 mm thick), which is cut to the desired size (~90 mm in diameter) before further use.

3.1.3 Marking Microgrooved Membranes

1. Using a needle attached to a syringe filled with Sanford waterproof black India ink, make a dotted line on the PDMS membrane surface perpendicular to the microgrooves.
2. Use these dots serve as reference points for measuring cell migration.

3.1.4 Plating Cells on Microgrooved Membranes

1. Sterilize the ink-marked silicone membranes in an autoclave and placed them in a 100 mm diameter petri dish.
2. Seed cells onto the microgrooved silicone membranes. Use cells at low concentration (usually 1×10^4 cells/cm^2 or 2×10^4 cells/cm^2, *see* **Note 2**) in DMEM containing 1 % fetal bovine serum, 50 U/mL penicillin, and 50 U/mL streptomycin. Culture at 37 °C and in 5 % CO_2.
3. Allow cells to attach for at least 6 h.
4. Monitor cell migration by creating cell-free areas on the membranes. Create cell-free areas by using a 2-mL pipette to scratch random areas perpendicular to the microgroove ridges (*see* **Note 3**). Be careful and make scratches adjacent to the ink markings so that cell movement can be precisely measured.
5. Remove cells that detach during this procedure along with the old medium and replenish with fresh media.

3.1.5 Obtaining Images of Cell Movement

1. Use a CCD camera equipped with image-capturing software to capture microphotographs of at least 50 cells at 200× magnification.
2. Move the microscope stage, and capture paired images at different *x*-axis coordinates while maintaining a constant *y*-axis coordinate. The direction parallel to the microgrooves is defined as the *x*-axis, while the direction perpendicular to the microgrooves is the *y*-axis.
3. Ensure that ink marks are included in at least one image in each pair. Use each ink mark as a reference to calculate the start point of cell migration (Fig. 1b).
4. Capture images every 2 h over a 12 h period, and name them sequentially: T2, T4, etc. to T12 (*see* **Note 4**).

3.1.6 Data Analysis

1. Analyze at least 50 cell images from each group using an image-processing software such as NIH Image J or Adobe Photoshop.
2. Cell movement is confined to a single direction, i.e., the *x*-axis; therefore, record the point where the ink mark crosses the ridge harboring the cell as the origin of the *x*-axis coordinate from where cell migration began (*see* **Note 5**).

3. Record the end point of cell migration as the *x*-coordinate at the center of a single cell calculated from its length.
4. Determine the distance traveled by the cell by measuring the difference between its *x*-coordinates at different time points.
5. Define positive cell movement as towards the scratched area (cell-free zone) and define negative movement as away from the scratched area and towards other attached cells.
6. Follow the above-described parameters, and calculate three types of movement: *net movement*, which is the distance traveled by a cell from its starting position (T0) to the position reached at 12 h (T12); *absolute net movement*, which is the absolute movement of a cell without considering the direction of movement; *total movement*, which is the absolute total distance traveled by the cell and is determined by adding the absolute values of the distance traveled between T0 and T4, T4 and T8, etc.
7. Evaluate the trends of all three movement types by using the median values instead of the means to avoid data that skew the means due to large variations in cell migration distances.

3.1.7 Applications of Microgrooved Membranes for Cell Migration Assay

Microgrooved membranes can be used to simultaneously monitor the migration of multiple cells. To show this, we used human patellar tendon fibroblasts (HPTFs) isolated from the tendon of a healthy male donor (21 years old). Since HPTFs move only in one direction in microgrooves, their migration·can be easily tracked.

1. Prepare the control group. At passage 6, plate HPTFs on three ProNectin-F-coated microgrooved silicone membranes in 6-well plates. Use cells at a concentration of 1×10^4 cells/cm^2 in DMEM containing 1 % fetal bovine serum, 50 U/mL penicillin, and 50 U/mL streptomycin.
2. Prepare the experimental group. Plate HPTFs and culture them as described in **step 2** except supplement the medium with 5 ng/mL of TGF-β1 to facilitate differentiation of fibroblasts to myofibroblasts [3, 4].
3. Confirm the presence of more myofibroblasts in the experimental group by the increase in the expression of α-smooth muscle actin (α-SMA) using western blot analysis (Fig. 2a).
4. After cell culture for 24 h, make random scratches on the membrane perpendicular to the microgrooves followed by the measurement of the three cell migration parameters, i.e., net movement, absolute net movement and total movement.
5. Using the low density of cells, identify and track the same fibroblasts from the beginning of experiment until 12 h later (Fig. 1a) (*see* **Note 6**).

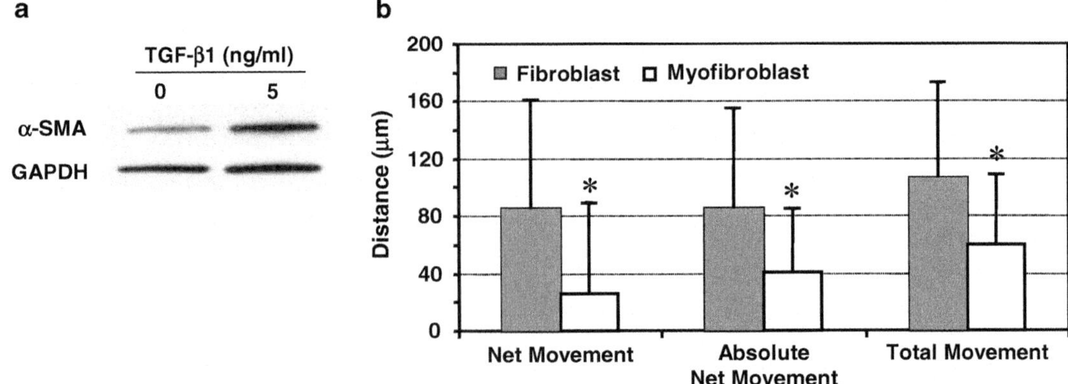

Fig. 2 Determining the differential mobility between fibroblasts and myofibroblasts on microgrooved membranes. (**a**) Western blot showing increased α-SMA protein expression in cells treated with 5 ng/mL TGF-β1. Treatment of fibroblast with TGF-β1 induced differentiation of fibroblasts into myofibroblasts. (**b**) Differences between myofibroblasts and fibroblasts in new movement, absolute net movement and total movement determined using microgrooved membranes. Myofibroblasts moved slower than fibroblasts. (Total number of cells $n > 50$; $*P < 0.05$). (Adopted with permission from Figs. 3, 4 in Thampatty et al. 2007, *Cell MotilCytoskeleton* 64(1):1–5)

4 Notes

1. The silicone precursor liquid should be thoroughly mixed by stirring, which inevitably creates a number of air bubbles within the liquid. Removing the air bubbles, ideally by vacuum, is critical for silicone membrane casting.
2. Depending on the cell type, different seeding density and fetal bovine serum concentration may be used.
3. Care should be taken to gently scratch the membrane so that only cells are removed but the silicone membrane surface is not damaged.
4. Measurement of cell movement in the microgrooved dishes has the following advantages: low cell density, frequent monitoring (2 h) of cell movement, high quality of the phase contrast images, and short distance (<200 μm) of cell migration along the microgrooves in a single direction. To avoid the possible influence of substrate topography (i.e., ridges vs. grooves) on cell migration, only cells on the ridges of the microgrooves are sampled.
5. The origin of the *x*-axis could be the point where the ink marking's edge crosses the ridge where the cell is migrating.
6. Data analysis of the acquired images revealed that myofibroblasts generally moved slower than fibroblasts. For example, the net movement of fibroblasts was 86 μm, while it was only 26 μm for myofibroblasts. Similarly, the absolute net movement

was 85 μm for fibroblasts and 52 μm for myofibroblasts. The total movement of fibroblasts and myofibroblasts was 107 and 60 μm, respectively (Fig. 2b). This example clearly shows that the migration of different types of cells can be effectively studied using microgrooved silicone membranes.

Acknowledgements

We gratefully acknowledge the funding support from NIH/NIAMS (AR065949, AR061395 and AR060920) for this work.

References

1. Rodriguez LG, Wu X, Guan JL (2005) Wound-healing assay. Methods Mol Biol 294:23–29
2. Thampatty BP, Wang JH (2007) A new approach to study fibroblast migration. Cell Motil Cytoskeleton 64:1–5
3. Ellis IR, Schor SL (1998) Differential motogenic and biosynthetic response of fetal and adult skin fibroblasts to TGF-beta isoforms. Cytokine 10:281–289
4. Song QH, Klepeis VE, Nugent MA, Trinkaus-Randall V (2002) TGF-beta1 regulates TGF-beta1 and FGF-2 mRNA expression during fibroblast wound healing. Mol Pathol 55:164–176

Chapter 16

The Study of Cell Motility by Cell Traction Force Microscopy (CTFM)

James H.-C. Wang, Guangyi Zhao, and Bin Li

Abstract

Migration is a vital characteristic of various cell types and enables various cellular functions during development and wound healing. Cell movement can be measured by monitoring cell traction forces, which are generated by individual cells and transmitted to the substrate below the migrant cells. This method, termed cell traction force microscopy (CTFM), has the advantage of directly measuring the "cause" (i.e., cell traction forces, CTFs) of cell movement rather than the "effect" (i.e., cell movement itself). This chapter details the methods involved in measuring cell traction forces. Several examples are also given to illustrate various applications of CTFM in cell biology research.

Key words Polyacrylamide gel, Cell movement, Traction force, Human fibroblast, Myofibroblast

1 Introduction

Cells move by applying traction forces to the underlying substrate [1–3] (Fig. 1). During these movements, CTFs are generated through the interaction of actin filaments with myosin-II (a cellular motor protein) and also by actin polymerization [4]. Once generated, CTFs are transmitted to the extracellular matrix (ECM) or substrate via focal adhesions (FAs), composed of structural and signaling proteins (e.g., integrins and kinases), which physically link the actin cytoskeleton with the ECM [5]. Therefore, determination of CTFs is useful not only in characterizing cell movement, but also to help define phenotype changes in cells caused by biochemical and biomechanical treatments.

Several methods have been developed to measure CTFs qualitatively or quantitatively. These include thin silicone membranes [3, 6, 7], micro-fabricated cantilevers [8], micropost force sensor arrays [9–11], and cell traction force microscopy (CTFM) [12–14]. Among these, CTFM is an efficient and reliable method to quantitatively determine traction forces of individual cells as well as a group of cells [4]. This method uses a polyacrylamide gel (PAG)

Ray H. Gavin (ed.), *Cytoskeleton: Methods and Protocols*, Methods in Molecular Biology, vol. 1365,
DOI 10.1007/978-1-4939-3124-8_16,

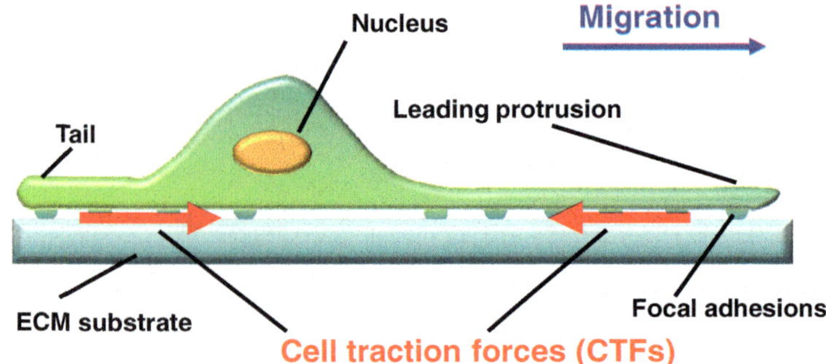

Fig. 1 Cell movement using traction forces on the underlying substrate

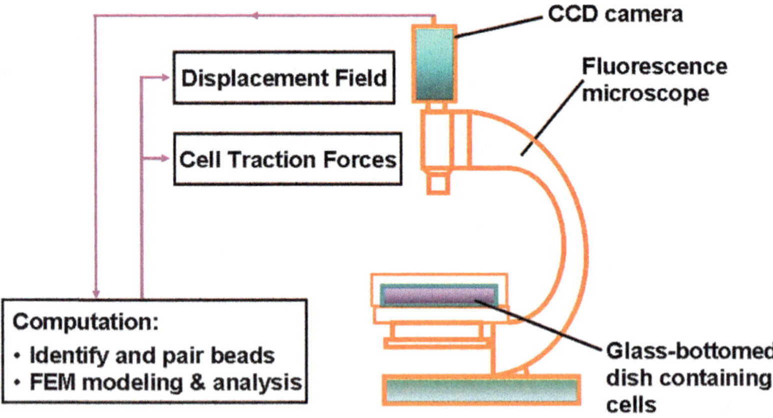

Fig. 2 General scheme of cell traction force microscopy (CTFM)

substrate that responds to a wide range of forces in a linear and elastic manner. More importantly, the substrate deformation is completely recovered after removal of the force [15]. In addition, the PAG stiffness can be easily adjusted to fit a broad range of cell types and its surface can be coated with ECM proteins of interest to cater to specific requirements when measuring traction forces of certain cell types [16].

The general scheme of CTFM construction is illustrated in Fig. 2 and involves four steps: (1) Fabricating the elastic PAG embedded with fine fluorescent beads (0.2–0.5 μm). After coating PAG with a layer of an ECM protein such as collagen type I, it is used as a substrate to grow cells; (2) Obtaining a pair of fluorescent bead images. These paired images are obtained when cells are attached to the PAG and after their removal, and are marked as "force-loaded" and "null-force," respectively; (3) Using the paired images to determine PAG substrate displacement field; and (4) "Converting" substrate displacements to CTFs. In this chapter, only steps 1 (fabrication of the ECM protein-coated PAG) and 2 (CTFM imaging) are described in detail. Readers interested in the details of steps 3 (determination of the substrate displacement field) and 4 (calculating the corresponding CTFs) are encouraged to refer to the references on CTFM technology [4, 12–14].

The PAG substrate preparation and CTFM determination involves the following steps: (1) activating the glass surface, (2) fabricating the PAG substrate, (3) collagen I conjugation to PAG, (4) measuring PAG thickness, (5) determining the Young's modulus and Poisson ratio of PAG substrate, (6) plating cells on PAG substrate, (7) acquiring cell and fluorescent microbead images, and (8) computing CTFs. Some examples of CTFM applications in determining traction forces of individual cells, detecting cell differentiation, determining traction forces of cells with controlled shapes, and determining traction forces of a group of cells are also described.

2 Materials

1. Glass-bottomed petri dish, Φ35 mm (with Φ14 mm glass bottom).
2. Circular cover glass, Φ12 mm.
3. Steel ball, Φ0.64 mm, 7.2 g/cm^3.
4. Sodium hydroxide: 0.1 M and 1 M NaOH solutions in water.
5. 3-aminopropyltrimethoxysilane.
6. Glutaraldehyde: 0.5 % solution in PBS.
7. Acrylamide: 10 % solution in water.
8. *N,N'*-methylenebisacrylamide (bis-acrylamide): 1 % solution in water.
9. Fluorescent beads (Molecular Probes, Eugene, OR) solution, Φ0.2 or 0.5 μm.
10. Ammonium persulfate (APS): 10 % solution in water.
11. *N,N,N',N'*-Tetramethylethylenediamine (TEMED).
12. *N*-sulfosuccinimidyl-6-[4′-azido-2′-nitrophenylamino] hexanoate (Sulfo-SANPAH): 0.2 mM solution in DMSO.
13. 200 mM HEPES buffer, pH 7.4.
14. Collagen type I (Angiotech Biomaterials, Palo Alto, CA): 100 μg/mL in PBS.
15. DMEM (GIBCO-BRL, Grand Inland, NY) containing 10 % fetal bovine serum (FBS), 20 ng/mL bFGF, 5 μg/mL heparin, 100 U/mL penicillin, and 100 μg/mL streptomycin.
16. Phosphate buffered saline (PBS).
17. Deionized water (DI water).
18. Transforming growth factor-β1 (TGF-β1).

3 Methods

3.1 Activating the Glass Surface

1. Treat a petri dish (35-mm) with a glass bottom (Φ 14-mm) with sodium hydroxide solution (0.1 M) for 1 day, and allow it to air-dry.
2. Add a few drops of 3-aminopropyltrimethoxysilane sufficient to cover the dried glass surface, and allow the drops to remain on the surface for 5 min. Wash the surface with DI water, and then treat with 0.5 % glutaraldehyde in PBS for 30 min.
3. Wash the dish thoroughly with copious DI water for 30 min and allow it to air-dry.

3.2 Fabricating the PAG Substrate

1. The procedure to fabricate PAG is described in the flow chart in Fig. 3.
2. Prepare a mixture of 5 % acrylamide and 0.1 % bis-acrylamide solution and subject it to vacuum suction for 20 min to remove air.
3. Add fluorescent microbeads at a ratio of 80/1 for 0.2 μm beads and 200/1 for 0.5 μm beads, 40 μL APS, and 4 μL TEMED and mix thoroughly (*see* **Note 1** for bead size).
4. Add 11 μL of this mix to the center of a pretreated glass-bottom petri dish and quickly covered with a piece of circular cover glass (Φ12 mm). Turn the dish upside down and maintain it at 4 °C for 2 h, which allows the microbeads to settle on the cover glass surface by gravity (*see* **Note 2**).

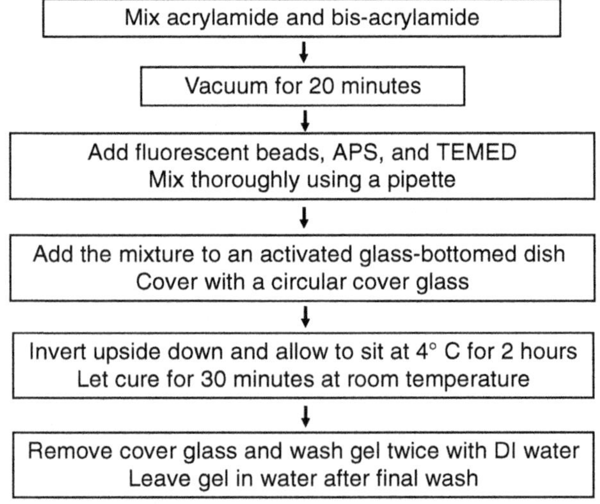

Fig. 3 Flowchart of the procedures involved in fabricating polyacrylamide gels (PAGs) embedded with fluorescent beads for CTFM

5. Subsequently, leave the dish at room temperature for 30 min to allow the formation of a PAG disk on the glass-bottom petri dish (*see* **Note 3**).
6. Use forceps to carefully remove the cover glass, and wash the PAG disk twice with DI water. Observe that the microbeads remain on the surface of the PAG.

3.3 Conjugating Collagen I to PAG

Prior to conjugation, the PAG substrate is first activated using a photo-linker, sulfo-SANPAH, with the aid of ultraviolet (UV) light.

1. Add 100 μL sulfo-SANPAH (5 mM) in 200 mM HEPES solution to the surface of the PAG previously fabricated as described above.
2. Expose the dish to UV light for 5 min, and then remove the sulfo-SANPAH solution.
3. Repeat **steps 1** and **2** two times, and then wash the PAG twice with PBS (*see* **Note 4**).
4. Add 130 μL collagen type I solution (100 μg/mL) to the PAG surface, and incubate overnight at 4 °C. Wash thoroughly three times with PBS prior to cell seeding.

3.4 Measuring PAG Thickness

1. Estimate the thickness of the activated PAG by using high power microscopy and sequentially changing the focus from the top to the bottom surfaces of the gel.
2. Record the vertical positions of the microscope stage in both positions. Determine PAG thickness from the difference between the two positions.
3. Alternatively, estimate the gel thickness from the ratio of the gel volume to the gel area.
4. Thickness of a typical gel used in CTFM experiments is 70–120 μm [17].

3.5 Determining the Young's Modulus and Poisson Ratio of PAG Substrate

1. Determine the Young's modulus of the PAG based on a previously published method [18]. First, place a steel ball (Φ0.64 mm) on the surface of the gel.
2. Use a microscope to measure the indentation caused by the weight of the steel ball by comparing the differences between the vertical positions of the fluorescent microbeads under the center of the steel ball.
3. Calculate the Young's modulus from the formula $\Upsilon = 3(1-\nu^2)f/4d^{3/2}r^{1/2}$, where ν is the Poisson ratio of the steel ball (0.3), f is the weight of the ball (neglecting the upward buoyancy force produced by the liquid surrounding the ball), d is the amount of indentation, and r is the radius of the steel ball (see **Note 5**).

3.6 Plating Cells on PAG Substrate

1. For CTFM experiments, seed approximately 2000–5000 cells on each PAG substrate (*see* **Note 6**).
2. Add cell suspensions to the collagen-coated PAG substrate, and allow the cells to spread on the gel and attach for 6–12 h before taking cell images (*see* **Note 7**).

3.7 Acquiring Cell and Fluorescent Microbead Images

1. Select a region on the PAG where a few individual cells are attached (*see* **Note 8**), and use a CCD camera attached to a fluorescence microscope to capture a phase contrast image of the cells.
2. Capture a fluorescence image of the microbeads. This image is the "force-loaded" image (*see* **Note 9**).
3. Apply 200 μL of 1 M NaOH to the medium to detach the cells, and capture an image of the fluorescent microbeads in the same view as in **step 2** (*see* **Note 10**). This image is the "null-force" image (*see* **Note 11**).

3.8 Computational Determination of CTFs

1. CTFs are determined using a general purpose FEM code on the ANSYS 8.0 or AMPS software (*see* **Note 12**).

3.9 CTFM Applications in Determining Traction Forces of Individual Cells

To illustrate that CTFM is an effective technology to determine traction forces of various cell types, we applied CTFM to determine the traction forces of individual human patellar tendon fibroblasts (HPTFs) as described below.

1. Fabricate a PAG substrate, and conjugate it to collagen type I following the protocol described above.
2. Seed HPTFs on the PAG substrates and allow them to spread for 6 h. After initial microscopic observation, choose cells with an elongated shape, and image them (*see* **Note 13**).
3. Use the "force-loaded" and "null-force" images to determine the substrate displacement field and CTFs based on previously published methods [14] (*see* **Note 14**).

3.10 Detecting Cell Differentiation by CTFM

Rabbit corneal stromal cells can be used to demonstrate the ability of CTFM to detect differentiated cells.

1. Induce rabbit corneal stromal cells for fibroblast differentiation in DMEM based on previously published protocols [19].
2. Confirm that the differentiated cells are fibroblasts and not myofibroblasts based on the minimal expression of α-smooth muscle actin (α-SMA), a specific marker for myofibroblasts [17, 20].
3. In order to obtain higher numbers of myofibroblasts, treat the fibroblasts with 2 ng/mL transforming growth factor-β1 (TGF-β1) for 24 h, and observe a high expression level of α-SMA, an indication of a high density of myofibroblasts [17].
4. Use CTFM to determine the CTFs of individual fibroblasts and differentiated myofibroblasts (Fig. 5a, b) (*see* **Note 15**).

3.11 Determining Traction Forces of Cells with Controlled Shapes (*See* Note 16)

1. Understanding the relationship between cell shape and cellular function is an important aspect of cell biology. Specifically, cell phenotypes play a critical role in maintaining tissue homeostasis and, if changed, can lead to pathological changes.
2. To facilitate the study of cell shapes by CTFM, we used differently shaped HPTFs obtained using micro-contact printing techniques.
3. Among the resulting cells, the rectangular cells had an aspect ratio of 4.9 and the circular cells had a diameter of 50 μm; however, both cells had the same area of 1960 μm^2.
4. These HPTFs with different shapes exhibited variations in actin cytoskeletal structure, spatial arrangement in focal adhesions, and spatial distribution and magnitude in cell traction forces (Fig. 6).
5. In rectangular cells, long and parallel actin filament bundles formed along the longitudinal direction. In circular cells, however, they formed along the circumference of the cells, and several bundles appeared to have formed from the nucleus area to the cell membrane.
6. Focal adhesions were visualized by staining the cytoskeletal protein, vinculin, at focal adhesions. Staining was concentrated at the two ends of the rectangular cells, but uniformly distributed along the circumference of the circular cells. These locations also had the largest CTFs in both rectangular and circular cells, respectively. Thus, CTFM can be effectively used to differentiate among the differently shaped same cell type.

3.12 Determining Traction Forces of a Group of Cells

CTFM can also be used to determine the traction forces of a cell group as described below.

1. Use micro-contact printing technology [21] to fabricate square fibronectin-coated micro-islands on PAG substrates.
2. Seed NIH 3T3 cells on the micro-island and culture until a square-shaped monolayer of cells (referred to as "Cell Island") forms on the micro-island, usually after 1 day in culture (Fig. 7a, c, e).
3. Using the protocol described for single cells, apply CTFM to determine the CTF distribution on the square cell island (*see* **Note 17**).

4 Notes

1. Depending on the purpose and resolution requirements, imaging can be performed at different magnifications for CTFM.

However, the bead size must be varied according to the magnification. It is recommended that 0.5 μm beads be used for a 10× objective and 0.2 μm beads for 20×. In addition, the bead concentration should be adjusted for 10× imaging and should be about half that for 20×.

2. Typically, it takes a few hours for the fluorescent microbeads to settle to the bottom during PAG fabrication. Maintaining the polymerizing mixture at a low temperature (4 °C) slows down the curing speed and allows enough time for the beads to settle down to the gel surface rather than being "frozen" in the bulk of the gel.
3. Although PAG fabrication is straightforward, care must be taken during the process. Failure to fabricate a uniform PAG substrate and achieve an even coating of collagen type I on the PAG surface will affect cell morphology and alter cell behavior, resulting in large variations and inconsistent CTF results from one experiment to another.
4. Washing the PAG after sulfo-SANPAH treatment should be completed as quickly as possible to avoid hydrolysis of the NHS groups in sulfo-SANPAH, which are the prime regions for protein conjugation.
5. A similar method [22] is used to determine the Poisson ratio of the PAG substrate. Typically, for a PAG substrate containing 5 % acrylamide and 0.1 % bis-acrylamide, the Young's modulus is ~3 kPa, and the Poisson ratio is ~0.48.
6. Cell seeding density should be low so that isolated individual cells are grown on PAG for CTF measurement.
7. Depending on the cell type, the exact time of cell spreading on collagen-conjugated PAG may need adjustment.
8. The same criteria should be applied when sampling individual cells for CTFM. Specifically, individual cells with similar spreading areas and shape should be selected to reduce variations in CTF values and thus increase statistical power to detect differences in CTF between treatment and control groups. In addition, the distance between cells should be at least twice the cell size so that influence from adjacent cells is minimized.
9. While an effort is made to let fluorescent beads settle near the surface of the PAG, it is normal for beads to exist slightly beneath the surface. When imaging fluorescent beads, the focus should be on the layer of beads that are closest to the PAG surface.
10. It is very important not to touch the dish when adding NaOH because even a slight displacement of the dish may result in a non-reversible shift or rigid body motion of the fluorescent bead image, making paired image analysis of beads a challenge.

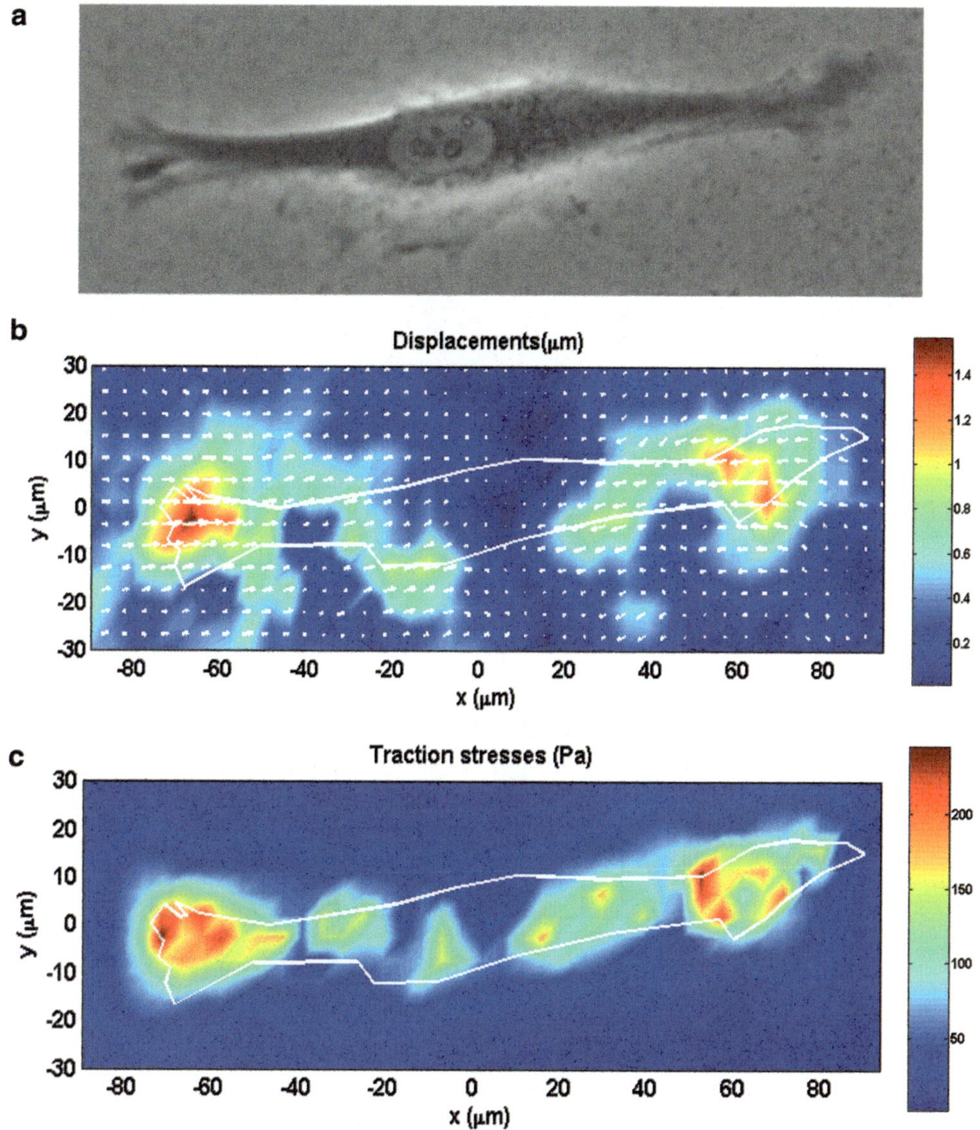

Fig. 4 Application of CTFM to determine traction forces of individual human patellar tendon fibroblasts (HPTFs). (**a**) HPTF on a PAG with embedded fluorescent beads (not shown); (**b**) substrate displacement field; (**c**) cell traction field

11. The "null-force" image must be taken at the same location and with the same focus as the "force-loaded" image. Where necessary, the microscope stage and focusing knob should be adjusted manually to ensure consistent location and focus.
12. In cases where single cells have small substrate deformations, Fourier transform traction cytometry, or FTTC [13], can be used.

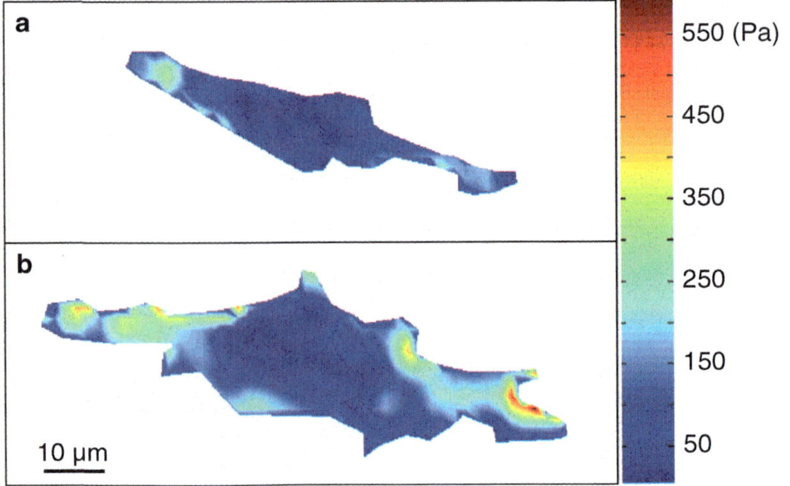

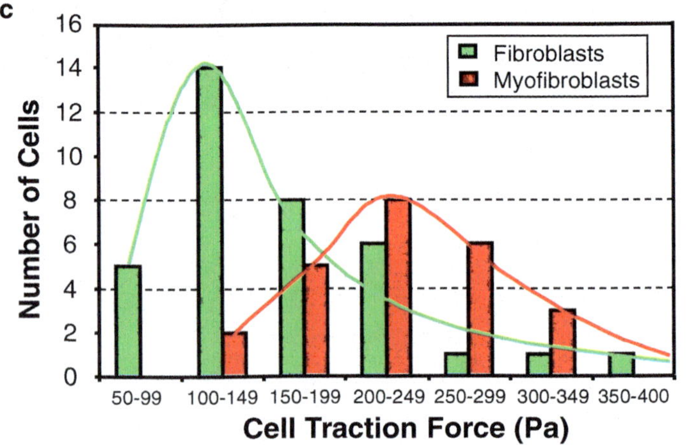

Fig. 5 Cell traction force fields of a typical fibroblast (**a**) and myofibroblast (**b**). The color within the cells represents the traction force magnitude. The myofibroblast is larger and generates a higher traction force than the fibroblast. (**c**) Evaluating cell differentiation by CTFM. After inducing fibroblast differentiation into myofibroblasts, the new CTF distribution shifted to the *right* because of the overall larger CTFs of myofibroblasts compared to fibroblasts. (**a** and **b** are adopted with permission from Fig. 1 in Chen et al. 2007, *Cell Motil Cytoskeleton* 64(4):248–257)

13. HPTFs can be elongated or rectangular in shape, although the elongated shape is more common.
14. The cell image, substrate displacement field, and cell traction stress distribution of the HPTFs are shown in Fig. 4. The maximum displacement obtained was ~1.2 μm, and the maximum traction stress was ~250 Pa. CTFs were concentrated at the front and rear of the cell and were sparse around the nucleus, which is consistent with general cell migration parameters.

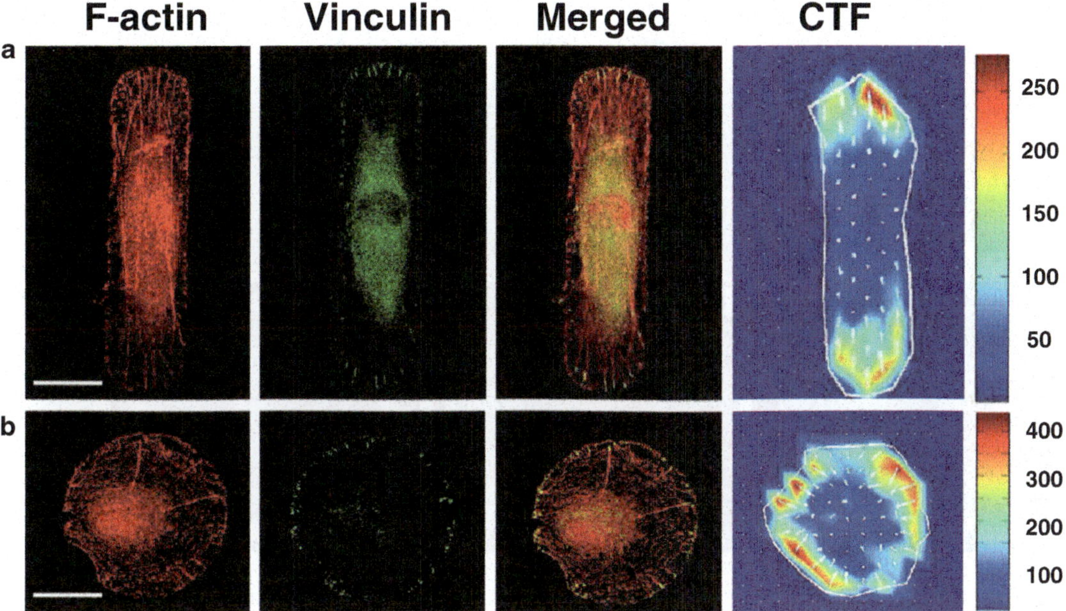

Fig. 6 Application of CTFM to determine traction forces of micropatterned cells. Immunofluorescence images of F-actin, vinculin, and their overlay, as well as CTFs in micropatterned HPTFs. (**a**) Rectangular HPTF. (**b**) circular HPTF. Development of focal adhesions and stress fibers is closely related to CTF distribution. Scale bars: 20 μm. (Adopted with permission from Fig. 6 in Li et al. 2008, *Cell Motil Cytoskeleton* 65(4):332–341)

15. The fibroblasts generated a mean CTF of about 100 Pa, while the mean CTF of the differentiated myofibroblasts was two-fold higher. It is evident from the CTF histogram (Fig. 5c) that the two cell populations differ, with the myofibroblasts shifted more to the right, indicating a higher CTF value compared to the fibroblast population. This example shows that CTFM may be employed to detect differentiation of cells, such as adult and embryonic stem cells, due to biochemical and biophysical stimuli [23].
16. Understanding the relationship between cell shape and cellular function is an important aspect of cell biology. Specifically, cell phenotypes play a critical role in maintaining tissue homeostasis and, if changed, can lead to pathological changes. To facilitate the study of cell shapes by CTFM, we used differently shaped HPTFs obtained using micro-contact printing techniques. Among the resulting cells, the rectangular cells had an aspect ratio of 4.9 and the circular cells had a diameter of 50 μm; however, both cells had the same area of 1960 μm^2. These HPTFs with different shapes exhibited variations in actin cytoskeletal structure, spatial arrangement in focal adhesions, and spatial distribution and magnitude in cell traction

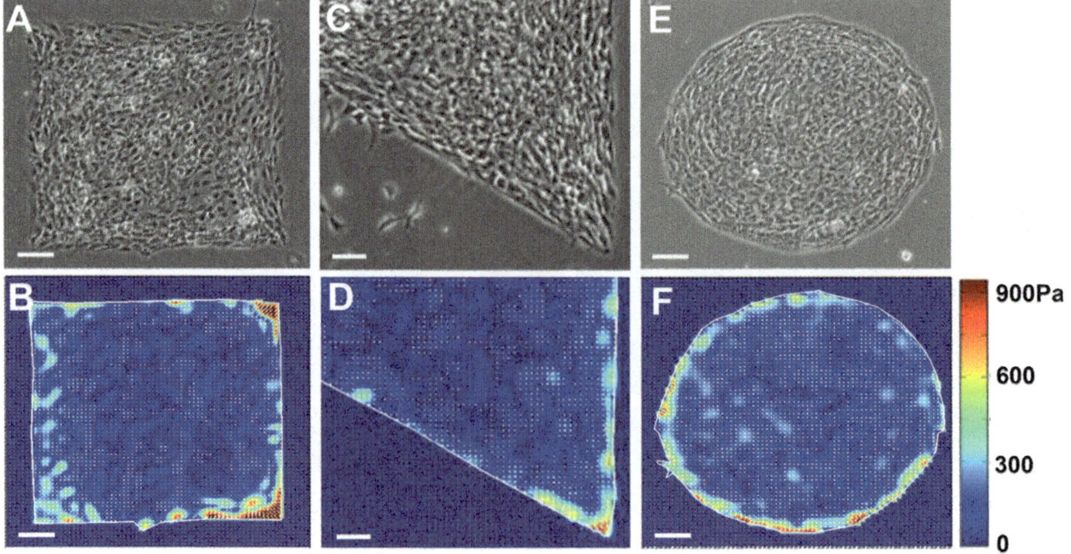

Fig. 7 Application of CTFM to determine traction forces of a micropatterned cell island. (**a**, **c**, **e**) Phase contrast micrographs of differently shaped NIH 3T3 cell islands. (**b**, **d**, **f**) CTF fields of the differently shaped NIH 3T3 islands (scale bar: 100 μm)

forces (Fig. 6). In rectangular cells, long and parallel actin filament bundles formed along the longitudinal direction. In circular cells, however, they formed along the circumference of the cells, and several bundles appeared to have formed from the nucleus area to the cell membrane. Focal adhesions were visualized by staining the cytoskeletal protein, vinculin, at focal adhesions. Staining was concentrated at the two ends of the rectangular cells, but uniformly distributed along the circumference of the circular cells. These locations also had the largest CTFs in both rectangular and circular cells, respectively. Thus, CTFM can be effectively used to differentiate among the differently shaped same cell type.

17. The results revealed the presence of larger CTFs at the corners and along the periphery of the cell island than in the interior (Fig. 7b, d, f).

Acknowledgements

We gratefully acknowledge the funding support from NIH/NIAMS (AR065949, AR061395, AR060920, and AR049921) for this work.

References

1. Ananthakrishnan R, Ehrlicher A (2007) The forces behind cell movement. Int J Biol Sci 3: 303–317
2. Lauffenburger DA, Horwitz AF (1996) Cell migration: a physically integrated molecular process. Cell 84:359–369
3. Oliver T, Dembo M, Jacobson K (1995) Traction forces in locomoting cells. Cell Motil Cytoskeleton 31:225–240
4. Wang, J. H. and J. S. Lin (2007). Cell traction force and measurement methods. Biomech Model Mechanobiol 6(6):361–371
5. Burton K, Park JH, Taylor DL (1999) Keratocytes generate traction forces in two phases. Mol Biol Cell 10:3745–3769
6. Harris AK, Wild P, Stopak D (1980) Silicone rubber substrata: a new wrinkle in the study of cell locomotion. Science 208:177–179
7. Balaban NQ, Schwarz US, Riveline D, Goichberg P, Tzur G, Sabanay I et al (2001) Force and focal adhesion assembly: a close relationship studied using elastic micropatterned substrates. Nat Cell Biol 3:466–472
8. Galbraith CG, Sheetz MP (1997) A micromachined device provides a new bend on fibroblast traction forces. Proc Natl Acad Sci U S A 94:9114–9118
9. du Roure O, Saez A, Buguin A, Austin RH, Chavrier P, Siberzan P et al (2005) Force mapping in epithelial cell migration. Proc Natl Acad Sci U S A 102:2390–2395
10. Tan JL, Tien J, Pirone DM, Gray DS, Bhadriraju K, Chen CS (2003) Cells lying on a bed of microneedles: an approach to isolate mechanical force. Proc Natl Acad Sci U S A 100:1484–1489
11. Li B, Xie L, Starr ZC, Yang Z, Lin JS, Wang JH (2007) Development of micropost force sensor array with culture experiments for determination of cell traction forces. Cell Motil Cytoskeleton 64:509–518
12. Dembo M, Wang YL (1999) Stresses at the cell-to-substrate interface during locomotion of fibroblasts. Biophys J 76:2307–2316
13. Butler JP, Tolic-Norrelykke IM, Fabry B, Fredberg JJ (2002) Traction fields, moments, and strain energy that cells exert on their surroundings. Am J Physiol Cell Physiol 282:C595–C605
14. Yang Z, Lin JS, Chen J, Wang JH (2006) Determining substrate displacement and cell traction fields – a new approach. J Theor Biol 242:607–616
15. Pelham RJ Jr, Wang Y (1997) Cell locomotion and focal adhesions are regulated by substrate flexibility. Proc Natl Acad Sci U S A 94: 13661–13665
16. Kandow CE, Georges PC, Janmey PA, Beningo KA (2007) Polyacrylamide hydrogels for cell mechanics: steps toward optimization and alternative uses. Methods Cell Biol 83:29–46
17. Chen J, Li H, Sundarraj N, Wang JH (2007) Alpha-smooth muscle actin expression enhances cell traction force. Cell Motil Cytoskeleton 64:248–257
18. Lo CM, Wang HB, Dembo M, Wang YL (2000) Cell movement is guided by the rigidity of the substrate. Biophys J 79:144–152
19. Anderson S, DiCesare L, Tan I, Leung T, SundarRaj N (2004) Rho-mediated assembly of stress fibers is differentially regulated in corneal fibroblasts and myofibroblasts. Exp Cell Res 298:574–583
20. Hinz B, Celetta G, Tomasek JJ, Gabbiani G, Chaponnier C (2001) Alpha-smooth muscle actin expression upregulates fibroblast contractile activity. Mol Biol Cell 12:2730–2741
21. Whitesides GM, Ostuni E, Takayama S, Jiang X, Ingber DE (2001) Soft lithography in biology and biochemistry. Annu Rev Biomed Eng 3:335–373
22. Li Y, Hu Z, Li C (1993) New method for measuring Poisson's ratio in polymer gels. J Appl Polym Sci 50:1107–1111
23. Wang JH, Thampatty BP (2008) Mechanobiology of adult and stem cells. Int Rev Cell Mol Biol 271:301–346

Chapter 17

Melanosome Motility in Fish Retinal Pigment Epithelial (RPE) Cells

Christina King-Smith

Abstract

Several model systems have been developed to investigate mechanism and regulation of intracellular organelle motility. The fish retinal pigment epithelial (RPE) cell represents a novel yet simple system for the study of organelle motility. Primary cultures of dissociated RPE cells are easily prepared and amenable to motility studies. In vivo, melanin-containing pigment granules (melanosomes) within fish RPE migrate distances up to 100 μm in response to light flux. When dissociated from the epithelial layer and cultured in vitro, RPE cells attach to the substrate with the apical projections extending radially from the central cell body. Melanosomes can be chemically triggered to aggregate or disperse throughout the projections, and are easily observed using phase contrast microscopy. Melanosome migration in RPE apical projections is dependent on actin filaments, and thus renders this model system useful for investigations of actin-dependent organelle motility.

Key words RPE, Actin, Melanosomes organelle transport, Microtubules, Motor proteins

1 Introduction

Retinal pigment epithelial (RPE) cells are located at the back of the vertebrate eye. In eyes of fish and other lower vertebrates, RPE melanosomes undergo light-dependent migrations, dispersing out into the cells' long apical projections in the light, and aggregating back into the cell body in darkness. Fish do not have dilatable pupils to control light that enters the retina. Rather, the moveable curtain of RPE melanosomes, together with elongation and contraction of rod and cone photoreceptors serve to modulate light flux, and are collectively called retinomotor movements [1].

RPE cells can be isolated from the eye, dissociated into single cells, and cultured for up to 24 h. In vitro, RPE cells lose their apical-basal polarity, become radially symmetrical, and attach to the substrate, with the formerly apical projections extending outward from the central, pigment-filled cell body (Fig. 1).

Ray H. Gavin (ed.), *Cytoskeleton: Methods and Protocols*, Methods in Molecular Biology, vol. 1365,
DOI 10.1007/978-1-4939-3124-8_17,

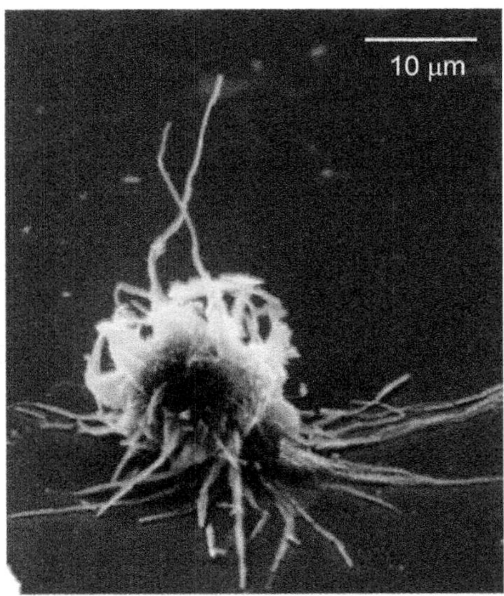

Fig. 1 Scanning electron micrograph of an isolated RPE cell. Formerly apical projections extend from the central cell body of dissociated cells

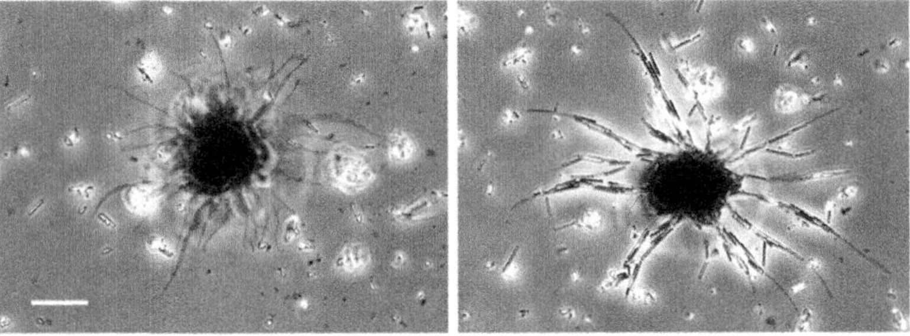

Fig. 2 Isolated RPE cells with aggregated and dispersed melanosomes. The cell on the *left* was treated with cAMP stimulating migration of melanosomes into the central cell body. Apical projections devoid of melanosomes appear phase-grey. The cell on the *right* has dispersed melanosomes; these appear phase-bright within the projections. Tips of projections are free of melanosomes because of the bidirectional shuttling of the melanosomes once full dispersion is achieved. Debris surrounding cells are melanosomes from lysed cells. Note melanosomes are mostly cylindrical, not round. Bar: 10 μm

Melanosome aggregation within the (formerly) apical projections can be triggered by application of underivatived cAMP [2]. Dispersion is stimulated by washout of cAMP in medium containing dopamine (Fig. 2).

Time-lapse videomicrography of melanosome motility within apical projections of dissociated RPE cells is easily carried out using a simple perfusion chamber on an upright, phase contrast

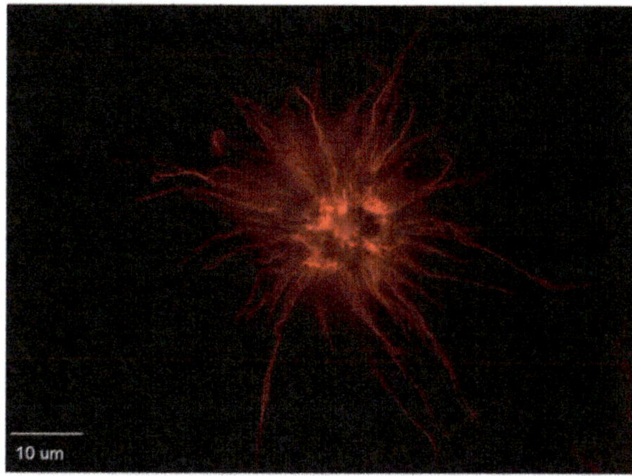

Fig. 3 Isolated RPE cell treated with cAMP to aggregate melanosomes, then fixed and labeled with rhodamine phalloidin revealing F-actin in the cell body and apical projections

microscope. Time-lapse videos reveal that patterns of melanosome motility differ during aggregation (retrograde motility) and dispersion (anterograde motility). During dispersion, melanosome motility is bidirectional and saltatory, at rates of 2–3 μm/min in the anterograde direction. In contrast, aggregation is continuous and unidirectional, also at 2–3 μm/min [3]. Studies using cytoskeletal inhibitors, time-lapse and fluorescent microscopy, and quantification of pigment position in fixed cells have demonstrated that while microtubules are plentiful in apical projections, they are not required for melanosome motility. Rather, actin filaments, also plentiful in the RPE cell body and apical projections (Fig. 3) are necessary and sufficient to support melanosome motility in the apical projections of isolated cells, with microtubules supplying structural support [3, 4]. Recent studies using myosin S1 labeling and platinum replica shadowing have demonstrated that actin filaments within RPE apical projections are uniformly oriented with barbed ends at the distal tips [5]. Previous work has suggested that melanosome aggregation may rely on retrograde actin flow, based on sensitivity of melanosome aggregation but not dispersion to treatments expected to interfere with actin retrograde flow, including (1) inhibition of myosin-II and Rho kinase (ROCK) [6], and (2) cross-linking of surface proteins using the lectin, ConA, [5] which has been demonstrated to interfere with cortical actin flow [7, 8]. Isolated RPE cells, therefore, represent a useful model system for study of intracellular organelle transport, especially actin-dependent motility.

2 Materials

2.1 Poly-L-Lysine-Coated Coverslips

1. 22 mm × 22 mm No. 1 coverslips.
2. Nitric acid.
3. Poly-L-lysine, (Sigma P-1524; >200 kD); 1 mg/mL in glass distilled, deionized water (ddH_2O). Store dry poly-L-lysine at −20 °C desiccated; store solution at −20 °C, it can be reused four times.

2.2 Dissection and Isolation of RPE Sheets

1. Green sunfish (*Lepomis cyanellus*) or bluegill (*Lepomis macrochirus*) 2–4 in. in length, commonly available at fish hatcheries. Maintain on 14 h:10 h light–dark cycle.
2. MS-222 (3-aminobenzoic acid ethyl ester, methane sulfonate salt, Sigma A-5040) anesthetic. Prepare 10 g/L solution in ddH_2O. Aliquot in volumes sufficient for dilution to 75 mg/L in fish tank water, and freeze at −20 °C.
3. Dark adaptation box with aeration.
4. Iris spring scissors (14 mm with curved blades; Fine Science Tools #15011-12) and forceps (Dumont #3).
5. HEPES-buffered Earles' Ringer (HER): 116.3 mM NaCl; 5.4 mM KCl; 1.8 mM $CaCl_2$; 0.8 mM $MgSO_4$; 1.0 mM Na_2HPO_4; 25.5 mM glucose; 4.0 mM $NaHCO_3$; 21 mM HEPES; pH 7.2. Made from 10× EBSS with phenol red in ddH_2O with additional 0.36 g/100 mL glucose; 0.034 g/100 mL $NaHCO_3$, and 0.50 g/100 mL HEPES. Make fresh or less than 24 h before use.
6. 35 mm petri dish lined with soft dental plastic or Parafilm.
7. 90 mm petri dish lined with white silicon rubber or Parafilm.

2.3 Preparation of Dissociated RPE Cells

1. Calcium- and magnesium-free HER (CMFHER) containing 1 mM EGTA: Made fresh from 10× CMFEBSS without phenol red. Additional glucose, $NaHCO_3$, and HEPES are added as described above for HER. pH 7.4 (*see* **Note 1**).
2. Papain digest solution: 10 U/mL papain (Worthington Biochemical, # LS003126) in 5 mL CMFHER activated with 0.14 mg/mL DNase and 3 μM L-cysteine. Papain suspension is aliquoted and stored at 4 °C. DNase and cysteine solutions are prepared fresh using 16.6 mg L-cysteine (Sigma C-8277) in 500 μL ddH_2O, and 10 mg DNase (Sigma DN25) in 500 μL ddH_2O, and kept on ice. Five milliliter papain digest solution is prepared per fish; added to that is enough papain for 10 U/mL, plus 64 μL L-cysteine solution and 36 μL DNase solution (*see* **Note 2**).
3. 35 mm petri dish lined with Parafilm.
4. 12–15 mL glass conical centrifuge tube.

5. HER supplemented with 0.4 % BSA (HERB).
6. HERB supplemented with 0.1 mg/mL DNase.
7. Poly-L-lysine coated coverslips in 35 mm petri dishes.
8. 100 cm glass petri dish with moistened filter paper (Whatman #1) in the cover.

2.4 Stimulation and Observation of Melanosome Motility

1. 1 mM cAMP (Sigma A-9501) in HERB made fresh from 10 mM cAMP/HERB stock, stored at −20 °C.
2. 10 μM dopamine (3-hydroxytyramine, Sigma H-8502) in HERB made fresh. Dopamine oxidizes easily; solutions should be remade after 1 h.
3. For perfusion chamber: standard microscope slides, double stick tape, wedges of filter paper (Whatman #3).

3 Methods

3.1 Preparation of Poly-L-Lysine Coated Coverslips

1. Soak coverslips in nitric acid for a few hours to overnight (*see* **Note 3**).
2. Wash coverslips in running tap water for 1 h, rinse 5× in deionized water.
3. Remove coverslips from racks and place in 10 cm petri dish with 1 mg/mL poly-L-lysine solution (PLL).
4. Rock or rotate for at least 1 h at room temperature.
5. Remove PLL and wash coverslips in ddH_2O at least 10× (*see* **Note 4**).
6. Rinse coverslips in 100 % ethanol and allow to air-dry on filter paper (Whatman # 1).
7. Store on filter paper; can keep for several months.

3.2 Dissection and Isolation of RPE Sheets

1. Thaw MS-222 and add to fish tank water in dark adaptation box (*see* **Note 5**).
2. Add fish and dark adapt 30 min to 2 h.
3. In dim light (0.002 $J/s/m^2$; *see* **Note 6**) enucleate eyes by blunt dissection using closed scissors; cut optic nerve. Place eyes on a paper towel lens side down.
4. Hemisect first eye; remove lens.
5. Holding edge of posterior eyecup with forceps (12 o'clock position) invert the eye cup retinal side down onto a paper towel. Gently press the back of the eyecup (sclera side) downwards onto the paper towel with the closed scissor blades so that the retina adheres to the paper towel; snip optic nerve to detach retina from the eyecup.

6. Transfer eyecup to dissection dish; hold edge of eyecup with forceps.
7. Gently irrigate the RPE/choroid with buffer (HER) using a Pasteur pipette to remove sheets of RPE (1–2 mm in diameter). RPE sheets look like black flecks in the medium. *See* **Note 7**.
8. Transfer RPE sheets to ca. 5 mL HER in wax-lined, 35 mm culture dish using a wide bore pipette or a glass 5 or 10 mL pipette (*see* **Note 8**).
9. Place 35 mm dish with sheets on a rotator at 30 rpm.
10. Repeat **steps 4–9** with the other eye; pool sheets into one dish.

3.3 Preparation of Dissociated RPE Cells

1. Transfer RPE sheets into papain solution in minimal volume of medium using a wide-bore or 10 mL serological pipette. Incubate for 30 min on rotator at 30 rpm.
2. After the digest period, transfer sheets to 3 mL HERB + DNase in a 12–15 mL glass conical centrifuge tube.
3. Allow sheets to settle to the bottom of the tube, then gently resuspend tissue using a Pasteur pipette. Aspirate off debris as tissue is resuspended. Remove solution and repeat in two more changes of 3 mL HERB + DNase (*see* **Note 9**).
4. Using a 5 mL serological pipette, transfer washed sheets to ca. 1 mL of HERB in a 1.5 mL microcentrifuge tube. Thoroughly wet a 1 mL syringe with 26 and 5/8 gauge needle by passing HERB in and out of the syringe a few times. Slowly dissociate sheets in the microcentrifuge tube by eight slow passes through the needle (four times in and out). Avoid air bubbles.
5. After dissociation, add 200–400 mL HERB-DNase to dilute the cell suspension. Optimal cell density is about 7.5×10^4 cells/mL (*see* **Note 10**).
6. Plate cells immediately after dissociation by adding 75–100 μL of the cell suspension to completely dry, poly-L-lysine coated coverslips in 35 mm dishes. Dishes should be placed in 100 cm glass petri plates with moistened filter paper in the cover before adding cells, and left undisturbed for 30 min–1 h, or until cells adhere (*see* **Note 11**).
7. After cells have adhered, change the medium to fresh HERB containing 10 μM dopamine to achieve full melanosome dispersion. Aspirate the old medium from the coverslip using a Pasteur pipette and immediately replace with 100 μL fresh medium. Repeat twice more for a total of three medium changes.

3.4 Stimulation and Observation of Melanosome Motility

1. To stimulate melanosome aggregation, change medium on coverslips to 1 mM cAMP in HERB using three changes of 100 μL. Full melanosome aggregation is achieved after 10–25 min. Melanosomes can be redispersed by washout of cAMP using 10 μM dopamine in HERB.

2. For time-lapse video observation, prepare a perfusion chamber using two strips of double-stick tape aligned along both sides of the long edge of a microscope slide. Add a drop of medium to the slide, and invert a coverslip with adherent cells onto the tape. View cells using a 40×–100× phase contrast objective on an upright microscope. Perfuse by pipetting medium on one side of the coverslip and absorbing it on the other side using a wedge of filter paper.

4 Notes

1. pH of CMFHER is adjusted to 7.4 initially; after addition of the cysteine solution pH will drop to 7.2.
2. Add papain to digest solution 10–15 min prior to dissection to allow 20–30 min activation of the enzyme before adding to RPE sheets. If multiple fish are to be dissected, add papain to separate aliquots of digest solution in Parafilm lined, 35 mm plates. Papain should be kept on ice until use.
3. Ceramic coverslip holders are available from Thomas Scientific (#8542E40); coverslips may be stored in nitric acid indefinitely until use.
4. Complete water rinse is critical as free PLL is toxic to cells.
5. We use plastic boxes (former rodent shoe-box style containers) that hold 3 L of water for up to six fish in each box. Boxes are placed in a custom made, light-tight wooden box with an aeration hose and bubbler connected to an aquarium pump outside the box. MS-222 will lower the pH of the water; measure pH before and after addition with pH paper and adjust pH up using a few drops of 10 N NaOH. MS-222 is a carcinogen; gloves should be worn.
6. We use a desk lamp with a 25 W incandescent bulb turned against the wall to let a small amount of light through.
7. After removal of RPE sheets, the back of the eye will remain black because of melanin pigment in the choroid.
8. Wet the inside of the pipette before use so tissue does not adhere to the glass. Do not let the sheets encounter the air–water interface. A wide-bore, fire polished pipette with a slight bend at the end also works well.
9. The washing step can be carried out in room light. Be sure not to introduce air bubbles into the solution during the washing step; likewise do not remove all medium between washes such that the tissue is exposed to the air. Each aliquot of wash solution should be added gently by slanting the tube and slowly trickling the wash solution down the side, so as not to

dissociate cells during washing. By holding the tube up to a black background, melanosomes from lysed RPE cells are visible in solution as the tissue is resuspended. Tissue should be washed in three changes of medium or until melanosomes from lysed cells are no longer visible.

10. Cells may be counted on a hemocytometer to gauge correct cell density, however cells must be plated immediately after dissociation or they will not adhere to the coverslips.
11. Be sure not to move cells after plating or they will not be distributed evenly on the coverslip. The moistened filter paper in the glass petri dish is important to maintain a humidified environment to discourage evaporation of the medium.

Acknowledgements

Thanks to Stephanie Santioleri for providing the scanning electron micrograph, and Jessica Oates for the fluorescence micrograph. This work was supported by an NIH AREA grant # R15 GM066961, a Howard Hughes Medical Institute Science Education Grant, and the Provost's Award for Funded Research from Saint Joseph's University.

References

1. Burnside B, King-Smith C (2010) Chapter 172: comparative eye: fish retinomotor movements. In: Dartt D, Besharse B, Dana R, Batelle B (eds) The encyclopedia of the eye. Elsevier, Amsterdam, The Netherlands
2. Garcia D, Burnside B (1994) Suppression of cAMP-induced pigment granule aggregation in RPE by organic anion transport inhibitors. Invest Ophthalmol Vis Sci 35:178–188
3. King-Smith C, Paz P, Lee C, Lam W, Burnside B (1997) Bi-directional pigment granule migration in isolated retinal pigment epithelial cells requires actin but not microtubules. Cell Motil Cytoskel 38:229–249
4. King-Smith C, Basciano P, Pham N (2001) Effect of the actin stabilizing drug, jasplakinolide, on pigment granule motility in the retinal pigment epithelium (RPE) of green sunfish, *Lepomis cyanellus*. Pigm Cell Res 14:14–22
5. King-Smith C, Vagnozzi RJ, Fischer NE, Gannon P, Gunnam M (2014) Orientation of actin filaments in teleost retinal pigment epithelial cells, and effect of the lectin, Concanavalin A, on melanosome motility. Vis Neurosci 31: 1–10
6. Barsoum IB, King-Smith C (2007) Myosin II and rho kinase are required for melanosome aggregation in fish retinal pigment epithelial (RPE) cells. Cell Motil Cytoskel 64:868–879
7. Canman JC, Bement WM (1997) Microtubules suppress actomyosin-based cortical flow in Xenopus oocytes. J Cell Sci 110(Pt 16): 1907–1917
8. Rosenblatt J, Cramer LP, Baum B, McGee KM (2004) Myosin II-dependent cortical movement is required for centrosome separation and positioning during mitotic spindle assembly. Cell 117:361–372

Chapter 18

Analysis of Stem Cell Motility In Vivo Based on Immunodetection of Planarian Neoblasts and Tracing of BrdU-Labeled Cells After Partial Irradiation

Junichi Tasaki, Chihiro Uchiyama-Tasaki, and Labib Rouhana

Abstract

Planarian flatworms have become an important system for the study of stem cell behavior and regulation in vivo. These organisms are able to regenerate any part of their body upon damage or amputation. A crucial cellular event in the process of planarian regeneration is the migration of pluripotent stem cells (known as neoblasts) to the site of injury. Here we describe two approaches for analyzing migration of planarian stem cells to an area where these have been ablated by localized X-ray irradiation. The first approach involves immunolabeling of mitotic neoblasts, while the second is based on tracing stem cells and their progeny after BrdU incorporation. The use of planarians in studies of cell motility is suitable for the identification of factors that influence stem cell migration in vivo and is amenable to RNA interference or pharmacological screening.

Key words Stem cells, Neoblasts, Planarian, Platyhelminthes, Mitosis, Regeneration, Cell migration, Cell motility

1 Introduction

An increasing number of researchers are turning to planarians for studies of stem cell regulation and function. Planarians present an exceptional opportunity to analyze stem cell dynamics in vivo due to the abundance of neoblasts present throughout their bodies, which are also the only proliferative cells in the soma of these organisms [1–3]. As such, neoblasts give rise to all differentiated cells during regeneration, growth, and homeostatic maintenance.

Neoblasts are classically defined as the cellular source for planarian regeneration, which can be lost after exposure to irradiation [4]. Normally, neoblasts comprise 20–30 % of all cells in these organisms [5], and are distributed throughout the mesenchyme of the planarian body [6, 7]. Upon amputation, neoblasts activate a proliferative response and accumulate at the site of injury, to form a mass of progenitor cells known as the blastema [8, 9].

Ray H. Gavin (ed.), *Cytoskeleton: Methods and Protocols*, Methods in Molecular Biology, vol. 1365, DOI 10.1007/978-1-4939-3124-8_18, © Springer Science+Business Media New York 2016

Cells in the blastema ultimately differentiate to compensate for lost structures in a process that requires ERK signaling and post-transcriptional regulation of gene expression [10, 11]. For decades, it has been postulated that blastema formation involves migration of neoblasts to the location where regeneration will take place. This concept was demonstrated by the observation that irradiated anterior portions of planarians were able to regenerate after irradiation and decapitation, as long as neoblasts in a distant location of the organism were shielded from irradiation [4]. Studies based on cellular morphology, tritium labeling, and grafting of planarian tissue, were able to confirm the motility of neoblasts during regeneration [12–21]. Most recently, Guedelhoefer and Sanchez Alvarado visualized the position of stem cells in partially irradiated planarians after head and/or tail amputation by neoblast-specific RNA in situ hybridization [22]. Using this approach, Guedelhoefer and Sanchez Alvarado were able to verify that neoblasts remain stationary during homeostasis but are actively recruited to the site of injury during regeneration [22].

The evidence for active recruitment of neoblasts to regenerating regions of planarians reinforces a number of questions regarding the mechanisms involved in somatic stem cell recruitment and motility. Fortunately, technical advancements in planarian research, such as RNA interference (RNAi) [23, 24], in situ hybridization [25–27], and RNAseq [28–36], will allow researchers to take advantage of this wonderful system to discover mechanisms of cellular migration present in stem cells and during regeneration. Here we present a simple step-by-step protocol to assess neoblast motility based on modifications to previously established partial irradiation approaches [22, 37] in combination with Histone H3 [pS10] immunodetection [7, 11] or 5-Bromo-2′-deoxyuridine (BrdU) labeling [7, 38] of neoblasts. Modifications presented in the current methodology offer a number of advantages: First, stem cell detection is based on labels that function across species, and can therefore be applied to clonal laboratory planarian lines, commercially purchased specimens, or samples caught in the wild; Secondly, the use of histone H3 [pS10] antibody requires only a small number of steps to infer planarian stem cell motility in vivo from whole-mount samples; Lastly, the use of BrdU for analysis of cell migration overcomes underestimations presented by the use neoblast markers to measure motility of stem cells during regeneration, being that neoblasts differentiate on transit to their final destination [22, 38, 39]. The described approach for the analysis of cell motility will facilitate the identification of factors modulating stem cell migration, such as signaling peptides and their receptors, components of phosphorylation cascades, or proteins involved in localized translation of mRNAs. This methodology could also be utilized for testing or screening chemicals capable of accelerating or inhibiting the recruitment of stem cells to sites of injury.

2 Materials

2.1 Planarians and Husbandry Media

1. Cultures of planarian species *Dugesia dorotocephala* are commercially available (Catalog No. 132970, Carolina Biological Supply Company, Burlington, NC). Clonal lines of species such as *Schmidtea mediterranea* and *Dugesia japonica* can be requested from academic laboratories in North America, Europe and Japan (*see* **Note 1**).
2. Husbandry media (0.75 × solution of Montjuïc salts). Montjuïc salts configuration as per [40, 41]): 1.6 mM NaCl, 1 mM $CaCl_2$, 1 mM $MgSO_4$, 0.1 mM $MgCl_2$, 0.1 mM KCl, 0.2 mM $NaHCO_3$. This solution is prepared with ultrapure water from filtered stocks of 5 M NaCl, 1 M $CaCl_2$, 1 M $MgSO_4$, 1 M $MgCl_2$, and 1 M KCl (*see* **Notes 2** and **3**).
3. Plastic or glass containers (e.g., Ziploc 40 or 72 oz plastic container, BPA-free).
4. Organic beef or chicken liver. Obtain fresh and store aliquots at −80 °C (*see* **Note 4**).
5. Plastic transfer pipettes.
6. 500 mL wash bottles.
7. Disposable paper towels.

2.2 X-Ray Irradiation

1. X-ray irradiation cabinet (e.g., Model 43855A [110 kVp, 3 mA], Faxitron Bioptics LLC, Tucson, AZ) (*see* **Note 5**).
2. Aluminum cool plate (e.g., CSAF-9600, Diversified Biotech, Dedham, MA).
3. Lead foil of 0.0156 in. minimal thickness cut in approximately 0.5 cm × 3 cm and 2 cm × 5 cm rectangles (Item. No. 871862, Carolina Biologicals, Burlington, NC) (*see* **Note 6**).
4. Whatman filter paper No. 3 (Catalog No. 1003-185, GE Healthcare Biosciences, Pittsburgh, PA).
5. Polystyrene box of approximately 10 in. length × 9 in. width × 6 in. height dimensions.
6. Laboratory ice.
7. Paint brush, small (e.g., Cra-Z-Art Artist Brushes, 7 ct).

2.3 Detection of Mitotic Stem Cells by Immunofluorescence

1. PBSTx: Phosphate-buffered saline (PBS) supplemented with 0.3 % Triton X-100. A 10× PBS stock solution is prepared by adding 80 g of NaCl, 2 g of KCl, 14.4 g of Na_2HPO_4, and 2.4 g of KH_2HPO_4, to reach a final volume of 1 L using ultrapure water. Adjust to pH 7.4 with HCl, and autoclave. Dilute 10× PBS stock into a 1× PBS working solution using ultrapure water and supplement each liter of 1× PBS with 3 mL of Triton X-100. Store at room temperature.

2. Husbandry media supplemented with 2 % HCl. Made fresh right before use by mixing 12 N HCl 1:19 (vol/vol) into husbandry media (*see* **Note 7**).
3. Methacarn fixative (6:3:1 methanol:chloroform:acetic acid). Make and store at −20 °C.
4. Methanol (99.5–99.9 % purity).
5. Bleaching solution: Methanol supplemented with 6 % H_2O_2. This sample is prepared fresh before application by mixing 4 parts of methanol to 1 part 30 % hydrogen peroxide (*see* **Note 7**).
6. Blocking solution #1: As per [38], made by supplementing PBSTx with 0.45 % gelatin from cold water fish skin (Catalog No. G7765-250ML, Sigma, St. Louis, MO) and 1 % bovine serum albumin. Stable stored at 4 °C for at least 2 weeks.
7. Histone H3 [pS10] rabbit polyclonal antibody (Catalog No. 44-1190G, Life Technologies, Carlsbad, CA).
8. Alexa Fluor® 568 goat anti-rabbit IgG (Catalog No. A-11011, Life Technologies, Carlsbad, CA) (*see* **Note 8**).
9. Mounting solution: Glycerol diluted to 80 % in PBS.
10. Microscope slides and covers (e.g., Catalog No. HD159871B and HEA159879G, Heathrow Scientific LLC, Vernon Hills, IL).
11. Fluorescence microscope (e.g., Zeiss AXIO Zoom.V16 with 578/603 nm excitation/emission capabilities) (*see* **Note 8**).

2.4 BrdU Pulse-Labeling and Detection

1. Low-melting-point agarose.
2. Liver solution: 1:2 (vol/vol) organic calf liver puree:ultrapure water.
3. 5-Bromo-2′-deoxyuridine.
4. Curwood Parafilm M wrapping film.
5. Disposable Petri dishes, 100 × 15 mm size.
6. Food color dye (e.g., Assorted NEON! Food colors & egg dye).
7. 5/8 Holtfreter's solution: 37 mM NaCl, 0.42 mM KCl, 0.57 mM $CaCl_2$, 1.5 mM $NaHCO_3$, pH 7.4. Prepare by adding 21.88 g of NaCl, 0.63 g of CaCl, 0.31 g of KCl and 1.25 g of $NaHCO_3$, and ultrapure water to reach a final volume of 10 L. Adjust to pH 7.4 by adding HCl. Store at room temperature.
8. 2 % HCl in 5/8 Holtfreter's solution. Make fresh right before use by diluting 12 N HCl 1:19 (vol/vol) into 5/8 Holtfreter's (*see* **Note 7**).
9. Formaldehyde fixative: 4 % paraformaldehyde and 5 % methanol in 5/8 Holtfreter's. To prepare 40 mL formaldehyde fixative, add 1.6 g paraformaldehyde to 30 mL of 5/8 Holtfreter's solution in a beaker and stir under a hood for 1 h at 60 °C.

Cool to room temperature and remove insoluble particles by filtration. Then, add 2 mL of methanol and adjust to a final volume of 40 mL with 5/8 Holtfreter's solution.

10. Bleaching solution: Methanol supplemented with 6 % H_2O_2. This sample is prepared fresh before application by mixing 4 parts of methanol to 1 part 30 % hydrogen peroxide (*see* **Note 7**).
11. Xylene-methanol solution: 1:1 (vol/vol) mixture of xylene and methanol.
12. Serial dilutions of 75, 50, and 25 % methanol (vol/vol) in 5/8 Holtfreter's solution.
13. PBS (*see* Subheading 2.3, **item 1**).
14. PBSTx: PBS containing 0.3 % Triton X-100.
15. Proteinase K solution: 20 mg/mL Proteinase K stock (Catalog No. AM2546, Invitrogen, Carlsbad, CA) diluted 1:10,000 into PBSTx to a final concentration of 2 μg/mL.
16. 2 N HCl solution: 12 N hydrochloric acid diluted 1:5 (vol/vol) with PBSTx.
17. Blocking solution #2: PBSTx containing 5 % horse serum (Catalog No. H1138-500ML, Sigma, St. Louis, MO) and 0.5 % Western Blocking Reagent (Catalog No. 11921673001, Roche, Mannheim, Germany). This solution is prepared by adding 2.5 mL of horse serum and 2.5 mL of Roche Western Blocking Reagent to 45 mL of PBSTx in a 50 mL conical tube. Stable at 4 °C for up to 2 weeks.
18. Purified anti-BrdU monoclonal antibody clone B44 (Catalog No. 347580, BD Biosciences, San Jose, CA).
19. Goat anti-mouse IgG (H + L), horseradish peroxidase (HRP)-conjugated secondary antibody (Catalog No. G-21040, Life Technologies, Carlsbad, CA).
20. Tyramide Signal Amplification (TSA) system (e.g., Cy3, Catalog No. NEL744001KT, PerkinElmer Inc, Waltham, MA).
21. Mounting solution: Glycerol diluted to 80 % in PBS.
22. Microscope slides and covers (e.g., Catalog No. HD159871B and HEA159879G, Heathrow Scientific LLC, Vernon Hills, IL).

3 Methods

3.1 Handling and Maintenance of Planarian Colonies

1. Upon collection or receipt of samples, transfer gradually into laboratory husbandry media. For the first week use a 1:1 dilution of the solution in which planarians are found and either 0.75× Montjuïc salts or Ice Mountain spring water. If planarian morphology looks normal after 5–7 days, then switch completely to husbandry media (*see* **Note 2**).

2. Clean plastic or glass containers sequentially with 70 % ethanol, tap water, deionized water, and ultrapure water. Dry containers using disposable paper towels. Avoid any possible contact with chemicals such as detergents. Planarians can be kept in the same container for long periods of time; however the containers must be cleaned once every 1–2 weeks or after every feeding session. Subsequent cleaning of containers does not require ethanol.
3. Maintain planarians under dark conditions as much as possible and do not seal containers shut to allow for aeration. Most planarians can be maintained between 18 and 24 °C. Temperature adjustments may be needed to mimic their natural habitat.
4. Planarians should be fed once every 2 weeks to maintain their size and number, starvation periods for longer than 2 weeks will lead to de-growth. Feed once to twice per week for growth in size and (in case of asexual strains) number. A piece of liver of approximately 1 cm^3 in size is enough to feed 50 planarians of 1 cm length. Place in the piece of liver into the bucket with planarians and allow 2 h in the dark for feeding. If food is completely eaten, then increase liver quantity in subsequent feedings. Planarians do not eat floating liver; so remove air bubbles in liver if it fails to sink to the bottom of the container. Clean containers (as in **step 2**) after feeding and replace husbandry media. A squirt bottle can be used to rinse out planarians, which can also be placed in a clean Petri dish while cleaning the container and replacing old media.
5. To increase population size under laboratory husbandry, cut planarians with a clean scalpel while immobilized on an iced aluminum plate (*see* immobilization procedure in Subheading 3.2; below). Return to clean husbandry media and bucket by rinsing of Whatman paper with transfer pipette or wash bottle. Return planarians to dark and change media again in the next couple of days. Allow regeneration to occur for 2 weeks before reinstating feeding regimen.
6. Planarians should not be fed for at least a week prior to experimentation in order to avoid artifacts.

3.2 Localized Depletion of Planarian Stem Cells by Irradiation

To analyze stem cell migration, the planarian stem cell population must be depleted in the anterior two thirds of the planarian anatomy. This is achieved by protecting stem cells in the posterior end with a lead foil during X-ray exposure as outlined in Figure 1.

1. To immobilize planarians, an irradiation stage must be assembled. To achieve this, place an aluminum cool plate on a polystyrene box full of ice and allow to reach ice temperature.
2. Cut a piece of Size 3 Whatman filter paper to fit on top of the aluminum cool plate without overhangs.

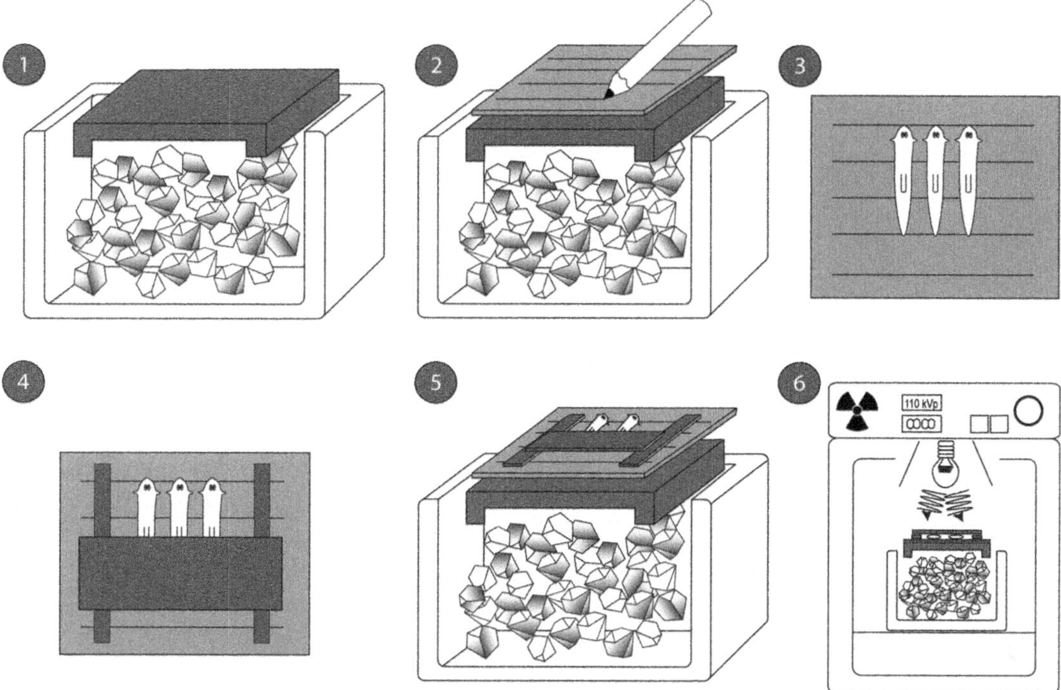

Fig. 1 Schematic illustration of stepwise assembly of irradiation stage and ablation of stem cells by exposure to X-rays. See text for details

3. Mark horizontal lines separated by 0.5 cm using a pencil and ruler.
4. Lightly moist the Whatman paper using husbandry media and place it on the aluminum cool plate, then place planarians on the Whatman paper using a transfer pipette. Arrange the samples by dragging gently using a small artist brush. Line up samples with the posterior end of their pharynx located on the same line and within 2–4 mm of each other (Fig. 1). Keep surface moist enough to allow the sliding of planarians around the stage without damaging them, but not so moist that planarians move around on their own. When ready, remove all excess liquid from the stage by pressing gently against the end of the Whatman paper with a dry paper towel to prevent samples from moving during irradiation.
5. Place lead foil spacers (example size: 0.5 cm × 3 cm) to the left and right of each line of samples. Then place a third lead foil (example size: 2 cm × 5 cm) on top of the spacers and lined up with the back end of the pharynx, covering the area of the samples that is to be protected from irradiation (Fig. 1).
6. Place the undisturbed polystyrene box, ice, stage, and samples inside the second stage of a Model 43855A Faxitron irradiation

cabinet (or equivalent) and apply a 10-min X-ray exposure at 110 kVp (*see* **Notes 5** and **9**).

7. Return planarians to husbandry container by rinsing off the Whatman paper with husbandry media with a wash bottle. Replace old husbandry media with fresh husbandry media and allow recovery for 12 h.
8. Analyze irradiated planarians and discard samples with damaged tissue (i.e., blisters, tissue loss).
9. Continue on to procedures described in Subheading 3.3 to analyze the position of mitotic stem cells migrating to irradiated tissue using a simplified immunodetection protocol. Alternatively, continue on to the methodology described in Subheading 3.4 for a more accurate calculation of cellular migration using more involved cell-tracing methodologies based on BrdU uptake by neoblasts in non-irradiated tissue.

3.3 Positional Analysis of Mitotic Stem Cell by Whole-Mount Immunofluorescence

To assess the migration of stem cells from protected to irradiated tissue, the position of a fraction of stem cells is visualized using antibodies with affinity to the M-phase-specific marker histone H3 [pS10]. Following a recovery period of at least 24 h after partial irradiation, samples are amputated posterior to photoreceptors and returned to husbandry conditions for 4–10 days until fixation for whole-mount immunofluorescence analysis.

1. For amputation, immobilize planarians on a damp size 3 Whatman paper placed on an ice-temperature irradiation stage (*as in* Subheading 3.2, **steps 1**, **2**, **4**) and amputate test samples posterior to photoreceptors, then return to husbandry conditions until fixation.
2. For fixation, transfer each group of planarians to a 15 mL conical tube (up to 15 individuals), or 50 mL conical tube (up to 50 individuals), and rinse using a wash bottle with husbandry media a couple of times.
3. Remove all but approximately 0.5 mL of media and allow samples to extend their anatomy by placing tubes in a rack. Then, quickly fill up the tube to 90–95 % capacity with Husbandry media supplemented with 2 % HCl. Proceed to rock samples immediately for 5 min at room temperature on a platform rocker or by hand (semi-gentle agitation; switching direction once per second).
4. Allow planarians to sink to the bottom of the tubes and decant most of the HCl solution, leaving just enough liquid to keep samples from drying. Then, fill up tubes between 90 and 95 % capacity with Methacarn fixative. Incubate for 2 h at room temperature with gentle agitation.
5. Allow planarians to sink to the bottom of the tubes, decant Methacarn fixing solution, and fill up tubes to 90–95 % capacity

with methanol pre-cooled at −20 °C. Immediately proceed to vortex samples for 1 min, and then rock for an additional 10 min at room temperature (*see* **Note 10**).

6. Replace methanol with bleaching solution and expose to fluorescent light at a distance of approximately 2 cm. Cover surfaces of the tube that are not directly exposed to light with aluminum foil to reflect fluorescent light to the entire sample surface. Incubate for 6 h to overnight. Samples should be white by the end of exposure.
7. Decant Bleaching solution and incubate samples for 15 min rocking at room temperature in 1:1 solution of PBSTx and methanol.
8. Decant Bleaching solution and incubate samples for 15 min rocking at room temperature in PBSTx.
9. Transfer samples to a microcentrifuge tube and replace PBSTx with blocking solution #1. Incubate samples in blocking solution #1 for 2 h at room temperature or overnight at 4 °C.
10. Replace Blocking solution with new blocking solution #1 containing a 1:1000 dilution of histone H3 [pS10] polyclonal antibody. Incubate by rocking at room temperature for 6 h or overnight at 4 °C.
11. Remove antibody solution and incubate samples in PBSTx while rocking at room temperature for 15 min. Repeat this washing step three more times.
12. Replace PBSTx with blocking solution #1 supplemented with a 1:500 dilution of Alexa Fluor® 568 conjugated anti-rabbit secondary. Incubate by rocking at room temperature for 6 h or overnight at 4 °C.
13. Remove secondary antibody solution and incubate samples in PBSTx while rocking at room temperature for 15 min. Repeat this washing step three more times.
14. Mount samples on a glass slide using modeling clay, electrical tape or commercial adhesive microscope slide spacers, in order to avoid damaging the samples. Remove excess PBSTx and replace with mounting solution. Seal slide cover with clear nail polish and allow to dry protected from light.
15. Proceed to fluorescence microscopy imaging and analysis as in (Fig. 2) or store at 4 °C for future analysis.

3.4 Comprehensive Analysis of Cell Motility by Tracing of BrdU-Labeled Neoblasts

To analyze migration of stem cells and differentiated progeny, planarians are subjected to a BrdU pulse 3 days after partial irradiation, amputated 2 days after BrdU labeling, and processed for cellular analysis 1 day after amputation.

1. Prepare a BrdU stock solution by dissolving 25 mg per mL of ultrapure water, heat at 70 °C for 5 min with inversion after

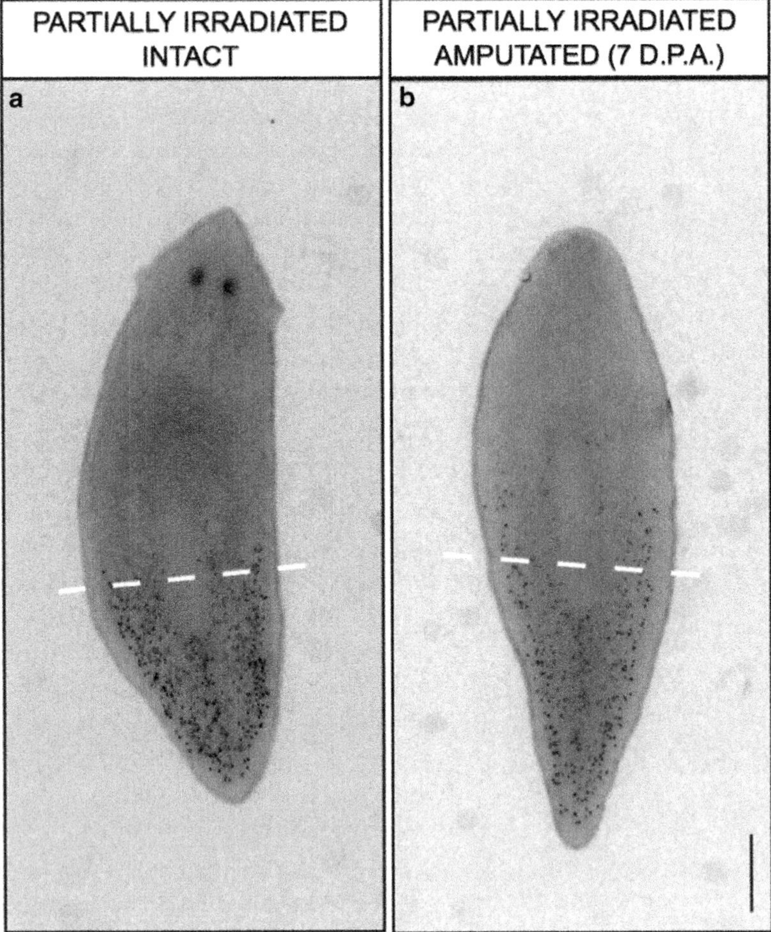

Fig. 2 Accelerated stem cell migration in partially irradiated *Dugesia dorotocephala* subjected to amputation. Mitotic cells (*black dots*) are detected by Histone H3 [pS10] immunofluorescence in intact (**a**) and amputated (**b**) planarians 7 days post-amputation (D.P.A.). *Dashed line* illustrates estimated border between irradiated and non-irradiated regions of the planarians. Scale bar: 500 μm

every minute, then use immediately or aliquot and store at −80 °C. BrdU stock is stable for a week when stored under −80 °C. Maintain protected from light.

2. Mix 60 μL of liver solution with 30 μL of 2 % Agar and 0.5 μL of food dye, then add 21 μL of BrdU stock solution and mix to homogeneity quickly. Aliquot BrdU-liver mix into 14 μL droplets on a piece of Parafilm wrapping film and allow to solidify by incubating at −20 °C for 15 min (Fig. 3).
3. Place the BrdU-containing droplets into the husbandry containers housing the partially irradiated planarians and allow planarians to feed from the droplets for 40 min at room temperature under dark conditions (Fig. 3).

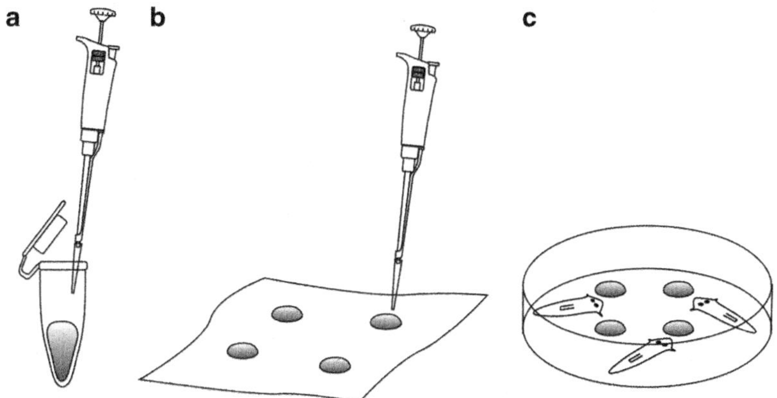

Fig. 3 BrdU feeding procedure. (**a**) BrdU stock [25 mg/mL] is mixed with liver solution, 2 % agarose and food dye. (**b**) Liver containing BrdU is aliquoted on Parafilm and solidified at −20 °C. (**c**) Planarians are allowed to eat from the droplets for 40 min to incorporate BrdU

4. Remove the droplets from the husbandry container and change culture media. Return planarians to dark husbandry conditions and allow 1 day for recovery before amputation.
5. Immobilize planarians over moist size 3 Whatman paper placed on an ice-temperature irradiation stage (as in Subheading 3.2, **steps 1** and **2**), amputate test samples posterior to photoreceptors, and then return to husbandry conditions until fixation.
6. Begin fixation at least 1 day of post-amputation to allow for the initiation of regeneration and cell motility. Then, remove culture media from the containers housing the BrdU-labeled planarians.
7. Pour 2 % HCl in Holtfreter's solution into the dish and immediately transfer planarians to 15 mL conical tubes (for up to 15 planarians). Fill conical to 90–95 % capacity with HCl solution and proceed to rock samples for 6 min at room temperature by hand or with a laboratory shaker.
8. Remove 2 % HCl in Holtfreter's solution and rock samples in formaldehyde fixative for 2 h at 4 °C.
9. Remove the formaldehyde fixative and replace with ice-cold methanol (*see* **Note 10**).
10. Remove methanol and replace with a 1:1 xylene:methanol solution. Incubate while rocking for 30 min at 4 °C.
11. Remove xylene:methanol solution and wash samples briefly with ice-cold methanol.
12. Rehydrate samples with gradual dilutions of methanol (75, 50, and 25 %) in 5/8 Holtfreter's. Incubate samples with gentle agitation 30 min at 4 °C for each dilution.

13. Incubate samples in PBSTx under gentle agitation for 30 min at 4 °C.
14. Transfer samples from conical tubes to 2 mL microcentrifuge tubes to reduce volumes used for antibody detection of incorporated BrdU.
15. Incubate planarians without agitation in Proteinase K solution for 8 min at 37 °C.
16. Remove the Proteinase K solution and rock samples for 30 min at room temperature in 2 N HCl solution (*see* **Note 7**).
17. Wash samples three times by rocking in PBSTx for 10 min at room temperature.
18. Incubate samples in blocking solution #2 for 1–2 h at 4 °C under gentle agitation.
19. Remove the blocking solution and treat samples with a 1:25 dilution of anti-BrdU purified monoclonal antibody (clone B44) in blocking solution #2 at 4 °C overnight with gentle agitation.
20. Remove the primary antibody solution and wash samples at least six times with PBSTx for 30 min at 4 °C.
21. Wash samples briefly with blocking solution #2 and incubate in a 1:100 dilution of HRP-conjugated anti-mouse secondary antibody in the blocking solution #2 overnight at 4 °C under gentle agitation.
22. Remove the secondary antibody solution and wash samples for at least six times with PBSTx for 30 min at 4 °C.
23. Develop samples for detection using PerkinElmer TSA system following the instructions from the manufacturer (*see* **Note 11**).
24. After the incubation, wash samples two times with PBSTx for 15 min at room temperature.
25. Mount samples on microscopic slides as instructed in Subheading 3.3, **step 14**, and analyze by fluorescent microscopy or store at 4 °C for future analysis as in (Fig. 4).

4 Notes

1. Planarians are broadly distributed in fresh water streams and lakes around the world. These can be brought to the laboratory and maintained in planarian husbandry media prepared in the laboratory or commercially available spring water (*see* **Note 3**). It is important to mimic source temperature and water quality. Maintain planarians in the dark as much as possible.
2. Use only ultrapure deionized water (resistivity of 18.2 MΩ cm at 25 °C) to prepare Montjuïc salts.

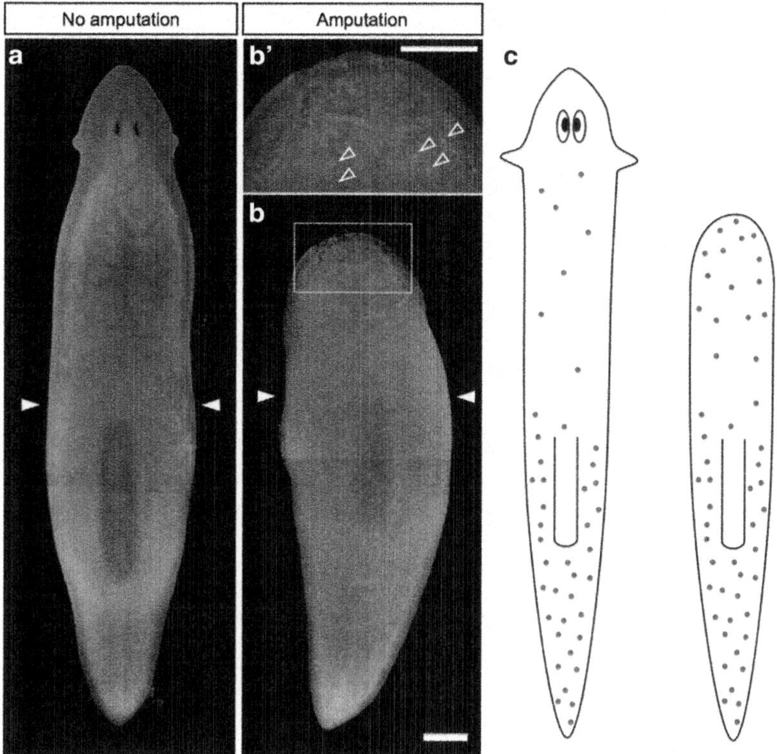

Fig. 4 Detection of BrdU-labeled stem cells and stem cell progeny in intact (**a**) and 1 day post-amputation (**b**) *Dugesia dorotocephala* planarians subjected to partial (anterior) irradiation. (**b′**) Magnified view of (**b**). BrdU-positive cells are visualized as *white dots*, some of which are detected in X-ray irradiated and amputated regions. (**c**) Schematic illustration of panels (**a**) and (**b**). *Non-filled arrowhead* shows border of area protected from X-ray irradiation. *Filled arrowhead* indicates BrdU-labeled cells located in the amputated end of irradiated planarian tissue. Scale bars: 500 μm

A 5× Montjuïc salts concentrated stock solution is prepared by adding 1.6 mL of 5 M NaCl, 5 mL of 1 M $CaCl_2$, 5 mL of 1 M $MgSO_4$, 0.5 mL of 1 M $MgCl_2$, 0.5 mL of 1 M KCl, and 0.504 g of $NaHCO_3$ per 1 L of ultrapure water. Stir the 5× stock for over 20 min and adjust pH to 7.0 with 2 N HCl. Stir overnight and confirm that the pH of the solution is within 6.8–7.2 after stirring overnight. Dilute 5× Montjuïc salts stock to working concentration of 0.75× (e.g., dilute 2.25 L of 5× Montjuïc salts with 12.75 L of ultrapure water in 15 L Nalgene Lowboy autoclavable carboys) mix and store at room temperature. Carboys are autoclaved empty once every 2–3 months to avoid contamination.

3. Ice Mountain natural spring water (Nestlé Waters North America, Stamford, CT) can be directly used as husbandry media for some planarian species.
4. Liver should be fresh and either minced or cut into small chunks within 24 h of purchase. Minced aliquots can be stored at −80 °C in small sterile Petri dishes and chunks wrapped in aluminum foil.
5. Laboratories have used different irradiation sources for ablation of neoblasts, both isotope and X-ray based. Settings for sufficient irradiation dosage must be established for individual X-ray source/planarian species combinations. A good starting dosage range is 3–8 Gy. Specific planarian species have been shown to display different minimal dosage sensitivity [42].
6. Lead is hazardous if swallowed and should be handled with care. Cut under a hood using a box-cutting knife and a metal ruler used exclusively for this purpose. Use disposable gloves and tweezers for handling.
7. To prepare 2 N HCl or other solutions, always add large solvent (e.g., husbandry media, methanol) first into receptacle and acid or solute (HCl, hydrogen peroxide) second to avoid a violent reaction.
8. A combination of Alexa Fluor® 488 goat anti-rabbit IgG (Catalog No. A-11008, Life Technologies, Carlsbad, CA) is also commonly used for this procedure. Choose appropriate secondary antibody depending on availability of fluorescent microscope excitation/emission capabilities.
9. Irradiation conditions must be empirically determined for each irradiation cabinet. This can be easily achieved by monitoring survival in groups of planarians subjected to different duration of irradiation exposure given a fixed position and voltage. Groups of planarians with sufficient irradiation exposure to ablate all neoblasts will begin to show signs of homeostatic loss during the second week following irradiation and eventually die. The lowest irradiation dosage that is sufficient to achieve complete lethality is usually chosen as baseline for future experiments.
10. Samples can be stored at −20 °C for 2 or 3 days after this step.
11. Development incubation periods longer than 25 min may lead to high background.

Acknowledgements

We thank Kimberly Morris for assistance with establishing irradiation conditions, Ricardo Pineda and Lucas Morton for help with initial partial irradiation trials, and Steve Sayson for revising early versions of the manuscript. This work was supported by startup funds to LR from the Department of Biological Sciences and the College of Science and Mathematics at Wright State University.

References

1. Elliott SA, Sanchez Alvarado A (2013) The history and enduring contributions of planarians to the study of animal regeneration. Wiley Interdiscip Rev Dev Biol 2(3):301–326. doi:10.1002/wdev.82
2. Newmark PA, Sanchez Alvarado A (2002) Not your father's planarian: a classic model enters the era of functional genomics. Nat Rev Genet 3(3):210–219. doi:10.1038/nrg759
3. Shibata N, Rouhana L, Agata K (2010) Cellular and molecular dissection of pluripotent adult somatic stem cells in planarians. Dev Growth Differ 52(1):27–41. doi:10.1111/j.1440-169X.2009.01155.x
4. Dubois F (1949) Contribution á l'ètude de la migration des cellules de règènèration chez les *Planaires dulcicoles*. Bull Biol Fr Belg 83: 213–283
5. Baguñá J, Romero R (1981) Quantitative analysis of cell types during growth, degrowth and regeneration in the planarians *Dugesia mediterranea* and *Dugesia tigrina*. Hydrobiologia 84: 181–194
6. Shibata N, Umesono Y, Orii H, Sakurai T, Watanabe K, Agata K (1999) Expression of vasa(vas)-related genes in germline cells and totipotent somatic stem cells of planarians. Dev Biol 206(1):73–87. doi:10.1006/dbio.1998.9130
7. Newmark PA, Sanchez Alvarado A (2000) Bromodeoxyuridine specifically labels the regenerative stem cells of planarians. Dev Biol 220(2):142–153. doi:10.1006/dbio.2000.9645
8. Wenemoser D, Reddien PW (2010) Planarian regeneration involves distinct stem cell responses to wounds and tissue absence. Dev Biol 344(2):979–991. doi:10.1016/j.ydbio.2010.06.017
9. Tasaki J, Shibata N, Sakurai T, Agata K, Umesono Y (2011) Role of c-Jun N-terminal kinase activation in blastema formation during planarian regeneration. Dev Growth Differ 53(3):389–400. doi:10.1111/j.1440-169X.2011.01254.x
10. Tasaki J, Shibata N, Nishimura O, Itomi K, Tabata Y, Son F, Suzuki N, Araki R, Abe M, Agata K, Umesono Y (2011) ERK signaling controls blastema cell differentiation during planarian regeneration. Development 138(12): 2417–2427. doi:10.1242/dev.060764
11. Rouhana L, Shibata N, Nishimura O, Agata K (2010) Different requirements for conserved post-transcriptional regulators in planarian regeneration and stem cell maintenance. Dev Biol 341(2):429–443. doi:10.1016/j.ydbio.2010.02.037
12. Stephan-Dubois F (1956) Migration and differentiation of neoblasts in anterior regeneration of Lumbriculus variegatus (Annelida, Oligochaeta). C R Seances Soc Biol Fil 150(6): 1239–1242
13. Stephan-Dubois F, Kolmayer S (1959) Migration and differentiation of the cells of regeneration in the planarian Dendrocoelum lacteum. C R Seances Soc Biol Fil 153: 1856–1858
14. Stephan-Dubois F, Gilgenkrantz F (1961) Transplantation and regeneration in the planarian, Dendrocoelum lacteum. J Embryol Exp Morphol 9:642–649
15. Wolff E (1961) Migrations et contacts cellulaires dans la régénération. Exp Cell Res 8:246–259
16. Cecere F, Grasso M, Urbani E, Vannini E (1964) Osservazioni autoradiografiche sulla rigenerazione di *Dugesia lugubris*. Rend Ist Sci Camerino 5:193–198
17. Lender T, Gabriel A (1965) Neoblasts labelled with tritiated uridine migrate and construct the regeneration blastema in fresh-water planaria. C R Hebd Seances Acad Sci 260:4095–4097
18. Sugino H, Okuno Y, Yoshinobu J (1970) Effect of transplanted pieces from non X-irradiated worms on irradiated ones in *Dugesia japonica*. Mem Osaka Kyoiku Univ 19:63–76
19. Salo E, Baguna J (1985) Cell movement in intact and regenerating planarians. Quantitation using chromosomal, nuclear and cytoplasmic markers. J Embryol Exp Morphol 89:57–70
20. Salo E, Baguna J (1989) Regeneration and pattern-formation in Planarians. 2. Local origin and role of cell movements in blastema formation. Development 107(1):69–76
21. Wagner DE, Wang IE, Reddien PW (2011) Clonogenic neoblasts are pluripotent adult stem cells that underlie planarian regeneration. Science 332(6031):811–816. doi:10.1126/science.1203983
22. Guedelhoefer OC, Sanchez Alvarado A (2012) Amputation induces stem cell mobilization to

sites of injury during planarian regeneration. Development 139(19):3510–3520. doi:10.1242/dev.082099

23. Sanchez Alvarado A, Newmark PA (1999) Double-stranded RNA specifically disrupts gene expression during planarian regeneration. Proc Natl Acad Sci U S A 96(9):5049–5054
24. Rouhana L, Weiss JA, Forsthoefel DJ, Lee H, King RS, Inoue T, Shibata N, Agata K, Newmark PA (2013) RNA interference by feeding in vitro-synthesized double-stranded RNA to planarians: methodology and dynamics. Dev Dyn 242(6):718–730. doi:10.1002/dvdy.23950
25. Umesono Y, Watanabe K, Agata K (1997) A planarian orthopedia homolog is specifically expressed in the branch region of both the mature and regenerating brain. Dev Growth Differ 39(6):723–727
26. Pearson BJ, Eisenhoffer GT, Gurley KA, Rink JC, Miller DE, Sanchez Alvarado A (2009) Formaldehyde-based whole-mount in situ hybridization method for planarians. Dev Dyn 238(2):443–450. doi:10.1002/dvdy.21849
27. King RS, Newmark PA (2013) In situ hybridization protocol for enhanced detection of gene expression in the planarian Schmidtea mediterranea. BMC Dev Biol 13:8. doi:10.1186/1471-213X-13-8
28. Rouhana L, Vieira AP, Roberts-Galbraith RH, Newmark PA (2012) PRMT5 and the role of symmetrical dimethylarginine in chromatoid bodies of planarian stem cells. Development 139(6):1083–1094. doi:10.1242/dev.076182
29. Blythe MJ, Kao D, Malla S, Rowsell J, Wilson R, Evans D, Jowett J, Hall A, Lemay V, Lam S, Aboobaker AA (2010) A dual platform approach to transcript discovery for the planarian *Schmidtea mediterranea* to establish RNAseq for stem cell and regeneration biology. PLoS One 5(12):e15617. doi:10.1371/journal.pone.0015617
30. Abril JF, Cebria F, Rodriguez-Esteban G, Horn T, Fraguas S, Calvo B, Bartscherer K, Salo E (2010) Smed454 dataset: unravelling the transcriptome of *Schmidtea mediterranea*. BMC Genomics 11:731. doi:10.1186/1471-2164-11-731
31. Adamidi C, Wang Y, Gruen D, Mastrobuoni G, You X, Tolle D, Dodt M, Mackowiak SD, Gogol-Doering A, Oenal P, Rybak A, Ross E, Sanchez Alvarado A, Kempa S, Dieterich C, Rajewsky N, Chen W (2011) De novo assembly and validation of planaria transcriptome by massive parallel sequencing and shotgun proteomics. Genome Res 21(7):1193–1200. doi:10.1101/gr.113779.110
32. Nishimura O, Hirao Y, Tarui H, Agata K (2012) Comparative transcriptome analysis between planarian *Dugesia japonica* and other platyhelminth species. BMC Genomics 13:289. doi:10.1186/1471-2164-13-289
33. Solana J, Kao D, Mihaylova Y, Jaber-Hijazi F, Malla S, Wilson R, Aboobaker A (2012) Defining the molecular profile of planarian pluripotent stem cells using a combinatorial RNAseq, RNA interference and irradiation approach. Genome Biol 13(3):R19. doi:10.1186/gb-2012-13-3-r19
34. Lapan SW, Reddien PW (2012) Transcriptome analysis of the planarian eye identifies ovo as a specific regulator of eye regeneration. Cell Rep 2(2):294–307. doi:10.1016/j.celrep.2012.06.018
35. Resch AM, Palakodeti D, Lu YC, Horowitz M, Graveley BR (2012) Transcriptome analysis reveals strain-specific and conserved stemness genes in *Schmidtea mediterranea*. PLoS One 7(4):e34447. doi:10.1371/journal.pone.0034447
36. Sikes JM, Newmark PA (2013) Restoration of anterior regeneration in a planarian with limited regenerative ability. Nature 500(7460):77–80. doi:10.1038/nature12403
37. Guedelhoefer OC, Sanchez Alvarado A (2012) Planarian immobilization, partial irradiation, and tissue transplantation. J Vis Exp (66). doi:10.3791/4015
38. Forsthoefel DJ, Park AE, Newmark PA (2011) Stem cell-based growth, regeneration, and remodeling of the planarian intestine. Dev Biol 356(2):445–459. doi:10.1016/j.ydbio.2011.05.669
39. Lapan SW, Reddien PW (2011) dlx and sp6-9 Control optic cup regeneration in a prototypic eye. PLoS Genet 7(8):e1002226. doi:10.1371/journal.pgen.1002226
40. Cebria F, Newmark PA (2005) Planarian homologs of netrin and netrin receptor are required for proper regeneration of the central nervous system and the maintenance of nervous system architecture. Development 132(16):3691–3703. doi:10.1242/dev.01941
41. Wang Y, Zayas RM, Guo T, Newmark PA (2007) nanos function is essential for development and regeneration of planarian germ cells. Proc Natl Acad Sci U S A 104(14):5901–5906. doi:10.1073/pnas.0609708104
42. Kobayashi K, Hashiguchi T, Ichikawa T, Ishino Y, Hoshi M, Matsumoto M (2008) Neoblast-enriched fraction rescues eye formation in eye-defective planarian 'menashi' *Dugesia ryukyuensis*. Dev Growth Differ 50(8):689–696. doi:10.1111/j.1440-169X.2008.01066.x

Chapter 19

Chemotaxis: Under Agarose Assay

Derrick Brazill

Abstract

The unicellular eukaryote *Dictyostelium discoideum* represents a superb model for examining chemotaxis. Under vegetative conditions, the amoebae are chemotactically responsive to pterins, such as folate. Under starved conditions, they lose their sensitivity to pterins, and become chemotactically responsive to cAMP. As an NIH model system, *Dictyostelium* offers a variety of advantages in studying chemotaxis, including its conservation of mammalian signaling pathways, its ease of growth, and its genetic tractability. In this chapter, we describe the use of the under agarose chemotaxis assay to identify proteins involved in controlling motility and directional sensing in *Dictyostelium discoideum*. Given the similarities between *Dictyostelium* and mammalian cells, this allows us to dissect the conserved pathways involved in eukaryotic chemotaxis.

Key words Chemotaxis, *Dictyostelium*, Amoeba, Folate, cAMP

1 Introduction

Efficient chemotaxis requires that the cell detect the direction of the chemotactic signal, polarize itself towards the source, and effectively move up the chemotactic gradient. In order to do this, cell surface receptors must recognize the presence of the chemoattractant and detect the direction of the highest concentration. Actin then must be polymerized and bundled at the front of the cell, leading to pseudopod formation [1, 2]. Finally, myosin must contract the actin cytoskeleton at the rear of the cell, providing the force required for motility [3]. These activities must be synchronized, or the cell will experience chemotactic defects. The molecular mechanisms underpinning these individual activities and their coordination can be dissected by using chemotaxis assays on the unicellular eukaryote *Dictyostelium discoideum*.

Dictyostelium and mammalian cells share a number of similarities in how they modify their cytoskeleton in response to chemoattractants. This evolutionary response conserves signaling mechanisms and biochemical interactions [4, 5]. In fact, much of what is known about mammalian chemotaxis was first discovered

Ray H. Gavin (ed.), *Cytoskeleton: Methods and Protocols*, Methods in Molecular Biology, vol. 1365,
DOI 10.1007/978-1-4939-3124-8_19,

in *Dictyostelium* [6]. *Dictyostelium* are haploid soil amoebae that consume bacteria and divide by binary fission. While in this vegetative state, the amoebae hunt bacteria by chemotaxing towards pteridines such as folate, which are released by the bacteria [7, 8]. This process is mediated by G protein-coupled receptors, which have yet to be identified [9, 10]. When the amoebae overgrow their bacterial food source and starve, they initiate development where 2×10^4–10^5 amoebae use cAMP as a chemoattractant to aggregate and form a multicellular organism [11]. Aggregation and development are mediated by four separate cAMP receptors, all of which are also G protein coupled [12]. Both chemotaxis to folate and chemotaxis to cAMP can be measured and analyzed using an under-agarose chemotaxis assay.

The under-agarose chemotaxis assay examines the movement of cells as they crawl beneath a layer of agarose towards a chemoattractant. It was first used to study migration in leukocytes [13, 14]. It was then later modified for use in *Dictyostelium* [15]. The assay is easy to set up and the results are simple to collect and analyze. Since the cells are required to crawl between the agarose and the plastic petri dish, defects in cortex structure or integrity can easily be identified. Slight defects which normally would not be detected can be exaggerated by using different concentrations of agarose. Because of its simplicity and flexibility, the under-agarose chemotaxis assay is incredibly useful for measuring chemotaxis in a variety of different eukaryotic cell types.

In this section, we explain how *Dictyostelium* mutants can be examined for efficient chemotaxis by using an under-agarose assay. The analysis of defective mutants can determine how and where the process of chemotaxis is disrupted. Examination of chemotaxis towards both folate and cAMP can uncover whether the defect is general to chemotaxis, or specific to one of the chemoattractants. Briefly, the assay consists of three steps: preparing the agarose plate, preparing and loading the cells and chemoattractant onto the agarose plate, and time-lapse video microscopy of cell movement. The videos are then used to measure three parameters of chemotaxis: chemotactic index, directionality, and speed. Chemotactic index is a measure of how well the cell moves up the concentration gradient. It is defined as $\cos\theta$, where θ is the angle of deviation between a direct line up the chemical gradient and the net path of a cell. It describes the ability of the cell to sense the chemoattractant and detect the direction of its source. Directionality is a measure of the deviation from a straight path, and is defined as the net cellular displacement divided by the total path length of the cell. It describes the ability to consistently produce a pseudopod in the same direction. Speed is a measure of how fast the cell moves. It is defined as the total path length of the cell divided by the time of travel. It describes the ability of the cell to generate force through contraction of the rear of the cell. These measurements can be used categorize the defective genes in chemotaxis mutants, providing a better understanding of the proteins and molecular mechanisms driving and regulating chemotaxis.

2 Materials

2.1 Agarose Components

1. SM Medium: 10 g Peptone, 10 g glucose, 10 g yeast extract, 1 g $MgSO_4 \cdot 7H_2O$, 1.9 g anhydrous KH_2PO_4, 1 g anhydrous K_2HPO_4. Dissolve in 1 L H_2O, and autoclave. Store at 4 °C.
2. PBM: 2.72 g anhydrous KH_2PO_4, 100 μL 0.1 M $CaCl_2$, 1 mL 1 M $MgCl_2$. Add to 1 L H_2O. Adjust to pH 6.1 and autoclave. Store at room temperature.
3. Agarose (*see* **Note 1**).
4. 100 mm × 15 mm polystyrene petri dish.

2.2 Chemo-attractants

1. Folate stock (50 mM): Dissolve 0.22 g folate in 10 mL 0.1 N NaOH. NaOH may need to be added to allow the folate to go into solution. This is stable for a month if kept in the dark at 4 °C.
2. Folate working solution (1 mM): Add 100 μL of 50 mM Folic acid stock solution to 5 mL of SM media. This is stable for a week if kept in the dark at 4 °C.
3. cAMP (10 mM): Make 10 mM of cAMP stock in PBM. Store at −20 °C.

2.3 Development (for Starving Cells for cAMP Chemotaxis)

1. Whatman #3 125 mm filter paper (GE Healthcare Life Sciences #1003-125).
2. Millipore nitrocellulose membrane AAWP (white) or AABP (black), 0.8 μm pore (Millipore AAWP04700 or AABP04700).
3. 150 mm × 20 mm polystyrene petri dish.
4. 50 mL Falcon tube.

2.4 Time-Lapse Imaging

1. An inverted microscope.
2. Image acquisition software.

3 Methods

3.1 Folate Chemotaxis: Preparation of the Agarose Plate

1. Add 200 mg of agarose to 20 mL of SM media (1 % agarose) in a 125 mL flask (*see* **Note 2**).
2. Cover the flask with a weigh boat, and heat in a microwave oven to the point where it the solution first begins to boil. While wearing a heat protective glove, carefully swirl the flask to gently mix the solution. Place the flask back in the microwave and heat to a boil again. Remove the flask and swirl as before. Repeat cycles of boiling and swirling until the agarose is completely melted.

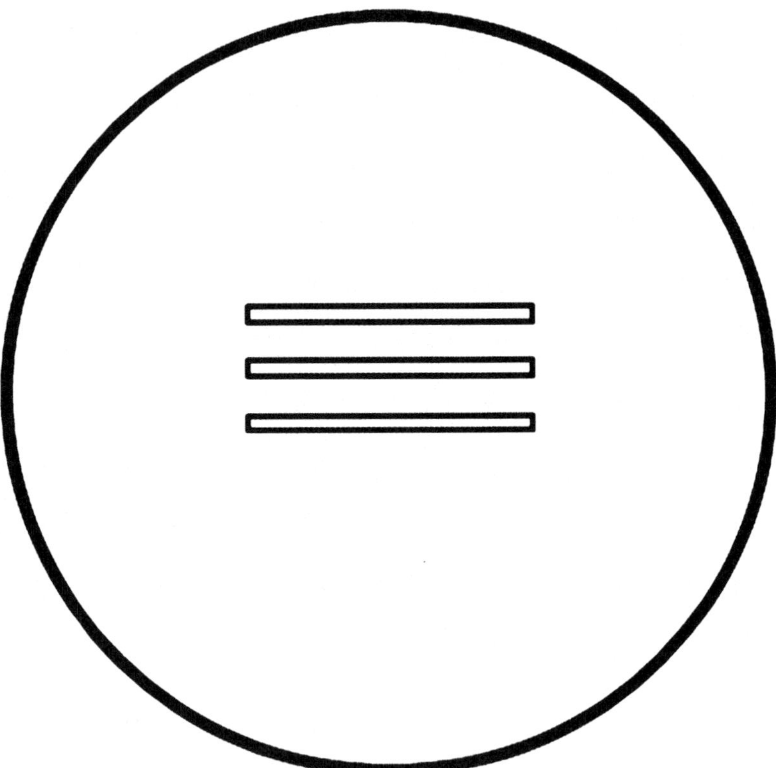

Fig. 1 Template for cutting troughs. Place the agarose dish on top of the template and use it as a guide to cut the troughs. The troughs should be 2 mm wide and 5 mm apart

3. Once the flask is cool enough to handle by hand, add 15 mL of agarose solution to a standard 100 mm Petri dish, and allow agarose to cool and harden (*see* **Note 3**). Plates can be poured up to 1 day before the experiment.
4. Loosen the agarose by carefully running a thin spatula between the walls of the dish and the edge of the agarose. Repeat this until the agarose is loose enough to visibly rotate. This separates the agarose from the bottom of the dish, allowing space for the amoebae to crawl through.
5. Place the dish on the template in Fig. 1. With the template as a guide, cut three troughs in the agarose (2 mm wide, 39 mm long, and 5 mm apart) by making precise, vertical cuts with a razor blade. It is important to avoid cutting the bottom of the petri dish, as this will impeded the motility of the amoebae.
6. To remove the agarose from the troughs, use a pointed spatula to cut the ends of each trough and carefully pry the agarose out. Avoid disturbing the surrounding agarose. The center trough will eventually contain folate (the chemoattractant). The outer wells will contain amoebae.

3.2 Folate Chemotaxis: Preparation and Loading of Amoebae and Chemoattractant

1. To create the folate gradient, pipette 1 mM folate to the center trough (*see* **Note 4**). Add enough folate to almost fill the trough, but not cause it to overflow (*see* **Note 5**). Allow a minimum of 30 min for the gradient to form. Once formed, the gradient is stable for several hours.
2. Once the folate has been added, prepare the amoebae. For optimal results, use axenically grown, log phase (1×10^6–2×10^6 cells/mL). Count and collect 1×10^7 cells by centrifugation at 500 g for 5 min. Remove the supernatant, and resuspend the cells in SM to a concentration of 5×10^6–1×10^7 cells/mL.
3. Immediately before placing the cells in the troughs, mix the cells thoroughly be vortexing. Pipette 100–200 μL of the amoebae to the outer troughs, making sure not to overflow the troughs (*see* **Notes 6** and 7). Once the amoebae have been added, do not disturb the dish for 30 min, in order to allow the cells to settle and attach to the bottom.
4. Once the amoebae have attached, place the dish on the stage of the inverted microscope for time-lapse imaging. Once attached, the amoebae will begin to move out of the troughs and underneath the agarose towards the center trough. The amoebae will continue to chemotax into the trough containing the folate. Amoebae that have exited the trough and are migrating under the agarose will appear flattened. This is usually seen once the cells have moved about 100 μm from the trough. Image these cells. If imaging is performed over the course of hours, it is important to make sure that the fluid in the troughs does not evaporate. Add SM to the troughs to avoid evaporation.

3.3 cAMP Chemotaxis: Preparation of the Agarose Plate

Chemotaxis to cAMP is performed similarly to that to folate with only a few differences.

1. Add 150 mg of agarose to 20 mL of PBM media (0.75 % agarose) in a 125 mL flask.

3.4 cAMP Chemotaxis: Preparation and Loading of Amoebae and Chemoattractant (Steps 1–8 Should Be Done at Least 6 h Before Performing the Assay)

1. Pour PBM into a 150 mm × 20 mm petri dish.
2. Float and then submerge a Whatman #3 125 mm diameter filter into PBM in dish.
3. Starting from one edge, use forceps to slowly lower a second Whatman #3 125 mm diameter filter directly on top of the first one. It is imperative that there be no bubbles between the two filters.
4. Discard excess PBM, leaving approximately 1 mL.
5. Starting from one edge, slowly lower a Millipore AAWP or AABP filter onto the Whatman filters.
6. Collect 1×10^7 log-phase amoebae at least 6 h before preparing the assay. Wash the amoebae twice in PBM to remove the growth medium and resuspend them to a concentration of 1×10^7 cells/mL.

7. Using a pipette, distribute the amoebae evenly over the Millipore filter, avoiding the edges. This is usually done with approximately 250 μL of cells at a time. Allow the liquid to wick into the filter before adding more cells.
8. After 15 min, use forceps to slide the filter containing the amoebae to another area of the Whatman paper. This ensures that the amoebae develop in a completely fresh environment.
9. Allow the amoebae to develop for 6–8 h at room temperature, in order to express genes required for cAMP chemotaxis.
10. To collect the starved amoebae, place the filter pad along the inside of a 50 mL Falcon tube, cell side up. Wash the cells off of the filter pad into the Falcon tube with 0.5 mL PBM. Repeatedly flush the filter for 30 s. Repeat with a fresh 0.5 mL PBM to remove the remainder of the amoebae.
11. Transfer amoebae to 15 mL Falcon tube and centrifuge at 500 g for 5 min. Remove the supernatant, and resuspend the amoebae in 500 μL PBM (5×10^6–1×10^7 cells/mL).
12. To create the cAMP gradient, pipette 5–10 μM cAMP to the center trough. Allow a minimum of 30 min for the gradient to form.
13. Add 100–200 μL of the amoebae to the outer troughs.

3.5 Imaging and Analysis

1. As soon as the cells have crawled beneath the agarose as determined by cell flattening, capture images every 30 s for a total of 20–30 min using a 20× objective lens (*see* **Note 8**), in order to make a movie.
2. Use the Manual Tracking plug-in in ImageJ to analyze cell movement in the acquired images (Fig. 2a). The specific instructions for Manual Tracking can be found online with the plug-in (*see* **Note 9**).

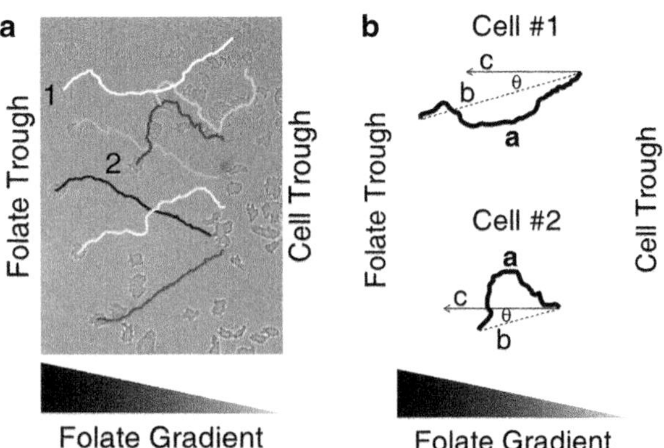

Fig. 2 An image of amoebae taken 3 h into a folate chemotaxis assay. (**a**) Several cells have been tracked, and their paths determined using the ImageJ Manual Tracking plug-in. (**b**) The paths of two selected cells. Directionality = *b*/*a*. Chemotactic index = cosθ

4 Notes

1. Molecular biology-grade agarose used for DNA gel electrophoresis works well.
2. Different types and brands of agarose will have slightly different stiffnesses. Therefore, it may be necessary to vary the agarose concentration. For folate chemotaxis, 1 % agarose works well. However, this concentration can be altered. We have used between 0.5 and 2 % agarose, and have found that different concentrations can have an impact on the efficacy of chemotaxis. Mutants that have a slight chemotactic defect may not be detected under the more permissive 0.5 % agarose, but become quite apparent at the less permissive 2 % agarose.
3. We use between 10 and 25 mL for a 100 mm Petri dish. We have found that we get better results with a thinner agarose layer, but the thinner agarose is more difficult to handle. So, we often prepare extra plates for our assays in case one or two of the plates do not turn out properly.
4. Folate concentrations can be varied anywhere between 0.5 and 5 mM.
5. The exact volume of chemoattractant added varies based on the thickness of the agarose and the exact width of the trough. Thus, we add chemoattractant to fill the well almost completely full without spilling over.
6. Normally, we place wild-type cells in one trough and mutant cells in the other. This allows the direct comparison of the two cell lines on the same plate under the exact same conditions.
7. Optimally, a near monolayer of cells is desired in the trough. This allows for a large number of cells to crawl under the agarose. If the cell density is low after the cells have settled, more cells can be added.
8. It is best to image cells at the leading edge of the group. These cells tend to be further spread out so that they do not bump into each other, complicating tracking. In addition, cells further back have poorer chemotaxis, most likely due to disruption of the gradient caused by the preceding cells.
9. Briefly, identify and mark individual cells in each frame of the movie. Given the time between frames, and distance traveled between each frame, the program then calculates the distance and speed of the cells between each frame. Thus, the overall speed of the cell, the total path length that the cell traveled, and the final displacement of the cell from its origin can be calculated. This allows the calculation of three basic chemotactic parameters. The first is simply the overall speed of the cell. This is just the average of the individual speed measurements taken between each frame. The second measurement is direc-

tionality (Fig. 2b). This is a measure of how directly the cell moved, and is defined as the ratio of the net cellular displacement (*b*) to the total path length of the cell (*a*). A value of 1 represents a direct path with no deviations, where a value of 0 represents a meandering path where the cell ends up back at its original position. In Fig. 2b, Cell #1 has a high directionality, close to 0.8. Cell #2 has a lower directionality, close to 0.5. The third measurement is the chemotactic index (Fig. 2b). This is a measure of how well the cell is able to sense and orient to the chemotactic gradient. It is defined as cosθ, where θ is the angle of deviation between a direct line up the chemical gradient (*c*) and the net cellular displacement of the cell (*b*). A value of 1 represents movement directly towards the source of the chemoattractant, while a value of 0 represents movement directly perpendicular to the source of the gradient. In Fig. 2b, both cells have similar chemotactic indices.

References

1. van Haastert PJM, Devreotes PN (2004) Chemotaxis: signalling the way forward. Nat Rev Mol Cell Biol 5:626–634
2. Diez S, Gerisch G, Anderson K, Muller-Taubenberger A, Bretschneider T (2005) Subsecond reorganization of the actin network in cell motility and chemotaxis. Proc Natl Acad Sci U S A 102:7601–7606
3. Tsujioka M et al (2012) Talin couples the actomyosin cortex to the plasma membrane during rear retraction and cytokinesis. Proc Natl Acad Sci U S A 109(32):12992–12997
4. Parent CA (2004) Making all the right moves: chemotaxis in neutrophils and Dictyostelium. Curr Opin Cell Biol 16:4–13
5. Artemenko Y, Lampert TJ, Devreotes PN (2014) Moving towards a paradigm: common mechanisms of chemotactic signaling in Dictyostelium and mammalian leukocytes. Cell Mol Life Sci 71(19):3711–3747
6. Jin T, Xu X, Fang J, Isik N, Yan J, Brzostowski JA, Hereld D (2009) How human leukocytes track down and destroy pathogens: lessons learned from the model organism Dictyostelium discoideum. Immunol Res 43:118–127
7. Pan P, Hall EM, Bonner JT (1975) Determination of the active portion of the folic acid molecule in cellular slime mold chemotaxis. J Bacteriol 122:185–191
8. Rifkin JL, Goldberg RR (2006) Effects of chemoattractant pteridines upon speed of D-discoideum vegetative amoebae. Cell Motil Cytoskeleton 63:1–5
9. Rifkin JL (2001) Folate reception by vegetative Dictyostelium discoideum amoebae: distribution of receptors and trafficking of ligand. Cell Motil Cytoskeleton 48:121–129
10. Rifkin JL (2002) Quantitative analysis of the behavior of Dictyostelium discoideum amoebae: stringency of pteridine reception. Cell Motil Cytoskeleton 51:39–48
11. Mahadeo DC, Parent CA (2006) Signal relay during the life cycle of Dictyostelium. Curr Top Dev Biol 73:115–140
12. Parent CA, Devreotes PN (1996) Molecular genetics of signal transduction in Dictyostelium. Annu Rev Biochem 65:411–440
13. Cutler JE, Munoz JJ (1974) A simple in vitro method for studies on chemotaxis. Proc Soc Exp Biol Med 147:471–474
14. Lauffenburger D, Rothman C, Zigmond SH (1983) Measurement of leukocyte motility and chemotaxis parameters with a linear under-agarose migration assay. J Immunol 131: 940–947
15. Laevsky G, Knecht DA (2001) Under-agarose folate chemotaxis of Dictyostelium discoideum amoebae in permissive and mechanically inhibited conditions. Biotechniques 31:1140–1149

Part IV

Genetic and Proteomic Protocols

Chapter 20

Functional Analysis of Actin-Binding Proteins in the Central Nervous System of *Drosophila*

Qi He and Christopher Roblodowski

Abstract

Using *Drosophila* actin-binding protein Dunc-115 as model system, this chapter describes a MARCM (mosaic analysis with a repressible cell marker)-based method for analyzing cytoskeletal components for their functions in the nervous system. Following a concise description about the principle, a step-by-step protocol is provided for generating the needed stocks and for histological analysis. Additional details and explanations have been given in the accompanying notes. Together, this should form a practical and sufficient recipe for performing at the single-cell-level loss-of-function and gain-of-function analyses of proteins associated with the cytoskeleton.

Key words *Drosophila*, Dunc-115, Actin-binding protein, MARCM, Confocal microscopy

1 Introduction

It has been well established that the growth cone, a highly sensitive and motile structure at the leading edge of a growing axon, receives and transduces guidance signals through its surface receptors to regulate actin cytoskeleton [1, 2]. Consequently, actin-binding proteins form a critical linkage to the cytoskeleton during the process of relaying guidance signals and studying the functional roles of these proteins holds promises of revealing the molecular underpinning of axon pathfinding. This chapter describes methods for generating genotypes, dissecting and staining the visual system, and analyzing the function of a *Drosophila* actin-binding protein Dunc-115. Other cytoskeletal components can be analyzed similarly.

A major step is to analyze axon pathfinding using a powerful technique, MARCM (mosaic analysis with a repressible cell marker) developed in *Drosophila* [3, 4]. As shown in Fig. 1, MARCM utilizes recombination mediated by the FLP/FRT system (*see* Fig. 1

Ray H. Gavin (ed.), *Cytoskeleton: Methods and Protocols*, Methods in Molecular Biology, vol. 1365,
DOI 10.1007/978-1-4939-3124-8_20, © Springer Science+Business Media New York 2016

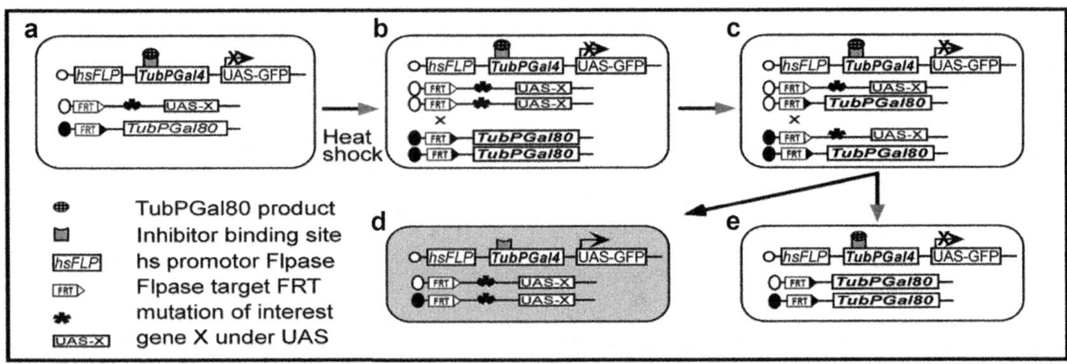

Fig. 1 The MARCM method. *Rectangles* represent individual cells. (**a**) Flippase (FLP) recombinase is under the control of a heat-shock promoter (hs), and both Gal4 and Gal80 are controlled by a universal tubulin 1 promoter (TubP). Transcription factor Gal4 binds to its unique upstream activator sequence (UAS) to drive the expression of GFP or gene X. Gal80 functions as a repressor of Gal4 activity and thus when Gal80 and Gal4 are coexpressed in the same cell, Gal80 (*hatched oval*) binds to a site on Gal4 to repress its activity. As a result, both GFP and gene X cannot be expressed in such a cell (*black cross and black arrow*, shown only for GFP). (**b**) A heat-shock activates FLP that catalyzes the exchange of two FRT targets during mitotic cell division after DNA duplication (*black X*), resulting in the exchange of chromosome arms distal to the FRT sites shown in (**c**). Following the completion of cell division, one daughter cell (**d**) did not inherit Gal80 gene and is also homozygous for the mutant (*black star*, Dunc-115 mutation in this chapter) (or transgene X such as Dunc-115 under UAS though for the purpose of expressing gene X, UAS-X does not need to be on the same chromosome arm carrying the FRT site). So in daughter cell (**d**) (and its offspring) Gal4 is active and drives the expression of GFP (and UAS-X). The other daughter cell inherited two copies of Gal80 and thus has no Gal4 activity or GFP expression, while the background will be cells that did not undergo recombination as in (**b**) and thus they will be like the cell in panel (**a**), also without Gal4 activity or GFP expression. Thus the only cells expressing GFP will be from panel (**d**). For simplicity, gene X (UAS-X) is depicted as on the same chromosome arm as the mutant (*black star*). In reality, the scheme can be used with a wild-type arm or with a mutant-carrying arm, and the UAS-X transgene can be on any chromosome arm except the one carrying Gal80. Heat-shock is carried out at 37 °C for a desired period of time and by controlling heat-shock time, it is possible to generate single-neuron clones. UAS-GFP is UAS-mCD8-GFP fusion protein (where mCD8 is mouse T cell surface protein CD8) that will be anchored in the plasma membrane. TubP is fused with Gal4 or Gal80 coding sequence. Schematics were inspired by and based on Lee and Luo (1999) and Wu and Luo (2006)

legend) to generate homozygous mutant clones for a gene of interest such as Dunc-115. Defects of Dunc-115 mutation in these clones during axon projection can then be analyzed by confocal microscopy. By creating the UAS-Dunc115 construct, ectopic Dunc-115 expression can be achieved in homozygous clones (Fig. 1) and thus gain of function of Dunc-115 can be analyzed. This approach will also allow the analysis of genes that generate lethality if mutated in the whole animal and thus provides a practical method for studying axon pathfinding functions of both viable and lethal genes.

2 Materials

2.1 Equipment

1. Standard incubator for culturing *Drosophila*.
2. Orbital shaker platform.
3. Dissecting microscope.
4. Confocal microscope.

2.2 Reagents

1. Standard cornmeal medium: Mix 850 mL of water with 7.9 g of agar, 27.5 g of torula yeast, 52 g of cornmeal, and 110 g of dextrose. Heat to a boil and cool to 80 °C. Add 2.4 g of tegosept dissolved in 92 mL ethanol and pour into culture vials and bottles (*see* **Note 1**).
2. G418 medium: Prepare standard cornmeal medium and cool to 50 °C. Add G418 (Fisher Scientific) to 1 mg/mL final concentration.
3. PBT: Add 0.3 % Triton X-100 to 1× PBS.
4. PBTN: Add 5 % normal goat serum to PBT.
5. 4 % formaldehyde (1 mL stock): Mix 0.792 mL H_2O, 0.1 mL 10× PBS and 0.108 mL of 37 % formaldehyde stock. The solution should be made fresh daily and unused portions should be discarded at the end of the day.
6. Primary Antibodies: Mouse anti-24B10, rat anti-N cadherin (both from Developmental Studies Hybridoma Bank), rabbit anti-GFP (Molecular Probes).
7. Secondary Antibodies: FITC-conjugated goat anti-mouse, Cy5-conjugated goat anti-rat (both from Jackson Lab), and Alexa647 conjugated donkey anti-rabbit (Molecular Probes).

2.3 *Drosophila* Strains and Culture

1. Dunc-115 mutant *dunc-115*^{KG03651} (85E4) [5].
2. The following strains are used and they are available at the Bloomington stock center:

   yw/Y; P[ry, FRT]82B.

   yw/Y; TM2/TM6B.

   yw hsFlp UAS-mCD8-GFP; Sp/CyO; TM2 Ubx/TM6B Hu.

   yw/Y; TubPGal4; FRT82B TubPGal80.

3 Methods

3.1 MARCM Analysis

3.1.1 Preparation of FRT dunc-115^{KG03651}, Gal4 and Gal80 Strains

1. *FRT dunc-115* strain (*see* **Note 2**):

   Set up the following cross:

   yw; dunc-115^{KG03651} × *yw/Y; P[ry, FRT]82B.*

Collect females and mate with *yw/Y; TM2/TM6B*:

yw; dunc-115^{KG03651}*/P[ry, FRT]82B*× *yw/Y; TM2/TM6B*.

Culture the cross in *neomycin* (G418) medium and select w^+ and cross to *yw/Y; TM2/TM6B* (single animal cross) (*see* **Note 3**):

yw; P[ry, FRT]82B dunc-115^{KG03651}/TM6B × *yw/Y; TM2/TM6B*.

Collect both males and females to establish the following stock:

yw; P[ry, FRT]82B dunc-115^{KG03651} or *yw; FRT82B dunc-115*^{KG03651}.

2. Gal4 and Gal80 strains (*see* **Note 4**):

   Set up the following cross:

   yw hsFlp UAS-mCD8-GFP; Sp/CyO; TM2 Ubx/TM6B Hu × *yw/Y; TubPGal4; FRT82B TubPGal80*.

   Collect males of *Sp* and *Hu* and set up:

   yw hsFlp UAS-mCD8-GFP/Y; TubPGal4/Sp; FRT82B TubPGal80/TM6B Hu × *yw hsFlp UAS-mCD8-GFP; Sp/CyO; TM2 Ubx/TM6B Hu*.

   Select both males and females of *Cy* and Hu^+ and set the following cross:

   yw hsFlp UAS-mCD8-GFP/Y; TubPGal4/CyO; FRT82B TubPGal80/TM2 Ubx × *yw hsFlp UAS-mCD8-GFP/yw hsFlp UAS-mCD8-GFP; TubPGal4/CyO; FRT82B TubPGal80/TM2 Ubx*.

   Select both males and females of Cy^+ and Ubx^+ to establish the following stock:

   yw hsFlp UAS-mCD8-GFP; TubPGal4; FRT82B TubPGal80.

3.1.2 Generation of MARCM Clones (*See* **Note 5**)

1. Set up the following cross:

   yw/Y; FRT82B dunc-115^{KG03651} × *yw hsFlp UAS-mCD8-GFP; TubPGal4; FRT82B TubPGal80*.

2. Collect eggs during a window of every 2 h; that is, switch the animals to a new vial every 2 h. Animals with the following genotype will be the target of analysis:

   yw hsFlp UAS-mCD8-GFP/Y or *yw; TubPGal4/+; FRT82B dunc-115*^{KG03651}*/FRT82B TubPGal80*.

3. Leave the collected eggs at 25 °C for 96 h.
4. Heat-shock at 37 °C for 1 h.
5. Dissect third instar larvae, and stain with antibodies as described below.

3.2 Histology

3.2.1 Larval Eye Disc/Brain Specimen Preparation

Dissect the visual system of the third instar larvae in PBS [5, 6] and fix in 4 % formaldehyde for 40 min at room temperature (RT) (*see* **Note 6**).

3.2.2 Antibody Staining

1. Following fixation, wash specimens three times for 5 min each time with PBT. Make certain all specimens have sunk to the bottom of the tube.
2. Block nonspecific binding sites on the specimens by incubating with PBTN for 30 min at RT.
3. Following the blocking step, add diluted primary antibodies, and incubate at 4 °C overnight (*see* **Note 7**).
4. Wash three times for 15 min each with PBT.
5. Add second antibodies (*see* **Note 8**). Incubate at 4 °C overnight.
6. Wash three times for 15 min each with PBT.
7. Perform 5-min serial washes with 25, 50, and 75 % glycerol in PBS. The 75 % glycerol solution contains 0.1 % *p*-phenylenediamine (anti-quenching agent to prolong the fluorescent intensity).
8. Mount the specimens on glass slides (*see* **Note 9**) and examine under a confocal microscope.

4 Notes

1. Pre-made fly food mix such as Jazzmix from Applied Scientific can be used by following the manufacturer's instruction, while 10× PBS can be purchased from many vendors such as Fisher Scientific.
2. This procedure is for loss-of-function analysis (***dunc-115*** mutant) using MARCM. For gain-of-function analysis (ectopic expression), a UAS construct (UAS-Dunc-115 fusion) will need to be built first using standard molecular biology methods and transgenic animals produced by embryo injection. The resulted UAS-Dunc-115 transgenic flies will be prepared for MARCM the same way as with the FRT strain. Similar analysis will reveal the effect of overexpressing Dunc-115 in a MARCM clone.
3. The structure P[ry FRT] carries two copies of the *neomycin* gene controlled by the heat-shock promoter (*hs-neo*) and thus confers the *neomycin* resistance [7], while the ***dunc-115***^{KG03651} strain carries the *mini-white* gene and thus has the w^+ phenotype. By selecting *neomycin* resistance and w^+, the only possible genotype is FRT ***dunc-115***^{KG03651}.

The chromosomal location of *dunc-115*^{KG03651} is 85E4 and the FRT used is at 82B. Given that the distance between the two is not long, it is necessary to set up several vials to increase the chances of recombination. For genes that have greater distances from the FRT site, the number of vials can be reduced. It would be informative to set up a pilot test to see the frequency of a given recombination.

4. The general principle behind MARCM is to bring all elements into the same animal after a final cross. Thus it is often necessary to prepare the two stocks needed for the final cross so that the needed constructs will be in place. For the FRT, it is required to have the Gal80 construct to be on the same chromosomal arm distal to the FRT site (Fig. 1), while for the other elements such as Gal4 or UAS-mCD8-GFP can be on any chromosomes. Traditionally, heat-shock-controlled flippase (hs Flp in Fig. 1) is linked with UAS-mCD8-GFP and located on the first chromosome. For a more detailed discussion, *see* [8].
5. Generate MARCM clones by crossing the strains created in Subheading 3.1.1. The genotype from this cross will be subjected to heat-shock to induce mitotic recombination. It is preferable to use females from the UAS-mCD8-GFP strain in the cross since both male and female offspring will have one copy of the GFP fusion protein and thus will be usable. Heat-shock time may vary depending on the actual setup. It is thus highly recommended to test the time first to make certain a desired result. The clone size in MARCM depends on at which developmental stage the heat-shock is given and for how long. A condition can be determined where single-cell clones can be generated (see for example [9]).
6. It is important to have all specimens submerged in the fixative to ensure proper fixation. Fixation and other incubation steps should be carried out on an orbital shaker platform.
7. Typically, three primary antibodies are used in this analysis: (1) anti-24B10 labels photoreceptor neurons, (2) anti-N cadherin stains neuropils in the optic lobes, and (3) anti-GFP labels GFP expressed in the MARCM clones (*see* Fig. 1). Dilute antibodies with PBTN. The dilution factor depends on the sensitivity of the antibody and will have to be determined empirically.
8. Dilute secondary antibodies in PBTN. The dilution factor is determined empirically; a typical range is between 200 and 500×. The three secondary antibodies are tagged with green (FITC-conjugated goat anti-mouse), red (Cy5-conjugated goat anti-rat) and blue (Alexa647-conjugated donkey anti-rabbit) fluorescent tags and will be detected separately by the confocal microscope according to the wavelength.

9. To examine along the dorsal-ventral axis, the visual system needs to be cleaned so that mouth hooks and other accessories are removed, and the whole specimen sits on its ventral base. For a lateral view, the two optic lobes need to be separated with the use of a surgical knife and each lobe should rest on the cut phase. Finally, one piece of cover slips should be placed on each side of the prepared specimen and a third cover slip should be placed on top of the first two so that the specimen will not be squashed by the cover slip to preserve its morphology.

Acknowledgments

I would like to thank the Bloomington Stock Center for providing fly strains and Ray Gavin for reviewing the manuscript. This work was supported in part by a PSC-CUNY grant.

References

1. Huber AB, Kolodkin AL, Ginty DD, Cloutier JF (2003) Signaling at the growth cone: ligand-receptor complexes and the control of axon growth and guidance. Annu Rev Neurosci 26: 509–563
2. Lewis TL Jr, Courchet J, Polle F (2013) Cellular and molecular mechanisms underlying axon formation, growth, and branching. J Cell Biol 202:837–848
3. Lee T, Luo L (1999) Mosaic analysis with a repressible cell marker for studies of gene function in neuronal morphogenesis. Neuron 22: 451–461
4. Hong W, Luo L (2014) Genetic control of wiring specificity in the fly olfactory system. Genetics 196:17–29
5. Garcia MC, Abbasi M, Singh S, He Q (2007) Role of *Drosophila* gene *dunc-115* in nervous system. Invert Neurosci 7:119–128
6. Kunes S, Wilson C, Steller H (1993) Independent guidance of retinal axons in the developing visual system of Drosophila. J Neurosci 13:752–767
7. Xu T, Rubin GM (1993) Analysis of genetic mosaics in developing and adult *Drosophila* tissues. Development 117:1223–1237
8. Wu JS, Luo L (2006) A protocol for mosaic analysis with a repressible cell marker (MARCM) in Drosophila. Nat Protoc 1:2583–2589
9. Morante J, Desplan C (2008) The color-vision circuit in the medulla of *Drosophila*. Curr Biol 18:553–565

Chapter 21

Proteomic Analysis of Cytoskeleton Proteins in Fish

Michael Gotesman*, Simon Menanteau-Ledouble*, and Mansour El-Matbouli

Abstract

In this chapter, we describe laboratory protocols for rearing fish and a simple and efficient method of extracting and identifying pathogen and host proteins that may be involved in entry and replication of commercially important fish viruses. We have used the common carp (*Cyprinus carpio* L.) and goldfish (*Cyprinus auratus*) as a model system for studies of proteins involved in viral entry and replication. The chapter describes detailed protocols for maintenance of carp, cell culture, antibody purification of proteins, and use of electrospray-ionization mass spectrometry analysis to screen and identify cytoskeleton and other proteins that may be involved in viral infection and propagation in fish.

Key words Cyprinid herpesvirus 3 (CyHV-3), Spring viremia of carp virus (SVCV), Carp, Cyprinidae fish, Electrospray-ionization mass spectrometry (ESI-MS) analysis

1 Introduction

Fish culture is an important component of the world food supply. Fish protein makes up 20 % or more of the total animal protein consumed in low-income, food-deficient countries, with aquaculture accounting for 40 % of the global food fish supply in 2010. Carp (*Cyprinidae*) makes up 71 % of the total freshwater supply of farmed fish and is an especially important food in China and the remainder of East Asia, which produce 61.3 % and 26.7 % of aquacultured carp, respectively [1]. However, the aquaculture industry is susceptible to outbreaks of infectious diseases that can deplete fish supplies. For example, cyprinid herpesvirus-3 (CyHV-3) and spring viremia of carp virus (SVCV) are important viruses that affect carp and other cyprinid fish and can cause greater than 80 % mortality during independent viral outbreak. Knowledge about protein interactions between pathogen and host can be used to understand how viruses enter host cells, propagate during infection, and ultimately cause mortality.

*Author contributed equally with all other contributors.

Ray H. Gavin (ed.), *Cytoskeleton: Methods and Protocols*, Methods in Molecular Biology, vol. 1365, DOI 10.1007/978-1-4939-3124-8_21, © Springer Science+Business Media New York 2016

Diseases of aquatic animals, such as viral infection of farmed fish, have become more problematic and are causing important economic losses to the aquaculture industry [2]. Among the diseases of great economic importance is koi herpesvirus disease (KHVD), which is caused by CyHV-3 and is a highly contagious viral disease in koi fish. Another serious disease is caused by SVCV and affects the common carp (*Cyprinus carpio* L.). The most common clinical signs of KHVD are white patches, sunken eyes, enlargement of the spleen and kidney, and necrosis at the gills in infected fish [3, 4]. CyHV-3 is exclusively a pathogen of freshwater fish, whereas SVCV, which is a member of the *Rhabdoviridae* family of viruses, infects multiple forms of freshwater fish and marine life [5]. In natural infection by SVCV, entry and early replication occurs through the gills, followed by dispersal via the bloodstream to the kidney, liver, spleen, and other major organs [6]. An outbreak of either virus can cause 80–100 % mortality under farming conditions and devastate any type of aquaculture or fishery practice.

Carp are especially amenable for aquaculture and laboratory maintenance because they tolerate diverse culture conditions. We have used the common carp and goldfish (*Cyprinus auratus*) as a model system for studies of proteins involved in viral entry and replication. This chapter describes detailed protocols for maintenance of carp, cell culture, and step-by-step immunochemical procedures for isolating viral-specific proteins from infected fish. The immunochemical protocols (Fig. 1) described in the chapter have been used to identify several fish cytoskeleton proteins including myosin, actin, integrin, and vimentin (Table 1). A similar immunochemical approach identified interactions between actin and tubulin within distinct regions of the MyTH4/FERM domains in a class XIV myosin in the *Tetrahymena thermophila* [7, 8]. Protocols in this chapter will be useful for future proteomic analyses and experimentation that aim to more fully understand how fish cytoskeletal proteins are involved in viral entry or replication.

2 Materials

2.1 Fish Tank

1. A polyethylene tank is generally preferable to a glass one because polyethylene is sturdier than glass. Moreover, the transparency and reflectivity of glass can be a source of stress on the fish. A square design (Fig. 2) leads to the apparition of a circular flow in the center of the tank, while the water remains static in the corners of the tank. Therefore, a round-shaped tank is preferable (*see* **Note 1**).

2.2 Cell Passaging

1. Pipette aid.
2. Pipette tips of various sizes (0.5–50 mL).

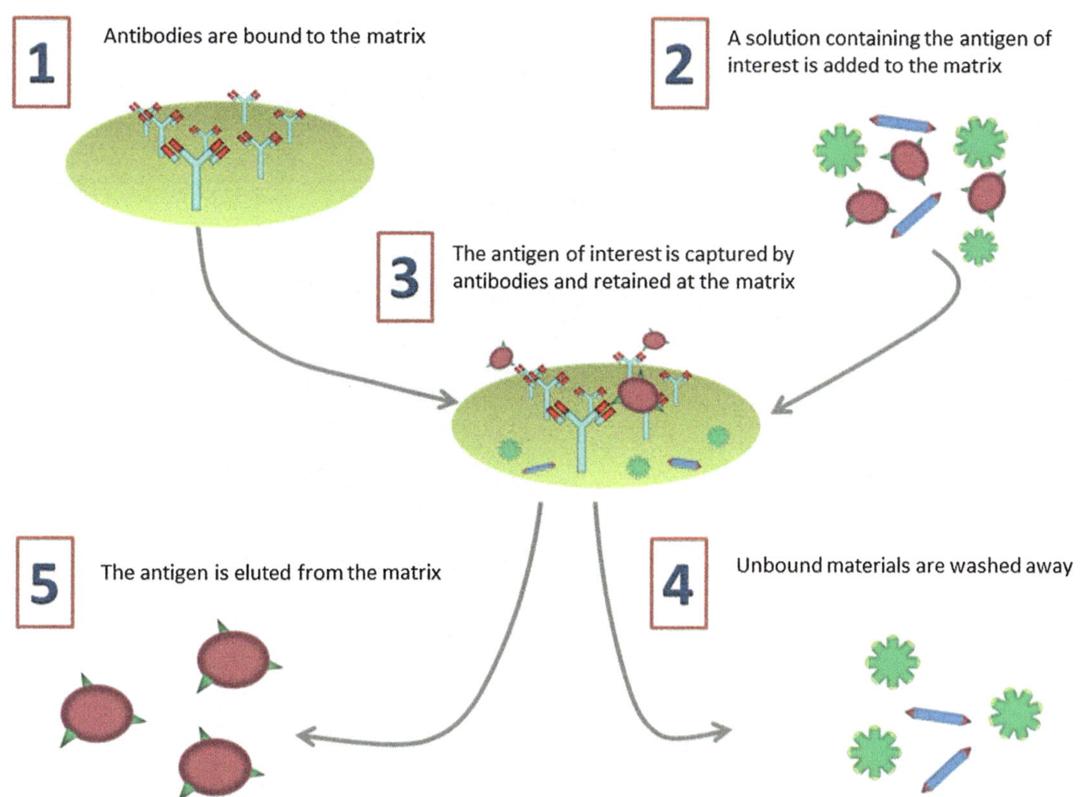
1
Antibodies are bound to the matrix
2
A solution containing the antigen of interest is added to the matrix
3
The antigen of interest is captured by antibodies and retained at the matrix
5
The antigen is eluted from the matrix
4
Unbound materials are washed away

Fig. 1 Schematic diagram of antibody-based affinity purification

Table 1
Cytoskeletal proteins that appeared in our analysis

Cytoskeletal proteins [18, 19]					
Accession no.	**Protein description**	**Score**	**Mass (Da)**	**Matches**	**Coverage (%)**
gi|2351221	Myosin heavy chain	45	222,051	1	0.8
gi|42560193	Beta-actin	496	42,068	24	25.1
gi|1703140	Alpha-actin-1	326	42,274	14	18.8
gi|13365501	Integrin beta-2 chain	173	87,597	5	7.3
gi|6226787	Vimentin	56	52,516	4	3.5

3. Sterile (100×) Na-pyruvate: Dissolve 1.2 g of Na-pyruvate in 100 mL of deionized water (dH_2O). Shake well, and filter using a 0.2 μm capacity filter.
4. 200 mM Glutamine (aqueous solution).
5. Commercially available 100× antibiotic/antimycotic mix.

Fig. 2 Photograph of a fish tank in use

6. Commercially available 100× nonessential amino acids mix.
7. Commercially available minimum essential medium (MEM).
8. Commercially available minimum essential medium with Earl's balanced salts (MEM-Earl's salts).
9. Commercially available minimum essential medium with Hank's balanced salts (MEM-Hanks' salts).
10. Fetal bovine serum.
11. Trypsin-EDTA: Cell culture grade between 0.05 and 0.25 %.
12. ZB4G medium: Add 10 mL of 200 mM l-glutamine, 10 mL of 100× antibiotic/antimycotic, 10 mL of 100× nonessential amino acids mix, and 100 mL of fetal bovine serum (FBS) to 870 mL MEM Earle's salts. Store at 4 °C (*see* **Note 2**).
13. ZB5 medium: Add 10 mL of 100× sterile Na-pyruvate, 10 mL of 100× antibiotic/antimycotic, 10 mL of 100× nonessential amino acids mix, and 100 mL of fetal bovine serum to a solution containing 385 mL MEM Earle's salts and 385 mL MEM Hanks' salts. Store at 4 °C (*see* **Note 3**).
14. Anesthetic solution: Benzocaine at 40 mg/L or tricaine methanesulfonate (MS-222) at 0.35 g/L.
15. RNA*later*® (Life technologies).

2.3 Spin Column

1. Antibody: Purchase or design appropriate antibody (*see* **Note 4**).
2. Spin columns: *N*-hydroxysuccinimide (NHS)-activated 33 mg capacity agarose spin column.

3. Phosphate-buffered saline (PBS): 13.7 mM NaCl, 0.27 mM KCl, 10 mM Na_2HPO_4, 0.2 mM KH_2PO_4 (pH 7.4). Make 10× PBS: Dissolve 80 g of NaCl, 2.0 g of KCl, 14.4 g of Na_2HPO_4, and 2.4 g of KH_2PO_4 in 800 mL dH_2O; adjust pH to 7.4, and adjust the volume to 1 L with additional dH_2O. Sterilize by autoclaving.
4. Ethanolamine (pH 7.4): Add 6 mL of reagent ethanolamine to 70 mL of dH_2O; adjust pH to 7.4 and adjust solution to 100 mL with additional dH_2O.

2.4 Tissue Lysate

1. Non-denaturing lysis buffer [9]: 50 mM Tris–HCl, pH 8.0, 150 mM NaCl, 20 mM ethylene diamine tetraacetic acid (EDTA), 1 % Na-deoxycholate, 1 % Triton X-100. To prepare the lysis buffer, make a 10× stock of 50 mM Tris–HCl, pH 8.0: Add 0.61 g Trizma base to 70 mL of sterile dH_2O; adjust the pH to 8.0 with HCl, and adjust the volume to 100 mL with additional dH_2O. Add 10 mL of the stock 10× 50 mM Tris–HCl solution to 50 mL dH_2O, then add 0.876 g NaCl, 0.744 g EDTA, 1 g Na-deoxycholate and 1 mL of Triton X-100, and adjust the volume to 100 mL with additional dH_2O to yield a working solution of non-denaturing lysis buffer. Add 50 μL of protease inhibitor cocktail mix per mL of lysis buffer.

2.5 Protein Purification

1. 0.2 M Glycine: Dissolve 0.75 g of glycine in 60 mL of dH_2O. Adjust the pH to 3.0 with HCl, and adjust the volume to 100 mL with additional dH_2O.
2. 0.1 M Tris–HCL, pH 8.0: Add 1.22 g Trizma base in 70 mL of sterile dH_2O; adjust the pH to 8.0 with HCl, and adjust the volume to 100 mL additional dH_2O.
3. Radioimmunoprecipitation assay buffer (RIPA buffer).

2.6 SDS-Polyacrylamide Gel

1. Materials for sodium dodecyl sulfate-polyacrylamide gel electrophoresis (SDS-PAGE) are described in Sambrook and Russel [10].

3 Methods

3.1 Setting Up the Fish Tank

1. Aeration: A schematic representation of a tank is shown in Fig. 3. The water outlet is at the top, and the water falls into the tank, creating a bubbling effect that contributes to oxygenating the tank. Fish should be kept in well-aerated water with a slow (about 1–1.5 % of the total volume per hour) but continuous flow. The permanent circulation of water allows for proper oxygenation and carries away dissolved metabolites that could otherwise build up to toxic levels (*see* **Note 5**).

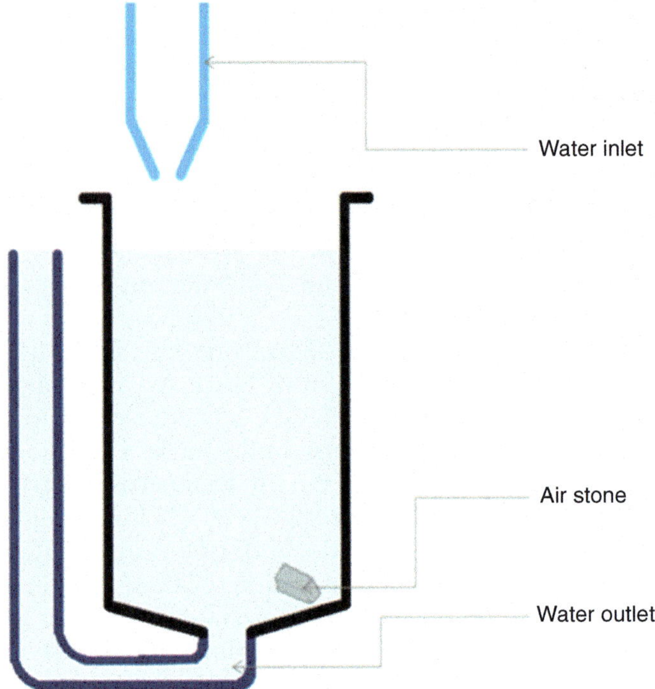

Fig. 3 Sketch of a fish-holding tank. The *slopped bottom* contributes to the removal of solid detritus as these will slide to the bottom and toward the outlet. The vertical pipe in the outlet should be removable so that pipes of different length can be used, allowing the height of the water in the tank to be controlled. This pipe should be removable and interchangeable, so that the tank can easily be modified to hold various volumes of water. A small plastic grid should be placed between the tank and the entry of the water outlet to prevent fish from being sucked in. Either a rigid plastic grid placed horizontally to cover the outlet or a flexible one, folded into the outlet

2. Temperature: Carp such as *Cyprinus carpio* L can tolerate temperatures ranging from 0 °C to above 35 °C and can be cultivated at temperatures between 10 and 30 °C (*see* **Note 6**).
3. Inorganic substrates: The most significant of these are ammonia (NH_3), nitrate (NO_3^-) and nitrite (NO_2^-). These three molecules are related because nitrifying bacteria will convert ammonia to nitrite while other bacteria will oxidize the nitrite into nitrate. Maintain nitrate levels below 25 mg/L. Determine nitrate levels by using a commercially available nitrate test strip, which can be used as an indicator for the accumulation of the other nitrogenous self-pollutants (*see* **Notes 7** and **8**).
4. Nutrition: Use a commercially available food source that has been specifically optimized for the farming industry and is available as extruded pellets. Use a pellet size appropriate for the age of the fish (Table 2) and the size of their mouth (*see* **Notes 9–11**).

Table 2
Feeding rate for cyprinid fish

Fish size (g)	Pellet size (mm)	Feeding ratio under 17 °C (%)	Feeding ratio at 20 °C (%)	Feeding ratio above 26 °C (%)	Number of feeding/day
<0.1	0.5	5	8	12	8–6
0.1–5	1.0–1.5	5	8	12	5
5–10	2.0–2.5	4	7	10.5	4
10–20	2.0–2.5	3.5	6	9	3
20–50	4–5	3	4	6	3
50–100	4–5	2.5	4	6	2
100–300	5–7	2	3	5	2
300–1000	7.5–9.5	1	2.5	3	2
>1000	10–12	0.8	1.5	2.5	2

5. General laboratory maintenance: Monitor fish regularly for any display of stress. In addition, record the water temperature daily. Weigh fish once per week, and calculate their average weight in order to adjust feeding as described in Table 2. As previously mentioned, the frequency of feeding is dependent of the size of the fish. A higher frequency of feeding allows a more homogeneous access to food and a better growth rate (*see* **Note 12**).

3.2 Preparing a Cell Culture for Viral Inoculation

1. Bring to room temperature the culture medium that will be used for the new flasks. Bring the trypsin to room temperature.
2. Sterilize the laminar hood.
3. Decant the medium from the culture flask after cells have reached a minimum of 70 % confluency (Fig. 4a).
4. For a 25 mL culture flask, wash out the residual medium with 0.5 mL trypsin.
5. Add an additional 1.0 mL trypsin, and allow the cells to incubate for up to 5 min. During the 5 min incubation, agitate the cells by mildly rocking the flask. Determine whether the cells have been detached (*see* **Note 13**).
6. Add sufficient medium to the detached cells to bring the total volume to 7.5 mL.
7. Ensure complete detachment of cells from each other by vigorously drawing cells up through a pipette tip and forcing them down against the surface of the flask 6–10 times.
8. Prepare three fresh culture flasks by adding 3 mL of culture medium (previously brought to room temperature as advised in **step 1**) to each.

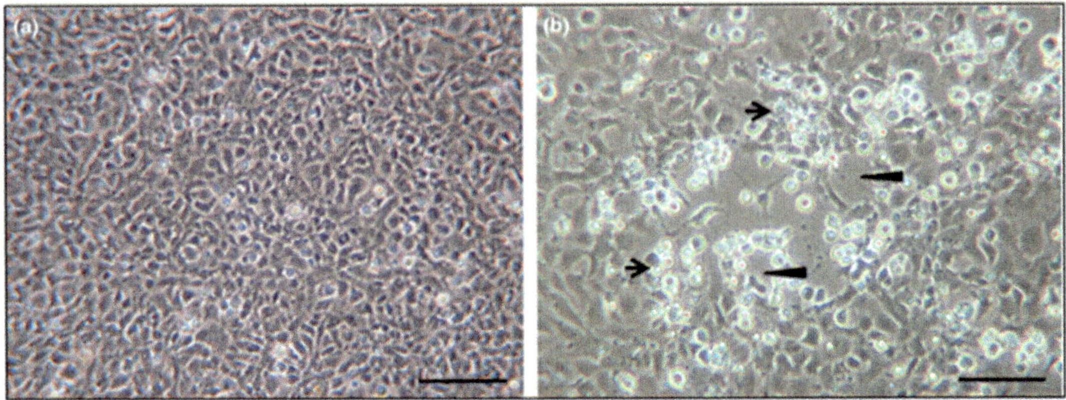

Fig. 4 EPC cells. Spring viremia of carp virus (SVCV)-induced cytopathic effect (CPE). Images of *epithelioma papulosum cyprini* (EPC) were taken 2 days post-incubation with SVCV. (**a**) Untreated, control cells (**b**) SVCV-infected cells. *Arrows* point to SVCV-infected rounded-up cells and *arrow heads* point to SVCV-induced cell clearance. Scale bar = 50 μm [17]

9. Dispense a 2.5 mL aliquot of the completely detached cells into each of the three fresh flasks prepared in **step 8**.
10. Incubate at appropriate temperature (15–25 °C) and duration (3–14 days) depending on the cell line and seeding density (*see* **Note 14**).

3.3 General Protocols for Viral Inoculation

3.3.1 Viral Passaging into Cultured Cells

1. Decant the medium from the culture flask 2 days post-seeding (described in Subheading 3.2) into a sterile beaker and immediately add 0.5–1.0 mL of previously propagated viral aliquots.
2. Incubate at 20 °C for 1 h.
3. Replenish the culture by adding an additional 5 mL of ZB5 medium, and incubate at 20 °C for an additional 2 days. Depending on the virus and initial starting titer, evidence of cytopathic effect can be seen 1–14 days after addition of virus (Fig. 4, [11]).

3.3.2 Viral Inoculation by Immersion of Fish

1. In the fish room, stop the water flow to the fish tank (Subheading 3.1). Reduce the volume of water to about 40 % of its original (*see* **Note 15**).
2. Add the volume of viral solution required to produce the desired final concentration of viral particles.
3. Every 2 h, add a volume of fresh water to the tank corresponding to approximately 20 % of the total tank volume, reaching the original volume in 6 h (*see* **Note 16**).
4. After the water level has been restored to its original volume, wait for two more hours, and then resume water flow (*see* **Note 17**).

3.3.3 Viral Inoculation by Injection of Fish

1. Immerse the carp for approximately 2–5 min in the anesthetic solution (*see* **Note 18**).
2. To inject the viral solution, introduce the needle in the abdominal cavity. The injection site should be slightly anterior to the anus. Position the needle to face the front of the fish at an angle of approximately 45° (*see* **Note 19**).

3.4 Organ Extraction

1. Euthanatize the fish by immersion in a anesthetic solution at a lethal concentration (for example, 1 part per thousand of tricaine methanesulfonate, MS-222; *see* **Note 20**).
2. Place fish on their left side, and disinfect the skin surface.
3. Open the brain cavity. Once sampled, place the brain into minimal essential medium (MEM) (*see* **Note 21**).
4. Use the tweezers to lift the gill operculum, and surgically remove it.
5. Remove the gill arches from the branchial cavity. Cut the cartilage away from the gill filaments and discard it. Transfer the gill filaments into MEM.
6. Use a scalpel to open the fish belly (*see* **Note 22**).
7. Use surgical scissors to cut through the remaining abdominal wall. Insert the scissors through the anterior part of the incision, vertically in relation to the body axis of the fish and interiorly, only slightly behind the pectoral fin.
8. Cut through the tissue while slowly positioning the scissors so as to follow the curve of the top of the peritoneal cavity. This cut should rejoin the first incision, allowing you to completely remove the side of the abdominal wall.
9. Use dissection scissors and tweezers to remove the spleen. Immediately place the organ in MEM (*see* **Note 23**).
10. Remove a section of the liver and place the liver section in MEM (*see* **Note 24**).
11. Use the tweezers to push the viscera out of the way (Fig. 5a). Remove the kidney, and place it in MEM (*see* **Note 25**).

3.5 Lysate Preparation

1. Freeze tissues in liquid nitrogen, and homogenize by carefully crushing with a pestle in a clean crucible to ensure sterile handling.
2. After homogenization, an aliquot of each homogenate can be used to propagate the virus on cell cultures. Common Carp brain is among the cell lines suitable for CyHV-3 (*see* **Note 26** and **27**).
3. Partition homogenate fractions in MEM and/or RNA*later*® depending on further use (*see* **Note 28**). Aliquot MEM partitioned fractions and prepare them for electrospray-ionization mass spectrometry (ESI-MS) analysis.

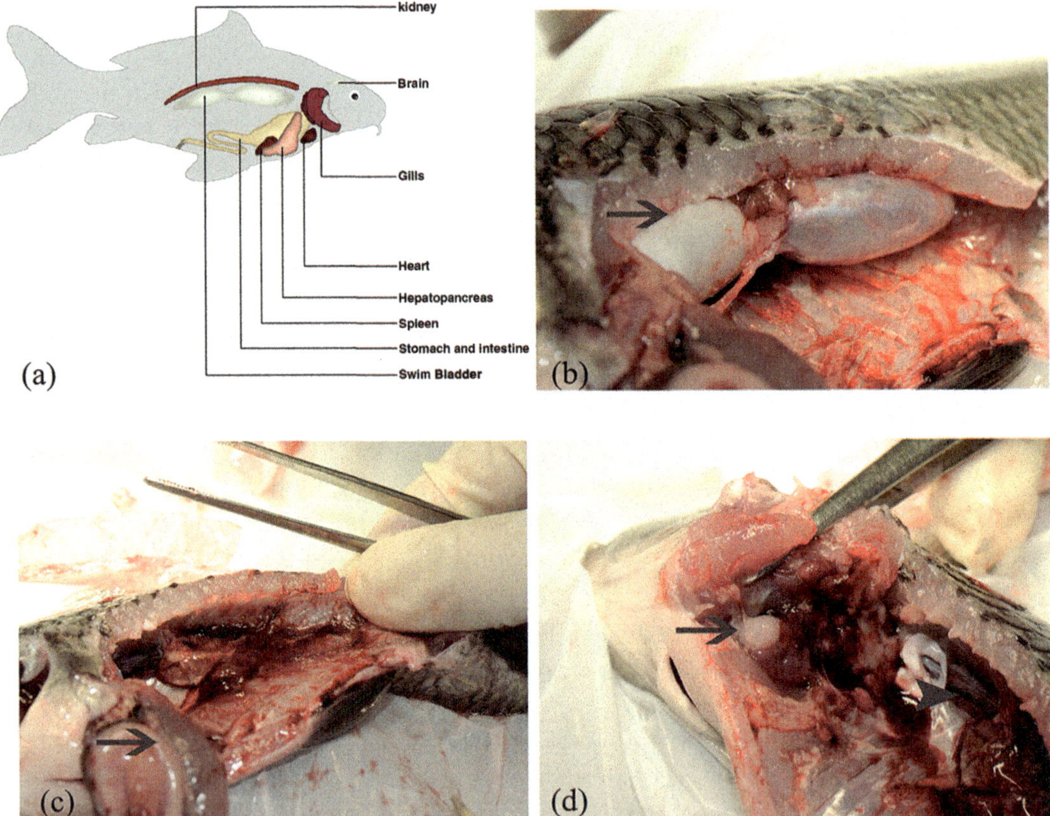

Fig. 5 Removing internal organs. (**a**) Schematic diagram of major fish organs including gills, liver, kidney, spleen and brain. (**b–d**) Dissected (**Cyprinus carpio**) fish. (**b**) *Arrow* points to Swim bladder. (**c**) *Arrow* points to intestine. (**d**) *Arrow* points to heart; *arrowhead* points to kidney

3.6 Preparation of Monoclonal Antibody-Linked Spin Columns

1. Depending on antibody avidity, resuspend 50–200 μg of antibody into 400 μL phosphate-buffered saline (PBS), pH 7.4, and incubate in an agarose spin column overnight at 4 °C with mild shaking (*see* **Note 29**).
2. Empty the spin column by gentle centrifugation (1500 × *g* for 2 min) and wash twice with PBS followed by gentle centrifugation (*see* **Note 30**).
3. Quench the column with 400 μL of 1 M ethanolamine (pH 7.4) by incubation for 1 h at 4 °C with mild shaking.
4. Clear spin columns by centrifugation at 1500 × *g* for 30 s and wash with PBS until OD_{280} of the final flow through is zero.

3.7 Protein Purification

1. Mix the homogenate from Subheading 3.5 in a 1:1 ratio with a non-denaturing lysis buffer containing protease inhibitor cocktail.
2. Put through a vigorously vortex, and centrifuge at 16,000 × *g* for 15 min.

3. Transfer the supernatant to a clean 1.5 mL microfuge tube, and centrifuge a second time at 16,000 ×*g* for 15 min.
4. Use the supernatant from the second centrifugation for affinity purification as described in **steps 5–10** (*see* **Note 31**).
5. Add the non-denatured whole cell extracts to the antibody-linked spin column and incubate overnight at 4 °C with mild shaking at 300 RPM.
6. Clear the spin column, and save the flow through for analysis by gel electrophoresis or mass spectrometry (MS).
7. Wash spin columns eight times with PBS, and or until the OD_{280} of the final flow through is zero.
8. Elute the spin column by incubation with 500 μL of 0.1 M glycine for 1 h at 4 °C with mild shaking. Neutralize the pH by adding 50 μL of 1 M Tris–HCl, pH 8.0 immediately after the incubation in glycine.
9. Wash the column with PBS. Use ESI-MS to analyze the eluate and the first wash fraction.
10. Alternatively, use SDS-PAGE followed by ESI-MS (Fig. 6) (*see* **Notes 32** and **33**).

4 Notes

1. Additional aeration with an aquarium aerator will be necessary if the density of fish or the water temperature do not remain very low throughout the experiments.
2. ZB4G medium is used to propagate the common carp brain (CCB) cell line.
3. ZB5 medium is used to propagate the *epithelioma papulosum cyprini* (epc) cell line.
4. From our experience, it appears that carp only recognize a very limited set of antigens from CyHV-3 [12]; therefore, the antibody should preferentially be raised in a different organism..
5. Carp tolerate low levels of oxygen and relatively poor water quality, which allow more latitude in their maintenance. For example, the fish will easily tolerate being stocked in static water with 20 % of the water volume being replaced every week.
6. The temperature of the fish's rearing water will depend primarily on the laboratory conditions or on the optimal conditions for the pathogen under investigation. The metabolism and food consumption of the fish will decrease at lower temperature. However, following Henry's law, the solubility of oxygen in the water is inversely correlated to the temperature of the water. Therefore, because feeding leads to an increase in the fish oxygen requirements, alimentation should only be performed at temperatures above 25 °C if effective aeration is available.

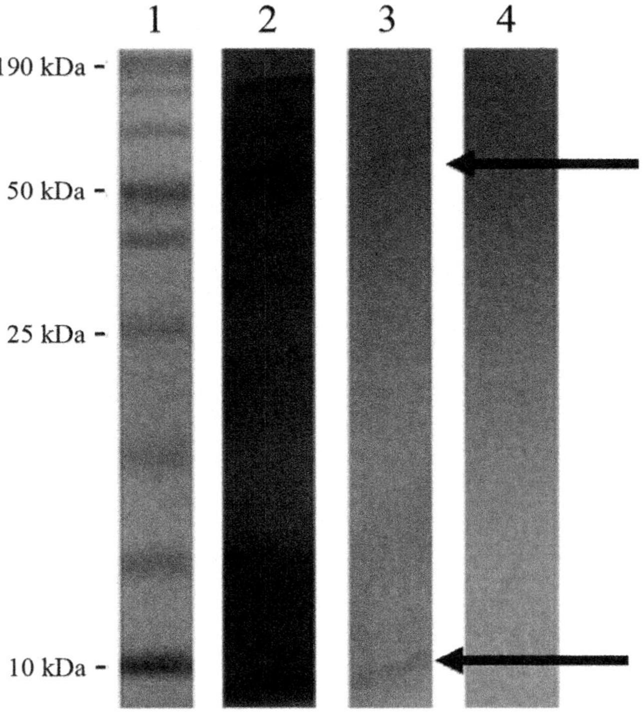

Fig. 6 Gel of immunoprecipitation. PAGE analysis of immunoprecipitation. *Lane 1*: 5 μL of the ProSieve™ Unstained Protein Marker II (by Lonza), 10–200 kDa was used, which consists of 11 bands: 10, 15, 20, 30, 40, 50, 70, 100, 120, 150, and 200 kDa. The 120, 150, and 200 kDa did not separate well and appear as 1 band. *Lane 2*: Flow-through of non-denatured lysate (*see* Subheading 3). *Lane 3*: Monoclonal antibody-purified peptides. Two bands are observed: one at ~10 kDa and another band corresponding to ~60 kDa. *Arrows* point to excised bands in *lane 3*. *Lane 4*: First phosphate-buffered saline wash after elution with glycine (pH 3.0). Forty microliters of the respective samples was run in *lanes 2–4*

7. Nitrite can enter the bloodstream and react with hemoglobin to form methemoglobin, which cannot transport oxygen. Failure of methemoglobin to transport oxygen causes a hypoxia, even under conditions in which the levels of dissolved oxygen in the water would usually be sufficient, a condition termed "brown blood disease."
8. Both nitrate (NO_3^-) and nitrite (NO_2^-) are by-products of the bacterial oxidation of ammonia excreted by fish as part of the digestion process of nitrogen compounds. Ammonia is first nitrified into the highly toxic nitrite that is then converted in nitrate. Increasing the flow of water or the frequency of water changes reduces excessive levels of nitrate. Alternatively, reducing the biomass and/or the quantity of food will also reduce the accumulation of self-pollutants.

9. Similar to other fish, protein consumption contributes a significant portion for their energetic need in addition to being a source of amino acids for these aforementioned species of carp. Therefore, an optimal diet should include from 30 to 38 % protein [13]. According to Watanabe et al. [14], the specific protein source has little impact on nutritional values, except for the notable exception of soybean meal, which contains phytoestrogens that have been well established to have a negative impact on the endocrine and immune system for cyprinidae fish [15, 16].
10. Lipids are also an efficient source of energy and can be added up to 15 % of the diet. A higher percentage of lipid results in a higher growth rate. The unsaturated fatty acids, *n*-6 and *n*-3, are a necessary requirement for the diet of carp. Inclusion of lipid in the diet results in a more efficient use of dietary proteins.
11. The ideal carbohydrate content in the diet is considered to be between 30 and 40 %. Unlike most other fish species, carp can efficiently digest carbohydrates. Inclusion of carbohydrate in the diet results in a more efficient use of dietary proteins.
12. Feeding rate is generally calculated as a proportion of the biomass in the tank, and regular calculation of this biomass, ideally once every week or every other week, is required to adjust this rate. The feeding rate is dependent on several factors (Table 2), mostly the size of the fish and the water temperature. Younger fish have less biomass than adult fish, but they eat more food than adults. The feeding rate can be adjusted to promote, or conversely delay, the growth of the animals.
13. An additional 0.5–1.0 mL of trypsin may be needed to completely separate the cells.
14. CCB and EPC cells can be split in an 1:3 ratio by trypsinization and seeded in fresh ~5 mL medium in 25 cm^2 cell culture flasks at initial concentrations of approximately 2.4×10^5 cells/mL at 20 °C.
15. If the design described in Fig. 2 is used, this can very easily be accomplished by replacing the outside pipe with a shorter one.
16. Adding water will dilute the various self-pollutants such as nitrite.
17. Because of the stress caused by the reduced water flow, it is important to monitor the fish closely during the whole procedure. If the fish show clear sign of stress such as piercing the surface to gulp air, immediately resume water flow.
18. Because fish can react differently to the anesthetic, it is important to keep monitoring the fish during the procedure. Immediately remove the fish from the anesthetic solution if respiratory movements cease.

19. A rule of thumb for the injection volume is to use 100–200 μL of solution per 10 mm of fish length.
20. To insure that the euthanasia is performed properly, keep fish in the anesthetic solution 5 min after cessation of respiratory movements.
21. In smaller fish, a scalpel blade can be inserted parallel to the fish body just above the level of the top of the eyes of the fish. Cutting through the width of the skull will allow removal of the skull cap and accessing the brain. In larger fish, the same procedure can be applied. However, the skull is much wider and thicker.
22. Care should be taken when incising the belly. The scalpel blade should not be too deep in order to avoid damaging the intestine behind the abdominal wall. Similar care should be taken whenever handling the intestine, as a damaged intestine will release a large number of bacteria that will contaminate samples.
23. The spleen appears as small dark-red triangular organ (Fig. 4). It may be hidden behind the intestine and difficult to see.
24. The liver is a large organ of pink to orange color and lies at the very end of the cavity. When manipulating the liver, avoid damaging the gall bladder, a small greenish, transparent organ that lies alongside the liver. The very fragile the gall bladder, if damaged, will release digestive enzymes that will damage the tissues.
25. The kidney is located at the very back of the cavity, hidden behind the swim bladder. The swim bladder appears as a whitish sack that may be more or less inflated at the time of the sampling. Behind the swim bladder, the kidney is a very thin dark red strip of tissue that lies against the spinal column. Because of its thinness, sampling of this organ can be difficult and it may be necessary to use a scalpel blade to scrape it away from the spine.
26. Alternatively, instead of **steps 1** and **2**, place the organs in lysis buffer and perform lysis by sonication or by using a bead-mill homogenizer.
27. Examination of the cytopathic effect on a cell line will allow confirmation of virus identity. In addition, DNA can also be extracted from these samples and used as template to perform PCR or qPCR.
28. PCR amplification on a fraction of the homogenate can be used to confirm many viral or bacterial infections. Extract DNA with DNeasy Blood Tissue Kit (Qiagen) and use standard protocols for conventional or quantitative real-time PCR (qPCR).

29. The antibody should be conjugated to the column on the previous day before the tissue lysate is prepared to allow for overnight conjugation of the antibody to the column.
30. The columns are sensitive to centrifugal force and should never be cleared by more than 2000 × *g*. Also, the columns should not be allowed to dry.
31. Two rounds of centrifugation ensure that particulates are not transferred to columns.
32. If the eluate is analyzed by SDS-PAGE, bands of interest can be sliced out for analysis by ESI-MS.
33. For our ESI-MS studies, the analysis was performed at the Proteomics Core Facility in the German Cancer Research Center (DKFZ) in Heidelberg, Germany.

Acknowledgements

Polyclonal antibodies were kindly provided by Dr Bergmann and Dr Fichtner, Friedrich Loeffler Institute (FLI), the national research centre for animal health in Greifswald, Germany. We thank Drs. Andrea Dressler, Ahmed Abd-Elfattah, and Julia Kattlun for laboratory assistance, and Dr. Mona Saleh and Wilfried-Matthias Stock for assistance with photography. We would also like to thank the Proteomics Core Facility of the German Cancer Research Centre Heidelberg for their excellent work regarding protein identification by ESI-MS. Funding for these projects were supported by the Austrian Science Fund (FWF) Grant Nos. P23550-B13 and P23850-B17.

References

1. FAO (2012) The State of World Fisheries and Aquaculture 2012. Rome, *FAO*, pp 3–36, Accessed on March 05, 2013, from http://www.fao.org/docrep/016/i2727e/i2727e.pdf
2. Xie J, Lu L, Deng M, Weng S, Zhu J, Wu Y, Gan L, Chan S, He J (2005) Inhibition of reporter gene and Iridovirus-tiger frog virus in fish cell by RNA interference. Virology 338: 43–52
3. Hedrick RP, Gilad O, Yun S, Spangenberg JV, Marty GD, Nordhausen RW, Kebus MJ, Bercovier H, Eldar A (2000) A herpesvirus associated with mass mortality of juvenile and adult koi, a strain of common carp. J Aquat Anim Health 12:44
4. Hedrick RP, Gilad O, Yun SC, Mcdowell TS, Waltzek TB, Kelley GO, Adkison MA (2005) Initial isolation and char acterization of a herpes-like virus (KHV) from koi and common carp. Bull Fish Res Agency Suppl 2:1–7
5. Talbi C, Cabon J, Baud M, Bourjaily M, de Boisseson C, Castric J, Bigarre L (2011) Genetic diversity of perch rhabdoviruses isolates based on the nucleoprotein and glycoprotein genes. Arch Virol 156:2133–2144
6. Ahne W (1978) Uptake and multiplication of spring viraemia of carp virus in carp, *Cyprinus carpio* L. J Fish Dis 1:265–268
7. Gotesman M, Hosein RE, Gavin RH (2010) A FERM domain in a class XIV myosin interacts with actin and tubulin and localizes to the cytoskeleton, phagosomes, and nucleus in *Tetrahymena thermophila*. Cytoskeleton 67: 90–101

8. Gotesman M, Hosein RE, Gavin RH (2011) MyTH4, independent of its companion FERM domain, affects the organization of an intramacronuclear microtubule array and is involved in elongation of the macronucleus in *Tetrahymena thermophila*. Cytoskeleton 68: 220–236
9. Williams NE (2000) Immunoprecipitation procedures. In: Asai DJ, Forney JD (eds) Methods in cell biology: Tetrahymena thermophila, vol 62. Academic, New York, NY, pp 449–453
10. Sambrook J, Russell D (2001) Molecular cloning. A laboratory manual, 3rd edn. Cold Spring Harbor Laboratory Press, Cold Spring Harbor, NY
11. Dishon A, Davidovich M, Ilouze M, Kotler M (2007) Persistence of cyprinid herpesvirus 3 in infected cultured carp cells. J Virol 81(9):4828–4836. doi:10.1128/JVI.02188-06
12. Kattlun J, Menanteau-Ledouble S, El-Matbouli M (2014) Non-structural protein pORF 12 of cyprinid herpesvirus 3 is recognized by the immune system of the common carp *Cyprinus carpio*. Dis Aquat Organ 111:269–273. doi:10.3354/dao02793
13. Takeuchi T, Satoh S, Kiron V (2002) Common Carp, *Cyprinus carpio*. In: Webster CD, Lim C (eds) Nutrient requirements and feeding of finfish in aquaculture. CAB International, Oxon
14. Watanabe T, Takeuchi T, Satoh S, Kiron V (1996) Digestible crude protein contents in various feedstuffs determined with four freshwater fish species. Fish Sci 62:278–282
15. Bagheri T, Imanpoor MR, Jafari V, Bennetau-Pelissero C (2013) Reproductive impairment and endocrine disruption in goldfish by feeding diets containing soybean meal. Anim Reprod Sci 139:136–144
16. Zhang J-X, Guo L-Y, Feng L, Jiang W-D, Kuang S-Y, Liu Y, Hu K, Jiang J, Li S-H, Tang L, Zhou X-Q (2013) Soybean β-conglycinin induces inflammation and oxidation and causes dysfunction of intestinal digestion and absorption in fish. PLoS One 8:e58115
17. Gotesman M, Soliman H, Besch R, El-Matbouli M (2014) Inhibition of spring viraemia of Carp virus replication in an *epithelioma papulosum cyprini* cell line by RNA*i*. J Fish Dis 38:197. doi:10.1111/jfd.12227
18. Gotesman M, Soliman H, El-Matbouli M (2013) Antibody screening identifies 78 putative host proteins involved in Cyprinid herpesvirus 3 infection or propagation in common carp, *Cyprinus carpio* L. J Fish Dis 36:721–733
19. Gotesman M, Abd-Elfattah A, Kattlun J, Soliman H, El-Matbouli M (2014) Investigating the interactions of Cyprinid herpesvirus-3 with host proteins in goldfish *Carassius auratus*. J Fish Dis 37:835. doi:10.1111/jfd.12172

Chapter 22

Using a Handheld Gene Gun for Genetic Transformation of *Tetrahymena thermophila*

Michael Gotesman and Selwyn A. Williams

Abstract

This chapter describes protocols for using a handheld gene gun to deliver transformation vectors for overexpression of genes or gene replacement in the macronucleus of *Tetrahymena thermophila*. The protocols provide helpful information for preparing *Tetrahymena* for biolistic bombardment, preparation of vector-coated microcarriers, and basic gene gun operating procedures.

Key words *Tetrahymena thermophila*, Biolistic bombardment, Gene gun, Overexpression, Gene replacement, Microcarriers

1 Introduction

Tetrahymena thermophila exhibits nuclear dimorphism, in which separate nuclei are used for vegetative and sexual processes. A macronucleus (MAC) provides for vegetative functions while a micronucleus (MIC), which provides for the sexual process of conjugation, is transcriptionally silent during vegetative growth. The MAC contains approximately 225 chromosomes [1] and approximately 45 copies of each gene and divides amitotically at cell division. The MIC contains five chromosomes and divides mitotically. *Tetrahymena* genome databases can be assessed at ciliate.org or at National Center for Biotechnology Information.

Biolistic bombardment is widely used as a means of delivering vector-coated microparticles into microorganisms, cultured cells, and tissues. The first particle delivery system contained a helium propulsion unit (the gun) mounted in a vacuum-controlled chamber. In contrast, the handheld gene gun does not operate within a chamber. It is completely handheld, easy and efficient to use, and it requires minimal space on the laboratory bench top. The basic operation of the gene gun involves adhering microcarriers coated

Ray H. Gavin (ed.), *Cytoskeleton: Methods and Protocols*, Methods in Molecular Biology, vol. 1365,
DOI 10.1007/978-1-4939-3124-8_22, © Springer Science+Business Media New York 2016

with the biological material of interest (DNA, RNA, or other substances) to the inner surfaces of small plastic tubes. High-pressure helium gas is used to dislodge the microcarriers ("bullets") and force them into the cell. Although the handheld gene gun has been used for several years in studies of cultured cells [2, 3], its apparent use in *Tetrahymena* has not been widespread [4, 5].

Biolistic bombardment has been successfully used for somatic, germline, and overexpression transformations in *Tetrahymena*. Somatic gene replacements target a gene of interest in the MAC. DNA from a transformation vector homologously recombines with the gene of interest. After many cycles of vegetative growth, a process known as phenotypic assortment [6] ensures that the MAC is homozygous for all alleles, and therefore all copies of the gene of interest have been replaced. However, if the replaced allele is nonfunctional and the gene is essential, phenotypic assortment may not be completed, and the MAC will retain a few copies of the endogenous allele, a condition referred to as a knockdown. Conjugation leads to destruction of the existing MAC and its replacement with a new MAC that is developed from the zygotic MIC. Therefore, somatic transformations, which occur solely in the MAC are lost during conjugation. In contrast, germline transformations target the MIC and are expressed in the newly formed MAC after conjugation. Overexpression transformations target the extrachromosomal rDNA at the time in conjugation when the new MAC is under formation.

Plasmid vectors for transformation of *Tetrahymena* contain a cassette consisting of a gene for antibiotic resistance, transcriptional start and stop sequences, and flanking regions that allow for homologous recombination with the endogenous gene of interest. Some of these vectors have been designed for either disruption or replacement of endogenous alleles [7–10]. The pIGF-1 vector [11] and its derivatives are frequently used for overexpression of genes in *Tetrahymena* (Fig. 1). This vector targets ribosomal genes located in an extrachromosomal segment that replicates autonomously. In current use, the pIGF-1 vector contains genes for antibiotic resistance, a metallothionein promoter that is inducible by ionic cadmium [12], a GFP gene, and a transcriptional stop sequence (Fig. 1). Genes of interest are cloned 3′ to the GFP site in pIGF-1 and therefore are expressed with an N-terminus GFP tag (Fig 1). Transformation vectors have been used for diverse recent studies involving overexpression or replacement of *Tetrahymena* genes [4, 5, 13–15]. A list of plasmid vectors is available from the National *Tetrahymena* Stock Center (https://tetrahymena.vet.cornell.edu).

This chapter describes protocols for using a hand-held gene gun to deliver transformation vectors for overexpression of genes or gene replacement in the macronucleus of *Tetrahymena thermophila*. Protocols provide helpful information for preparing

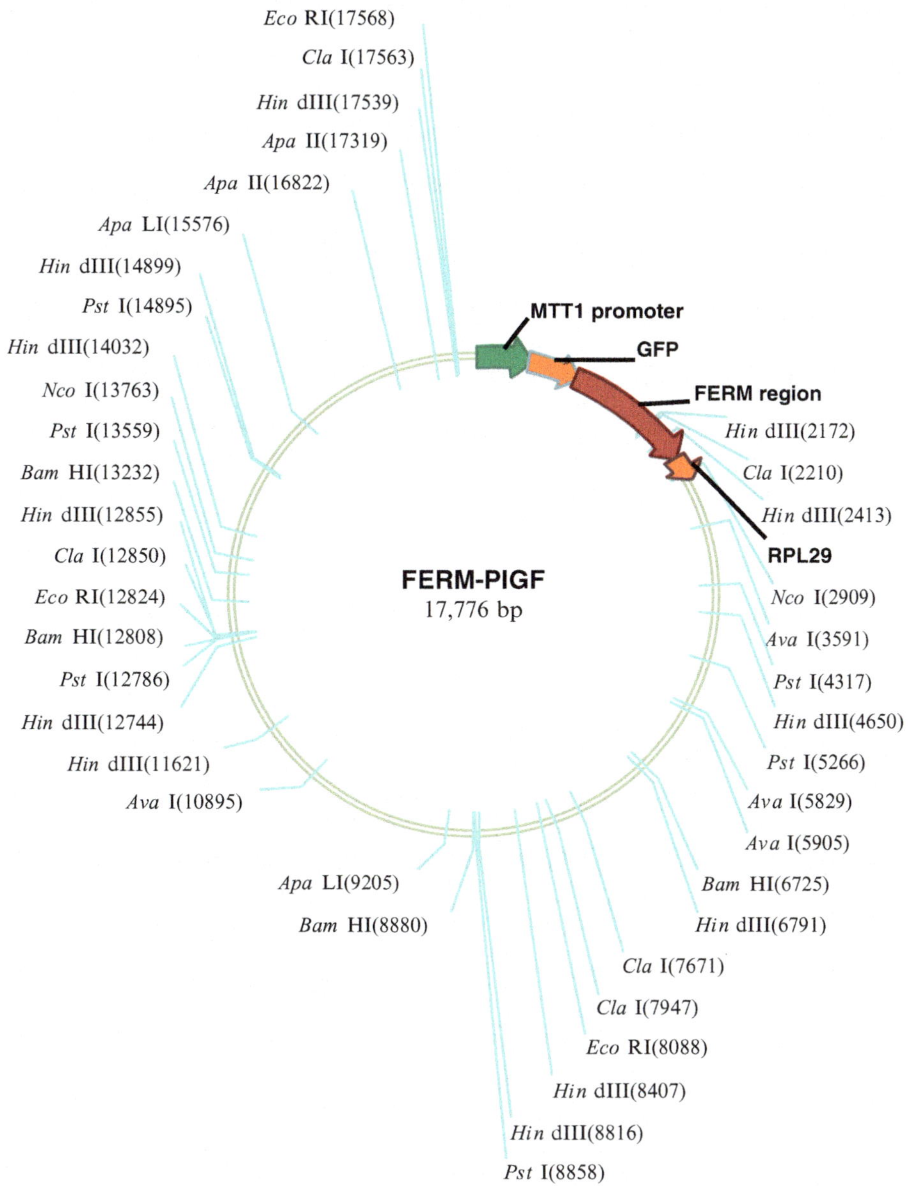

Fig. 1 A pIGF-1 vector was used to prepare a construct for transforming *Tetrahymena thermophila* for overexpression of the FERM domain in a myosin gene [4]. Unique restriction endonuclease cloning sites MTT1 (*green*) GFP (*orange*) FERM insert (*red*) and the transcriptional termination site RPL (*darker orange*) are pictured from $5' \rightarrow 3'$

Tetrahymena for biolistic bombardment, preparation of DNA-coated gold particle "bullets" as adapted from Woods and Zito [16], and basic gene gun operating procedures for the Bio-Rad Helios Hand-Held Gene Gun based on the Bio-Rad Handbook for the Helios Gene Gun. We are not aware of any other maker of a handheld gene gun. However, the protocols described in this chapter should be adaptable to other versions of the gene gun.

2 Materials

2.1 Growth and Starvation Media

1. Modified Neff's medium [17]: 0.25 % Proteose peptone, 0.25 % yeast extract, 0.5 % glucose, 33.3 μM $FeCl_3$ (*see* **Note 1**).
2. Dryl's starvation medium [18]: 2 mM $C_6H_5Na_3O_7$, 1 mM NaH_2PO_4, 1 mM Na_2HPO_4, and 15 mM $CaCl_2$ (*see* **Notes 2** and **3**).

2.2 Antibiotics

1. Paromomycin: 120 μg/mL aqueous solution. Make 100× stock solution: Dissolve 120 mg of paromomycin in 10 mL of sterile of dH_2O.
2. Cycloheximide: 30 μg/mL aqueous solution. Make a 100× stock solution: Dissolve 30 mg of cyclocheximide in 10 mL of sterile dH_2O.
3. Commercially available 100× antibiotic/antimycotic mix.

2.3 Other Reagents

1. $CdCl_2$: 10 μg/mL aqueous solution. Make a 1000× stock solution: Dissolve 10 mg of $CdCl_2$ into 10 mL of sterile dH_2O.

2.4 DNA Coating

1. Vector DNA: ~1.0 μg/μL in dH_2O (*see* **Note 4**).
2. 100 % ethanol.
3. Spermadine: Add 1.45 g of spermadine to 100 mL pure ethanol to make a 0.10 M solution.
4. Polyvinylpyrrolidine (*see* **Notes 5** and **6**).
5. Gold (Au): 0.6–1.0 μm gold particles (*see* **Note 7**).

2.5 Gene Gun

1. Helios Gene Gun System (Bio-Rad 165-2431) includes the helium hose assembly with regulator tubing prep station, syringe kit, Tefzel tube cutter, and optimization kit.
2. Tank of compressed helium.
3. Tank of compressed nitrogen.

3 Methods

3.1 Coating Vector DNA to Gold (AU) Microcarriers

1. Add 50 μg (0.6–1.0 μm) gold to a sterile microfuge tube containing 50 μL of ethanol and 50 μL of 0.10 M spermidine. Sonicate for 1–2 min, and put the mixture through a vortex for 2–3 s. Next, add between 50 and 100 μg of 1 μg/μL vector DNA to the solution, and gently sonicate (*see* **Note 8**).
2. Drop-wise, add 100 μL of 1.0 M calcium chloride to the previously prepared solution, and sonicate for 2–3 min. With the microfuge tube cap open, allow the mixture to remain at room temperature for 10 min.

3. After the 10-min incubation, briefly sonicate, and pulse-centrifuge the mixture for 15 s. Carefully remove and discard the supernatant (*see* **Note 9**).
4. Repeat three washes by adding 1 mL of fresh ethanol and pulse-centrifuge the mixture for 15 s. Carefully remove and discard the supernatant (*see* **Note 9**).
5. After the final wash, use 3.5 mL PVP ethanol working solution to transfer the DNA-coated gold particles to a sterile 10 mL screw-cap container by repeatedly adding 200 μL of PVP (0.01 mg/mL) aliquots.
6. Complete the transfer of the DNA-coated gold particles, and add the remaining PVP working solution to the container, cap and invert the container several times to ensure an even suspension (Fig. 2a).

3.2 Coating the Plastic Tubing with Gold Microcarriers (See Note 10)

1. Cut a 30-in. (80 cm) Tefzel tube (*see* **Note 11**), and dry its inner surface by allowing nitrogen to flow through the tube at 0.35–0.4 L per min for at least 15 min.
2. Use a 5 mL syringe to quickly transfer the DNA-coated gold particles to the dried tubing (Fig. 2b) (*see* **Notes 12** and **13**).
3. Allow the micro-carriers to settle for 3–5 min. Then remove the mixture at a rate of 0.5–1.0 in./s (the tube should be emptied within 30–45 s) (*see* **Note 13**).
4. Rotate the tube 180° (by holding the switch to position II), and allow the gold to coat the inside of the tubing for 3–4 s.

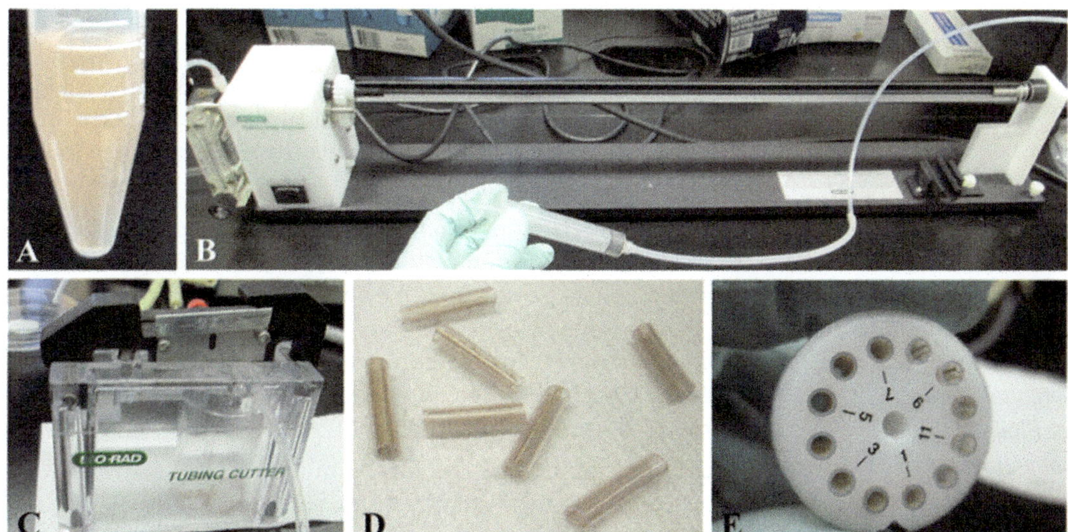

Fig. 2 Preparation of "bullets" for use in the gene gun. (**a**) DNA-Gold mixture suspended in solution. (**b**) Loading the mixture in the drying chamber. (**c**) Cutting "bullets" in the tube cutter. (**d**) Cut bullets. (**e**) Loading the bullets into the barrel of the gene gun

5. Turn on (I) the tubing prep station for 20–30 s to rotate the tube and smear it with gold.
6. Allow nitrogen to flow through the tube at a rate of 0.35–0.4 L per min for 3–5 min.
7. Turn off the rotator (0), and close the nitrogen valve.

3.3 Tube Cutting and Barrel Preparation

1. Transfer the gold-smeared tube from the tubing prep station to the tube cutter. Ensure that the end of the tubing contacts the rear end of the prep station. Cut the tubing into sections approximately 1 in. (2.5 cm) in length (Fig. 2c).
2. Load cut tubes (Fig. 2d) into the barrel of the gun (Fig. 2e) or store the tubes at −20 °C until cells have been prepared.

3.4 Preparing Cells for Transformation Using Overexpression Vectors

The pIGF-1 vector and its derivatives target developing macronuclei at conjugation. Therefore, two different mating-competent strains are required for this type of transformation. Mating-competent cells can be purchased from the National *Tetrahymena* Stock Center (https://tetrahymena.vet.cornell.edu).

1. Separately grow wild-type *Tetrahymena* cells (strains 427 and 428) in bottles containing 30 mL of modified Neff's medium at 30 °C overnight, usually 12–18 h dependent upon the concentration of the inoculum.
2. Initiate a starvation regime while cells are still in exponential phase of growth. Gently centrifuge cells at 300 × *g* for 5 min in a sterile glass centrifuge tube. Carefully decant the Neff's medium without disturbing the soft pellet of cells. Resuspend the pellet in Dryl's starvation medium and wash three times using the gentle centrifugation regime with fresh Dryl's starvation medium for each wash.
3. After the third wash, resuspend each strain in 30 mL of Dryl's starvation medium to yield an optical density of 0.2 as measured at 540 nm. Incubate the cells overnight (approximately 18 h) at 30 °C (*see* **Note 14**).
4. The following day add ~2.5 mL of each starved *Tetrahymena* strain to a sterile 100 × 15 mm plastic Petri dish, and incubate at 30 °C. Expect conjugal pairing in 2–3 h after the initial cell mixing.

3.5 Preparing Conjugal Pairs for Biolistic Bombardment

1. Concentrate ~30 mL of conjugal *Tetrahymena* pairs (8–10 h after mixing) by gentle centrifugation at 300 × *g* for 5 min in a sterile glass centrifuge tube.
2. Gently decant half of the Dryl's medium, and resuspend the conjugal pairs in the small volume (approximately 2–3 mL) of Dryl's that remains in the centrifuge tube.
3. Decant the cell suspension on to the surface of sterile 24 cm GF/A Whatman filter paper placed in 100 mm × 15 mm plastic Petri dishes.

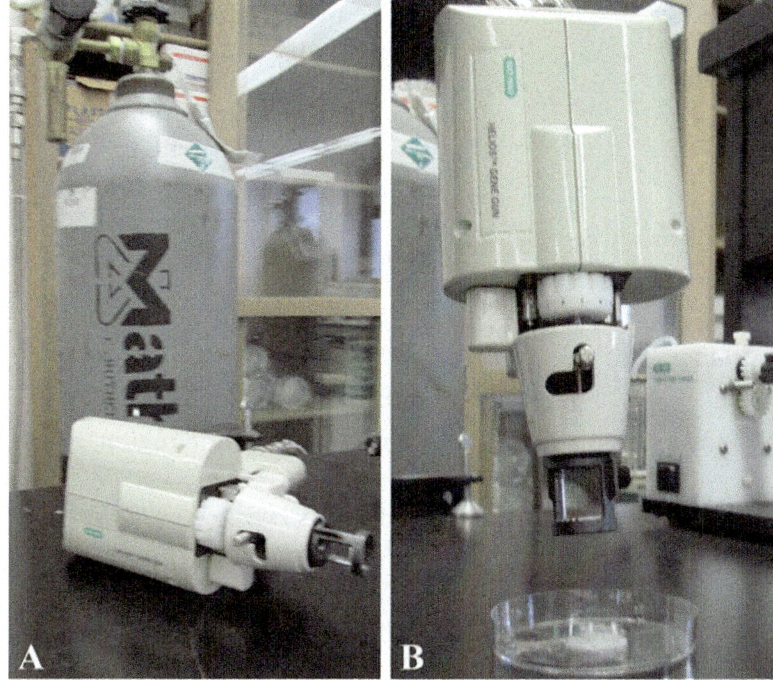

Fig. 3 Assembled gene gun. (**a**) Gene gun attached to a helium tank. (**b**) Firing position of gun

3.6 Using the Helios Handheld Gene Gun for Particle Bombardment

1. Load the stored bullets (Fig. 2d) into the barrel of the gun (Fig. 2e); load the barrel into the gun, and reassemble the remaining components (Fig. 3a).
2. Pressurize the gun to 200 psi with helium.
3. Hold the gene gun approximately 20 cm above the petri dish containing the conjugal pairs (Fig. 3b) and discharge the gene gun with a pressure of 120–180 psi (*see* **Note 15**).

3.7 Screening for Overexpression Transformants

1. Wash the bombarded cells from the petri dish into Neff's medium containing antibiotic/antimycotic mix, and incubate at 30 °C overnight.
2. Add the appropriate selection antibiotic either paromomycin (120–300 μg/mL) or cycloheximide (30–120 μg/mL) depending on the vector of choice (*see* **Note 16**). Vectors that contain a metallothionein promoter can be activated for expression of antibiotic resistance by addition of 1.0 μg/mL $CdCl_2$ (*see* **Note 17**). Next, dispense 250 μL aliquots of cell culture to each well of a 96-well plate, and incubate at 30 °C.
3. Examine each well for antibiotic resistant clones after 3–5 days. Antibiotic sensitive cells (non-transformers) will die within 3–5 days. Transformants will display rapid movement, and over time the population density will increase within each well (*see* **Note 18**).

4. Remove a 100 μL aliquot of culture from each well containing transformants, and put each aliquot into a tube containing 3 mL of Neff's medium. Add the appropriate selection antibiotic at a concentration of 120 μg/mL paromomycin or 30 μg/mL cycloheximide.
5. Repeat **step 4** except use a higher concentration of the appropriate selection antibiotic (180–240 μg/mL paromomycin or 45–60 μg/mL cycloheximide) (*see* **Note 18**).

3.8 Somatic Transformations

Somatic transformations are achieved by using plasmid vectors that target genes in the macronucleus of vegetative cells. Therefore, conjugal pairs are not used for this technique.

1. Perform the starvation regime as described in Subheading 3.4, **steps 1–3**.
2. Transfer concentrated cells on to the surface of sterile 24 cm GF/A Whatman filter paper placed in 100 mm × 15 mm plastic Petri dishes.
3. Prepare the gene gun and use it as described in Subheading 3.6, **steps 1–3**.

3.9 Screening for Somatic Transformants

Screen for transformants using the protocols described in Subheading 3.7, **steps 1–4**.

4 Notes

1. Modified Neff's medium [17]: 0.25 % Proteose peptone, 0.25 % yeast extract, 0.5 % glucose, 33.3 μM $FeCl_3$. Add 500 mL of dH_2O to an autoclavable bottle Weigh and add 2.50 g proteose peptone, 2.50 g yeast extract, 5.0 g glucose, and 1 mL of 1000× 333 μM $FeCl_3$. Bring the solution up to 1 L with dH_2O, and mix vigorously. Autoclave. Store at 4 °C.
2. Dryl's starvation medium [18]: 2 mM $C_6H_5Na_3O_7$, 1 mM NaH_2PO_4, 1 mM Na_2HPO_4, and 15 mM $CaCl_2$. Add 50 mL of dH_2O to an autoclavable bottle. Weigh and add 590 mg Na citrate · $2H_2O$, 140 mg $NaH_2PO_4 \cdot H_2O$, 140 mg Na_2HPO_4. Bring the solution up to 100 mL with dH_2O, and mix vigorously. Add 50 mL of dH_2O to a separate autoclavable bottle. Weigh and add 130 mg $CaCl_2$. Bring the solution up to 100 mL with dH_2O, and mix vigorously (*see* **Note 3**). Autoclave each solution for 20 min, and allow it to cool before using. Mix the two solutions under aseptic conditions by allowing each solution to pass through the same filter sterilization column, and adjust the volume to 1 L with dH_2O.
3. Alternatively Dryl's can be prepared as 10× aliquots.

4. Prepare a suitable transformation vector for either overexpression or replacement of genes in the macronucleus. A list of plasmid vectors is available from the National *Tetrahymena* Stock Center (https://tetrahymena.vet.cornell.edu).
5. Prepare a stock solution of polyvinylpyrrolidine (PVP): Dissolve 200 mg PVP into 10 mL of dH_2O to make a 20 mg/mL stock solution. Prepare a working solution of PVP: Add 175 μL of 20 mg/mL PVP stock solution to a sterile screw cap container and bring the volume up to 3.5 mL with 100 % ethanol to yield a concentration of 0.01 mg/mL PVP.
6. Polyvinylpyrrolidone (PVP) serves as an adhesive during the cartridge preparation process. The working concentration of PVP should be between 0.01 and 0.1 mg/mL. We found that 0.01 mg/mL of PVP worked well for us.
7. We have experienced variability in the effectiveness of some commercially available gold particles. However we achieved good results with gold microcarriers from Seashell Tech (http://www.seashelltech.com/dnadel.shtml).
8. We used the Branson Model B200 Ultrasonic Cleaner 120 V, which has only on and off settings. For gentle sonication, the unit was turned on for 20–30 s.
9. A standard bench top microcentrifuge was used for pulse centrifugations. Do not disturb the pellet when decanting the supernatant; it is acceptable to leave a small fraction of the supernatant behind.
10. It is advisable to begin drying the tube shortly before coating the gold with DNA.
11. Tefzel tubing is supplied with the Bio-Rad Gene Gun.
12. Care should be taken to work rapidly when coating the plastic with gold.
13. A vacuum pump can be used to fill and to empty the Tefzel tube with the gold-DNA mixture. Alternatively, with a bit of practice, this step can be performed manually with a syringe.
14. In order to prevent low mating efficiency, the Neff's medium must be completely washed from the culture and replaced with Dryl's starvation medium in order to ensure adequate starvation of cells. Incompletely starved cells will continue to feed and will not mate. Too much centrifugal force will lyse some of the cells, and the cell debris becomes nutrient material for the cells and consequently they are not starved. The optimal starvation time is 18 h assuming no nutrient medium is present in the culture.
15. *Tetrahymena* cells are very sensitive to biolistic force and will splatter out of the Petri dish when subjected to bombardment at pressures greater than 200 psi. We observed very good

results at psi range between 120 and 180. For higher discharge pressures, preparing cartridges with higher PVP concentrations can increase the total number of particles delivered.

16. Begin screening with a low concentration of antibiotic (120 μg/mL paromomycin or 30 μg/mL cycloheximide). Increase the antibiotic concentration until cells can no longer tolerate a further increase in concentration.
17. We observed basal expression of GFP or GFP-tagged epitopes with the pIGF-1 vector in the absence of added $CdCl_2$ consistent with another observation [19]. It is possible that the $FeCl_3$ in Neff's medium mildly activates the MTT1 promoter.
18. It is advisable to replenish the 96-well plates with fresh Dryl's medium containing antibiotic/antimycotic mix and the appropriate selection antibiotic after 2–3 days. $CdCl_2$ should not be added to the replenishment medium.

Acknowledgements

The authors thank Roland E. Hosein for laboratory assistance. Protocols in this chapter are based on research performed at Brooklyn College-CUNY and supported by Grants MCB 0517083 and DBI 0619460 from National Science Foundation awarded to Dr. Ray H. Gavin.

References

1. Eisen JA, Coyne RS, Wu M et al (2006) Macronuclear genome sequence of the ciliate *Tetrahymena thermophila* a model eukaryote. PLoS Biol 4(9):e286. doi:10.1371/journal.pbio.0040286
2. O'Brien JA, Lummis SCR (2006) Biolistic transfection of neuronal cultures using a hand-held gene gun. Nat Protoc 1(2):977–981. doi:10.1038/nprot.2006.145
3. O'Brien JA, Lummis SCR (2007) Diolistics: incorporating fluorescent dyes into biological samples using a gene gun. Trends Biotechnol 25(11-2):530–534. doi:10.1016/j.tibtech.2007.070.14
4. Gotesman M, Hosein RE, Gavin RH (2010) A FERM domain in a class XIV myosin interacts with actin and tubulin and localizes to the cytoskeleton phagosomes and nucleus in Tetrahymena thermophila. Cytoskeleton 67:90–101
5. Gotesman M, Hosein RE, Gavin RH (2011) MyTH4 independent of its companion FERM domain affects the organization of an intramacronuclear microtubule array and is involved in elongation of the macronucleus in *Tetrahymena thermophila*. Cytoskeleton 68:220–236
6. Merriam EV, Bruns PJ (1988) Phenotypic assortment in tetrahymena thermophila: assortment kinetics of antibiotic-resistance markers tsA death and the highly amplified rDNA locus. Genetics 120:389–395
7. Gaertig J, Gorovsky MA (1992) Efficient mass transformation of *Tetrahymena thermophila* by electroporation of conjugants. Proc Natl Acad Sci 89(19):9196–9200
8. Gaertig J, Gu L, Hai B, Gorovsky MA (1994) High frequency vector-mediated transformation and gene replacement in *Tetrahymena*. Nucleic Acids Res 22(24):5391–5398
9. Cassidy-Hanley D, Bowen J, Lee JH, Cole E, VerPlank L, Gaertig A, Gorovsky M, Bruns PJ (1997) Germline and somatic transformation of mating Tetrahymena thermophila by particle bombardment. Genetics 146:135–147

10. Bruns PJ, Cassidy-Hanley D (2000) Biolistic transformation of macro- and micronuclei. In: Asai DJ, Forney JD (eds) Tetrahymena thermophila, vol 62, Methods in cell biology. Academic, San Diego, CA, pp 501–512
11. Yao MC, Yao CH (1991) Transformation of Tetrahymena to cycloheximide resistance with a ribosomal protein gene through sequence replacement. Proc Natl Acad Sci 88: 9493–9497
12. Shang Y, Song X, Bowen J et al (2002) A robust inducible-repressible promoter greatly facilitates gene knockouts conditional expression and overexpression of homologous and heterologous genes in *Tetrahymena thermophila*. Proc Natl Acad Sci U S A 99(6):3734–3739. doi:10.1073/pnas.052016199
13. Cole ES, Cassidy-Hanley D, Pinello JF, Zeng H, Hsueh M, Kolbin D, Clark TG (2014) Function of the male gamete-specific fusion protein HAP2 in a seven-sexed ciliate. Curr Biol 24(18):2168–2173. doi:10.1016/j.cub.2014.07.064
14. Horrell SA, Chalker DL (2014) LIA4 encodes a chromoshadow domain protein required for genomewide DNA rearrangements in *Tetrahymena* thermophila. Eukaryot Cell 13: 1300–1311
15. Turkewitz AP, Bright LJ (2011) A Rab-based view of membrane traffic in the ciliate *Tetrahymena thermophila*. Small GTPases 2:222–226. doi:10.4161/sgtp.2.4.16706
16. Woods G, Zito K (2008) Preparation of gene gun bullets and biolistic transfection of neurons in slice culture. J Vis Exp (12): e675. doi: 10.3791/675
17. Orias E, Hamilton EP, Orias J (2000) Tetrahymena as a laboratory organism: useful strains cell culture and cell line maintenance. In: Asai DJ, Forney JD (eds) Tetrahymena thermophila, vol 62, Methods in cell biology. Academic, New York, NY, pp 189–211
18. Dryl S (1959) Antigenic transformation in paramecium aurelia after homologous antiserum treatment during autogamy and conjugation. J Protozool 6:25
19. Couvillion MT, Collins K (2012) Biochemical approaches including the design and use of strains expressing epitope-tagged proteins. Methods Cell Biol 109:347–355

Chapter 23

Proteomic Tools for the Analysis of Cytoskeleton Proteins

Michael Scarpati, Mary Ellen Heavner, Eliza Wiech, and Shaneen Singh

Abstract

Proteomic analyses have become an essential part of the toolkit of the molecular biologist, given the widespread availability of genomic data and open source or freely accessible bioinformatics software. Tools are available for detecting homologous sequences, recognizing functional domains, and modeling the three-dimensional structure for any given protein sequence. Although a wealth of structural and functional information is available for a large number of cytoskeletal proteins, with representatives spanning all of the major subfamilies, the majority of cytoskeletal proteins remain partially or totally uncharacterized. Moreover, bioinformatics tools provide a means for studying the effects of synthetic mutations or naturally occurring variants of these cytoskeletal proteins. This chapter discusses various freely available proteomic analysis tools, with a focus on in silico prediction of protein structure and function. The selected tools are notable for providing an easily accessible interface for the novice, while retaining advanced functionality for more experienced computational biologists.

Key words Proteomics, Homology modeling, Comparative modeling, Threading, Sequence similarity, Multiple sequence alignment, Protein domains, Secondary structure prediction, Structure analysis

1 Introduction

The cytoskeleton is made up of a diverse family of structural proteins that operate in conjunction with interaction partners to form a conspicuous and dynamic internal superstructure within a cell. The proteins comprising the cytoskeletal system in any particular cell type represent a bewildering array of distinct molecular architectures: combinations of various modular domains, the function of which may or may not be known. Although a wealth of structural and functional information is available for representative members of the various classes of cytoskeletal proteins, many more remain partially or fully uncharacterized. However, this dearth of experimental data may be supplemented by predictive analyses. Computational tools that are freely and easily accessible to the scientific community currently provide an excellent starting point for prediction of the structural and functional properties of these

Ray H. Gavin (ed.), *Cytoskeleton: Methods and Protocols*, Methods in Molecular Biology, vol. 1365,
DOI 10.1007/978-1-4939-3124-8_23, © Springer Science+Business Media New York 2016

uncharacterized protein sequences. These predictions may in turn lead to well-designed experiments that can further probe the hypothesized function.

Proteomic tools are available for a variety of tasks, ranging from sequence analysis and characterization to the generation of high-resolution 3D models. For example, tools exist for finding sequentially similar sequences and putative homologs (e.g., BLAST, HMMER), recognizing and characterizing functional protein domains and residues (e.g., ScanProsite, Pfam), and generating tertiary structure models for a given sequence (e.g., MODELLER, I-TASSER, HHpred) (*see* **Note 1**, Table 1). These tools may be used alone as a platform for studying uncharacterized proteins or to map experimentally observed functionality onto a structural model. Alternatively, proteomic analyses may be carried out in conjunction with traditional wet-lab techniques as a means to design more meaningful experiments. For example, a preliminary bioinformatics analysis may identify targets that are more likely to be pertinent to a given study. This chapter discusses various proteomic analysis tools with a focus on protein structure prediction and functional characterization. These in silico experiments are designed to begin with a raw sequence, which may have no functional annotation associated with it, and in a step-wise manner derive sequence and structural information leading up to a three-dimensional model of the sequence or its constituent domain(s), which can be analyzed for its biophysical properties.

2 Materials

2.1 Computer with Internet Access

A computer with any underlying operating system and access to the Internet is required for the computational analyses described in the methods. For certain tools, familiarity with a command line interface or the UNIX operating system may be useful.

2.2 Visualization Software

In order to visualize and manipulate three-dimensional protein structures a molecular visualization system is required. Chimera is a comprehensive software package for rendering and manipulating three-dimensional protein structures [1]. In addition to providing an expansive toolset related to protein structural analyses, Chimera also includes limited functionality with respect to sequence analysis and alignment, protein–ligand docking, and molecular dynamics. While many software packages are available for each of these discrete functions, having the toolset available in a single operating environment may result in a more efficient workflow.

Chimera binaries are available for the Windows platform (32 and 64-bit) as well as UNIX-based operating systems such as Mac OS X platform, and all are freely accessible for noncommercial use. The most recent version of Chimera, 1.10.1, was released on December 24, 2014.

Table 1
List of alternative software choices

Function	Program or database	URL	Primary reference
Sequence and Structure Databases	GenBank (proteins)	http://www.ncbi.nlm.nih.gov/protein	[2]
	RefSeq	http://www.ncbi.nlm.nih.gov/refseq/	[5]
	UniProt	http://www.uniprot.org/	[6]
	EMBL-EBI	http://www.ebi.ac.uk/	[3]
	DDBJ	http://www.ddbj.nig.ac.jp/	[4]
	RCSB PDB	http://www.rcsb.org/	[26]
	neXtProt	http://www.nextprot.org/	[34]
	PIR	http://pir.georgetown.edu/	[35]
Sequence Search	BLAST	http://blast.ncbi.nlm.nih.gov/Blast.cgi *Note*: PSI-BLAST, Delta-BLAST, and other variants are available at this address.	[7]
	CS-BLAST	http://toolkit.tuebingen.mpg.de/cs_blast	[36]
	HHblits	http://toolkit.tuebingen.mpg.de/hhblits	[37]
	HMMER	http://hmmer.janelia.org/ *Note*: JackHMMer, HMMsearch, and other HMMer variants are available at this address.	[9]
	FASTA Suite	http://www.ebi.ac.uk/Tools/sss/fasta/ *Note*: searches are available with SSEARCH (local), GGSEARCH (global), and GLSEARCH (global query, local database).	[38]
Multiple Sequence Alignment	CLUSTAL Omega	http://www.ebi.ac.uk/Tools/msa/clustalo/	[10]
	T-Coffee	http://tcoffee.crg.cat/apps/tcoffee/do:regular	[11]
	MAFFT	http://mafft.cbrc.jp/alignment/server/	[12]
	MUSCLE	http://www.ebi.ac.uk/Tools/msa/muscle/	[13]
	PRALINE	http://www.ibi.vu.nl/programs/pralinewww/	[39]
	COMPASS	http://prodata.swmed.edu/compass/compass.php	[40]
	PROMALS3D	http://prodata.swmed.edu/promals3d/promals3d.php	[41]
	Probalign	http://probalign.njit.edu/probalign/login	[42]
	PROBCONS	http://probcons.stanford.edu/	[14]

(continued)

Table 1 (continued)

Function	Program or database	URL	Primary reference
Secondary Structure Prediction	PSIPRED	http://bioinf.cs.ucl.ac.uk/psipred/	[23]
	PSSpred	http://zhanglab.ccmb.med.umich.edu/PSSpred/	[25]
	Jpred4	http://www.compbio.dundee.ac.uk/jpred4/index.html	[43]
	NetSurfP	http://www.cbs.dtu.dk/services/NetSurfP/	[44]
	PredictProtein	https://www.predictprotein.org/	[45]
	Porter	http://distill.ucd.ie/porter/	[46]
	SOPMA	https://npsa-prabi.ibcp.fr/cgi-bin/npsa_automat.pl?page=npsa_sopma.html	[47]
	RAPTOR-XSS8	http://raptorx.uchicago.edu/StructurePrediction/predict/	[24]
Sequence Analysis (domain, motif, active site annotation, etc.)	ScanProsite	http://prosite.expasy.org/	[15]
	Pfam	http://pfam.sanger.ac.uk	[17]
	PRATT	http://www.ebi.ac.uk/Tools/pfa/pratt/	[16]
	CDART	http://www.ncbi.nlm.nih.gov/Structure/lexington/lexington.cgi	[18]
	SMART	http://smart.embl-heidelberg.de/	[19, 20]
	HHrepID	http://toolkit.tuebingen.mpg.de/hhrepid	[21].
	REPRO	http://www.ibi.vu.nl/programs/reprowww/	[22]
	WebLogo3	http://weblogo.threeplusone.com/	[48]
Model Generation (homology modeling, threading)	MODELLER	http://salilab.org/modeller/	[30]
	I-TASSER	http://zhanglab.ccmb.med.umich.edu/I-TASSER/	[49]
	RAPTOR-X	http://raptorx.uchicago.edu/	[24]
	HHpred	http://toolkit.tuebingen.mpg.de/hhpred	[50]
	LOMETS	http://zhanglab.ccmb.med.umich.edu/LOMETS/	[51]
	Phyre2	http://www.sbg.bio.ic.ac.uk/phyre2/	[52]
	pGenThreader	http://bioinf.cs.ucl.ac.uk/psipred/?pgenthreader=1	[53]
	FUGUE	http://tardis.nibio.go.jp/fugue/	[54]
	SPARKS-X	http://sparks-lab.org/yueyang/server/SPARKS-X/	[55]

Model Quality Assessment	ProSA	http://prosa.services.came.sbg.ac.at/prosa.php	[32]
	MODFOLD	http://www.reading.ac.uk/bioinf/ModFOLD/	[56]
	QMEAN	http://swissmodel.expasy.org/qmean/cgi/index.cgi	[57]
	TM-score	http://zhanglab.ccmb.med.umich.edu/TM-score/	[58]
	PSVS	http://psvs-1_5-dev.nesg.org/	[59]
	PROCHECK	http://services.mbi.ucla.edu/SAVES/	[33]
	ProQ	http://www.sbc.su.se/~bjornw/ProQ/ProQ.cgi	[60]
	ModEval	http://modbase.compbio.ucsf.edu/evaluation/	[61]
	VERIFY3D	https://genesilico.pl/toolkit/unimod?method=Verify3D	[31]
Miscellaneous Online Servers	ESPript3	http://espript.ibcp.fr/ESPript/ESPript/ *Note*: generates publication quality images from multiple sequence alignment data.	[62]
	BoxShade	http://www.ch.embnet.org/software/BOX_form.html *Note*: generates publication quality images from multiple sequence alignment data.	None
	Pro-Origami	http://munk.csse.unimelb.edu.au/pro-origami/ *Note*: generates 2D representations of protein domain architecture from a given PDB.	[63]
Downloadable Software Packages	CHIMERA	http://www.cgl.ucsf.edu/chimera/	[1]
	Pymol	http://www.pymol.org	N/A
	Jalview	http://www.jalview.org/	[64]
	Illustrator for Biological Sequences (IBS)	http://ibs.biocuckoo.org/	None

2.2.1 Installation of Chimera

1. Chimera binaries are available at www.cgl.ucsf.edu/chimera/download.html
2. Download the binary installer that is appropriate for your operating system of choice, e.g., 32 or 64-bit Windows, Mac OS X, Linux.
3. Run the installer (on Windows) to install the program. You will be prompted to select an installation directory. On Max OS X, load the installer's ".dmg" file and drag the enclosed executable to your applications folder.

2.2.2 Alternative Molecular Visualization Tools

Pymol is another popular open source visualization package (www.pymol.org; The PyMOL Molecular Graphics System, Version 1.5.0.4 Schrödinger, LLC.). Pymol and Chimera each provide a perfectly adequate graphical user interface (GUI) for molecular visualization and annotation. However, Chimera's manipulation toolset is far more advanced.

Chimera also has the advantage of being freely available as an up to date, ready-to-use binary. Pymol is released as three different versions; (1) a ready to use commercial version; (2) a free, ready to use but outdated academic version; and (3) as source code that must be pre-compiled by the end user, but free for noncommercial use. That being said, the cost of a 1-year academic use license for the commercial version is fairly inexpensive. For lightweight modeling tasks, Pymol may be a perfectly adequate option.

3 Methods

3.1 Retrieval of Protein Sequences

GenBank [2], a highly accessed sequence repository maintained by the National Center for Biotechnology Information (NCBI), is composed of individual categorical databases, including one for protein sequences. GenBank is redundant with similar data structures, including those administered by European (EMBL) [3] and Japanese consortiums (DDBJ) [4]. GenBank's nonredundant, curated database, RefSeq [5], is often used as a primary resource for obtaining sequences of interest, as its collection provides a comprehensive, integrated, nonredundant, well-annotated set of sequences.

NCBI sequence entries are assigned curated, cross-referenced accession numbers in conjunction with EMBL and DDBJ. For proteomic studies, the Universal Protein Resource (UniProt) administered by EMBL, SIB, and PIR, is another excellent repository [6]. UniProt differs slightly in its annotation and accession number system, as compared to the NCBI. However, cross-referencing is provided and the variations in annotations are frequently useful.

Protein sequences, and their annotations, from both NCBI and UniProt are available in several formats (*see* **Note 2**). The following steps explain how to retrieve a specific protein sequence record from the NCBI's GenBank protein database:

1. Point your browser to www.ncbi.nlm.nih.gov/ for the main portal entrance (Entrez), or to www.ncbi.nlm.nih.gov/guide/proteins/ for immediate access to all of the NCBI's protein-specific databases.
2. Search using keyword(s) (*see* **Note 3**) or accession number(s) (e.g., in NCBI or UniProt format). If you choose to start at the protein-specific entry point, select "Protein" from the drop-down list, left of the "Search" field. A BLAST search can also be used to query the database(s) (*see* Subheading 3.2 for details).
3. Select a search result of interest by clicking on its check box. The viewing options include: "GenPep," a detailed annotation record; "FASTA," a minimally annotated sequence in FASTA format (*see* **Note 2**); and "Graphics," for a schematic illustration of the gene locus that encodes the selected protein.
4. The protein's sequence, in FASTA format, can be copied and pasted or saved to a text file, as follows. From the "Send to" drop-down list, choose "File" and "FASTA" format. A FASTA formatted text file containing the sequence will be downloaded to your browser's default downloads folder.
5. This ".fasta" file contains the selected sequence as plain text. Depending on your operating system, you may need to replace the ".fasta" extension in order for your text editor application to recognize the file. For example, on Windows, you may need to replace ".fasta" with ".txt" in order for the Notepad application to recognize and open the file for viewing/editing.
6. It is highly recommended that a simple plain text reader be used for sequence manipulation such as Microsoft's Notepad application. Programs that allow more advanced document formatting options, such as Microsoft Word, are likely to add invisible formatting characters (e.g., an invisible tab or newline symbol) that may interfere with subsequent analyses by other programs.

3.2 Searching for Similarities and Inferring Homologies

The BLAST algorithm [7] quickly finds sequences within a given database (e.g., NCBI's GenBank) that are sequentially similar to a query sequence (*see* **Note 4**). Sequences identified as high-scoring matches in a BLAST search presumably share a functional and/or evolutionary relationship (i.e., putative homology) to the query. However, in order to correctly assign homology, potential relationships must be thoroughly investigated, rather than assumed, and the types of inferences necessary will not be addressed here.

Since the time of its inception, the basic BLAST algorithm has spawned multiple specialized spin-offs, especially for hard to characterize sequences. Methods to find remote similarities for novel, or less widely conserved proteins and their constitutive domains, have been particularly welcomed by researchers. Exemplary algorithms include: Position-Specific Iterative BLASTs (PSI-BLAST, [8]) and hidden Markov model based approaches (HMMER, [9]).

The PSI-BLAST method allows the user to dynamically create a customized position-specific scoring matrix ("PSSM," *see* **Note 5**) based on an iterative series of BLAST searches. The search series is repeated until convergence (i.e., no change in the "hit" list) is observed. During these iterations, the user can adjust the alignment scoring function for the current and subsequent BLAST rounds by setting a cutoff value that determines whether to include low-scoring sequence hits into the PSSM. The addition of low-scoring hits gradually reduces the stringency of the search, but may lead to the detection of distant homologs. In contrast, with a standard BLAST search these hits might not be found. The relevance of data uncovered in a PSI-BLAST is highly dependent on optimized parameters (e.g., selection of an appropriate cutoff value). It is recommended that novice users review the NCBI's PSI-BLAST tutorial, located at http://www.ncbi.nlm.nih.gov/books/NBK2590/

The HMMER method [9] is another useful tool for identifying homologous proteins. Like PSI-BLAST, HMMER relies on a position-specific scoring profile. However, HMMER constructs this profile using a hidden Markov model approach: a dynamic Bayesian network to model the likelihood of each of the 20 standard amino acids occurring at each position in a given query based on a multiple sequence alignment of the query and sequentially similar proteins. In its simplest form, an implementation of the HMMER method follows a two-step process: a profile-HMM is constructed based on the multiple sequence alignment of the query and similar sequences; and (2) the profile-HMM is queried against a database of protein sequences, with results limited based on a selected cutoff value. By using the profile-HMM in place of the original sequence as the search query, HMMER is able to identify sequentially distant homologs that may have escaped detection in a BLAST or PSI-BLAST search.

3.2.1 Conducting a BLASTP (Protein-Specific BLAST) of the Nonredundant NCBI RefSeq Database

1. Point your browser to the NCBI BLAST page (http://blast.ncbi.nlm.nih.gov/Blast.cgi).
2. Select "protein blast" by clicking on its hyperlink.
3. Cut and paste your sequence of interest (in FASTA format) into the "Enter Query Sequence" textbox (*see* **Note 6**). You can also browse and upload the sequence from a file (in FASTA format).

4. In the "Databases" field select the nonredundant (nr) database from the pull-down list (*see* **Note 7**).
5. In the "Program Selection" field select the "blastp" algorithm (*see* **Note 8**).
6. Click on "BLAST" to initiate the search and wait for the results page to load. The Blast output page has graphics at the top and textual details at the bottom. At the very top you will see the program details such as the version of the program used, the reference for BLAST, the name and length of the query sequence, the database searched, and the RID number (*see* **Note 9**). As you scroll down, a graphical output, with color-coded horizontal lines that represent query–hit alignments in descending order of significance, will become visible. If you mouse-over the lines, you can see summary information for each hit sequence. In the next section, a hyperlinked description and accession number is provided for each hit, along with its BLAST scores and *E*-value (*see* **Note 10**). Pairwise alignments between the query and similar regions in the related sequences are given below this. All of these sections display the BLAST hits in the same order, from smallest to largest *E*-values (*see* **Note 10**).
7. In the hyperlinked description section, select the protein(s) of interest by clicking on the checkbox to the left of each sequence.
8. Click on "Get selected sequences" to download a file containing the selected sequences. The options for downloading are similar to those discussed in Subheading 3.1, above.

3.2.2 Conducting a PHMMER (Profile-HMM) Search of the Nonredundant NCBI RefSeq Database

1. The following steps describe how to perform a phmmer search (i.e., a protein sequence HMMER search of a protein database). To begin, navigate to the Janelia Farms HMMER home page (http://hmmer.janelia.org/).
2. From the menu bar, select "Search" and then "phmmer."
3. Input your sequence of interest into the query textbox by (a) pasting (in FASTA format); (b) uploading the sequence in a file (in FASTA format); or (c) entering an NCBI or UniProt accession number.
4. Scroll to the databases selection area and choose the "NR" database, which refers to the NCBI nonredundant protein sequence database (*see* **Note 11**).
5. Other options and advanced modifications of the search algorithm are available. Click the question mark symbol above each section for a description of each parameter and the available options.
6. Click on the "submit" button below the query textbox in order to initiate the search. Wait for the results page to load.

7. The phmmer output page displays a menu bar at the top of the screen that allows the user to view the results sorted by (a) score, (b) taxonomy, or (c) protein domains. The entire set of results can also be downloaded in multiple formats via the (d) download button link. If you wish to save multiple hit sequences at once, visit the "Download" section and select the desired output format. The "significant hits" FASTA format option is recommended as starting point (left column, third button). Returning to the main output page, a graphical display of the sequence features identified in the search is provided (immediately below the menu bar). This section illustrates structural features, domains, and a summary of the query–hit alignment profile. A simple taxonomic hit distribution is displayed next, but this information can be explored in more detail via the "Taxonomy" button at the top of the page. At the bottom of the main output page, a customizable listing of the hits, ranked by *E*-value is displayed (*see* **Note 10**). The NCBI accession numbers of the hits, with hyperlinks to the full annotations, are listed under the "Target" column.
8. You can download any hit sequence by visiting its NCBI annotation page link. After selecting this link, click on "FASTA" to download the full-length hit sequence in FASTA format.
9. The main output section provides additional information for each of the hit sequences, arranged by column. By default, the "Description," "Species" (hyperlinked), and *E*-values for each hit are listed in the columns to the right of each accession number. The "Customize" button provides the user with additional options. The hit list can be expanded to show the query–hit pairwise alignment by (a) clicking on the ">" to the left of the accession number or (b) clicking on "(show all) alignments" at the very bottom of the hit list. This expanded view displays the query–hit sequence similarity and identity, and the lengths of each alignment.
10. Variations on the basic protein HMMER algorithm (phmmer) are available. For example, the iterative JackHMMer algorithm may be preferable for detecting distant homologs. A link to this method, and other alternatives, is available from the main HMMER page (http://hmmer.janelia.org/).

3.3 Multiple Sequence Alignment

Multiple sequence alignments (MSAs) assist in identifying evolutionary relationships between a set of sequences descended from a common ancestor. From the resulting MSA, sequence homology can be inferred and phylogenetic analysis can be conducted to assess the aligned sequences' shared evolutionary origins. MSAs also assist in identifying sequence conservation of protein sequence motifs and patterns, structural and functional domains, and secondary structures elements.

In a basic sense, MSAs are constructed by aligning a given set of sequences in a manner that maximizes a given scoring function. In order to account for residues present in only one (or a subset) of the aligned sequences, gaps may be inserted into the alignment, though gaps normally carry a penalty that reduces the scoring function more so than a residue mismatch. For pairwise alignments, and small MSAs, it may be possible for a program to sample effectively all of the potential alignment permutations in order to identify the highest-scoring MSA. However, as the set of sequences increased the computational requirements scale up fast enough that this strategy becomes prohibitive for medium and large sets. This problem has spurred the development of MSA algorithms that can efficiently sample the available solution space in order to generate a top-scoring MSA with a reasonable degree of confidence. Some of the most popular MSA generation tools include: CLUSTAL Omega [10], T-Coffee [11], MAFFT [12], MUSCLE [13] and PROBCONS [14]. An explanation of the underlying algorithm employed by these servers is available in the papers cited above. Generally speaking, the output from each of these tools is likely to be highly similar, if not identical, for most datasets. However, discrepancies may arise due to biases in the algorithms implemented by these tools. As a result, it is often useful to submit a given query to multiple servers and to evaluate the result manually. The user may also wish to apply corrections to the MSA to account for information available to the user (e.g., knowledge of a particular protein family's phylogeny) that was likely not factored into the algorithm.

3.3.1 Creation of a MSA Using CLUSTAL Omega

To create a MSA from a set of protein sequences:

1. Visit the CLUSTAL Omega site hosted by EMBL at the following address: http://www.ebi.ac.uk/Tools/msa/clustalo/
2. Paste your set of protein sequences in FASTA format into the text box or upload a file containing your set of sequences (*see* **Note 12**).
3. Click on "Submit" and wait for the results page to load. The MSA is displayed on the results page under the "Alignments" tab. A consensus line is provided below the alignment to indicate residue positions that are invariant, conserved or semi-conserved (*see* **Note 13**). The MSA may be colored coded by selecting the "Show Colors" option.
4. To view the MSA interactively click on the "Results Summary" tab and then press the "Start Jalview" button (*see* **Note 14**).
5. To download the MSA select the "Results Summary" tab and click on the hyperlink below the label "Alignment in CLUSTAL format." The resulting page will display the alignment in CLUSTAL format, which should be saved as a file with the

".ali" or ".clustal" extension. Depending on the browser, you may also be able to shift-click on this link to save the file directly to your default downloads folder.

3.4 Protein Motifs, Patterns and Domains

3.4.1 PROSITE

PROSITE is a database of protein motifs, patterns, and profiles that represent known functional sites, families, and domains, which may be searched using the ScanProsite tool [15]. To identify these regions/sites in a given protein sequence:

1. Point your browser to PROSITE: http://prosite.expasy.org/
2. Paste your protein sequence in the text box provided under "Quick Scan mode of ScanProsite".
3. Click "Scan" (*see* **Note 15**).
4. The results page provides a visual schematic and detailed summary of protein domains and families identified in your target sequence. For additional options, click the "ScanProsite" hyperlink (under "Quick Scan mode of ScanProsite").
5. By default, ScanProsite will scan the given protein sequence against the PROSITE collection of motifs (Option 1). In addition, the user may submit a motif and have it scanned against the PROSITE collection of proteins (Option 2) or submit one or more protein sequences and motifs to scan them against each other (Option 3). For motif-based searches, one may wish to consider generating a pattern to be searched using PRATT [16] (*see* "Other Tools") or to submit a particular (known) pattern of interest.
6. To create a cartoon to represent the identified motifs/domains in your query sequence through PROSITE or any other protein domains/motifs detection method, the MyDomains Image Creator can be used (*see* "Other tools"). Alternatively, one may use the Illustrator for Biological Sequences (IBS), a standalone package for visualizing and annotating protein or nucleotide sequences (http://ibs.biocuckoo.org/).

3.4.2 PFAM

PFAM [17] is a hidden Markov model (HMM) profile based method for the detection of protein families and domains. To analyze your protein sequence for Pfam matches:

1. Point your browser to PFAM database hosted by the Sanger Institute (http://pfam.sanger.ac.uk).
2. Click "Sequence Search."
3. Paste your protein sequence into the text box.
4. To search against the Pfam-A families (a set of manually curated families), click "Go."
5. To search against both the Pfam-A and Pfam-B families (a set of automatically generated families), click "here" underneath the text box and check "Search for PfamBs" (this is a useful option

if no Pfam-A matches were identified). The results page will display the graphical output from the search, showing domains identified in your protein sequence as well as a table with statistically significant matches. This table provides additional details regarding the hits: descriptions of domains/families, the alignment between the target sequence and the HMM profile match, and relevant *E*-values.

3.4.3 CDART

The Conserved Domain Architecture Retrieval Tool (CDART) [18] can be used to search for the query protein conserved domains against the NCBI GenBank Protein Database [2] based on protein domain profiles. To identify conserved domains in a sequence of interest, follow these steps:

1. Point your browser to (http://www.ncbi.nlm.nih.gov/Structure/lexington/lexington.cgi).
2. Enter a query protein sequence in the text box.
3. Click "Submit." The results page, normally generated within 2 min, displays conserved domains detected within the query sequence and a list of proteins with similar domain architectures that include at least one domain from the query sequence.

3.4.4 SMART

The Simple Modular Architecture Research Tool (SMART) is a database of protein domains that allows for the analysis of protein domain architectures by providing identification and annotation tools for the constituent domains in protein sequences [19, 20]. To analyze the domain architecture of your protein sequence:

1. Point your browser to SMART: http://smart.embl-heidelberg.de/ (*see* **Note 16**). Click on the box to the left, which has "Normal Mode" highlighted.
2. In the "Sequence Analysis" section on the following page, paste your protein sequence into the "Protein sequence" text box or provide the UniProt accession number of your sequence (the "Sequence ID or ACC" textbox).
3. Click on "Sequence SMART." The results page will display the domain architecture of your sequence in cartoon form. Click on any of the identified domain to get the sequence details and functional annotation ascribed to that domain.
4. Click on the "Information," "Interactions," "PTMs" (post-translational modifications), or "Orthology" tabs to reveal predictions associated with each of these topics.

3.4.5 HHrepID

HHrepID is a de novo repeat identification method that uses hidden Markov model (HMM) profiles [21]. To detect repeats in your protein sequence:

1. Point your browser to HHrepID (http://toolkit.tuebingen.mpg.de/hhrepid).

2. Copy and paste a single query sequence or a MSA into the "Input" field, or upload a file containing one or more query sequences in FASTA format.
3. Click "Submit job." The search results are presented as a graphical output with all identified repeats mapped onto the query. In addition, this output is provided in tabular format and as a colored alignment. Probability values (*P*-values) are provided for each individual repeat and for the whole group of homologous repeat groups found together.
4. The default parameters may be modified (see the "Options" tab below the search box). For example, the MSA generation algorithm may be selected from either PSI-BLAST or HHblits and the *P*-value thresholds for repeat family detection and self-alignment may be adjusted based to increase or decrease stringency. In addition, secondary structure prediction and domain boundary detection may be enabled.

3.4.6 REPRO

REPRO is de novo repeat identification method [22]. To detect repeats in your protein sequence using this method:

1. Point your browser to REPRO: http://www.ibi.vu.nl/programs/reprowww/
2. Paste your protein sequence of interest into the text box in FASTA format or upload a file containing a sequence to be analyzed.
3. Click "Run Repro." An initial list of "top-scoring non-overlapping local alignments" will be displayed when the search is complete.
4. Specify the number of alignments that would be clustered to detect the repeat sets.
5. Click "Run Repro." The results are presented under four categories: "Top alignments," "Stacked alignments," "Evaluation," and "Fragments." The final repeat sets are listed under "Fragments."

3.5 Protein Secondary Structure Prediction

Secondary structure refers to the localized regularly occurring structure of a given stretch of protein sequence. Secondary structure is dominated by hydrogen bonding between nearby residues and so presents a far less challenging task than tertiary structure prediction, which must account for additional parameters such as long distance interactions. As a result, high confidence secondary structure prediction can be accomplished quickly for the majority of protein sequences.

The canonical secondary structure motifs include alpha helices, beta strands and loops. Surveys of the available crystal structures indicate that each of the proteogenic amino acids has a particular propensity to occur as one of these three secondary structural motifs. These biases result from the unique physiochemical properties of each amino acid and its surrounding residues.

Modern secondary structure prediction algorithms typically operate by assigning a secondary structure to each residues in given sequence based upon a propensity table as well as knowledge-based potentials extracted from analyses of crystalized structures. A particular algorithm's accuracy may vary based upon the type of protein (globular or fibrous) or based upon the protein family. As a result, it may be worthwhile to submit a sequence to several different algorithms. PSIPRED [23], Raptor-XSS8 [24], and PSSpred [25] are examples of commonly used secondary structure prediction programs that generally provide results with high accuracy.

3.5.1 PSIPRED

1. PSIPRED utilizes a neural network to analyze the results from a PSI-BLAST search of the query sequence. The current version of PsiPred (v3.3) can be accessed at http://bioinf.cs.ucl.ac.uk/psipred/
2. Copy and paste your sequence of interest into the "input" field, in FASTA format. Alternatively, a multiple sequence alignment can be submitted (e.g., several aligned homolog sequences).
3. In the "Submission Details" section, you must provide a "Short identifier" for the job and may optionally provide an email address to receive a job completion alert.
4. Select "Predict" to begin the analysis. Results will be available on the following page, which will periodically refresh. Results are normally complete within 30–60 min. If you provided an email address in **step 3**, you can navigate away from this page; a link to the results will be sent to that address.
5. The results page displays the query sequence, with alpha helices and beta strands colored purple and yellow, respectively. Download links are provided for the raw output and for a PDF file containing a visual illustration of the results.

3.5.2 PSSpred

1. PSSpred utilizes multiple back-propagation neural networks to analyze the results from a PSI-BLAST search of the query sequence. The current version of PSSpred (v2) is available online at http://zhanglab.ccmb.med.umich.edu/PSSpred/ A downloadable package is also available at this address.
2. Copy and paste your sequence of interest into the "PSSpred on-line" field, in FASTA format.
3. You must provide an email address in the "email" field below the submission box. A job name is optional.
4. After submitting the job, a page will be generated indicating the address where results will be posted upon completion.

5. The results page displays the query sequence, with a second row "SS:" showing the predicted secondary structure at each position (H: alpha helix, E: beta strand, C: coil) and a row "conf:" displaying the confidence score (1–9: low-high).

3.5.3 RaptorX-SS8

1. RaptorX-SS8 predicts both 3-state and 8-state secondary structure using conditional neural fields generated from PSI-BLAST profiles. This differs from PSIPRED and PSSpred, which predict secondary structure using a 3-state model (helix, strand, coil). The current version of RaptorX-SS8 is available online at http://raptorx.uchicago.edu/StructurePrediction/predict/
2. Copy and paste 1–20 sequences of interest into the "Sequence for Prediction" field, in FASTA format, then press "submit."
3. The following page will provide a link to the address where results will be stored, as well as a progress indicator. Note that you can create an account in order to organize and store past and current jobs.
4. The results page will provide additional information besides secondary structure, including a complete three-dimensional model of the full sequence and any individual domains that were detected. Secondary structure predictions are available under the tab, "Section III. Detailed Prediction Results." Click on the "+" sign to expand this tab.
5. Initially, the "SS3" tab will be active, and the secondary structure prediction for each residue will be displayed graphically as a 3-state system (helix: blue, strand: yellow, coil: red). Higher bars represent higher confidence.
6. Click on the "SS8" tab to switch the results to an 8-state system. The 8-state model includes five additional secondary structure motifs beyond the three basic elements. The 8-state system is colored as follows: helix: blue, isolated beta bridge: green, extended strand in a beta ladder: light yellow, 3-helix: dark yellow, 5-helix (pi-helix): tan, hydrogen bonded turn: red, bend: orange, coil: brown. Again, higher bars represent higher confidence.
7. A link to download the "detailed prediction results" is displayed on the right under the heading "Section II. Summary Prediction Results."

3.6 Retrieval of Three-Dimensional Coordinates of Known Protein Structures

The column-based Protein Data Bank format (PDB), and newer chemical-based formats such as the macromolecular chemical interchange format (mmCIF) and molecular modeling database format (MMDB), are different standards for encoding the atomic coordinates of amino acid chains in three-dimensional space so that this information can be recognized and manipulated by software. Solved coordinates, obtained from X-ray crystallography and NMR spectrometry, can be downloaded from repositories such as the RCSB Protein Data Bank (described below) or built using molecular modeling software (*see* Subheading 3.7). In the RCSB PDB repository [26], each structure file has a unique four-character alphanumeric code, called a PDB ID code, for example, the PH domain of *Drosophila* β-spectrin has a PDB ID code 1DRO. To obtain a coordinate file from the RCSB PDB:

1. Point your browser to the RCSB PDB repository (http://www.rcsb.org/).
2. You can query the repository by keyword (e.g., "cytoskeleton") or by PDB ID (e.g., "1DRO" or by author) (*see* **Note 17**).
3. Enter a search term into the search box provided at the top of the screen and click "Go." If you searched for a specific PDB ID, the associated record page will be displayed (proceed to **step 4**). If you searched by molecule or author name, the results page will display database hits associated with this search term. Click on the "Structure Hits" tab to display PDBs associated with the given search term and then select a particular record.
4. The record page displays information about the structure, such as the regions of the protein that were crystalized and the resolution of the structure. To visualize the structure click on the "3D View" link, which will launch the JMol viewer (*see* **Note 14**).
5. To download the record, click the "Download PDB file." When a new window appears, press "save" and select the download location.

3.7 Modeling the Three-Dimensional Structure of a Protein or Domain Sequence

3.7.1 Modeling Algorithms

Currently, the best route to create a three-dimensional model for a full-length protein or isolated domain is by "homology modeling" (comparative modeling), which involves predicting the tertiary structure of a given protein sequence by inferring constraints from a sequentially similar template structure. These algorithms operate on the assumption that similar protein sequences will result in a similar tertiary structure. As a result, the quality of homology models depends on the closeness of the evolutionary relationship on which they are based [27]. In this sense, homology modeling is distinguishable from ab initio or de novo methods, which generate models based purely on the underlying physical principles. Homology modeling methods are limited in that in order to model a given sequence, a suitable experimentally crystalized template must be available, with the threshold for suitability often considered to be ≥30 % pairwise sequence identity across the aligned target–template region [28]. In this case of multi-domain targets, this limitation may be overcome by selecting templates for individual domains [29].

The second major family of modeling algorithms relies on threading (fold recognition) in order to predict the tertiary structure of a target sequence. Threading algorithms operate by extracting spatial constraints from a given template structure. However, threading methods differ from homology modeling in that templates are selected based upon a sequence–structure alignment rather than a sequence–sequence alignment. For example, a simple threading algorithm might operate by predicting the secondary structure of a given query sequence, searching the PDB databank for crystal structures with a similar secondary structure architecture, and extracting 3D coordinates from the aligned template region in order to predict where the aligned residues in the query should be placed. That being said, modern threading algorithms typically employ much more sophisticated scoring algorithms that account for additional physiochemical parameters and energy minimization. Given that structure is more conserved than sequence, threading methods may be able to generate high-quality models for queries that fail to satisfy the sequence identity threshold for homology modeling.

Homology modeling involves the use of four basic steps: (1) identification of a suitable structural template, (2) alignment of the template and target sequences, (3) model building, and (4) model evaluation (*see* **Note 18**). Structural templates may be identified using the sequence similarity based algorithm, PSI-BLAST to search for matches in the protein databank (i.e., the "nr PDB" database option). In cases where a target protein shares low sequence identity with previously crystalized proteins, homology modeling may need to be coupled with fold recognition methods, which identify the likely protein fold even in cases where there is no clear sequence homology. To that end, many modern prediction servers employ a mixture of homology modeling and threading algorithms ("multi-servers"). Examples of multi-servers include: HHpred, Phyre2, LOMETS, I-TASSER, Raptor-X, pGenTHREADER, FUGUE, and SPARKS-X (*see* **Notes 1** and **19**). With respect to desktop software, MODELLER is perhaps the most widely used homology modeling tool [30]. MODELLER is free for academic use and provides a full-featured modeling package that rivals commercial modeling software suites. However, MODELLER functions as a command line tool, which may be daunting to the novice, though easy-to-follow tutorials are provided by the Sali lab (http://salilab.org/modeller/tutorial/). Users unfamiliar with command line software may wish to access MODELLER through a GUI provided by third party tools (e.g., Chimera) and online servers (e.g., HHpred). Several novice-level methods for model generation are discussed below.

3.7.2 Using HHpred to Generate a Homology Model Structure

1. HHpred is accessible through your web browser at the follow address: http://toolkit.tuebingen.mpg.de/hhpred
2. Paste your protein sequence in the designated "Input" field.

3. Select the format of your sequence, e.g., FASTA.
4. Click "Submit job" button and wait for the results.
5. When the results appear in the browser window, click on "Create model" in the "Results" panel. Proceed to either **step 6** or **8**, as desired.
6. *Manual selection of templates*: Check the box(es) next to the top structure and/or any other listed structure(s) that you wish to use as a template for the target sequence. By default, HHpred selects the top hot as a single template for the sequence. Select the "Create model from manually selected template(s)" button to initiate model generation.
7. HHpred will then proceed to align your query against the one or more templates that were selected and generate an alignment file in ".pir" format.
8. *Automatic selection of templates*: Click "Automatically Select Best template" to have HHpred select the optimal template and initiate model building.
9. On the "results" page select "optimal single template" or "optimal multiple templates," followed by "Generate alignment for MODELLER." HHpred will then select the best single template or multiple templates, and generate an alignment file in ".pir" format.
10. In either case, when the results appear in the browser window, enter your MODELLER license key (*see* **Note 20**). Click the "Submit job" button and wait for your results.
11. When the results appear in the browser window, click "Save" button in the "PDB-file" panel to save the model on your local computer.
12. When the "File Download" window appears, click "save" and select the directory, where want to store the file.
13. Click "Save."
14. Click "View 3D structure" to obtain a quick cartoon rendition of the model.
15. Click "Quality by VERIFY3D" to assess the quality of the model using VERIFY3D [31]. Alternative model validation tools may also be used to analyze the quality of the resulting structure (*see* Subheading 3.7.4).

3.7.3 Using I-TASSER to Generate a Threading-Based Model

1. Visit http://zhanglab.ccmb.med.umich.edu/I-TASSER/registration.html in order to request an academic license. Submit a registration form using a ".edu" address.
2. Access the main I-TASSER submission page at http://zhanglab.ccmb.med.umich.edu/I-TASSER/
3. Copy and paste your target sequence (in FASTA format) into the input form.

4. Select "Run I-TASSER" to initiate the model building process. The resulting page will provide a link to the modeling output (record this address). Model building and optimization often requires 1–3 days to complete.
5. Results will be available at the address provided in **step 4**. I-TASSER's modeling process involves the generation of several hundred structures for a given sequence, which are then clustered based on three-dimensional similarity. The top-scoring structure from each of the five largest clusters is selected for further refinement (e.g., energy minimization) and provided as output to the user.
6. Results may be downloaded as a single archive file (".bz2") via a link at the top of the results page. This archive should be downloaded by the user, as the results page will be deleted after a short period of time.
7. The results page includes several useful sections in addition to the structural modeling output: "Predicted secondary structure," "Predicted solvent accessibility," "Predicted normalized B-factor" (a measure of flexibility), "Function prediction using COFACTOR," and "Predicted binding sites." A description is provided for each of these functions and the associated metrics for measuring the confidence of these predictions.
8. The most important output from an I-TASSER run is provided in the section titled "Top 5 Models predicted by I-TASSER." These models are ranked according to the clustering frequency (i.e., Model 1 is from the cluster with the most members). Confidence scores (C-Score and TM-Score), an estimated RMSD and a graph of the estimated RMSD at each residue position are provided for each of the top 5 models.
9. Evaluating the initial output: a TM-Score of <0.17 indicates a model is likely incorrect, whereas a TM-Score of >0.5 indicates that a model is likely to have the correct topology. The estimated RMSD and target–template alignments should be inspected; a low overall TM-Score may result from poor modeling at the highly variable distal ends of a protein while intervening functional domains are modeled accurately.

3.7.4 Validation of Protein Structures

I-TASSER and HHpred provide options for assessing model validation internally, through the former's calculation of TM-Score and the latter's incorporation of VERIFY3D. In addition to these validation methods, one may also wish to consider submitting any generated models to external servers, such as ProSA [32] or PROCHECK [33]. To submit a model for evaluation, simply visit either server using your browser (*see* **Note 1**) and upload a PDB file to be evaluated. For example, ProSA evaluates overall model quality based on the C-alpha atoms of a given structure and outputs a z-score, plotted on a graph containing the z-scores of experimentally crystalized structures of various lengths. A homology or threading based model should ideally display a z-score comparable to those of the displayed similar-length experimentally solved structures. For a more detailed explanation of ProSA's z-score calculation and the metrics used by the PROCHECK, see the respective help sections provided by each server.

3.8 Tools for Analyzing Structures

3.8.1 Visualization of Protein Structure

1. Open Chimera.
2. To open a protein structure saved locally, from the menu bar, select "File," then "Open" and search for the directory containing a locally saved ".pdb" file. Select "Open" and proceed to **step 4**.
3. Alternatively, to download and open a protein structure from one of the major online structure repositories, select "File" followed by "Fetch by ID." A popup menu will appear. By default the PDB (Protein Databank) repository is selected. In the "ID" field, enter the four-character structural identifier, e.g., "1MK4."
4. In the main window, the protein structure will appear, depicted in "ribbon" format.
5. *Setting up the workspace*: From the menu bar, select "TOOLS," followed by "General Controls" and finally "Model Panel." This will generate the Model Panel, a small window that provides access to many of Chimera's functions for model analysis. From the menu bar, if you select "TOOLS," followed by "Sequence" and finally "Sequence" a dialog box will appear with the request, "Show sequence for" and a list of all polypeptide chains identified in the PDB file. Select the chain of interest, and press "Show." This will generate the Sequence panel, which shows the amino acid sequence of the selected chain. Alpha helices and beta strands, if identified in the file, are colored yellow and green, respectively.
6. *Basic visualization techniques*: Alternative depiction schemes are available (e.g., ribbon, ball and stick, sphere) by selecting "Action" from the menu bar and then selecting a depiction option listed under the "Atoms/Bonds" or "Ribbon" submenu. To calculate and display the surface of protein, select "Action" from the menu bar, followed by "Surface" and "Show." This surface can be depicted as a solid, mesh or dot representation, or alternatively hidden, by selecting these options on the "Surface" submenu. The "Select" panel on the menu bar provides various options for selecting individual atoms, residues or polypeptides chains. This submenu also

provides options for selecting residues by time, property or for finding and selecting a given sequence. The depiction schemes discussed above can be applied selectively to different subsets of residues or atoms.

7. *Coloring the protein*: To color the protein based upon secondary structure, select "Tool" from the menu bar, followed by "Depiction" and "Color Secondary Structure." A dialog box will appear with options for the color scheme to use for helices, strands and coils. Select "Apply" to color the protein while leaving this menu open, or "OK" to apply the color scheme and close this menu. To color the protein as a rainbow, (blue to red from the N-terminus to the C-terminus), select "Tools" from the menu bar, followed by "Depiction" and "Rainbow." To color the protein surface based upon residue hydrophobicity according to the Kyte-Doolittle scale, select "Tools" from the menu bar, followed by "Structure Analysis" and "Render by Attribute." A dialog box will appear. Set the "Attributes of" field to "residues" and select "kdHydrophobicity" from the "Attribute" dropdown menu. Choose "Apply" to color the molecule while leaving the dialog box open or "OK" to color the molecule and close the dialog box. The protein surface will be calculated and displayed, with blue indicating hydrophilic regions and red indicating hydrophobic regions. To color the protein surface based upon electrostatic potential (Coulombic), select "Tools" from the menu bar, followed by "Surface Binding/Analysis" and "Coulombic Surface Coloring." A dialog box will appear with options for fine-tuning the representation. Select "Apply" to color the molecule according to the selected settings while leaving the menu open, or "OK" to apply the changes and close the menu. The calculation should be complete within 30 s. Under the default color scheme, red indicates negatively charged regions and blue indicates positively charged regions. To reset the color scheme for a protein, select "Actions" from the menu bar, followed by "Color" and "None."

8. *Structure comparison*: It is often useful to compare the three-dimensional structure of two or more protein structures in order to highlight conserved regions or differences. Chimera provides this functionality. Open two PDB files, as described above. The models will both appear within the main workspace. Select "Tools" from the menu bar, followed by "Structure Comparison" and "Matchmaker." A dialog box will appear. Select a "Reference structure" and a "Structure to match" from the menu, followed by "Apply" or "OK." Matchmaker will calculate the best fit of the two polypeptide chains and then move the "Structure to match" so that it is overlaid on the "reference structure". Chimera can also be

used to generate a multiple sequence alignment of two or more proteins based upon a structural superposition. In other words, two or more structures can be overlaid to maximize the best fit and then the sequence of the residues in the aligned structural regions can be extracted to generate a pairwise or multiple sequence alignment. To generate a sequence alignment based on structural superposition, align two structures using the Matchmaker tool described above. However, before selecting "Apply" or "OK," check the box titled "After superposition, compute structure-based multiple sequence alignment."

9. *Ramachandran plots*: Chimera can also be used to generate a Ramachandran plot for a given structure. This functionality is useful for comparing homology models and evaluating structural refinement. Load a protein structure and activate the "Model Panel." On the Model Panel menu, highlight a protein structure of interest and select the "Ramachandran Plot" button. A Ramachandran plot will be generated for the structure. The "Show region for" field provides options for configuring the plot.

3.8.2 Manipulation of Protein Structures

1. *Energy minimization*: Chimera has implemented limited energy minimization functionality and can be used to refine small errors in homology models (e.g., steric clashes) by moving the model towards a local energy minimum. To minimize a structure, select "Tools" from the menu bar, followed by "Structure Editing" and "Energy Minimization." A dialog box will appear. By default, Chimera's energy minimization routine applies 100 steps of steepest descent followed by 100 steps of conjugate gradient minimization. The "Fixed atoms" field may be used to define regions of the structure that will be held fixed in place during minimization. This feature may be used, for example, to refine only a loop region within a protein.
2. *Refining rotamer mutating individual residues*: Chimera can be used to refine the side chains of specific residues or to mutate individual residues in a protein, using the Dunnbrack or Richardson rotamer libraries. To access this functionality, select "Tools" from the menu bar, followed by "Structure Editing" and "Rotamers." A dialog box will appear. Select a residue (e.g., by control-clicking on the model). Choose a "Rotamer type" on the dialog box and the "Rotamer library" and press "Apply." If you intend to refine a side chain, the "rotamer type" should be the same as the amino acid being refined, whereas if mutation is intended the "rotamer type" should be the replacement amino acid. A second dialog box will appear showing a table of various side chain conformations (Chi angles and associated probabilities). Select a side chain

conformation from the table. The 3D model will be updated to display the selected conformation. By default, the "Existing side chain(s):" field on the second dialog box is set to "replace." Thus, pressing "Apply" will replace the selected residue with the active amino acid rotamer, refining or mutating the given residue. If a residue has been mutated, it is recommended that the energy minimization routine discussed above be applied to the structure.

4 Notes

1. All analyses described can be carried out using more than one program. Listed below are some alternative software choices for each type of analysis described earlier. This list is not exhaustive, but refers the reader to some of the more popular tools available.
2. FASTA is the most popular format for protein sequence data and is widely used in most sequence based proteomic software tools. It is characterized by a comment line which is a single-line description, followed by lines of sequence data. The description line is distinguished from the sequence data by a greater-than (">") symbol in the first column. It is recommended that all lines of text be shorter than 80 characters in length. The sequence data is single letter codes in uppercase with no white spaces between them. An example sequence in FASTA format is:

```
>gi|5031877|ref|NP_005564.1| lamin B1 [Homo sapiens]
MATATPVPPRMGSRAGGPTTPLSPTRLSRLQEKEELRELNDRLAVYIDKVRSLETENSALQLQVTEREEV
RGRELTGLKALYETELADARRALDDTARERAKLQIELGKCKAEHDQLLLNYAKKESDLNGAQIKLREYEA
ALNSKDAALATALGDKKSLEGDLEDLKDQIAQLEASLAAAKKQLADETLLKVDLENRCQSLTEDLEFRKS
MYEEEINETRRKHETRLVEVDSGRQIEYEYKLAQALHEMREQHDAQVRLYKEELEQTYHAKLENARLSSE
MNTSTVNSAREELMESRMRIESLSSQLSNLQKESRACLERIQELEDLLAKEKDNSRRMLTDKEREMAEIR
DQMQQQLNDYEQLLDVKLALDMEISAYRKLLEGEEERLKLSPSPSSRVTVSRASSSRSVRTTRGKRKRVD
VEESEASSSVSISHSASATGNVCIEEIDVDGKFIRLKNTSEQDQPMGGWEMIRKIGDTSVSYKYTSRYVL
KAGQTVTIWAANAGVTASPPTDLIWKNQNSWGTGEDVKVILKNSQGEEVAQRSTVFKTTIPEEEEEEEEA
AGVVVEEELFHQQGTPRASNRSCAIM
```

3. You can increase the specificity of your searches by combining your keywords or accession numbers and the Boolean operators "AND," "OR," and "NOT."
4. BLAST, a heuristic sequence alignment and search algorithm, is used to query a database and uncover potentially related sequences. Alignments are scored and used to rank the similarity of the "hits" with respect to the query. Scoring is based on

the alignment length, gaps/inserts, residue similarities, and statistical significance with respect to the database size. Residue similarities, at each position of the query–hit alignment, are calculated based on biochemical natures and adjusted according to evolutionary parsimony. The user can customize the BLAST scoring, but the standard parameters are frequently sufficient. Many variations on the BLAST search method and its scoring have been developed and can, in some cases, uncover additional hits.

5. A Position Specific Scoring Matrix (PSSM) is a scoring system applied to each position of a paired sequence alignment that corresponds to the likelihood of a substitution from one residue (of the query) to another (of the hit). Various PSSMs exist and are based on alignments of known protein families. The scores imposed by a PSSM are derived from the amount of sequence variation present within the family and reflects evolutionary relatedness/divergence. BLOSUM (BLOck Substitution Matrix, reference) and PAM (Point Accepted Mutation, reference) are two commonly used families of scoring matrices.
6. As an alternative option you can paste the accession number of your protein sequence or limit the search to a sub-region of your provided sequence.
7. You can select "PDB" as the database of choice if you are looking for only those related sequences that have solved structures associated with the protein sequence.
8. Customized versions of the BLAST algorithm, such as PSI-BLAST (see above), PHI-BLAST, and DELTA-BLAST, can be selected at this point.
9. By clicking on the RID (Reference ID) hyperlink a new, identical tab will open, but the HTML will now include the RID for your search. You can return to your search results using this HTML for up to 48 h, after which time the RID will expire.
10. The *E*-value gives a statistical measure of the significance of the alignment and describes the number of hits that would be "expected" to occur by chance when searching a sequence database of a particular size; the lower the *E*-value, the more significant the match. An *E*-value of 1 means that it would be expected to find one match with a similar score simply by chance in a database of the same size. Exact and closely matched hits will have *E*-values of 0. If the query sequence is short (less than 100 nucleotides or amino acids long), the top *E*-values may be larger than 10–50 even if there is an exact match. Therefore it is a good idea to check the similarities and identities (given by percent) as well as the *E*-value.

11. The databases available in phmmer searches are similar to the NCBI options, but also include proteomic databases from UniProt. You can again search for proteins that have solved structures by selecting "PDB" or for proteins that have Pfam annotated domains by selecting Pfamseq.
12. Make sure that your sequences have different names as the first 30 characters of the name are significant, and if they are not unique, the program will fail. Also, remove any white space or empty lines from the beginning of your input
13. "*" denotes identity, ":" denotes a conserved residue, and "." denotes a semi-conserved residue in the MSA.
14. You will need a Java-enabled browser for this step.
15. The default scan excludes patterns with a high probability of occurrence to reduce the number of false positive hits. You can uncheck this option if you are interested in doing a more complete but less stringent scan of your sequence. The "ScanProsite" advanced form allows you to further tweak the scan.
16. When your browser first encounters the SMART website, you will be asked to choose between the "normal" or "genomic" mode, after which the mode setting will be saved for future visits. The models differ in the underlying protein database used and only the proteomes of completely sequenced genomes are used in the genomic mode. For most routine purposes "normal" mode is chosen but the mode can be changed by accessing the "Setup" link on the top of the page at any time.
17. Use the advanced search form if you are looking for very specific information and do not know the PDB ID for that data record.
18. These steps often need to be repeated and tweaked until a satisfactory model is obtained. The second step is crucial to making a reliable model since the alignment dictates the relationship between your protein sequence and the template sequence.
19. You may want to parse your sequence into its constituent domains and submit each domain sequence individually since structural templates will be identified in most cases separately for each domain. Rarely, there will be structural templates that can be used to model entire multi-domain proteins
20. This key is freely available for academic users (with academic e-mail accounts) and easily obtainable at: http://salilab.org/modeller/registration.shtml

References

1. Pettersen EF et al (2004) UCSF Chimera--a visualization system for exploratory research and analysis. J Comput Chem 25(13): 1605–1612
2. Benson DA et al (2014) GenBank. Nucleic Acids Res 41:D36
3. Kulikova T et al (2007) EMBL nucleotide sequence database in 2006. Nucleic Acids Res 35(Database issue):D16–D20
4. Tateno Y et al (2002) DNA Data Bank of Japan (DDBJ) for genome scale research in life science. Nucleic Acids Res 30(1):27–30
5. Pruitt KD et al (2014) RefSeq: an update on mammalian reference sequences. Nucleic Acids Res 42(Database issue):D756–D763
6. UniProt C (2014) Activities at the Universal Protein Resource (UniProt). Nucleic Acids Res 42(Database issue):D191–D198
7. Altschul SF et al (1990) Basic local alignment search tool. J Mol Biol 215(3):403–410
8. Altschul SF et al (1997) Gapped BLAST and PSI-BLAST: a new generation of protein database search programs. Nucleic Acids Res 25(17):3389–3402
9. Finn RD, Clements J, Eddy SR (2011) HMMER web server: interactive sequence similarity searching. Nucleic Acids Res 39(Web Server issue):W29–W37
10. Sievers F et al (2011) Fast, scalable generation of high-quality protein multiple sequence alignments using Clustal Omega. Mol Syst Biol 7:539
11. Di Tommaso P et al (2011) T-Coffee: a web server for the multiple sequence alignment of protein and RNA sequences using structural information and homology extension. Nucleic Acids Res 39(Web Server issue):W13–W17
12. Katoh K et al (2002) MAFFT: a novel method for rapid multiple sequence alignment based on fast Fourier transform. Nucleic Acids Res 30(14):3059–3066
13. Edgar RC (2004) MUSCLE: multiple sequence alignment with high accuracy and high throughput. Nucleic Acids Res 32(5): 1792–1797
14. Do CB et al (2005) ProbCons: probabilistic consistency-based multiple sequence alignment. Genome Res 15(2):330–340
15. de Castro E et al (2006) ScanProsite: detection of PROSITE signature matches and ProRule-associated functional and structural residues in proteins. Nucleic Acids Res 34(Web Server issue):W362–W365
16. Jonassen I, Collins JF, Higgins DG (1995) Finding flexible patterns in unaligned protein sequences. Protein Sci 4(8):1587–1595
17. Finn RD et al (2014) Pfam: the protein families database. Nucleic Acids Res 42(Database issue):D222–D230
18. Geer LY et al (2002) CDART: protein homology by domain architecture. Genome Res 12(10):1619–1623
19. Schultz J et al (2000) SMART: a web-based tool for the study of genetically mobile domains. Nucleic Acids Res 28(1):231–234
20. Letunic I, Doerks T, Bork P (2014) SMART: recent updates, new developments and status in 2015. Nucleic Acids Res 43:D257
21. Biegert A, Soding J (2008) De novo identification of highly diverged protein repeats by probabilistic consistency. Bioinformatics 24(6):807–814
22. George RA, Heringa J (2000) The REPRO server: finding protein internal sequence repeats through the Web. Trends Biochem Sci 25(10):515–517
23. Buchan DW et al (2013) Scalable web services for the PSIPRED Protein Analysis Workbench. Nucleic Acids Res 41(Web Server issue):W349–W357
24. Kallberg M et al (2012) Template-based protein structure modeling using the RaptorX web server. Nat Protoc 7(8):1511–1522
25. Yang J et al (2015) The I-TASSER Suite: Protein structure and function prediction. Nature Methods 12: 7–8
26. Berman HM, The Protein Data Bank et al (2000) Nucleic Acids Res 28(1):235–242
27. Moult J (2005) A decade of CASP: progress, bottlenecks and prognosis in protein structure prediction. Curr Opin Struct Biol 15(3): 285–289
28. John B, Sali A (2003) Comparative protein structure modeling by iterative alignment, model building and model assessment. Nucleic Acids Res 31(14):3982–3992
29. Fernandez-Fuentes N et al (2007) Comparative protein structure modeling by combining multiple templates and optimizing sequence-to-structure alignments. Bioinformatics 23(19): 2558–2565
30. Sali A et al (1995) Evaluation of comparative protein modeling by MODELLER. Proteins 23(3):318–326
31. Eisenberg D, Luthy R, Bowie JU (1997) VERIFY3D: assessment of protein models with three-dimensional profiles. Methods Enzymol 277:396–404
32. Wiederstein M, Sippl MJ (2007) ProSA-web: interactive web service for the recognition of errors in three-dimensional structures of pro-

teins. Nucleic Acids Res 35(Web Server issue):W407–W410

33. Laskowski RA et al (1993) PROCHECK - a program to check the stereochemical quality of protein structures. J App Cryst 26: 283–291
34. Lane L et al (2012) neXtProt: a knowledge platform for human proteins. Nucleic Acids Res 40(Database issue):D76–D83
35. Barker WC et al (2001) Protein Information Resource: a community resource for expert annotation of protein data. Nucleic Acids Res 29(1):29–32
36. Biegert A, Soding J (2009) Sequence context-specific profiles for homology searching. Proc Natl Acad Sci U S A 106(10):3770–3775
37. Remmert M et al (2012) HHblits: lightning-fast iterative protein sequence searching by HMM-HMM alignment. Nat Methods 9(2): 173–175
38. Pearson WR, Lipman DJ (1988) Improved tools for biological sequence comparison. Proc Natl Acad Sci U S A 85(8):2444–2448
39. Bawono P, Heringa J (2014) PRALINE: a versatile multiple sequence alignment toolkit. Methods Mol Biol 1079:245–262
40. Sadreyev RI et al (2009) COMPASS server for homology detection: improved statistical accuracy, speed and functionality. Nucleic Acids Res 37(Web Server issue):W90–W94
41. Pei J, Grishin NV (2014) PROMALS3D: multiple protein sequence alignment enhanced with evolutionary and three-dimensional structural information. Methods Mol Biol 1079:263–271
42. Chikkagoudar S, Roshan U, Livesay D (2007) eProbalign: generation and manipulation of multiple sequence alignments using partition function posterior probabilities. Nucleic Acids Res 35(Web Server issue):W675–W677
43. Cole C, Barber JD, Barton GJ (2008) The Jpred 3 secondary structure prediction server. Nucleic Acids Res 36(Web Server issue): W197–W201
44. Petersen B et al (2009) A generic method for assignment of reliability scores applied to solvent accessibility predictions. BMC Struct Biol 9:51
45. Yachdav G et al (2014) PredictProtein--an open resource for online prediction of protein structural and functional features. Nucleic Acids Res 42(Web Server issue): W337–W343
46. Pollastri G, McLysaght A (2005) Porter: a new, accurate server for protein secondary structure prediction. Bioinformatics 21(8): 1719–1720
47. Geourjon C, Deleage G (1995) SOPMA: significant improvements in protein secondary structure prediction by consensus prediction from multiple alignments. Comput Appl Biosci 11(6):681–684
48. Crooks GE et al (2004) WebLogo: a sequence logo generator. Genome Res 14(6): 1188–1190
49. Zhang Y (2008) I-TASSER server for protein 3D structure prediction. BMC Bioinform 9:40
50. Soding J, Biegert A, Lupas AN (2005) The HHpred interactive server for protein homology detection and structure prediction. Nucleic Acids Res 33(Web Server issue): W244–W248
51. Wu S, Zhang Y (2007) LOMETS: a local meta-threading-server for protein structure prediction. Nucleic Acids Res 35(10):3375–3382
52. Bennett-Lovsey RM et al (2008) Exploring the extremes of sequence/structure space with ensemble fold recognition in the program Phyre. Proteins 70(3):611–625
53. Lobley A, Sadowski MI, Jones DT (2009) pGenTHREADER and pDomTHREADER: new methods for improved protein fold recognition and superfamily discrimination. Bioinformatics 25(14):1761–1767
54. Shi J, Blundell TL, Mizuguchi K (2001) FUGUE: sequence-structure homology recognition using environment-specific substitution tables and structure-dependent gap penalties. J Mol Biol 310(1):243–257
55. Yang Y et al (2011) Improving protein fold recognition and template-based modeling by employing probabilistic-based matching between predicted one-dimensional structural properties of query and corresponding native properties of templates. Bioinformatics 27(15):2076–2082
56. McGuffin LJ, Buenavista MT, Roche DB (2013) The ModFOLD4 server for the quality assessment of 3D protein models. Nucleic Acids Res 41(Web Server issue):W368–W372
57. Benkert P, Kunzli M, Schwede T (2009) QMEAN server for protein model quality estimation. Nucleic Acids Res 37(Web Server issue):W510–W514
58. Zhang Y, Skolnick J (2004) Scoring function for automated assessment of protein structure template quality. Proteins 57(4):702–710
59. Bhattacharya A, Tejero R, Montelione GT (2007) Evaluating protein structures determined by structural genomics consortia. Proteins 66(4):778–795
60. Wallner B, Elofsson A (2003) Can correct protein models be identified? Protein Sci 12(5):1073–1086

61. Shen MY, Sali A (2006) Statistical potential for assessment and prediction of protein structures. Protein Sci 15(11):2507–2524
62. Robert X, Gouet P (2014) Deciphering key features in protein structures with the new ENDscript server. Nucleic Acids Res 42(Web Server issue):W320–W324
63. Stivala A et al (2011) Automatic generation of protein structure cartoons with Pro-origami. Bioinformatics 27(23):3315–3316
64. Waterhouse AM et al (2009) Jalview Version 2--a multiple sequence alignment editor and analysis workbench. Bioinformatics 25(9): 1189–1191

Chapter 24

Homology Modeling Procedures for Cytoskeletal Proteins of *Tetrahymena* and Other Ciliated Protists

Giovanni J. Pagano, Linda A. Hufnagel, and Roberta S. King

Abstract

In recent years there has been an explosive increase in the number of annotated protein sequences available through genome sequencing, as well as an accumulation of published protein structural data based on crystallographic and NMR methods. When taken together with the development of computational methods for the prediction of protein structural and functional properties through homology modeling, an opportunity exists for prediction of properties of cytoskeletal proteins in a suitable model organism, such as *Tetrahymena thermophila* and its ciliated protist relatives. In particular, the recently sequenced genome of *T. thermophila*, long a model for cytoskeletal studies, provides a good starting point for undertaking such homology modeling studies. Homology modeling can produce functional predictions, for example regarding potential molecular interactions, that are of great interest to the drug industry and Tetrahymena is an attractive model system in which to follow up computational predictions with experimental analyses. We provide here procedures that can be followed to gain entry into this promising avenue of analysis.

Key words Ciliates, Ciliophora, Cytoskeleton, Molecular modeling, Protein structure, Structural bioinformatics

1 Introduction

Three-dimensional analysis of proteins based on data from X-ray crystallography [1], high resolution electron microscopy on frozen samples [2], and NMR spectroscopy [3] has been essential in developing models for their functional properties and interactions with other cellular molecules (c.f. [4]). For proteins that are not yet able to be prepared for X-ray analysis, electron microscopy, or NMR analysis, another approach has become popular: homology modeling (c.f. [5]). Homology modeling of proteins depends on the availability of reliable amino acid sequence data; however, as more and more gene sequences are validated and annotated, in many different species, the needed starting material is rapidly becoming available [6].

Ray H. Gavin (ed.), *Cytoskeleton: Methods and Protocols*, Methods in Molecular Biology, vol. 1365,
DOI 10.1007/978-1-4939-3124-8_24, © Springer Science+Business Media New York 2016

Homology modeling is a computational approach to the construction of three-dimensional, atomic resolution topology models of macromolecules, such as proteins and nucleic acids, by use of comparisons with experimentally modeled homologues [5]. Homology modeling of proteins makes use of information coded in genes and expressed RNA sequences that code for these proteins. For this reason, genes that have been well annotated, both structurally and functionally, can provide a basis for three-dimensional modeling when experimentally derived models are not feasible or are too time-consuming or costly to produce. Homology modeling can provide functional predictions, such as potential molecular interactions, that are proving particularly attractive to the drug industry (c.f. [7]).

The ciliated protists, and in particular species of the genera *Tetrahymena* and *Paramecium*, have long served as model systems in which to investigate the cell cytoskeleton (c.f. [8–10]). In part this is because, before the advent of modern fluorescence labeling procedures, their highly developed cytoskeletal structures were readily seen by light microscopy and later by means of a variety of electron microscopy methods. Tetrahymena and Paramecium, in particular, have long served as model systems for purification of cytoskeletal proteins and other interacting proteins, for further structural and functional analysis, due to the fact that both species can be grown very cheaply in large quantities [11, 12]. The genetics, cell biology and molecular properties of both species have been well described (for *T. thermophila* c.f. http://www.lifesci.ucsb.edu/~genome/Tetrahymena/genetics.htm#Tetrahymena and http://faculty.jsd.claremont.edu/ewiley/about.php; for Paramecium, see http://www.genoscope.cns.fr/spip/Paramecium-a-model-ciliate.html).

The genome of *Tetrahymena thermophila* was initially sequenced in 2006 [13] and subsequently has been undergoing refinement and annotation [14, 15]. Genes of *T. thermophila* can be accessed at the Tetrahymena Genome Database Wiki (www.ciliate.org) and their expression profiles are available at the Tetrahymena Functional Genomics Database (TetraFGD; tfgd.ihb.ac.cn/; [16]). Many of its genes have been shown to share strong homology with human genes and genes of important human pathogens, in particular those of the parasitic apicomplexan protist, Plasmodium, which causes malaria [13]. Genomes of several other ciliated protists are also now available as sources of potentially useful genes, transcribed RNA and protein sequences; these include *T. borealis*, *T. elliotti*, and *T. malaccensis* (www.ciliate.org), *Paramecium tetraurelia* (Paramecium DB, available at http://paramecium.cgm.cnrs-gif.fr), *P. biaurelia*, *sexaurelia*, and *caudatum* (genes available at Paramecium DB), *Oxytricha trifallax* (oxy.ciliate.org/index.php/home/welcome), and the fish parasite *Ichthyophtherius multifiliis* (www.ciliate.org/gb2/gbrowse/ich/).

Other resources for those particularly interested in the genes of Tetrahymena include the Ciliate Genomics Consortium (http://faculty.jsd.claremont.edu/ewiley/) and the Tetrahymena Stock Center (https://tetrahymena.vet.cornell.edu).

A search via PubMed for published atomic structures of proteins of Tetrahymena reveals that such structures have been reported for relatively few proteins. These include most of the proteins found in the large and small ribosomal subunits [17, 18], as well as proteins associated with chromosomal telomeres (c.f. [19]). In both cases, experimentally derived data on interactions with other proteins and factors was obtained and made available to others via the Protein Data Bank (PDB). As far as we know, the predictive power of homology modeling of proteins has been applied in Tetrahymena in just one case, to characterize the conformation of phosphoprotein P0, a protein found in the GTPase center of the large ribosomal subunit of *T. thermophila*, and to explore its association with the 26S ribosomal RNA [20]. Therefore, the field appears to be wide open for the application of homology modeling to cytoskeletal proteins of Tetrahymena. The procedures described below are meant to provide a pathway for entry into this promising avenue of analysis.

2 Materials

Homology modeling is a process that requires a significant amount of computer memory. Therefore, the building of models should be carried out on a computer with a large amount of available RAM and hard disk space (*see* **Note 1**); computers with inadequate capacity are likely to crash when trying to carry out homology modeling.

2.1 Target Sequence

1. There are various starting points for choosing the target sequence. Generally, a protein with a known and experimentally verified amino acid sequence is the best starting point for homology modeling. A gene with an experimentally verified sequence, or a gene with a structurally annotated sequence derived from expressed messenger RNA [21] are also acceptable as sources for a derived target sequence (*see* **Note 2**). Least desirable would be an amino acid sequence derived from a gene sequence that has not been structurally annotated.
2. If a gene sequence is used, it should be translated into a likely amino acid sequence before use. One way to translate the sequence is to use a tool like ExPASy Translate (http://web.expasy.org/translate/) with the amino acid substitution table appropriate for the organism from which the target sequence is derived. If the target sequence has known orthologs in other

organisms, the orthologs can be compared to the predicted amino acid sequence of the target to help validate the accuracy of the translation.

2.2 Programs and Web Servers

While there are many available programs and databases for carrying out homology modeling, the programs and websites listed below have been found to produce reliable, high-quality homology models in the case of *Tetrahymena thermophila* [20].

1. NCBI BLAST (Web Server): http://blast.ncbi.nlm.nih.gov/Blast.cgi
2. MCoffee/Expresso (Web Server/Download): http://www.tcoffee.org
3. UCSF Chimera (Download): http://www.cgl.ucsf.edu/chimera/
4. Protein Data Bank (Web Site): http://www.rcsb.org/pdb/home/home.do
5. Accelrys Discovery Studio Client (Download): http://accelrys.com/products/discovery-studio/ (*see* **Note 3**).

3 Methods

3.1 Identifying Existing Protein Structures to Serve as Templates for the Homology Model

3.1.1 Gather Homologous Structures and Sequences

While there are many different methods to collect sequence data, this method is generally reliable for collecting high-quality data. For definitions of target, template, structure, sequence, and other terms, *see* **Note 4** below.

1. Visit the BLAST search page (http://blast.ncbi.nlm.nih.gov/Blast.cgi) and click "Protein BLAST". Under "Enter Query Sequence", enter the accession number, gi number or FASTA sequence, or upload a text file with the protein sequence of interest. A FASTA-format sequence begins with a header line consisting of a ">" symbol, followed by the name of the sequence. Any name will do, either an already published name or a user-generated name (*see* **Note 5**). The sequence begins on a new line following the header. For example:
   >Ribosomal protein P0, Tetrahymena thermophila
   MPPAKVD…
2. Under "Choose search set", select "Protein data bank proteins (pdb)" as the "Database". This will allow you to search for protein sequences with available structures that match your protein. If you wish to limit your search to a particular group of organisms, enter the name of the group in the "Organism" box. Otherwise, leave it blank. Leave the other parameters at their default settings.
3. Run the BLAST program by clicking the "BLAST" button.

4. When the BLAST is complete, look through the top approximately 20 results, including only those with an *E*-value of less than 1E−5. The *E*-value is a parameter that indicates the likelihood of a valid match between two sequences. Proteins with a higher *E*-value are less likely to be useful template structures.
5. Clicking on the name of the hit will take you to an alignment showing where the query sequence and the subject sequence share homology. Clicking on the link under the "Accession" column will take you to a page with more information on the protein, such as which organism it comes from, references to the sequence in the literature, and annotation information for the gene, if available. Under the title of the Genbank entry, note the alphanumeric sequence next to the "PDB" header (i.e., 3u5i_q). This is the PDB ID, and is used to identify a structure in the Protein Data Bank (*see* **Note 6**).
6. Go to the Protein Data Bank homepage (http://www.pdb.org/pdb/home/home.do). Type the first four characters of the PDB ID (i.e., 3U5I) in the search box and hit Enter to be taken to the entry for that structure. Go to the "Experimental Details" section, and record the method by which the structure was determined, as well as its resolution (*see* **Note** 7). If the structure was derived from an X-ray diffraction experiment, record the *R* and R-Free values (*see* **Note** 7) as well. Finally, record which, if any, regions are structurally unresolved and which are not reported in the crystal's structural coordinates (*see* "Unresolved 3D Structure" in **Note** 7).
7. To download FASTA-formatted sequences for the best hits, check the boxes next to the matches you wish to download. Under the "Download" button above the matches, make sure the "FASTA (complete sequence)" option is selected, and click the "Continue" button. The FASTA sequences will be downloaded as a text file (*see* **Note 8**). As the file will have the default name "seqdump", we recommend that you rename the downloaded file to something more descriptive.
8. In a text editor (*see* **Note 8**), view the file you just downloaded. Some of the sequences will have multiple headers above them, indicating the different names of the protein in the NCBI databases. Remove all but one of the headers. Rename each of the sequences using a short (no more than 20 characters) and descriptive name (i.e., the protein name, or the organism the sequence comes from). You may want to note the old and new FASTA names in a separate text file, to avoid confusion.

*3.1.2 Aligning Multiple Sequences (**See Note 9**) and Identifying Potential Templates*

1. Go to the TCoffee site (www.tcoffee.org) and select the MCoffee mode [22]. MCoffee is an alignment tool that combines the outputs of several different alignment tools into a single "meta-alignment". In the "sequences input" section, paste the text of the FASTA sequences obtained in **step** 7 from

Subheading 3.1.1 into the text box, or upload the text file directly under the "Click here to upload a file" link. Leave the other options at their default settings.

2. Run MCoffee. After the alignment is complete, download all the output files using ZIP file. Look at the "score/HTML" file to determine how well certain areas are aligned. Regions with a red background are better aligned than regions with a blue background, and are more reliable when selecting what regions to model. Make note of the residue numbers of these inconsistently aligned (blue) regions for use in **step 5** of Subheading 3.2.2.
3. The "Expresso" mode of TCoffee is able to build an alignment that incorporates structural data by automatically searching for structures to associate with each sequence in the FASTA file [23]. Go to the TCoffee website (www.tcoffee.org) and select the Expresso mode. In the "sequences input" section, paste the text of the FASTA sequences from Subheading 3.1.1, **step 8** into the text box, or upload the text file directly under the "Click here to upload a file" link. Leave the other options at their default settings.
4. Run Expresso and analyze the output as in **step 2**. Compare the MCoffee and Expresso alignments, and keep the alignment that results in the highest alignment scores (*see* **Note 10**).

3.1.3 Selecting the Best Template

1. Using the information collected in Subheadings 3.1.1 and 3.1.2, select the template structure or structures to be used. First, try to select a structure from an organism that is as closely related as possible to the organism the target sequence is from. Then, consider the structures that have sequences that are most similar to your target sequence. If there are multiple possible templates from the same organism and protein, select the template with the highest resolution (lower numbers) data. Generally, X-ray diffraction and NMR structures are higher resolution than structures derived from cryo-electron microscopy (Cryo-EM).
2. Once you have identified the template structure(s) to be used, download their PDB files from the PDB website. On the page for each structure, click the "Download Files" link and select "PDB File (text)" to automatically download the full PDB file. These will be trimmed down for ease of use in a later step.
3. If you have only a few template structures, it may be necessary to gather more sequence data by running another protein BLAST (Subheading 3.1.1) by leaving the "Database" selection at its' default setting. Download the FASTA-formatted protein sequences, using the protocols in **step 7** of Subheading 3.1.1. Copy the FASTA-formatted sequences into the same file as the best PDB matches.

3.2 Making the Homology Model (See Note 3 Regarding Discovery Studio Software)

3.2.1 Trimming PDB File for Ease of use (See Note 11)

1. Open the PDB file from Subheading 3.1.3, **step 2** in the program UCSF Chimera [24] (*see* **Note 12**). Alternatively, Chimera can automatically fetch the structure from PDB ("File" → "Fetch by ID", select the "PDB" option in the Database column, enter the first four characters of the PDB ID and click "Fetch"). Depending on the size of the file, it may take a while for the structure to download.
2. Using the "Select" → "Chain" option, select the chain matching the PDB chain ID you noted in Subheading 3.1.1, **step 5**.
3. Select "Actions" → "Write PDB". In the window that pops up, check the "Save selected atoms only" option. Choose a descriptive name for the new PDB file, and click "Save". This will save a PDB file that contains only the structural coordinates corresponding to the template structure. Repeat this section for each template structure.

3.2.2 Building the Homology Model Backbone

1. Before the template structure can be used for homology modeling, it must be modified for use. In the program Discovery Studio (*see* **Note 13**), go to "Edit" → "Preferences", and under "Protein Utilities", select "Clean Protein". Check all of the boxes under "Correct Problems", the "Modify Hydrogens" box, the "Modify termini" box and the "Display report" box. Under the "Tools" → "Prepare Protein" tab, click "Clean Protein". After the cleaning is complete, read the "Protein Clean Report" and record any missing residues and other remaining problems with the PDB file.
2. Open the alignment from Subheading 3.1.2. If there are regions with deletions or insertions (*see* **Note 14**) between the template and target sequences, delete the corresponding amino acids from the alignment. Save this edited alignment under a separate file name, so the full alignment can be returned to later if necessary.
3. In order to make the alignment and structure work together, they must be "linked" in Discovery Studio using the "Sequence" → "Link Sequence and Structure" dialog. The name of the template structure and sequence must match exactly in order for them to be linked. If they do not match, edit the sequence name in the FASTA alignment file as necessary, and retry the "Link Sequence and Structure" dialog.
4. Build the initial models using the "Build Homology Models" protocol in Discovery Studio Client [25], under the "Tools → Create Homology Models" tab (license needed; *see* **Note 3**). We recommend using the following parameters:

   Input Sequence Alignment: Select the sequence alignment from **step 3**.

   Input Template Structures: Select the name of the template structure from **step 3**.

Input Model Sequence: Select the name of the target sequence in the alignment window.

Protein → Optimize Sidechains: True.

Waters: False.

Number of Models: 25.

Optimization Level: None.

Cut Overhangs: True.

Disulfide Bridges: False.

Cis-Prolines: False.

Refine Loops: False.

5. After building 25 models, visually inspect them to find areas of variability in the structures. These variable areas are indicated by the lack of overlap between different models. Based on the inspection, manually adjust the input alignment in the Alignment Window by adding and shifting gaps and residues in regions with poor overlap in the models, using the space and backspace keys. In general, adjustments are often needed at the areas noted as inconsistently aligned in Subheading 3.1.2. If there are sequences present in the alignment that might interfere with the quality of the alignment between the target and template sequence, they may be removed from the alignment entirely.
6. When the alignment has been refined, build 25 new homology models as in **step 4**. Use the refined alignment as the new input alignment, and the same template structure used in the first round of modeling.

3.2.3 Refinement of the Homology Model

1. Read the report file produced after the Homology Modeling is complete. This report provides details about the quality of the homology models, including the PDF and DOPE (*see* **Note 15**) energy scores for each model. Generally, models with more negative energy scores are better to use for the rest of the homology modeling process. Additionally, you may use the "Check Structure" tool under "X-Ray" → "Validate Protein Structure" to check the models for any other errors (refer to the Discovery Studio help for more information on the types of errors you may find).
2. Select the best-scoring (lowest energy) model from the second homology modeling. Go to the "Forcefield" section (under "Simulation" → "Change Forcefield"). From the "Forcefield" dropdown menu, select "CHARMm Polar H". From the "Partial Charge" dropdown menu, select "Momany-Rone". Click "Apply Forcefield" to apply the CHARMm forcefield to the homology model.

3. Use the "side-chain refinement" tool ("Minimize and Refine Protein" → "Side-Chain Refinement") in Discovery Studio to rotate the side chains of the model into more favorable positions [26]. To determine if the side chains are in more favorable positions after refinement, use the "Calculate Energy" tool (under "Simulation" → "Run Simulations") on the refined model to compare the energy between the models before and after refinement. If the side chain refinement was successful, then the energy score of the refined model should be more negative than the energy score of the original model. Repeat **step 3** until there is little to no difference in the energy of the previous model and the current model. This is a sign that the model has reached a side-chain refinement minimum, and no further side-chain refinement is useful.
4. Use the "minimization" tool of Discovery Studio ("Run Simulations" → "Minimization") to refine the entire protein. In the "Algorithm" dropdown menu, select "Smart Minimizer", which incorporates steepest descent and conjugate gradient approaches. Under the "Implicit Solvent Model" dropdown menu, select "Generalized Born". Once these options have been chosen, click "Run".
5. Record the initial and final potential energy from the report generated by Discovery Studio after the minimization.
6. Repeat **steps 4** and **5** until there is little to no difference in the initial and final Potential Energy. This is a sign that the model has reached an energy minimum, and no further minimization is useful at this point.
7. Assess the quality of the refined model using the "Check Structure" tool, to make sure no additional problems were introduced into the homology model by side-chain refinement or minimization. If the quality of the model is sufficient for the intended use [27], then the model is complete.

3.2.4 Homology Modeling of Sequence Inserts and Unresolved 3D Structure Regions

If there were regions of amino acids which were deleted in **step 2** of Subheading 3.2.2, they will now be added back to the sequence alignment and modeled using de novo (without a template structure) techniques. To insert the deleted amino acids, type in the single-letter codes into the proper place in the target sequence. Use the space key to add gaps to the same place in the template sequence. Save this revised alignment under a separate file name. Use this revised alignment with the model from **step 7** of Subheading 3.2.3 to generate models containing the insert(s).

1. Create 25 new homology models using the "Build Homology Models" function under the same parameters used for the initial models. Use the "Check Structure" option to assess each of the models.

2. Select the lowest-energy model for further refinement. Refine the side chains of the model as above, and verify the energy of the protein using "Calculate Energy".
3. Minimize the entire model using the same steps as above, until the energy reaches a plateau.
4. Using the "Loop Refinement" protocol, minimize the inserts present in the model with CHARMm minimization [28]. Use the "CHARMm Polar H" forcefield if it is not already present. Make 50 models of the insert, and compare the energies of each to determine the most likely conformation for each one. Select the lowest-energy model for future use.
5. Assess the quality of the completed model as in **step 7** of Subheading 3.2.3.

4 Notes

1. Version 4.1 of Discovery Studio requires a computer with at least 4 GB of RAM and at least 1 GB of free disk space. Depending on what version you use, the system requirements may vary.
2. Expressed RNA libraries, including EST libraries, can reveal alternative splice variants and avoid complications due to introns.
3. The Discovery Studio Client requires a license in order to use its homology modeling features. Your research center or university may have access to Discovery Studio. If Discovery Studio is unavailable, there are other programs for homology modeling such as I-TASSER, SWISS-MODEL (swissmodel.expasy.org), and MODELLER (salilab.org/modeller/). It may be useful to create homology models using more than one program and compare the results.
4. Target: The protein for which one is making a model three-dimensional structure; the goal of the homology modeling process.

   Template: A protein with a known 3D structure and amino acid sequence, usually obtained from a public database like the Protein Data Bank (PDB).

   Sequence: Unfolded chain of amino acids, also known as the primary structure of the protein.

   Structure: Folded chain of amino acids, also known as the tertiary structure of the protein.
5. A short user-generated name is often helpful for easily identifying a sequence from a particular species in an alignment of sequences from many species.

6. The code has two parts; the PDB code of the structure the match is from (the first four characters), followed by the ID of the chain the match is from (the remaining characters). Single letters (like 3u5i_q) indicate lowercase chain IDs, and double letters (like 3izs_ss) indicate uppercase chain IDs.
7. Resolution: A quantitative measurement of the amount of detail that will be seen in a map of a crystal, measured in Angstroms. "Higher resolution" models will have resolution values closer to zero (http://www.rcsb.org/pdb/101/static101.do?p=education_discussion/Looking-at-Structures/resolution.html).

   R value: A quantitative measurement of the quality of an atomic model, based on a comparison between a model's simulated diffraction pattern versus the experimentally derived diffraction pattern (http://www.rcsb.org/pdb/101/static101.do?p=education_discussion/Looking-at-Structures/Rvalue.html). An *R*-value of around 0.20 is typical for a crystal structure.

   R-Free value: A quantitative measurement similar to the *R* value. The R-free value is based on removing a portion of the experimental observations prior to refinement, then how well the refined model predicts these removed observations. An R-Free value of around 0.26 is typical for models (http://www.rcsb.org/pdb/101/static101.do?p=education_discussion/Looking-at-Structures/Rvalue.html).

   Unresolved 3D Structure Region: Part of the template where 3D structural coordinates were not reported (even if they are a known portion of the crystallized protein).
8. When the FASTA file is opened in your operating system's basic text editor (e.g., Notepad), the headers and sequences may be squashed together on the same line of text. Opening the file in a word processor (i.e., Word) or in an advanced text editor like Notepad++ should correctly place the line returns, and allow you to read the FASTA sequences clearly.
9. Alignment of three or more sequences.
10. Alignment Score: A numerical value given by the T-Coffee suite of alignment programs that indicates the reliability of the alignment for a sequence. Higher values indicate more reliable alignments.
11. The PDB files contain more information than needed, while there are many methods to extract only the relevant structural information, we describe here a simple method using the UCSF Chimera software.
12. These methods were written using Version 1.6.1 of UCSF Chimera. Newer versions of the program will have similar menus.
13. These methods were written using Version 3.1 of Discovery Studio. Newer versions of the program will have similar menus.

14. Sequence Insert: A general term describing a sequence of amino acids (generally up to 20) that occurs in one amino acid sequence, but not another to which it is being compared. It could apply to sequences that have been aligned with one another, or a template used in homology modeling.
15. PDF: Probability Density Function. A function that is used to evaluate the "energy" of a homology model, in the form of a numerical score. More negative energy scores generally indicate a better homology model.

    DOPE: Discrete Optimized Protein Energy. A calculation that is used by the MODELLER algorithm to score the "energy" of a homology model. More negative energy scores generally indicate a better homology model.

Acknowledgements

Computational resources provided in part by the RI-INBRE Centralized Research Core Facility which is supported by an Institutional Development Award (IDeA) from the National Institute of General Medical Sciences of the National Institutes of Health under grant number 2 P20 GM103430.

References

1. Wlodawer A, Minor W, Dauter Z, Jaskolski M (2013) Protein crystallography for aspiring crystallographers or how to avoid pitfalls and traps in macromolecular structure determination. FEBS J 280(22):5705–5736
2. Milne JL, Borgnia MJ, Alberto Bartesaghi A, Erin EH, Tran EE, Earl LA, Schauder DM, Lengyel J, Jason Pierson J, Subramaniam S, Patwardhan A (2013) Cryo electron microscopy—a primer for the non-microscopist. FEBS J 280(1):28–45
3. Markley JL, Bahrami A, Eghbalnia HR, Peterson FC, Tyler RC, Ullrich EL, Westler WM, Volkman BF (2009) Macromolecular structure determination by NMR spectroscopy. In: Gu J, Bourne PE (eds) Structural bioinformatics, 2nd edn. Wiley, Hoboken, NJ, pp 93–128
4. Yalcin EB, Stangl H, Pichu S, Mather TN, King RS (2011) Monoamine neurotransmitters as substrates for novel tick sulfotransferases, homology modeling, molecular docking, and enzyme kinetics. ACS Chem Biol 6(2):176–184
5. Venselaar H, Krieger E, Vriend G (2009) Homology modeling. In: Gu J, Bourne PE (eds) Structural bioinformatics, 2nd edn. Wiley, Hoboken, NJ, pp 715–735
6. Mills CL, Beuning PJ, Ondrechen MJ (2015) Biochemical functional predictions for protein structures of unknown or uncertain function. Comput Struct Biotechnol J 13:182–191
7. Schmidt T, Bergner A, Schwede T (2014) Modelling three-dimensional protein structures for applications in drug design. Drug Discov Today 19(7):890–897
8. Gaertig J (2000) Molecular mechanisms of microtubular organelle assembly in Tetrahymena. J Eukaryot Microbiol 47(3):185–190
9. Beisson J (2007) Preformed cell structure and cell heredity. In: Chernoff Y (ed) Protein-based inheritance. Landes Biosciences, Austin, TX, pp 106–118
10. Beisson J, Bétermier M, Bré MH, Cohen J, Duharcourt S, Duret L, Kung C, Malinsky S, Meyer E, Preer JR Jr, Sperling L (2010) Paramecium tetraurelia: the renaissance of an early unicellular model. Cold Spring Harb Protoc 2010:pdb.emo140. doi:10.1101/pdb.emo140
11. Orias E (1997) Introduction to the genetics of Tetrahymena. Tetrahymena Genome Project.

Available from https://www.lifesci.ucsb.edu/genome/Tetrahymena/

12. Beisson J, Bétermier M, Bré MH, Cohen J, Duharcourt S, Duret L, Kung C, Malinsky S, Meyer E, Preer JR Jr, Sperling L (2010) Mass culture of Paramecium tetraurelia. Cold Spring Harb Protoc 2010:pdb.prot5362. doi:10.1101/pdb.prot5362
13. Eisen JA, Coyne RS, Wu M, Wu D, Thiagarajan M, Wortman JR, Badger JH, Ren Q, Amedeo P, Jones KM, Tallon LJ, Delcher AL, Salzberg SL, Silva JC, Haas BJ, Majoros WH, Farzad M, Carlton JM, Smith RK Jr, Garg J, Pearlman RE, Karrer KM, Sun L, Manning G, Elde NC, Turkewitz AP, Asai DJ, Wilkes DE, Wang Y, Cai H, Collins K, Stewart BA, Lee SR, Wilamowska K, Weinberg Z, Ruzzo WL, Wloga D, Gaertig J, Frankel J, Tsao CC, Gorovsky MA, Keeling PJ, Waller RF, Patron NJ, Cherry JM, Stover NA, Krieger CJ, del Toro C, Ryder HF, Williamson SC, Barbeau RA, Hamilton EP, Orias E (2006) Macronuclear genome sequence of the ciliate Tetrahymena thermophila, a model eukaryote. PLoS Biol 4(9):e286
14. Coyne RS, Thiagarajan M, Jones KM, Wortman JR, Tallon LJ, Haas BJ, Cassidy-Hanley DM, Wiley EA, Smith JJ, Collins K, Lee SR, Couvillion MT, Liu Y, Garg J, Pearlman RE, Hamilton EP, Orias E, Eisen JA, Methé BA (2008) Refined annotation and assembly of the *Tetrahymena thermophila* genome sequence through EST analysis, comparative genomic hybridization and targeted gap closure. BMC Genomics 9:562–579. doi:10.1186/1471-2164-9-562
15. Stover NA, Krieger CJ, Binkley G, Dong Q, Fisk DG, Nash R, Sethuraman A, Weng S, Cherry JM (2006) Tetrahymena Genome Database (TGD): a new genomic resource for Tetrahymena thermophila research. Nucleic Acids Res 34(Database issue): D500–D503
16. Xiong J, Lu X, Lu Y, Zeng H, Yuan D, Feng L, Chang Y, Bowen J, Gorovsky MA, Fu C, Miao W (2011) Tetrahymena Gene Expression Database (TGED): a resource of microarray data and co-expression analyses for Tetrahymena. Sci China Life Sci 54(1):65–67
17. Rabl J, Leibundgut M, Ataide SF, Haag A, Ban N (2011) Crystal structure of the eukaryotic 40S ribosomal subunit in complex with initiation factor 1. Science 331:730–736
18. Klinge S, Voigts-Hoffmann F, Leibundgut M, Arpagaus S, Ban N (2011) Crystal structure of the eukaryotic 60S ribosomal subunit in complex with initiation factor 6. Science 334: 941–948
19. Zeng Z, Min B, Huang J, Hong K, Yang Y, Collins K, Lei M (2011) Structural basis for *Tetrahymena* telomerase processivity factor Teb1 binding to single-stranded telomeric-repeat DNA. Proc Natl Acad Sci U S A 108(51):20357–20361. doi:10.1073/pnas.1113624108
20. Pagano GJ, King RS, Martin LM, Hufnagel LA (2015) The unique N-terminal insert in the ribosomal protein, phosphoprotein P0, of Tetrahymena thermophila: bioinformatic evidence for an interaction with 26S rRNA. Proteins 83(6):1078-1090. doi:10.1002/prot.24800
21. Reid AJ, Yeats C, Lees J, Orengo CA (2009) Structural annotation of genomes. In: Gu J, Bourne P (eds) Structural bioinformatics, 2nd edn. Wiley, Hoboken, NJ, pp 539–558
22. Moretti S, Armougom F, Wallace IM, Higgins DG, Jongeneel CV, Notredame C (2007) The M-Coffee web server: a meta-method for computing multiple sequence alignments by combining alternative alignment methods. Nucleic Acids Res 35(Web Server issue):W645–W648
23. Armougom F, Moretti S, Poirot O, Audic S, Dumas P, Schaeli B, Keduas V, Notredame C (2006) Expresso: automatic incorporation of structural information in multiple sequence alignments using 3D-Coffee. Nucleic Acids Res 34(Web Server issue):W604–W608
24. Pettersen EF, Goddard TD, Huang CC, Couch GS, Greenblatt DM, Meng EC, Ferrin TE (2004) UCSF Chimera--a visualization system for exploratory research and analysis. J Comput Chem 25(13):1605–1612
25. Eswar N, Marti-Renom MA, Webb B, Madhusudhan MS, Eramian D, Shen M, Pieper U, Sali A (2006) Comparative protein structure modeling with MODELLER. Curr Protoc Bioinformatics 15:5.6.1–5.6.30
26. Spassov VZ, Yan L, Flook PK (2007) The dominant role of side-chain backbone interactions in structural realization of amino acid code. ChiRotor: a side-chain prediction algorithm based on side-chain backbone interactions. Protein Sci 16:494–506
27. Wlodawer A, Minor W, Dauter Z, Jaskolski M (2008) Protein crystallography for non-crystallographers, or how to get the best (but not more) from published macromolecular structures. FEBS J 275(1):1–21
28. Spassov VZ, Flook PK, Yan L (2008) LOOPER: a molecular mechanics-based algorithm for protein loop prediction. Protein Eng Des Sel 21:91–100

Index

A

B

C

D

E

F

Ray H. Gavin (ed.), *Cytoskeleton: Methods and Protocols*, Methods in Molecular Biology, vol. 1365,
DOI 10.1007/978-1-4939-3124-8, © Springer Science+Business Media New York 2016

N

O

P

R

S

T

X

Y

Z

Printed by Printforce, the Netherlands